Advances in Modal Logic

volume 5

Advances in Modal Logic

volume 5

Edited by

Renate Schmidt, Ian Pratt-Hartmann, Mark Reynolds and Heinrich Wansing

ISBN 0-904987-22-2

King's College Publications
Department of Computer Science
Strand, London WC2R 2LS, UK.

http://www.dcs.kcl.ac.uk/kcp

Printed by Lightning Source, UK
Cover design by Richard Fraser, Avalon Arts, UK

CONTENTS

Preface

The present volume contains papers from the fifth conference on *Advances in Modal Logic* (AiML 2004), including an invited paper by Valentin Goranko. After previous conferences in Berlin (1996), Uppsala (1998), Leipzig (2000), and Toulouse (2002), AiML 2004 took place at the University of Manchester from September 9th to September 11th, 2004. The conference was organized by Renate Schmidt and Ian Pratt-Hartmann. It brought together a total of sixty delegates from all over the world to present and discuss recent results and developments in all ramifications of modal logic and its applications in mathematics, computer science, artificial intelligence, and philosophy.

As in previous years, AiML 2004 could boast a glittering array of invited speakers:

- Philippe Balbiani (Toulouse, France),
- Keith Devlin (Stanford, USA),
- Valentin Goranko (Johannesburg, Rep. South Africa),
- Wiebe van der Hoek (Liverpool, UK),
- Maarten Marx (Amsterdam, The Netherlands), and
- Robert Stalnaker (MIT, USA).

Moreover, the programme committee selected 28 out of 62 submitted papers for presentation in Manchester.

It is well-known that the impetus for the study of modal logic has shifted in recent years. While the origins of the subject mainly lie in the philosophical analysis of specific modalities—alethic, epistemic, deontic, or temporal—many of its more recent developments stem from the fact that systems of modal logic can often be alternatively viewed as decidable fragments of first-order (or occasionally higher-order) logic enjoying interesting computation-theoretic properties. On this view, the most salient issues include, for example, expressive power, decidability, computational complexity and practicable implementation. Unsurprisingly, many of the talks at AiML 2004 reflected this computational bent. Nevertheless, the more traditional concerns of modal logic were still very much in evidence, with papers on logics of knowledge, agency, time and—an intriguing recent development—space. Last but not least, the conference featured a small number of talks which were primarily philosophical in character or motivation.

We would like to thank our colleagues in the programme committee for the great amount of work they invested in putting together the programme of AiML 2004. The programme committee consisted of:

- Patrick Blackburn (Nancy, France),
- Alexander Chagrov (Tver, Russia),
- Vincent Hendricks (Roskilde, Denmark),
- Ian Pratt-Hartmann (Manchester, UK),
- Mark Reynolds (co-chair; Perth, Australia),
- Maarten de Rijke (Amsterdam, The Netherlands),
- Ulrike Sattler (Manchester, UK),
- Holger Schlingloff (Berlin, Germany) ,
- Renate Schmidt (Manchester, UK),
- Nobu-Yuki Suzuki (Shizuoka, Japan),
- Heinrich Wansing (co-chair; Dresden, Germany),
- Frank Wolter (Liverpool, UK), and
- Michael Zakharyaschev (London, UK).

We are grateful to the following additional referees for their kind assistance: Adrianna Alexander, Philippe Balbiani, Howard Barringer, Jochen Burghardt, Melvin Fitting, Tim French, Paul Gochet, Rajeev Goré, Ian Hodkinson, John Horty, Ullrich Hustadt, Rosalie Iemhoff, Manfred Jaeger, Ryo Kashima, Oliver Kutz, Gerhard Lakemeyer, Christoph Lüth, Carsten Lutz, Maarten Marx, Till Mossakowski, Larry Moss, Marco Ragni, Markus Roggenbach, Takafumi Sakurai, Katsumi Sasaki, Ken Satoh, Karsten Schmidt, Lutz Schroeder, Yoshihito Tanaka, Dmitry Tishkovsky, Stefan Wölfl, and other anonymous reviewers.

Moreover, it is our pleasure to acknowledge generous support by the following institutions:

- TARSKI EU COST Action 274,
- CologNet (Areas 'Automated Reasoning, Deduction, Theorem Proving, and Model-Checking' and 'Logic and Multi-Agent Systems'),
- British Logic Colloquium,
- School of Information Technology, Murdoch University,
- Language and Inference Technology Group, University of Amsterdam, and
- Department of Computer Science, University of Manchester.

Information about the Advances in Modal Logic initiative is available at: `http://www.aiml.net` .

Renate Schmidt, Manchester
Ian Pratt-Hartmann, Manchester
Mark Reynolds, Perth
Heinrich Wansing, Dresden

Complexity of Strict Implication

FÉLIX BOU

ABSTRACT. The aim of the present paper is to analyze the complexity of strict implication (together with falsum, conjunction and disjunction). We prove that Ladner's Theorem remains valid when we restrict the language to the strict implication fragment, and that the same holds for Hemaspaandra's Theorem. As a consequence we have that the validity problem for most standard normal modal logics is the same one than the validity problem for its strict implication fragment. We also prove that the validity problems for Visser's basic propositional logic and Visser's formal propositional logic are **PSPACE** complete. Finally, a polynomial reduction from most standard normal modal logics into its strict implication fragment is presented.

Strict implication is defined in the modal language as

$$\varphi_0 \rightarrow \varphi_1 := \quad \Box(\varphi_0 \supset \varphi_1) \qquad \textit{strict implication},$$

where $\supset$ refers to material implication. Strict implication was already considered by Lewis in the birth of modal logic. Nevertheless, there is almost no complexity result in the literature for proper fragments of the modal language that are based on strict implication. One of the few exceptions is intuitionistic propositional logic [22].

On the other hand, there are a lot of complexity results for logics in the modal language. Two of the most outstanding results are Ladner's Theorem [15, Theorem 3.1] and Hemaspaandra's Theorem [21, 13]. They state, respectively, that all modal subsystems of **S4** have a **PSPACE** hard validity problem and that all modal extensions of **S4.3** have a co-**NP** complete validity problem.

In this paper we analyze the complexity of the language based on strict implication ($\rightarrow$) together with falsum ($\bot$), conjunction ($\wedge$) and disjunction ($\vee$). We notice that true ($\top$) is easily definable in this language, while neither material implication ($\supset$) nor classical negation ($\sim$) are definable. The main results of the paper say that Hemaspaandra's Theorem and Ladner's Theorem also hold when we consider the restriction to the strict implication fragment. Hence, the complexity of the validity problem for most standard

Advances in Modal Logic, Volume 5. © *2005, Félix Bou.*

normal modal logics is the same one than the complexity of the validity problem for its strict implication fragment[1].

The first section is devoted to fix the notation and to remind the complexity results for the modal language. In this section we also sketch a proof of Ladner's Theorem since we will follow the same strategy for the strict implication fragment. In Section 2 we prove Hemaspaandra's Theorem and Ladner's Theorem for the strict implication fragment, and we state some consequences of them. The aim of the third section is to show a map that is a polynomial reduction from most standard normal modal logics into its strict implication fragment. Finally, the last section is devoted to state some open problems. We notice that all the results of this paper are contained in the author's Ph.D. Dissertation [3].

1 Preliminaries

First of all, let us fix our notation. Throughout the paper we will consider two languages. One is the well-known modal language and the other is its strict implication fragment (together with falsum, conjunction and disjunction). Let Prop be an infinite set of propositions. The *modal formulas* and the *strict implication formulas* are defined, respectively, by

$$\varphi ::= p \mid \bot \mid \varphi_0 \wedge \varphi_1 \mid \varphi_0 \vee \varphi_1 \mid \varphi_0 \supset \varphi_1 \mid \Box\varphi$$

and

$$\varphi ::= p \mid \bot \mid \varphi_0 \wedge \varphi_1 \mid \varphi_0 \vee \varphi_1 \mid \varphi_0 \rightarrow \varphi_1,$$

where p ranges over elements of Prop. The set of modal formulas will be denoted by $\mathcal{L}^{mod}(\mathsf{Prop})$, and the set of strict implication formulas by $\mathcal{L}^{s}(\mathsf{Prop})$. The set of subformulas of a formula φ is denoted by $Sub(\varphi)$, and the *modal degree* $\deg(\varphi)$ of a modal formula φ is the number of nested modalities.

These formulas are to be interpreted in Krikpe models. Given a Kripke model $\mathcal{M} = \langle M, R, V\rangle$, a world $m \in M$ and a formula φ the *satisfiability relation* $\mathcal{M}, m \Vdash \varphi$ is defined as follows:

$\mathcal{M}, m \Vdash p$	iff	$m \in V(p)$
$\mathcal{M}, m \nVdash \bot$		
$\mathcal{M}, m \Vdash \varphi_0 \wedge \varphi_1$	iff	$\mathcal{M}, m \Vdash \varphi_0$ and $\mathcal{M}, m \Vdash \varphi_1$
$\mathcal{M}, m \Vdash \varphi_0 \vee \varphi_1$	iff	$\mathcal{M}, m \Vdash \varphi_0$ or $\mathcal{M}, m \Vdash \varphi_1$
$\mathcal{M}, m \Vdash \varphi_0 \supset \varphi_1$	iff	$\mathcal{M}, m \nVdash \varphi_0$ or $\mathcal{M}, m \Vdash \varphi_1$
$\mathcal{M}, m \Vdash \Box\varphi$	iff	$\forall m'(mRm' \Rightarrow \mathcal{M}, m' \Vdash \varphi)$
$\mathcal{M}, m \Vdash \varphi_0 \rightarrow \varphi_1$	iff	$\forall m'(mRm' \;\&\; \mathcal{M}, m' \Vdash \varphi_0 \Rightarrow \mathcal{M}, m' \Vdash \varphi_1)$.

[1] Besides the complexity results there are a lot of other results suggesting that the expressive power of the strict implication fragment is very close to the one of the modal language. A thorough study can be found in [3].

If $\mathcal{M}, m \Vdash \varphi$ then we say that φ *is satisfied in* $\mathcal{M}$ *at* a. It is said that φ *is valid in* $\mathcal{M}$ (notation: $\mathcal{M} \Vdash \varphi$) if $\mathcal{M}, m \Vdash \varphi$ for every world $m \in M$. Given two formulas φ_0 and φ_1 we will write $\varphi_0 \equiv \varphi_1$ whenever both formulas are satisfied in the same worlds for every Kripke model. We will use the following abbreviations in the strict implication language: $\top := \bot \rightarrow \bot$, $\neg\varphi := \varphi \rightarrow \bot$ and $\Box\varphi := \top \rightarrow \varphi$. In the modal language we write $\sim\varphi := \varphi \supset \bot$, $\varphi_0 \supset\subset \varphi_1 := (\varphi_0 \supset \varphi_1) \wedge (\varphi_1 \supset \varphi_0)$ and $\Diamond\varphi := \sim\Box\sim\varphi$. Given $n \in \omega$ we denote by $\Box^n\varphi$ and $\Box^{(n)}\varphi$ the formulas

$$\overbrace{\Box\ldots\Box}^{n \text{ times}}\varphi \quad \text{and} \quad \varphi \wedge \Box\varphi \wedge \ldots \wedge \Box^n\varphi.$$

It is clear that the semantics of the non-primitive connectives is the expected one[2]. In particular,

$$\mathcal{M}, m \Vdash \sim\varphi \qquad \text{iff} \quad \mathcal{M}, m \nVdash \varphi.$$

We notice that if we add the material implication $\supset$ or the classical negation $\sim$ to $\mathcal{L}^s(\mathsf{Prop})$ then the language that we obtain has the same expressive power than $\mathcal{L}^{mod}(\mathsf{Prop})$. We also single out that the semantics on the strict implication language is precisely the standard semantics given for intuitionistic propositional logic except for the fact that we range over arbitrary Kripke models (and not only over intuitionistic models[3]).

The interest in the strict implication language comes from the fact that there is a wide range of well-known logics that can be formalized in it. As we have already pointed out the more famous example is *intuitionistic propositional logic* **IPL**, which is the set of strict implication formulas that are valid in all reflexive and transitive frames under persistent valuations [14, 5]. We recall that a valuation V is *persistent* when it satisfies that for every $p \in \mathsf{Prop}$ and every worlds m and m', if mRm' and $m \in V(p)$ then $m' \in V(p)$. Another example of logic that can be obtained in this way is *classical propositional logic* **CPL**. It is the set consisting of all strict implication formulas that are valid in all reflexive, transitive and symmetric frames under persistent valuations (or simply the set of all strict implication formulas that are valid in the reflexive frame that consists of a single point). Other examples in the literature are all superintuitionistic logics [5], some subintuitionistic logics [8, 9, 17, 25, 4], Visser's *formal propositional logic* **FPL** [24] and Visser's *basic propositional logic* **BPL** [24, 18, 19]. Let us explain which logics are the last two. We define the *Gödel translation* $\mathsf{T} : \mathcal{L}^s(\mathsf{Prop}) \longrightarrow \mathcal{L}^{mod}(\mathsf{Prop})$ using the following clauses:

[2] We notice that the semantics of $\Box$ for the modal case coincides with its semantics as a defined connective for the strict implication case. Therefore, there is no ambiguity.

[3] For this sake we do not call our language the intuitionistic propositional language. The author's opinion is that it is better to use a different name to point out this fact.

$$\text{i) } \mathsf{T}(p) = \Box p, \quad \text{ii) } \mathsf{T}(\bot) = \bot, \quad \text{iii) } \mathsf{T}(\varphi_0 \wedge \varphi_1) = \mathsf{T}(\varphi_0) \wedge \mathsf{T}(\varphi_1),$$
$$\text{iv) } \mathsf{T}(\varphi_0 \vee \varphi_1) = \mathsf{T}(\varphi) \vee \mathsf{T}(\varphi_1), \quad \text{v) } \mathsf{T}(\varphi_0 \rightarrow \varphi_1) = \Box(\mathsf{T}(\varphi_0) \supset \mathsf{T}(\varphi_1)).$$

It is well-known that T is an embedding of **IPL** into both the normal modal logic **S4** and the normal modal logic **Grz**, i.e.,

$$\varphi \in \mathbf{IPL} \quad \text{iff} \quad \mathsf{T}(\varphi) \in \mathbf{S4} \quad \text{iff} \quad \mathsf{T}(\varphi) \in \mathbf{Grz}.$$

It is interesting to notice that the same map is also an embedding of **CPL** into **S5**. The logics **FPL** and **BPL** were introduced by Visser in [24] as the logics such that T is an embedding of them into **GL** and **K4**, respectively. That is,

$$\mathbf{FPL} = \{\varphi \in \mathcal{L}^s(\mathsf{Prop}) : \mathsf{T}(\varphi) \in \mathbf{GL}\},$$

and

$$\mathbf{BPL} = \{\varphi \in \mathcal{L}^s(\mathsf{Prop}) : \mathsf{T}(\varphi) \in \mathbf{K4}\}.$$

It is not hard to check that **FPL** is precisely the set of strict implication formulas that are valid in all frames that are Noetherian strict orders under persistent valuations[4], and that **BPL** is the set of strict implication formulas that are valid in all transitive frames under persistent valuations. In [24] Visser was interested in **FPL** for provability reasons, and **BPL** was simply suited for technical reasons. However, nowadays **BPL** has acquired an interest of its own since Ruitenburg has argued its philosophical interest as a constructive logic [18, 19].

We notice that there is a general method to associate a strict implication logic with a normal modal logic. The idea is to consider its strict implication fragment. Given a normal modal logic L we define its strict implication fragment L^s as the set $\{\varphi \in \mathcal{L}^s(\mathsf{Prop}) : \sigma(\varphi) \in L\}$ where σ is the translation defined by:

$$\begin{array}{lcl}
\sigma(p) & := & p \\
\sigma(\bot) & := & \bot \\
\sigma(\varphi_0 \wedge \varphi_1) & := & \sigma(\varphi_0) \wedge \sigma(\varphi_1) \\
\sigma(\varphi_0 \vee \varphi_1) & := & \sigma(\varphi_0) \vee \sigma(\varphi_1) \\
\sigma(\varphi_0 \rightarrow \varphi_1) & := & \Box(\sigma(\varphi_0) \supset \sigma(\varphi_1)).
\end{array}$$

The map σ has been considered several times in the literature. Its first appearance under this name is in [9], and since then it has been also used by other authors, e.g., Celani and Jansana [4].

We have previously announced that we will prove in the next section that in most cases the complexity of the validity problem for a normal modal

[4] If we replace 'Noetherian' with 'finite' we also obtain **FPL**.

logic coincides with the complexity of the validity problem for its strict implication fragment. The following easy example shows that this is not true for all normal modal logics (as far as $\mathbf{P} \neq \mathbf{NP}$).

EXAMPLE 1. Let **Verum** be the normal modal logic given by the frame $\mathfrak{F}$ that is a single irreflexive point. Since the problem "$\varphi \in$ **CPL**?" is co-**NP** complete [7] it easily follows that "$\varphi \in$ **Verum**?" is also co-**NP** complete. On the other hand, it is not hard to prove that "$\varphi \in \mathbf{Verum}^s$?" is in **P**. It follows from the fact that the model checking for modal formulas (in particular for strict implication formulas) is in **P** together with the fact that for every $\varphi \in \mathcal{L}^s(\mathsf{Prop})$, $\varphi \in \mathbf{Verum}^s$ iff φ holds in the Kripke model over $\mathfrak{F}$ such that all propositions fail in its unique state.

In the rest of the section we recall two of the main results concerning complexity of normal modal logics. The first one is due to Hemaspaandra (*née* Spaan) [21, 13].

THEOREM 2 (Hemaspaandra). *Let* Prop *be a countable set and let L be a normal modal logic extending* **S4.3**. *Then, the problem "$\varphi \in L$?" is co-***NP** *complete.*

The other result is due to Ladner [15] and gives a lower bound of the complexity of normal modal logics that are subsystems of **S4**.

THEOREM 3 (Ladner). *Let* Prop *be a countable set and let L be a normal modal logic such that* $\mathbf{K} \subseteq L \subseteq \mathbf{S4}$. *Then, the problem "$\varphi \in L$?" is* **PSPACE** *hard.*

Using that for the validity problem of the normal modal logics **K**, **T**, **K4**, **S4**, **D**, **GL** and **Grz** it is known the existence of algorithms running in **PSPACE**, it follows that the validity problem for these normal modal logics is **PSPACE** complete.

To finish the section we review a proof of Ladner's Theorem because we will follow the same strategy in the next section. Since in the next section we want to deal with persistent valuations we cannot consider the proof given by Ladner (see [2] for a more accessible presentation of this proof). The argument that we detail is very close to the one exhibited by Halpern and Moses in [12].

Let L be a normal modal logic such that $\mathbf{K} \subseteq L \subseteq \mathbf{S4}$. Now we show a polynomial reduction from the complementary of the validity problem of the logic **QBF** of quantified Boolean formulas. It is enough because it is well known that this last problem is **PSPACE** complete [23]. Let us describe what is ***QBF***. The set of *quantified Boolean formulas*[5] consists of

[5]Sometimes these formulas have been called prenex quantified Boolean formulas. Then

expressions of the form

$$Q_0p_0 \dots Q_{n-1}p_{n-1}\varphi(p_0, \dots, p_{n-1})$$

where each $Q_i \in \{\forall, \exists\}$, and $\varphi(p_0, \dots, p_{n-1})$ is a Boolean formula (called the *matrix*) with propositions among $p_0, \dots, p_{n-1}$. The quantifiers range over the truth values 1 (true) and 0 (false), and a quantified Boolean formula without free variables is *true* if and only if it evaluates to 1, i.e., the subformulas $\forall p\, \varphi(p)$ and $\exists p\, \varphi(p)$ are regarded to be true iff $\varphi(\top) \wedge \varphi(\bot)$ and $\varphi(\top) \vee \varphi(\bot)$ are true, respectively. For instance, $\exists p_0 \forall p_1 (p_0 \vee p_1)$ is true, while $\forall p_0 \exists p_1 \forall p_2 (p_0 \wedge p_1 \wedge p_2)$ is not true. The logic $\boldsymbol{QBF}$ is the set of quantified Boolean formulas that are true. Hence, the $\boldsymbol{QBF}$ problem is the problem of deciding, given an arbitrary quantified Boolean formula β, whether $\beta \in \boldsymbol{QBF}$.

As far as we are interested in a **PSPACE** complete problem we can restrict[6] the definition of quantified Boolean formulas to the case that $Q_0 = \exists$ and $n \geq 2$. This is obvious by the following trivial polynomial reduction:

$$\begin{gathered} Q_0p_0 \dots Q_{n-1}p_{n-1}\varphi(p_0, \dots, p_{n-1}) \text{ is true} \quad \text{iff} \\ \exists q_0 \exists q_1 Q_0p_0 \dots Q_{n-1}p_{n-1}\big(q_0 \wedge q_1 \wedge \varphi(p_0, \dots, p_{n-1})\big) \text{ is true,} \end{gathered}$$

where q_0 and q_1 are two new propositions. From now on we will assume that all quantified Boolean formulas satisfy the requirements $Q_0 = \exists$ and $n \geq 2$.

Now we present the promised polynomial reduction from a **PSPACE** complete problem to L. Let β be a quantified Boolean formula

$$Q_0p_0 \dots Q_{n-1}p_{n-1}\varphi(p_0, \dots, p_{n-1})$$

with $Q_0 = \exists$ and $n \geq 2$. We consider new propositions $q_0, \dots, q_{n+1}$, and we define $g(\beta)$ as the conjunction of formulas displayed in Figure 1, where

$$\gamma_{in} := (q_i \wedge {\sim} q_{i+1}) \supset \big((q_0 \wedge \dots \wedge q_{i-1}) \wedge ({\sim} q_{i+2} \wedge \dots \wedge {\sim} q_{n+1}) \wedge ({\sim} p_i \wedge \dots \wedge {\sim} p_{n-1})\big),$$

$$\theta_i := (q_i \wedge {\sim} q_{i+1}) \supset \Diamond(q_{i+1} \wedge {\sim} q_{i+2}),$$

$$\delta_i := (q_i \wedge {\sim} q_{i+1}) \supset \big(\Diamond(q_{i+1} \wedge {\sim} q_{i+2} \wedge p_i) \wedge \Diamond(q_{i+1} \wedge {\sim} q_{i+2} \wedge {\sim} p_i)\big),$$

and

$$\psi_i := \big((q_i \wedge p_{i-1}) \supset \Box p_{i-1}\big) \wedge \big((q_i \wedge {\sim} p_{i-1}) \supset \Box {\sim} p_{i-1}\big).$$

---

the set of quantified Boolean formulas is defined by

$$\varphi ::= \bot \mid \top \mid p \mid {\sim}\varphi \mid \varphi_0 \wedge \varphi_1 \mid \forall p \varphi \mid \exists p \varphi$$

where p ranges over elements of Prop.

[6]Indeed, there is no reason to adopt this restriction in the modal case, but in the strict implication case this restriction simplifies the argument.

(i) q_0

(ii) $\sim(q_1 \vee q_2 \vee \ldots \vee q_n \vee q_{n+1} \vee p_0 \vee p_1 \vee \ldots \vee p_{n-1})$

(iii) $\Diamond(q_1 \wedge \sim q_2)$

(iv) $\Box^1\gamma_{1n} \wedge \Box^2\gamma_{2n} \wedge \ldots \wedge \Box^n\gamma_{nn}$

(v) $\Box^1\theta_1 \wedge \Box^2\theta_2 \wedge \ldots \wedge \Box^{n-1}\theta_{n-1}$

(vi) $\bigwedge_{i\in\{j<n:Q_j=\forall\}} \Box^i\delta_i$

(vii) $\Box^1\psi_1\wedge\Box^2(\psi_1\wedge\psi_2)\wedge\Box^3(\psi_1\wedge\psi_2\wedge\psi_3)\wedge\ldots\wedge\Box^{n-1}(\psi_1\wedge\psi_2\wedge\psi_3\wedge\ldots\wedge\psi_{n-1})$

(viii) $\Box^n(q_n \supset \varphi)$

Figure 1. The modal formula $g(\beta)$

It is clear that $\deg(g(\beta)) = n$, and it is easy to check that for every quantified Boolean formula β we can compute in polynomial time the modal formula $g(\beta)$. Hence, to obtain a polynomial reduction from $\boldsymbol{QBF}^c$ to L it is enough to prove that

$$\beta \in \boldsymbol{QBF} \quad \text{iff} \quad \sim g(\beta) \notin L,$$

for every quantified Boolean formula β. This equivalence is a trivial consequence of the following lemma.

LEMMA 4. *Let β be a quantified Boolean formula*

$$Q_0p_0 \ldots Q_{n-1}p_{n-1}\varphi(p_0, \ldots, p_{n-1})$$

with $Q_0 = \exists$ and $n \geq 2$. The following statements are equivalent:

- *β is true.*
- *$g(\beta)$ is satisfiable in a Kripke model.*
- *$g(\beta)$ is satisfiable in a finite strict order with a persistent valuation.*
- *$g(\beta)$ is satisfiable in a finite partial order with a persistent valuation.*

Sketch of the proof. Ladner's idea is that when we are evaluating β we are essentially generating a finite number of binary 'trees' $\mathcal{T}_0, \ldots, \mathcal{T}_{k-1}$ of height n such that each one of their branches gives us a Boolean valuation

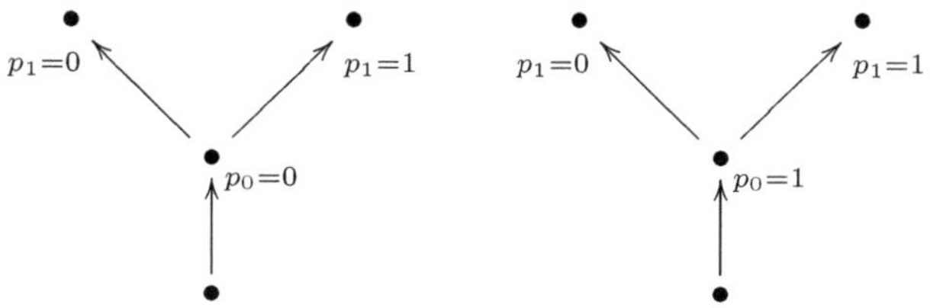

(a) The binary 'trees' generated by β

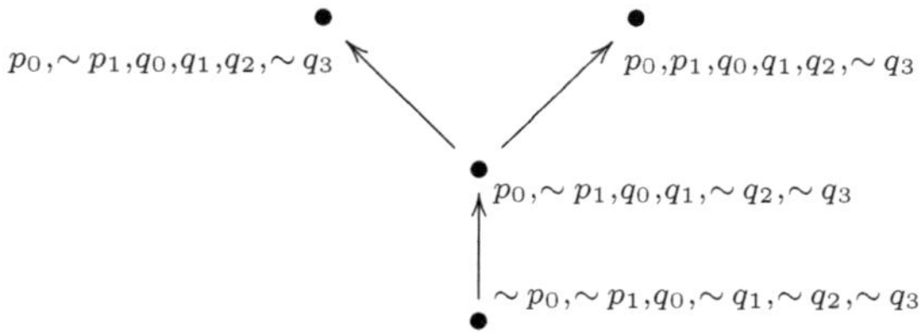

(b) A Kripke model witnessing that β is true

Figure 2. Let β be the quantified Boolean formula $\exists p_0 \forall p_1 (p_0 \vee p_1)$

in $\{p_0, \dots, p_{n-1}\}$. These 'trees' consists of the root node, and then — working inwards along the quantifier string — each existential quantifier extends it by adding a single branch, and each universal quantifier extends it by adding two branches (In Figure 2(a) the reader can find an example). For every $j < k$, we associate a Kripke model $\mathcal{M}_j$ over the propositions $\{p_0, \dots, p_{n-1}, q_0, \dots, q_{n+1}\}$ with the binary 'tree' $\mathcal{T}_j$. Let us explain how to define $\mathcal{M}_j$: the universe and the accessibility relation are given by $\mathcal{T}_j$, and the behaviour of its valuation at a node of height i ($\leq n$) is the following one:

- propositions in $\{q_0, \dots, q_i\}$ are true.
- propositions in $\{p_i, \dots, p_n, q_{i+1}, \dots, q_{n+1}\}$ are false.
- propositions in $\{p_0, \dots, p_{i-1}\}$ behaves according to the Boolean valuation given by a branch containing the node.

It holds that $\mathcal{M}_j$ is a tree, and that it has a persistent valuation (In our previous example the Kripke model associated with the second 'tree' in Figure 2(a) is the one depicted in Figure 2(b)).

The connection of the previous ideas with the problem that we are interested in is that β is true iff there is $j < k$ such that the matrix φ is evaluated to 1 in all Boolean valuations recorded by the binary 'tree' $\mathcal{T}_j$ (In our previous example the witnessing 'tree' is the second one in Figure 2(a)). Now let us analyze the different implications that we have in the present lemma.

$(1 \Rightarrow 3)$: Assume that β is true. Then, there is $j < k$ such that the matrix φ is evaluated to 1 in all Boolean valuations recorded by the binary 'tree' $\mathcal{T}_j$. Now we define $\mathcal{N}_j$ as the Kripke model $\mathcal{M}_j$ except for the fact that we take the $R^{\mathcal{N}_j}$ as the transitive closure of $R^{\mathcal{M}_j}$. It is clear that $\mathcal{N}_j$ is a finite strict order with a persistent valuation; and from the fact that φ is evaluated to 1 in all Boolean valuations recorded by the binary 'tree' $\mathcal{T}_j$ it is simple to check that $\mathcal{N}_j$ satisfies $g(\beta)$ in the state that is the root of $\mathcal{M}_j$.

$(1 \Rightarrow 4)$: It is proved as the previous implication, except for the fact that now we take the reflexive-transitive closure.

$(3 \Rightarrow 2)$: Trivial.

$(4 \Rightarrow 2)$: Trivial.

$(2 \Rightarrow 1)$: We suppose that $g(\beta)$ is satisfiable in a certain Kripke model $\mathcal{M}$ at a world m. Replacing it with its unravelling [20] we can assume that $\mathcal{M}$ is a tree with root m. Using that $\deg(g(\beta)) = n$ we can assume that $\mathcal{M}$ is a tree of length n: simply remove all states with length $> n$. Using all clauses except (viii) of the definition of $g(\beta)$, it is easily verified that $\mathcal{M}$ is isomorphic to a (perhaps not generated) submodel of $\mathcal{M}_j$ for some $j < k$. By clause (viii) of the definition of $g(\beta)$, it follows that φ is evaluated to 1 in all Boolean valuations recorded by the binary 'tree' $\mathcal{T}_j$. Therefore, β is true. ■

2 New Complexity Results

In this section we prove that Hemaspaandra's Theorem and Ladner's Theorem are also true when we restrict ourselves to the strict implication fragment.

THEOREM 5. *Let* Prop *be a countable set and let* L *be a normal modal logic extending* **S4.3**. *Then, the problems "$\varphi \in L$?" and "$\varphi \in L^s$?" are both co-***NP** *complete.*

Proof. By Hemaspaandra's Theorem it is enough to show that "$\varphi \in L^s$?" is co-**NP** hard. It is known that the equational logic of distributive lattices is co-**NP** hard (see [10]). Given two formulas φ_0 and φ_1 using only the connectives $\wedge$ and $\vee$, it is clear that

$$\begin{array}{cr}
\varphi_0 \approx \varphi_1 \text{ holds in all distributive lattices} & \text{iff} \\
\varphi_0 \approx \varphi_1 \text{ holds in all Boolean algebras} & \text{iff} \\
\varphi_0 \supset\subset \varphi_1 \in \mathbf{CPL} & \text{iff} \\
\varphi_0 \supset\subset \varphi_1 \in L & \text{iff} \\
\Box(\varphi_0 \supset\subset \varphi_1) \in L & \text{iff} \\
(\varphi_0 \to \varphi_1) \wedge (\varphi_1 \to \varphi_0) \in L^s. &
\end{array}$$

Hence, we have obtained a polynomial reduction that allows us to conclude that "$\varphi \in L^s$?" is co-**NP** hard. ■

Now we state our version of Ladner's Theorem. In the proof we will use that if we restrict ourselves to quantified Boolean formulas β with matrix in conjunctive normal form, then the problem "$\beta \in \boldsymbol{QBF}$?" is also **PSPACE**-complete [1, Corollary 1.36].

THEOREM 6. *Let* Prop *be a countable set and let* L *be a set of* $\mathcal{L}^s$*-formulas such that either* $\boldsymbol{K}^s \subseteq L \subseteq \boldsymbol{FPL}$ *or* $\boldsymbol{K}^s \subseteq L \subseteq \boldsymbol{IPL}$*. Then,* L *is* **PSPACE** *hard.*

Proof. Let L be a set of $\mathcal{L}^s$-formulas such that either $\boldsymbol{K}^s \subseteq L \subseteq \boldsymbol{FPL}$ or $\boldsymbol{K}^s \subseteq L \subseteq \boldsymbol{IPL}$. The method of the proof is to show a polynomial reduction from a known **PSPACE** complete problem to L^c. The **PSPACE** complete problem considered is the logic $\boldsymbol{QBF}$ of quantified Boolean formulas in conjunctive normal form.

Let β be a quantified Boolean formula

$$Q_0 p_0 Q_1 p_1 \ldots Q_{n-1} p_{n-1} \varphi(p_0, \ldots, p_{n-1})$$

such that $Q_0 = \exists$, $n \geq 2$ and φ is in conjunctive normal form. Therefore, φ is of the form $(\nu_0 \supset \pi_0) \wedge \ldots \wedge (\nu_{k-1} \supset \pi_{k-1})$ where the ν's are finite (maybe empty) conjunctions of propositions and the π's are finite (maybe empty) disjunctions of propositions. Hence, the ν's and the π's are $\mathcal{L}^s$-formulas. We consider new propositions $q_0, \ldots, q_{n+1}$, and we define the following $\mathcal{L}^s$-formulas:

$$\gamma'_{in} := \left(q_i \to (q_{i+1} \vee \bigwedge_{j \in \{0,\ldots,i-1\}} q_j) \right) \wedge \bigwedge_{j \in \{i+2,\ldots,n+1\}} ((q_i \wedge q_j) \to q_{i+1}) \wedge$$

$$\wedge \bigwedge_{j \in \{i,\ldots,n-1\}} ((q_i \wedge p_j) \to q_{i+1}),$$

$$\theta'_i := (q_i \wedge (q_{i+1} \to q_{i+2})) \to q_{i+1},$$

$$\delta'_i := \left(\left(q_i \wedge ((q_{i+1} \wedge p_i) \to q_{i+2}) \right) \to q_{i+1} \right) \wedge$$

(i) q_0

(ii) $q_1 \vee q_2 \vee \ldots \vee q_n \vee q_{n+1} \vee p_0 \vee p_1 \vee \ldots \vee p_{n-1}$

(iii) $q_1 \rightarrow q_2$

(iv) $\gamma'_{1n} \wedge \Box^1 \gamma'_{2n} \wedge \ldots \wedge \Box^{n-1} \gamma'_{nn}$

(v) $\theta'_1 \wedge \Box^1 \theta'_2 \wedge \ldots \wedge \Box^{n-2} \theta'_{n-1}$ [7]

(vi) $\bigwedge_{i \in \{j<n : Q_j = \forall\}} \Box^{i-1} \delta'_i$ [8]

(vii) $\psi'_1 \wedge \Box^1(\psi'_1 \wedge \psi'_2) \wedge \Box^2(\psi'_1 \wedge \psi'_2 \wedge \psi'_3) \wedge \ldots \wedge \Box^{n-2}(\psi'_1 \wedge \psi'_2 \wedge \psi'_3 \wedge \ldots \wedge \psi'_{n-1})$

(viii) $\Box^{n-1}\Big(\big((q_n \wedge \nu_0) \rightarrow \pi_0\big) \wedge \big((q_n \wedge \nu_1) \rightarrow \pi_1\big) \wedge \ldots \wedge \big((q_n \wedge \nu_{k-1}) \rightarrow \pi_{k-1}\big)\Big)$

Figure 3. A list of $\mathcal{L}^s$-formulas

$$\wedge \left(\left(q_i \wedge \left(q_{i+1} \rightarrow (p_i \vee q_{i+2})\right)\right) \rightarrow q_{i+1} \right),$$

and

$$\psi'_i := \left(q_i \wedge p_{i-1}\right) \rightarrow \Box p_{i-1}\right) \wedge \left(q_i \rightarrow (p_{i-1} \vee \neg p_{i-1})\right).$$

It is easy to check that $\gamma'_{in} \equiv \Box\gamma_{in}$, $\theta'_i \equiv \Box\theta_i$, $\delta'_i \equiv \Box\delta_i$ and $\psi'_i \equiv \Box\psi_i$, where the formulas without $'$ are the ones considered in the proof of Ladner's Theorem (see p. 6). Now we consider the list of $\mathcal{L}^s$-formulas displayed in Figure 3. We notice that each one of them is either equivalent to the corresponding one in Figure 1 or equivalent to the classical negation $\sim$ of it. We define $f_0(\beta)$ as the conjunction of (i),(iv),(v),(vi),(vii),(viii); and we define $f_1(\beta)$ as the disjunction of (ii),(iii). Let $f(\beta)$ be the $\mathcal{L}^s$ formula $f_0(\beta) \rightarrow f_1(\beta)$. A moment of reflection shows that

$$\Diamond g(\beta) \equiv \sim f(\beta),$$

where $g(\beta)$ is the formula considered in the proof of Ladner's Theorem. Using this together with Lemma 4 it easily follows that the following statements are equivalent:

- β is true.
- $\sim f(\beta)$ is satisfiable, i.e., $f(\beta) \notin \boldsymbol{K}$.

[7] The assumption $n \geq 2$ guarantees that it is a strict implication formula.

[8] Since $Q_0 = \exists$ we know that $0 \notin \{j < n : Q_j = \forall\}$. Hence, $i - 1 \geq 0$ whenever $i \in \{j < n : Q_j = \forall\}$.

- $\sim f(\beta)$ is satisfiable in a finite strict order with a persistent valuation, i.e., $f(\beta) \notin \boldsymbol{FPL}$.
- $\sim f(\beta)$ is satisfiable in a finite partial order with a persistent valuation, i.e., $f(\beta) \notin \boldsymbol{IPL}$.

Therefore,

$$\beta \in \boldsymbol{QBF} \quad \text{iff} \quad f(\beta) \notin \boldsymbol{K} \quad \text{iff} \quad f(\beta) \notin \boldsymbol{FPL} \quad \text{iff} \quad f(\beta) \notin \boldsymbol{IPL}.$$

Thus, using the fact that either $\boldsymbol{K}^s \subseteq L \subseteq \boldsymbol{FPL}$ or $\boldsymbol{K}^s \subseteq L \subseteq \boldsymbol{IPL}$ we obtain that

$$\beta \in \boldsymbol{QBF} \quad \text{iff} \quad f(\beta) \notin L.$$

It is easy to check that the map f is a polynomial time reduction, what allows us to conclude that L is **PSPACE** hard. ∎

By the last theorem it trivially follows that $\boldsymbol{K}^s$, $\boldsymbol{T}^s$, $\boldsymbol{K4}^s$, $\boldsymbol{S4}^s$, $\boldsymbol{D}^s$, $\boldsymbol{GL}^s$ and $\boldsymbol{Grz}^s$ are **PSPACE** complete. Another easy consequence of Theorem 6 is that the logics $\boldsymbol{BPL}$ and $\boldsymbol{FPL}$ are **PSPACE** complete. We know that they are in **PSPACE** by the Gödel translation. As far as the author is aware this is the first time that the complexity classes for $\boldsymbol{BPL}$ and $\boldsymbol{FPL}$ have been calculated. From this theorem it also follows that all subintuitionistic logics defined using Kripke models are **PSPACE** hard, e.g., all logics considered in [8, 9, 17, 25, 4].

3 A Polynomial Reduction from Normal Modal Logics

Up to now we have proved that most standard normal modal logics have the same complexity class as their strict implication fragment, but we have not given any polynomial reduction from the normal modal logic into its strict implication fragment. Now we present a reduction of this type that works well under certain (general) assumptions.

Let Prop be the set $\{p_n : n \in \omega\}$, and let Prop' be $\{p_n : n \in \omega\} \cup \{q_n : n \in \omega\} \cup \{r_\varphi : \varphi \in \mathcal{L}^{mod}(\mathsf{Prop})\}$. Then, we simultaneously define two translations $^+$ and $^-$ from $\mathcal{L}^{mod}(\mathsf{Prop})$ into $\mathcal{L}^s(\mathsf{Prop}')$:

$$\begin{array}{lllllll}
\bot^+ & := & \bot & \qquad & \bot^- & := & \top \\
\top^+ & := & \top & & \top^- & := & \bot \\
p_n^+ & := & p_n & & p_n^- & := & q_n \\
(\sim\varphi)^+ & := & \varphi^- & & (\sim\varphi)^- & := & \varphi^+ \\
(\varphi_0 \wedge \varphi_1)^+ & := & \varphi_0^+ \wedge \varphi_1^+ & & (\varphi_0 \wedge \varphi_1)^- & := & \varphi_0^- \vee \varphi_1^- \\
(\Box\varphi)^+ & := & \Box\varphi^+ & & (\Box\varphi)^- & := & r_\varphi.
\end{array}$$

By a straightforward simultaneous induction it is easily verified that for every $\mathcal{L}^{mod}(\mathsf{Prop})$-formula φ with propositions among $p_0, \ldots, p_{n-1}$ and with modal degree k, it holds that

$$(1)\quad \left(\Box^{(k)}\Big(\bigwedge_{0\leq i<n}(p_i \supset\subset \sim q_i) \wedge \bigwedge_{\phi\in Sub(\varphi)}(r_\phi \supset\subset \sim \Box\phi^+)\Big)\right) \supset (\varphi \supset\subset \varphi^+) \in \boldsymbol{K}$$

and

$$(2)\quad \left(\Box^{(k)}\Big(\bigwedge_{0\leq i<n}(p_i \supset\subset \sim q_i) \wedge \bigwedge_{\phi\in Sub(\varphi)}(r_\phi \supset\subset \sim \Box\phi^+)\Big)\right) \supset (\sim\varphi \supset\subset \varphi^-) \in \boldsymbol{K}.$$

It is clear that the box of the modal formula

$$\bigwedge_{0\leq i<n}(p_i \supset\subset \sim q_i) \wedge \bigwedge_{\phi\in Sub(\varphi)}(r_\phi \supset\subset \sim \Box\phi^+)$$

is equivalent to the formula

$$\bigwedge_{0\leq i\leq n}\Big(((p_i\wedge q_i) \to \bot)\wedge\Box(p_i\vee q_i)\Big)\wedge \bigwedge_{\phi\in Sub(\varphi)}\Big(((r_\phi\wedge\Box\phi^+) \to \bot)\wedge\Box(r_\phi\vee\Box\phi^+)\Big).$$

Let us call this $\mathcal{L}^s(\mathsf{Prop}')$-formula $t(\varphi)$. We define $h(\varphi)$ as the $\mathcal{L}^s(\mathsf{Prop}')$-formula

$$\Box^{(k)}t(\varphi) \to \Box\varphi^+.$$

It is easy to check that h is computable in polynomial time. Before stating the main theorem of this section we need a new definition.

DEFINITION 7. A normal modal logic L is *closed under extensions by a predecessor* if it is characterized by a class C of frames such that

> for every $\mathcal{F} \in \mathsf{C}$ and every world m in the frame $\mathcal{F}$ there is a frame $\mathcal{F}' \in \mathsf{C}$ with a world m' such that (i) m' is not an initial world in $\mathcal{F}'$ (i.e., it has a predecessor), and (ii) the subframe of $\mathcal{F}$ generated by m and the subframe of $\mathcal{F}'$ generated by m' are isomorphic.

It is obvious that if the state m is not an initial world in $\mathcal{F}$, then the previous condition trivially holds since we can take $\mathcal{F}'$ as $\mathcal{F}$, and m' as m. In particular all Kripke frame complete extensions of **T** are closed under extensions by a predecessor. An easy fact to check is that the normal modal logics **K**, **T**, **K4**, **S4**, **D**, **GL** and **Grz** are closed under extensions by a predecessor.

THEOREM 8. *Let L be a normal modal logic such that is closed under extensions by a predecessor. For every modal formula φ, it holds that*

$$\varphi \in L \quad \textit{iff} \quad h(\varphi) \in L^s.$$

Proof. $(\Rightarrow)$: Suppose that $\varphi \in L$. By (1) we obtain that

$$\Box^{(k-1)}t(\varphi) \supset \varphi^+ \in L.$$

From here it follows that $\Box^{(k)}t(\varphi) \to \Box\varphi^+ \in L$, i.e., $h(\varphi) \in L$.

$(\Leftarrow)$: Let us assume that $h(\varphi) \in L$, and let C be the class of frames given by the closure under extensions by a predecessor condition. Let $\mathcal{F} \in \mathsf{C}$, $m \in F$, and V a valuation in Prop for $\mathcal{F}$. We want to prove that $\mathcal{F}, V, m \Vdash \varphi$. Applying twice the property that C satisfies we can assume that there are states m_0 and m_1 such that $\langle m_0, m_1 \rangle \in R^{\mathcal{F}}$ and $\langle m_1, m \rangle \in R^{\mathcal{F}}$. Now we extend the valuation V to a valuation V' in Prop' for $\mathcal{F}$ according to the following conditions:

- for every $n \in \omega$, $V'(q_n) := \{x \in F : x \notin V(p_n)\}$,
- for every $\phi \in \mathcal{L}^{mod}(\mathsf{Prop})$, $V'(r_\phi) := \{x \in F : \mathcal{F}, V, x \nVdash \Box\phi\}$.

By a straightforward induction it is easily verified that for every modal formula $\phi \in \mathcal{L}^{mod}(\mathsf{Prop})$ and every $x \in F$,

(3) $\quad \mathcal{F}, V, x \Vdash \phi$ iff $\mathcal{F}, V', x \Vdash \phi$ iff $\mathcal{F}, V', x \Vdash \phi^+$ iff $\mathcal{F}, V', x \nVdash \phi^-$.

By (3) and the definition of V' it is not hard to verify that $\mathcal{F}, V' \Vdash t(\varphi)$. Using now that $\mathcal{F}, V', m_0 \Vdash h(\varphi)$ (because $h(\varphi) \in L$) and that $\mathcal{F}, V', m_1 \Vdash \Box^{(k)}t(\varphi)$ we deduce that $\mathcal{F}, V', m_1 \Vdash \Box\varphi^+$. Therefore, $\mathcal{F}, V', m \Vdash \varphi^+$. By (3) it follows that $\mathcal{F}, V, m \Vdash \varphi$. ■

We have already seen that f is a polynomial time reduction for all normal modal logics that are closed under extensions by a predecessor. In particular it holds for **K**, **T**, **K4**, **S4**, **D**, **GL** and **Grz**.

4 Open Problems

Halpern proved in [11] that Ladner's Theorem for normal modal logics holds even in the case that there is a single proposition (see also [6]). On the other hand, in this paper we have seen that the strict implication language satisfies a version of Ladner's Theorem. So, a natural question is whether in the case that there is a single proposition the strict implication language also has a certain version of Ladner's Theorem. We cannot give the same version than in Theorem 6 because it is known that intuitionistic propositional logic with one variable is decidable in linear time [16]. But perhaps it is possible to give the same version when there are only two propositions. Indeed, as far as the author knows it is still open the characterization of the complexity of **IPL** with two variables (see [5, p. 564]). The only result in this direction already proved is in [3] and claims that $\mathbf{K}^s$ is **PSPACE** hard even when Prop is empty.

Another interesting open question is what happens when the strict implication is alone (i.e., without falsum, conjunction and disjunction). One of the few results known for the pure strict implication fragment is that in the case of **IPL** this fragment is also **PSPACE** complete [22].

Acknowledgments

I would like to thank Nick Bezhanishvili for some previous discussions. I also owe my gratitude to Valentin Shehtman for some remarks during the conference. The research presented in this paper was partially supported by Catalan grant 2001SGR-00017 and by Spanish grant BFM2001-3329.

BIBLIOGRAPHY

[1] J. L. Balcázar, J. Díaz, and J. Gabarró. *Structural complexity. I.* Texts in Theortical Computer Science. An EATCS Series. Springer-Verlag, Berlin, second edition, 1995.

[2] P. Blackburn, M. de Rijke, and Y. Venema. *Modal logic.* Number 53 in Cambridge Tracts in Theoretical Computer Science. Cambridge University Press, Cambridge, 2001.

[3] F. Bou. *Strict-Weak Languages. An analysis of strict implication.* Ph. D. Dissertation, University of Barcelona, 2004.

[4] S. Celani and R. Jansana. A closer look at some subintuitionistic logics. *Notre Dame Journal of Formal Logic*, 42(4):225–255, 2001.

[5] A. Chagrov and M. Zakharyaschev. *Modal Logic*, volume 35 of *Oxford Logic Guides.* Oxford University Press, 1997.

[6] A. V. Chagrov and M. N. Rybakov. How many variables does one needs to prove PSPACE-hardness of modal logics. volume 4 of *Advances in Modal Logic*, pages 71–82. 2003.

[7] S. A. Cook. The complexity of theorem-proving procedures. In *ACM, Proc. 3rd ann. ACM Sympos. Theory Computing, Shaker Heights, Ohio 1971, 151-158.* 1971.

[8] G. Corsi. Weak logics with strict implication. *Zeitschrift für Mathematische Logik und Grundlagen der Mathematik*, 33:389–406, 1987.

[9] K. Došen. Modal translations in K and D. In M. de Rijke, editor, *Diamond and Defaults*, pages 103–127. Kluwer Academic Publishers, 1993.

[10] R. Freese. Algorithms in finite, finitely presented and free lattices. Preprint, July 1999.

[11] J. Y. Halpern. The effect of bounding the number of primitive propositions and the depth of nesting on the complexity of modal logic. *Artificial Intelligence*, 75(2):361–372, 1995.

[12] J. Y. Halpern and Y. Moses. A guide to completeness and complexity for modal logics of knowledge and belief. *Artificial Intelligence*, 54:319–379, 1992.

[13] E. Hemaspaandra. The price of universality. *Notre Dame Journal of Formal Logic*, 37:174–203, 1996.

[14] S. Kripke. Semantical analysis of intuitionistic logic I. In J. M. Crossley and M. A. E. Dummet, editors, *Formal Systems and Recursive Functions*, pages 92–130. North-Holland, Amsterdam, 1965.
[15] R. E. Ladner. The computational complexity of provability in systems of modal propositional logic. *SIAM J. Comput.*, 6(3):467–480, 1977.
[16] I. Nishimura. On formulas of one variable in intuitionistic propositional calculus. *The Journal of Symbolic Logic*, 25:327–331 (1962), 1960.
[17] G. Restall. Subintuitionistic logics. *Notre Dame Journal of Formal Logic*, 35(1):116–129, 1994.
[18] W. Ruitenburg. Constructive logic and the paradoxes. *Modern Logic*, 1:207–301, 1991.
[19] W. Ruitenburg. Basic logic and Fregean set theory. In H. Barendregt, M. Bezem, and J. W. Klop, editors, *Dirk van Dalen Festschrift, Questiones Infinitae*, volume 5, pages 121–142. Department of Philosophy, Utrecht University, 1993.
[20] H. Sahlqvist. Completeness and Correspondence in First and Second Order Semantics for Modal Logic. In S. Kanger, editor, *Proceedings of the third Scandinavian logic symposium*, pages 110–143. North-Holland, Amsterdam, 1975.
[21] E. Spaan. *Complexity of modal logics.* Ph. D. Dissertation, Institute for Logic, Language and Computation, 1993.
[22] R. Statman. Intuitionistic propositional logic is polynomial-space complete. *Theoretical Computer Science*, 9(1):67–72, 1979.
[23] L. J. Stockmeyer and A. R. Meyer. Word problems requiring exponential time: preliminary report. In *Fifth Annual ACM Symposium on Theory of Computing (Austin, Tex., 1973)*, pages 1–9. Assoc. Comput. Mach., New York, 1973.
[24] A. Visser. A propositional logic with explicit fixed points. *Studia Logica*, 40:155–175, 1981.
[25] H. Wansing. Displaying as temporalizing. Sequent systems for subintuitionistic logic. In S. Akama, editor, *Logic, Language and Computation*, pages 159–178. Kluwer, 1997.

Félix Bou
JAIST, Nomi, Ishikawa, 923-1292, Japan
bou@jaist.ac.jp

Elementary Canonical Formulae: A Survey on Syntactic, Algorithmic, and Model-theoretic Aspects

WILLEM CONRADIE, VALENTIN GORANKO AND
DIMITER VAKARELOV

ABSTRACT. In terms of validity in Kripke frames, a modal formula expresses a universal monadic second-order condition. Those modal formulae which are equivalent to first-order conditions are called *elementary*. Modal formulae which have a certain persistence property which implies their validity in all canonical frames of modal logics axiomatized with them, and therefore their completeness, are called *canonical*. This is a survey of a recent and ongoing study of the class of elementary and canonical modal formulae. We summarize main ideas and results, and outline further research perspectives.

1 Introduction

1.1 Elementary canonical formulae

We study modal formulae ϕ which are:

(i) elementary (first-order definable):

- *locally,* if there is a first-order formula $\alpha(x)$ such that for every frame F and $w \in F$:

$$F, w \models \phi \text{ iff } F \models_{FO} \alpha(w).$$

- or, *globally,* if there is a first-order sentence α such that for every frame F:

$$F \models \phi \quad \text{iff} \quad F \models_{FO} \alpha.$$

  Clearly, every locally elementary formula is globally so, too. The converse does not hold, as we will see later.

(ii) canonical: informally, that means valid in the canonical frames of all modal logics in which such formula is an axiom. This property is important, because it implies *frame completeness* of logics axiomatized with such formulae.

Formally, we define canonicity in a somewhat stronger, but more uniform and precise way, as *persistence* with respect to a suitable class of general frames containing all canonical general frames of the language. In the standard (polyadic) modal languages, these are the *descriptive frames* (see e.g. [2]), and we identify canonicity with persistence with respect to such frames ($\mathcal{D}$-persistence). However, we note that if the language contains special sorts, such as nominals, or the logics admit special inference rules, the notion of canonicity accordingly changes.

While the class[1] of globally elementary and canonical formulae properly extends the class of locally elementary and canonical ones (see examples in [3]), these two classes behave very similarly, and hereafter we will concentrate mainly on the latter one. By the Fine-van Benthem theorem (see [3]), an elementary modal formula is canonical *iff* the modal logic axiomatized by that formula is complete, so the elementary and canonical formulae are precisely the elementary and complete ones.

The elementary and canonical formulae axiomatize many important modal logics and are of particular practical interest, as they lend themselves to the computational tools developed for first-order logic.

Examples of locally elementary canonical formulae include:

- every valid modal formula;
- the axioms for most well-known logics, incl. T, B , $K4$, $S4$, $S5$,...
- all Sahlqvist formulae [39], and many more...

Non-examples include:

- any formula axiomatizing an incomplete logic, e.g. van Benthem's formula $\Box\Diamond\top \to \Box(\Box(\Box p \to p) \to p)$, which is elementary, but not complete, hence not canonical.
- Fine's formula $\Diamond\Box(p \vee q) \to \Diamond(\Box p \vee \Box q)$, which is canonical, but not elementary.

[1] Here and further we often use the term 'class' when referring to sets of specially defined formulae, not in set-theoretic sense, but as a stylistic way of emphasizing their importance and internal structure.

- McKinsey's formula $\Box\Diamond p \to \Diamond\Box p$ and the Gödel-Löb formula $\Box(\Box p \to p) \to \Box p$, which, although complete, are neither first-order definable, nor canonical.

Hereafter, 'elementary' modal formula will usually mean a globally elementary one, unless otherwise specified in the context.

The sets of locally and globally elementary canonical formulae are not recursive (see [3]) hence the problem arises to establish useful characterizations and to identify rich natural subclasses of effectively recognizable such formulae. Our study follows three main threads of obtaining such characterizations: *syntactic, algorithmic, and model-theoretic,* which are discussed in the subsequent sections. Below we summarize the main issues and results.

1.2 Syntactic classes of elementary canonical formulae

The best-known class of elementary canonical formulae, which was also the starting point of this study, is the class of *Sahlqvist formulae* [39]. While bearing a clear semantic motivation, these formulae are defined purely syntactically, and that syntactic definition is only a lower approximation of the underlying semantic idea (of *minimal valuations*, see [3]). The syntactic definition is extremely fragile, as it does not withstand even simple boolean transformations, or even substitutions changing the polarity of propositional variables. It has, therefore, become customary to tacitly consider the Sahlqvist formulae closed under such simple transformations. On the other hand (see [7]), axiomatic equivalence to a Sahlqvist formula is not decidable, and hence it would be unreasonable to close the class of Sahlqvist formulae under such equivalences. Thus, the notion of Sahlqvist formulae has become fuzzy, and the question '*What is a Sahlqvist formula?*' has gained increasing pertinence

In [23] and [25] we have extended the Sahlqvist formulae to the class of *inductive formulae* in arbitrary polyadic languages, also generalized for hybrid modal languages in [24]. These are still syntactically defined elementary canonical formulae, and their first-order equivalents are still computed by means of minimal, first-order definable valuations which enable elimination of the second-order predicate variables. These minimal valuations are defined inductively, in an order determined by certain syntactic dependencies between the propositional variables within the formula. Like Sahlqvist formulae, the syntactic shape of inductive formulae is rather vulnerable to otherwise inessential transformations, and thus the question 'What is an inductive formula?' remains actual.

In our study we analyze the possibilities to extend the class of syntactically determined elementary canonical formulae:

- by extending further the syntactic definition of inductive formulae;

- by adding and refining a *pre-processing* phase, in attempt to transform the formula into an inductive formula, while preserving its frame condition.
- by closing the class of inductive formulae under suitable equivalences preserving the elementary canonical formulae, to larger, still effectively recognizable classes.

Also, we develop purely syntactic procedures for computation of the first-order equivalents of effectively defined classes of elementary canonical formulae. For instance, in [24] we present such method for inductive formulae in temporal (more generally, reversive) languages with nominals.

1.3 Algorithmic approach to elementary canonical formulae

A natural extension of the syntactic approach aims at development of algorithms which identify elementary canonical formulae, and thus produce *effectively enumerable* classes of such formulae. Such algorithmically definable classes need not be decidable, but they are much less tied-up with the syntactic shape of the formulae, and the algorithmic approach penetrates deeper into the semantic nature of the elementary canonical formulae.

To establish first-order definability of a modal formula amounts to elimination of the monadic second-order quantifiers occurring in its standard translation. Therefore, prime candidates for algorithms producing elementary canonical formulae are the two currently developed and implemented algorithms for second-order quantifier elimination, viz. SCAN and DLS (see [34]). Both are provably correct and incomplete and, while seemingly based on different ideas and with quite distinct computational behavior, none of them is stronger than the other for that task. It has been proved in [27] that SCAN succeeds for all Sahlqvist formulae. Moreover, this holds for all inductive formulae, and the same applies to DLS (see [12]).

In [10] we have developed a new algorithm, SQEMA, for computing first-order equivalents of modal formulae and have proved the canonicity of all formulae on which it succeeds.

1.4 Model-theoretic aspects of elementary canonical formulae

This direction of research aims at characterizing the elementary canonical formulae of a given modal language in practically more useful model-theoretic terms. A typical such characterization (as a sufficient condition) is a suitable notion of persistence, e.g.: with respect to all descriptive frames, in the case of standard polyadic modal languages; with respect to all discrete frames, in the case of hybrid modal languages with nominals; with respect to all refined frames, in the case of logics with additional non-orthodox rules of inference.

Various other persistence properties have emerged as useful tools for model theoretic analysis and classification of elementary canonical, and related, formulae. For instance, as established by van Benthem in [3], the modal formulae amenable to the method of substitutions turn out to be precisely those persistent with respect to the general frames in which all parametrically first-order definable sets are admissible. Also, in [25] we have introduced a new notion of persistence which separates the Sahlqvist formulae from the inductive ones, and have proved that, up to local equivalence in all discrete general frames in reversive languages with nominals, the inductive, pure (not containing propositional variables), and locally discrete-persistent formulae, coincide, thus delineating a very large and natural class of elementary and 'discretely-canonical' formulae in such languages.

The persistence properties of the elementary canonical formulae have a distinct topological nature, first identified in [40] and used in [41] to give a uniform proof of first-order definability and canonicity of Sahlqvist formulae. We have continued and extended that analysis, and used topological arguments to establish first-order definability and canonicity of inductive formulae in [25], and of the formulae on which the algorithm SQEMA succeeds in [10].

2 Preliminaries

In this paper we assume that the reader has basic familiarity with syntax and semantics of modal logic, some useful references on which include [3], [2], and [6]. For the reader's convenience, we briefly recall some important facts related to general frames and persistence.

2.1 General frames

For technical simplicity, we will only consider a basic monadic modal language. For treatment of general polyadic languages see [23], [25], as well as [24] for hybrid polyadic languages. For general background on model theory of modal logic see [2], [6], [8], and [28].

Given a Kripke frame $F = \langle W, R\rangle$, a **general frame over** F is a structure $\mathfrak{F} = \langle W, R, \mathbb{W}\rangle$ expanding F with a modal algebra $\mathbb{W}$ of **admissible subsets** of W, closed under all Boolean and modal operators, i.e., $\mathbb{W}$ is a modal subalgebra of $\langle \mathcal{P}(W); \cap, -, \Box, \varnothing\rangle$ where $\Box X = \{x \in W \mid \forall y(Rxy \rightarrow y \in X)\}$. The operator $\Diamond$ is defined dually: $\Diamond X = \{x \in W \mid \exists y(Rxy \wedge y \in X)\}$.

A valuation V in the frame F is **admissible** in $\mathfrak{F}$ if $V(p) \in \mathbb{W}$ for every variable p. A **Kripke model over** $\mathfrak{F}$ is any Kripke model $M = \langle F, V\rangle$ with a valuation V admissible in $\mathfrak{F}$.

Local (at a state) and **global validity of a modal formula in a general frame** is defined as truth at the state (resp. validity) in every admissible model over that general frame.

Every general frame $\mathfrak{F} = \langle W, R, \mathbb{W} \rangle$ defines a topology $T(\mathfrak{F})$ on W with $\mathbb{W}$ as a base of clopen sets, i.e. the closed sets of the topology are precisely all intersections of admissible sets.

2.2 Some important classes of general frames

A general frame $\mathfrak{F} = \langle W, R, \mathbb{W} \rangle$ is:

- **differentiated** if for every $x, y \in W$, if $x \neq y$ then $x \in X$ and $y \notin X$ for some $X \in \mathbb{W}$ or, equivalently, if $T(\mathfrak{F})$ is Hausdorff.
- **tight** if for any $x, y \in W$, Rxy iff $\forall Y \in \mathbb{W}(y \in Y \Rightarrow x \in \Diamond(Y))$, or, equivalently, if R is point-closed, i.e. $R(\{x\})$ is closed for every $x \in W$, where $R(X)$ denotes the set of all successors of points in X.
- **refined** if it is differentiated and tight.
- **compact** if every family of admissible sets in $\mathfrak{F}$ with FIP has a non-empty intersection, or, equivalently, if $T(\mathfrak{F})$ is compact.
- **discrete**, if $\{u\} \in \mathbb{W}$ for every $u \in W$.
- **elementary**, if every subset of W, which is parametrically first-order definable in the first-order language for Kripke frames, is admissible.
- **descriptive** if it is refined and compact.

The class of all differentiated (resp. tight, refined, discrete, elementary, descriptive) general frames will be denoted by $\mathcal{DF}$ (resp. $\mathcal{T}, \mathcal{R}, \mathcal{DI}, \mathcal{E}, \mathcal{D}$).

Some relationships between these classes (see [28]):

$$\mathcal{E} \subsetneq \mathcal{DI} \subsetneq \mathcal{R} = \mathcal{DF} \cap \mathcal{T}; \quad \mathcal{D} \subsetneq \mathcal{R}; \quad \mathcal{D} \not\subseteq \mathcal{DI} \not\subseteq \mathcal{D}.$$

2.3 Persistence and canonicity

Let $\mathcal{C}$ be any class of general frames. A modal formula is **locally $\mathcal{C}$-persistent**, if for every general frame $\mathfrak{F} = \langle F, \mathbb{W} \rangle \in \mathcal{C}$, and $w \in F$: $\mathfrak{F}, w \models \phi$ implies $F, w \models \phi$; ϕ is **$\mathcal{C}$-persistent**, if for every such general frame, $\mathfrak{F} \models \phi$ implies $F \models \phi$.

Clearly, local persistence implies persistence, but the converse does not always hold. If we denote by $\mathcal{C}^p$ the set of all $\mathcal{C}$-persistent formulae, we have the following (see [28]):

$$\mathcal{DF}^p \cap \mathcal{T}^p = \mathcal{R}^p \subsetneq \mathcal{DI}^p \subsetneq \mathcal{E}^p; \quad \mathcal{R}^p \subsetneq \mathcal{D}^p; \quad \mathcal{DI}^p \not\subseteq \mathcal{D}^p \not\subseteq \mathcal{E}^p.$$

The same relationships hold for the local persistences.

How are persistence and canonicity related? Descriptive frames typically appear as the *canonical general frames* of every normal modal logic without any special inference rules. Thus, all $\mathcal{D}$-persistent formulae are valid in the underlying canonical Kripke frames, and hence they axiomatize Kripke complete logics. For that reason the $\mathcal{D}$ -persistent formulae are often (incl. in this study) identified with canonical formulae.

However, in hybrid logics with nominals, or in logics with special additional ('context', or 'non-orthodox') rules of inference, $\mathcal{D}$ -persistent formulae *need not* be canonical, because the canonical general frames for such logics are only *discrete* (for hybrid logics, see [2]) or *refined* (in logics with additional rules).

Furthermore, $\mathcal{DI}$-persistent formulae have the important property of remaining canonical when added as axioms to hybrid logics with nominals, while $\mathcal{R}$-persistent formulae remain canonical not only in the presence of other axioms, but even if additional rules of inference of the type mentioned above are added to the axiomatic system. Thus, the right notion of canonicity in such languages is $\mathcal{DI}$-persistence, resp. $\mathcal{R}$-persistence.

3 Syntactic classes of elementary canonical formulae

3.1 The starter: Sahlqvist formulae

After the introduction of Kripke semantics for modal logics, a quest for general completeness results ensued, which culminated in Sahlqvist's theorem [39]. Sahlqvist proved two notable facts about a large, syntactically defined class of modal formulae, now called Sahlqvist formulae: the *first-order correspondence:* that they all define first-order conditions on Kripke frames and these conditions can be effectively "computed" from the modal formulae; and the *canonicity:* that all these formulae are valid in their respective canonical frames, and hence axiomatize completely the classes of frames satisfying their corresponding first-order conditions.

DEFINITION 1. In a fixed standard modal language ML we define the following syntactic classes of formulae.

- **Positive and negative formulae** are defined as usual.
- A **boxed atom** is a formula $\Box_1 \ldots \Box_n p$ where $\Box_1, \ldots, \Box_n$ is a (possibly empty) string of unary boxes and p is a propositional variable.
- A **Sahlqvist antecedent** is a formula constructed from boxed atoms and negative formulae by applying conjunctions, disjunctions, and diamonds.
- A **definite Sahlqvist antecedent** is a Sahlqvist antecedent obtained without applying disjunctions.

- A **(definite) Sahlqvist implication** is a formula $A \to P$ where A is a (definite) Sahlqvist antecedent and P is a positive formula.
- A **Sahlqvist formula** is a formula obtained from Sahlqvist implications by freely applying conjunctions, disjunctions, and boxes.

  We note that every Sahlqvist implication is tautologically equivalent to a formula of the type $\neg A$ where A is a Sahlqvist antecedent, and therefore every Sahlqvist formula is semantically equivalent to a negated Sahlqvist antecedent, too.

  The set of Sahlqvist formulae of ML will be denoted by SF(ML), or just SF if the language is clear from the context.

Some examples: $\Diamond\Box p \to \Box p$ and $\Box((\Box(\Diamond\neg p \vee \Diamond\Box\neg q) \wedge \Diamond\Box p) \to \Box\Diamond\Box(p \vee \Diamond\Box q))$ are Sahlqvist formulae, but not $\Box\Diamond p \to p$ or $\Box(p \vee q) \to (p \vee q)$. Even the K-axiom $\Box(p \to q) \to (\Box p \to \Box q)$, or its equivalent $\Box p \wedge \Box(\neg p \vee q) \to \Box q$ are (syntactically) not Sahlqvist formulae.

THEOREM 2 (Sahlqvist, 1973). *All Sahlqvist formulae are elementary and canonical.*

For more on Sahlqvist's theorem, including a proof and related results, see [3], [41], [2], [6], [33], [30], [16]. A generalization of Sahlqvist's theorem for polyadic languages has been proposed in [38].

What makes Sahlqvist formulae tick? The characteristic semantic feature of the Sahlqvist formulae, which is in the heart of *Sahlqvist-van Benthem substitution method* (see [3], [2]), is the existence of a *minimal valuation* for the occurring propositional variables, which makes the antecedent true. The minimal valuations have the following property:

If a Sahlqvist formula is valid for the minimal valuation in any given frame, then it is valid for every valuation on that frame.

Thus, the idea of the method of substitutions, applied to Sahlqvist formulae, is to compute the minimal valuations from the antecedent of the standard translation, and then to substitute them in the consequent. The result of that substitution is an equivalent formula where the second-order predicates in the standard translation are eliminated.

Example: Take $\theta = \Box p \to \Box\Box p$.

Then $ST(\theta)(x) = \forall y(Rxy \to Py) \to \forall y(Rxy \to \forall z(Ryz \to Pz))$. The minimal valuation of P satisfying the antecedent is $P = \{y \mid Rxy\}$. Substituting in the consequent yields $\forall y(Rxy \to \forall z(Ryz \to Rxz))$, i.e. transitivity.

Furthermore, the minimal valuations for Sahlqvist formulae are:

- *first-order definable,* whence the first-order definability of Sahlqvist formulae;
- *closed sets* in the topological spaces generated by the admissible sets in descriptive frames, whence the canonicity of Sahlqvist formulae can be derived.

How far does the class of Sahlqvist formulae stretch? On one hand, it is quite large but on the other, being syntactically defined, it is unstable even under Boolean transformations, so usually some *pre-processing* is needed to transform (if possible) a formula into that shape. For instance:

- $\neg p \to \neg\Box p$ is not a Sahlqvist formula but becomes one after a trivial Boolean transformation. Likewise for the contradictory formula $(\Box(\Box p \to p) \to \Box p) \wedge \neg(\Box(\Box p \to p) \to \Box p)$.
- $\Box(p \to q) \to (\Box p \to \Box q)$ becomes a Sahlqvist formula after a Boolean transformation and substitution of $\neg q$ for q. Likewise for $\Box\Diamond p \to p$ and $\Box(p \vee q) \to (p \vee q)$.
- $p \wedge \Box(\Diamond p \to \Box q) \to \Diamond\Box q$ is not a Sahlqvist formula, nor is it reducible to one with such 'simple' syntactic transformations. Yet, it is an elementary canonical formula and determines the same frame condition as the Sahlqvist formula $p \to \Diamond(\Diamond p \vee \Box\bot)$.

Sahlqvist formulae do not cover, in any reasonable sense, all elementary canonical formulae. For instance:

- $p \wedge \Box(\Diamond p \to \Box q) \to \Diamond\Box\Box q$ is an elementary canonical formula but not a Sahlqvist formula, nor is it reducible to one. In fact, it does not determine the same frame condition as any Sahlqvist formula in the basic modal language, as proved in [25]. Still, in the basic tense language it is equivalent to the Sahlqvist formula $p \to FGGP(Fp \wedge Pp)$.
- Likewise, $(\Box\Diamond p \to \Diamond\Box p) \wedge (\Box p \to \Box\Box p)$ is an elementary canonical formula [3], but not frame equivalent to any Sahlqvist formula.
- $\Diamond\Box(p \vee q) \to \Diamond(\Box p \vee \Box q)$, $\Box\Diamond p \to \Diamond\Box p$, and $\Box(\Box p \to p) \to \Box p$ are not Sahlqvist formulae, because they are neither elementary nor canonical [3].

A formula which is not in SF can possibly still be reduced to an equivalent Sahlqvist formula, in one or another sense. So, *what should we call a Sahlqvist formula?* To address this question, we first need to digress from it and discuss in more detail the various natural notions of equivalence arising in modal logic.

3.2 A hierarchy of equivalences

DEFINITION 3. Modal formulae A and B are:

- **tautologically equivalent (TAU)** if $A \leftrightarrow B$ is a propositional tautology.
- **semantically equivalent (SEM)** if $A \leftrightarrow B$ is a valid modal formula, i.e. A and B are valid at the same states of every Kripke model.
- **model-equivalent (MOD)** if valid in the same Kripke models.
- **locally equivalent (LOC)** if valid at the same states of every general frame.
- **algebraically equivalent (ALG)** if valid in the same modal algebras, equivalently, in the same general frames.
- **locally frame-equivalent (LFR)** if valid at the same states in every frame.
- **frame-equivalent (FR)** if valid in the same Kripke frames.
- **axiomatically equivalent (AX)** if the logics $\mathbf{K}+A$ and $\mathbf{K}+B$ have the same theorems. Equivalently, if $\mathbf{K}+A \vdash B$ and $\mathbf{K}+B \vdash A$.

We want to close the class of Sahlqvist formulae under as strong as possible equivalences, so as to preserve its effectiveness. Note, that **AX**, **LFR** and **FR** are not decidable [3]. Moreover, **AX** is not decidable even on the class SF [7]. Thus, we cannot safely close the class of Sahlqvist formulae under **AX**, if we want to preserve its effectiveness. Decidable equivalence closures of SF are currently under investigation.

3.3 Beyond Sahlqvist formulae: monadic inductive formulae

Here we present an extension of the class of Sahlqvist formulae, introduced for arbitrary polyadic languages in [23], [25].

DEFINITION 4. Let ML be a fixed monadic (multi-)modal language and # be a symbol not in ML. We define **box-forms of #** as follows:

- # is a box-form of #.
- If $\mathbf{B}(\#)$ is a box-form of #, then $\Box\mathbf{B}(\#)$ is a box-form of #, for any box $\Box$ in ML.
- If $\mathbf{B}(\#)$ is a box-form of #, and A is a positive formula, then $A \rightarrow \mathbf{B}(\#)$ is a box-form of #;

Thus, box-forms are, up to semantic equivalence, of the type

$$\Box_1(A_1 \to \Box_2(A_2 \to \dots \Box_n(A_n \to \#)\dots))$$

where $\Box_1, \dots, \Box_n$ are compositions of boxes in ML(τ) and $A_1, \dots, A_n$ are positive formulae.

A **box-formula of** p is the result $\mathbf{B}(p)$, of substitution of p for $\#$ in any box-form $\mathbf{B}(\#)$. The last occurrence of the variable p is the **head** of $\mathbf{B}(p)$ and every other occurrence of a variable in $\mathbf{B}(p)$ is **inessential** there.

DEFINITION 5. A **(monadic) regular formula** is any modal formula built from positive formulae and negated box-formulae by applying conjunctions, disjunctions, and boxes.[2]

The **dependency digraph** of a set of box-formulae $\mathcal{B} = \{\mathbf{B}_1(p_1), \dots, \mathbf{B}_n(p_n)\}$ is a digraph $G = \langle V, E\rangle$, where $V = \{p_1, \dots, p_n\}$ is the set of heads in $\mathcal{B}$, and $p_i E p_j$ iff p_i occurs as an inessential variable in a box from $\mathcal{B}$ with a head p_j; in such case we say that p_j **depends** on p_i. A digraph is called **acyclic** if it does not contain oriented cycles (including loops).

A **monadic inductive formula** is a monadic regular formula for which the dependency digraph of the set of all box-formulae occurring in it as subformulae, is acyclic.

EXAMPLE 6. The formula

$$D = p \land \Box(\Diamond p \to \Box q) \to \Diamond\Box\Box q \equiv \neg p \lor \neg\Box(\Diamond p \to \Box q) \lor \Diamond\Box\Box q$$

is an inductive formula, obtained as a disjunction of the negated box-formulae $\neg p$ and $\neg\Box(\Diamond p \to \Box q)$, and the positive formula $\Diamond\Box\Box q$. The dependency digraph of D over the set of heads $\{p, q\}$ has only one edge, from p to q.

Sahlqvist formulae are a simple particular case of inductive formulae, where all box-formulae are just boxed atoms $\Box_1 \dots \Box_n p$, and hence the dependency digraph has no arcs at all. In fact, the class SF can be substantially generalized simply by replacing in the definition of classical monadic (multi-)modal Sahlqvist formulae 'boxed atoms' by 'box-formulae', and further requiring that the set of all box-formulae occurring as subformulae in the antecedent is *independent*, i.e. they all have different heads, and no head occurs inessentially in any of them. For instance,

$$\Diamond(\Box(\Box\Diamond q \to \Box\Box p_1) \land \Box\Box(\Diamond\Box q \to \Box(\Diamond q \to p_2))) \to \Diamond(p_1 \land \Box(\Diamond p_2 \lor q))$$

is not a Sahlqvist formula, but a simply generalized one.

[2]Just like a Sahlqvist formula, where the boxed atoms are now any box-formulae.

THEOREM 7 ([23, 25]). *All monadic inductive formulae are locally elementary and canonical.*

Proof. (Sketch) The first-order equivalents of inductive formulae can be computed by the method of substitutions, just like for Sahlqvist formulae, but inductively, following a partial ordering $\prec$ induced by the dependency digraph. More specifically, we first compute the minimal valuations of the variables which are not heads of box-subformulae. They only occur positively in the formula, so their minimal valuations are $\varnothing$. Then we proceed with the head variables in the box-subformulae, beginning with those which do not depend on any variables (i.e. the sources in the dependency graph). Thus, step by step we compute the minimal valuations of all head variables which only depend on variables whose valuations have already been computed. The acyclicity of the dependency graph of the inductive formula guarantees the successful completion of that procedure.

The canonicity follows similar lines, but needs some topological arguments. Let $A = A(q_1, \ldots, q_n)$ be a monadic inductive formula, $\mathfrak{F} = \langle \mathbf{F}, \mathbb{W} \rangle$ be a descriptive general frame such that $\mathfrak{F} \models A$, and V_m be the minimal valuation for $q_1, \ldots, q_n$. It suffices to prove that $\mathbf{F}, V_m \models A$.

Problem: *the minimal valuation need not be admissible* in $\mathfrak{F}$, so we cannot claim that $\mathfrak{F}, V_m \models A$. However, it suffices to show the following:

(C1) V_m is closed i.e. an intersection of admissible valuations.

(C2) For every closed valuation U in $\mathfrak{F}$ and a positive formula P, $U(P) = \bigcap_{U \preceq V} V(P)$ where the intersection ranges over all admissible valuations V which extend U.

(C1) is proved by $\prec$ -induction for every $V_m(q_j)$. (C2) is proved by structural induction on positive formulae, where the crucial step is Esakia's lemma (see e.g. [6], [2]) which essentially claims that infinite intersections of admissible sets in descriptive frames distribute over $\Diamond$, thus implying that $\Diamond$ is a closed operator in every topology over a descriptive frame. ■

REMARK 8. The conditions (C1) and (C2) in the proof above hold trivially for Kripke frames (i.e. full general frames), which allows for simultaneous treatment of first-order definability and canonicity of inductive formulae, in the spirit of Sambin and Vaccaro's proof [41].

EXAMPLE 9. The local first-order correspondent of the formula $D = \neg p \vee \neg\Box(\Diamond p \rightarrow \Box q) \vee \Diamond\Box\Box q$ is computed as follows. Since $p \prec q$, we first compute $V_m(p) = \{w\}$ where w denotes the current state in a frame with domain W. Then $V_m(q)$ is the minimal subset $Q(w)$ of W such that

$w \in \Box(\Diamond\{w\} \to \Box Q(w))$. This is equivalent to $\Diamond^{-1}\{w\} \in \Diamond\{w\} \to \Box Q_w$, i.e. $\Diamond^{-1}\{w\} \cap \Diamond\{w\} \subseteq \Box Q(w)$, i.e. $\Diamond^{-1}(\Diamond^{-1}\{w\} \cap \Diamond\{w\}) \subseteq Q(w)$. Thus, $V_m(q) = \Diamond^{-1}(\Diamond^{-1}\{w\} \cap \Diamond\{w\})$ and the (set-theoretic record of the) local first-order equivalent of D at w is

$$w \in \Diamond\Box\Box\Diamond^{-1}(\Diamond^{-1}\{w\} \cap \Diamond\{w\}).$$

This condition corresponds to the local first-order formula

$$\mathrm{FO}(D)(w) = \exists y(Rwy \wedge \forall z(R^2yz \to \exists u(Rwu \wedge Ruw \wedge Ruz))).$$

As mentioned earlier, the formula D is not frame equivalent to any Sahlqvist formula in the basic modal language (see section 5.1). In particular, we note that FO(D) is not a Kracht formula (see [32]). For more details on computing first-order equivalents of inductive formulae, see [23], [25].

3.4 Inductive formulae in polyadic modal languages

We now outline the generalization of monadic inductive formulae to arbitrary polyadic languages, introduced in [23], [25].

First, note that the inductive formulae *are not implications (like Sahlqvist implications), but composite polyadic boxes of special shape.* In order to extend the inductive formulae to polyadic modal languages we will adopt a somewhat non-orthodox view on these languages, by treating conjunctions and disjunctions as modal operators, and allowing compositions of modal operators, in PDL style. This treatment flattens the structure of polyadic modal formulae and makes their syntactic classification simpler.

Purely modal polyadic languages.

DEFINITION 10. A purely modal polyadic language $\mathcal{L}_\tau$ contains a countably infinite set propositional variables VAR, negation $\neg$, and a **modal similarity type** τ consisting of a set of **basic modal terms** (modalities) with pre-assigned finite arities, including a 0-ary modality ι_0, a unary one ι_1, and a binary one ι_2.

The intuition behind the 3 distinguished modalities above: ι_0 will be interpreted as the constant $\top$ and its dual as $\bot$; ι_1 will be the self-dual identity; ι_2 will be $\vee$ and its dual — $\wedge$.

DEFINITION 11. By simultaneous mutual induction we define the set of **modal terms** $MT(\tau)$ and their **arity function** ρ, and the set of **(purely) modal formulae** $MF(\tau)$ as follows:

(MT i) Every basic modal term is a modal term of the predefined arity.

(MT ii) Every constant formula (having no variables) is a 0-ary modal term.

(MT iii) If $n > 0$, $\alpha, \beta_1, \ldots, \beta_n \in MT(\tau)$ and $\rho(\alpha) = n$, then $\alpha(\beta_1, \ldots, \beta_n) \in MT(\tau)$ and $\rho(\alpha(\beta_1, \ldots, \beta_n)) = \rho(\beta_1) + \ldots + \rho(\beta_n)$.

Modal terms of arity 0 will be called **modal constants**.

(MF i) Every propositional variable is a modal formula.

(MF ii) Every modal constant is a modal formula.

(MF iii) If A is a formula then $\neg A$ is a formula;

(MF iv) If $A_1, \ldots, A_n$ are formulae, α is a modal term and $\rho(\alpha) = n > 0$, then $[\alpha](A_1, \ldots, A_n)$ is a modal formula.

Note that all formulae in a purely modal language are literals, boxes, or diamonds (negations of boxes). For technical purposes we extend the series of ι's with n-ary modalities ι_n: inductively as follows: $\iota_{n+1} = \iota_2(\iota_1, \iota_n)$ for $n > 1$.

Some notation on formulae:
$\langle\alpha\rangle(A_1, \ldots, A_n) := \neg[\alpha](\neg A_1, \ldots, \neg A_n)$; $\top := \iota_0, \bot := \neg\iota_0$;
$A \vee B := [\iota_2](A, B)$, $A \wedge B := \langle\iota_2\rangle(A, B)$, and respectively
$A_1 \vee \ldots \vee A_n := [\iota_n](A_1, \ldots, A_n)$, $A_1 \wedge \ldots \wedge A_n := \langle\iota_n\rangle(A_1, \ldots, A_n)$;
$A \to B := \neg A \vee B$; $A \leftrightarrow B := (A \to B) \wedge (B \to A)$.

For instance, the formula $D = p \wedge \Box(\Diamond p \to \Box q) \to \Diamond\Box\Box q$, after elimination of $\to$ and $\wedge$ becomes $\neg p \vee \neg\Box(\Box\neg p \vee \Box q) \vee \Diamond\Box\Box q$, which is represented in the polyadic language as:

$$D = [\iota_3](\neg p, \neg[\alpha(\iota_2(\alpha, \alpha))](\neg p, q), \langle\alpha\rangle[\alpha][\alpha]q),$$

where $[\alpha]$ corresponds to $\Box$.

Positive and negative occurrences of variables and positive and negative formulae are defined as usual.

Let us fix an arbitrary purely modal language $\mathcal{L}_\tau$. The semantics of $\mathcal{L}_\tau$ is a straightforward combination of the standard Kripke semantics for polyadic modal languages and PDL-type polymodal languages, taking into account the fact that conjunctions and disjunctions are now treated as modalities. This is accomplished by using the $(n+1)$-ary *identity relation* as the accessibility relation corresponding to $[\iota_n]$. Also, the notions of general frames and truth and validity in them generalize in a predictable way. (For details, see [23], [25].) The **standard translation** ST extends to polyadic languages with the clauses:

- $ST(\sigma) = R_\sigma(x)$ for every modal constant σ;
- $ST([\alpha](A_1, \ldots, A_n) = \forall\overline{y}(R_\alpha xy_1 \ldots y_n \to \bigvee_{i=1}^n ST(A_i)(y_i/x))$

Note that the propositional logical connectives $\wedge, \vee, \to$, as defined above, have their standard semantic interpretation. Therefore, the purely modal polyadic languages and the traditional ones are equally expressive.

Polyadic inductive formulae.

Given a purely modal polyadic language $\mathcal{L}_\tau$, an **essentially box-formula** in it is a modal formula of one of the following two types:

- **Headless boxes**, of the form $B = [\beta](N_1, \ldots, N_m)$, where β is any m-ary (composite) modal term, for $m \geq 1$, and $N_1, \ldots, N_m$ are negative formulae.
- **Headed boxes**, of the form $B = [\beta](p, N_1, \ldots, N_m)$, where β is any $(m+1)$-ary (composite) modal term, for $m \geq 0$, and $N_1, \ldots, N_m$ are negative formulae. The variable p is called the **head** of the box (here the head is put on the first place only for convenience of notations). In particular, p and $[\beta]p$ for any unary modal term β, are headed boxes.

All variables in an essentially box-formula except for the head of the formula (if any) are called **inessential variables** in that formula.

A **regular (polyadic) formula** is any modal constant (a 0-ary modal term) or a formula $A = [\alpha](\neg B_1, \ldots, \neg B_n)$ where α is an n -ary modal term and $B_1, \ldots, B_n$ are essentially box-formulae. The **dependency digraph** of A is a digraph $G = \langle \mathbf{V}_A, \mathbf{E}_A \rangle$ where $\mathbf{V}_A = \{p_1, \ldots, p_n\}$ is the set of heads in A, and $p_i \mathbf{E}_A p_j$ iff p_i occurs as an inessential variable in a formula from $B_1, \ldots, B_n$ with a head p_j.

A **(polyadic) inductive formula** is any regular formula A with an acyclic dependency digraph.[3] Note that the class of polyadic inductive formulae contains all monadic ones. In particular,
$D = [\iota_3](\neg p, \neg[\alpha(\iota_2(\alpha, \alpha))](\neg p, q), \langle\alpha\rangle[\alpha][\alpha]q)$ is a polyadic inductive formula. The class of polyadic inductive formulae can be further closed under conjunctions.

THEOREM 12 ([23],[25]). *All polyadic inductive formulae are locally elementary and canonical.*

The proof extends the one for monadic inductive formulae with the due technical overhead, but without essential conceptual complications.

[3] In [23] these were called 'polyadic Sahlqvist formulae'.

EXAMPLE 13. Computing the first-order equivalent to the polyadic inductive formula $B = [\mathbf{3}](\neg[\mathbf{1}]p, \neg[\mathbf{2}](\neg p, q), \langle\mathbf{1}\rangle[\mathbf{1}]q)$: the dependancy graph has one arc, $p \prec q$, so we first compute $V_m(p) = R_1(y_1)$. Then $V_m(q) = \{z | \exists s(R_2 y_2 s z \wedge R_1 y_1 s)\}$. Finally, $FO(B) =$
$\forall x y_1 y_2 y_3 (R_\mathbf{3} x y_1 y_2 y_3 \rightarrow \exists v(R_\mathbf{1} y_3 v \wedge \forall w(R_\mathbf{1} v w \rightarrow \exists s(R_\mathbf{2} y_2 s w \wedge R_\mathbf{1} y_1 s))))$.

Note that, once $V_m(p)$ is determined, $[\mathbf{2}](\neg p, q)$ can be regarded as a *unary box*: $[\alpha](q) = [\mathbf{2}](\neg V_m(p), q)$ where $\alpha = \mathbf{2}(\neg V_m(p), \iota_1)$ is a *unary parametrized modal term*, the relation of which can be accordingly computed: $R_\alpha xy$ iff $\exists s(R_2 xsy \wedge V_m(p)(s))$. This trick is essential in the proof of canonicity.

3.5 Inductive formulae in hybrid and reversive modal languages

Given a modal similarity type τ we extend the modal language $\mathcal{L}_\tau$ by adding *nominals* and the *universal modality* (see e.g. [18] or [2]), as well as inverse (residual) modalities, to obtain $\mathcal{L}_{\tau,r}^{u,\mathbf{n}}$. We now briefly consider inductive formulae in such languages, as developed in [24].

A **pure formula** in (a sublanguage of) $\mathcal{L}_{\tau,r}^{u,\mathbf{n}}$ is a formula that contains no propositional variables. Note that that *every pure formula is locally first-order definable.*

The definition of modal terms in $\mathcal{L}_{\tau,r}^{u,\mathbf{n}}$ extends the original one with the clause: *Every pure formula is a* 0*-ary modal term*, i.e. modal terms can be parameterized with pure formulae.

Inductive polyadic formulae in $\mathcal{L}_{\tau,r}^{u,\mathbf{n}}$ are defined as in purely modal polyadic languages, but on the extended set of modal terms.

A modal polyadic language is **reversive**[4] if, together with every n-ary modal term α it contains its 'inverses' $\alpha^1, \ldots, \alpha^n$ where for each $k = 1, \ldots, n$:

$$xR_{\alpha^k} y_1 \ldots y_k \ldots y_n \text{ iff } y_k R_\alpha y_1 \ldots x \ldots y_n.$$

In fact, it suffices to require this condition for the *basic* modal terms from the signature. An example of a reversive language is the basic tense language.

THEOREM 14 ([24]). *Every inductive formula in a reversive language with nominals,* $A = [\alpha](\neg H_1, \ldots, \neg H_n, P_1, \ldots, P_k)$, *is axiomatically equivalent*[5] *to a pure formula*

$$A^\circ = [\alpha](\neg c_1, \ldots, \neg c_n, Q_1, \ldots, Q_k)$$

where $c_1, \ldots, c_n$ *are nominals and* $Q_1, \ldots, Q_k$ *are obtained by means of effectively computable pure substitutions.*

[4]These are similar to Venema's *versatile* languages [45], but not quite the same.

[5]In an axiomatic system with additional rules for the nominals (see [18, 2]).

Note that the corresponding pure formula of a given Sahlqvist formula A codes the intended first-order equivalent of A.

Recall, that the right notion of canonicity in languages with nominals is $\mathcal{DI}$-persistence, so we will hereafter refer to this notion as *discrete canonicity.*

COROLLARY 15. *Every inductive formula in a reversive language with nominals is elementary and discretely canonical.*

Thus, proving the analogue of Sahlqvist's theorem for inductive formulae in reversive hybrid language with universal modality becomes merely a syntactic exercise.

3.6 Pushing the limits of the syntactic approach

Sahlqvist and inductive formulae do not exhaust the shapes of elementary canonical formulae. Other syntactic classes of such formulae include:

- All formulae of modal depth 1: van Benthem has classified them and proved their FO definability in [3]. Their canonicity can be verified by considering all cases.
- Consider *modal reduction principles* $M_1p \to M_2p$, where M_1 and M_2 are strings of boxes and diamonds. Again, van Benthem [3] has identified the first-order definable ones, and they are all easily seen to be canonical.
- *All* modal reduction principles on transitive frames.
- *Complex formulae,* see [44]. Example:

  $$\Diamond\Box(p \vee q) \wedge \Diamond\Box(p \vee \neg q) \wedge \Diamond\Box(\neg p \vee q) \to POS(p \to q, p \leftrightarrow q, p \wedge q)$$

  for any positive formula $POS(p_1, p_2, p_3)$.

  These are not inductive, but can be converted into inductive formulae by means of rather intricate substitutions.

All these syntactic classes are unstable under inessential transformations. For instance, the inductiveness can foolishly fail, e.g. in $\Box(p \leftrightarrow q) \to q$. What more can be done to extend the scope of the syntactic approach? Here are some further ideas:

- *Pre-processing* (see [24],[25]), using Boolean and modal equivalences, suitable substitutions, e.g. changing polarities or the special substitutions for complex formulae, normal forms, etc.

- In the definitions of Sahlqvist and inductive formulae, '*positive*' and '*negative*' formulae can be replaced respectively by *upwards* and *downwards monotone*. By Lyndon's monotonicity theorem for modal logic (see e.g. [37]), such replacements preserve the formulae up to semantical equivalence, and hence preserve first-order definability and canonicity. Note that testing monotonicity of a modal formula is *decidable:* $B(p)$ is upwards monotone iff $\models B(p \wedge q) \rightarrow B(p)$ for q not occurring in $B(p)$. So, the definitions can be amended without loss of effectivity (though, at the expense of increased complexity).

- The definitions of both Sahlqvist and inductive formulae can be further extended by closing under effective equivalences, e.g. under semantic equivalence.

All these techniques push the limits of the syntactic approach farther. Still, it has firm boundaries, as it only produces *decidable* (usually, of fairly low complexity) classes of elementary canonical formulae. So, let us try something stronger...

4 Algorithmic approach to elementary canonical formulae

A natural strengthening of the syntactic approach is to develop algorithms that generate or identify elementary canonical formulae. Such algorithms need not be complete, i.e. successful for all elementary canonical formulae, but should always produce a correct result, if any, and thus define (recursively enumerable) classes of elementary canonical formulae. The roots of such an algorithm can be found in the *method of substitutions*, which originated from Sahlqvist's paper and was independently developed by van Benthem [4], see also [3]. That method was further sophisticated and extended by Simmons [43]. In particular, Simmons presented it in an explicitly algorithmic form which can be regarded as the first algorithm for producing elementary canonical formulae[6]. Simmons' algorithm works on a larger set of formulae, including non-elementary ones which have equivalents in FOL + Henkin quantifiers, such as all modal reduction principles. Note that the Sahlqvist-van Benthem substitution method works successfully only on formulae with fixed syntactical shape, such as Sahlqvist's formulae. In [23, 25] the method has been extended to work on inductive modal formulae, defining the appropriate substitutions by induction on the order determined by

[6]Though, Simmons' algorithm uses Skolemization, but does not involve a mechanism for unskolemization.

the dependency graph, but it still works only on formulae of the precise syntactic shape. The same applies for Simmons' method.

REMARK 16. The substitution method and Simmons' algorithm only establish first-order definability of modal formulae, but not their canonicity. However, if properly restricted and precisely specified, they can be shown to produce canonical formulae, by suitably modifying the proof of Sahlqvist's theorem.

Later in this section we show how the Sahlqvist-van Benthem substitution method can be considerably extended, by introducing the algorithm SQEMA [10] which works on arbitrary modal formulae and when it renders a successful result, it can be obtained by a sequence of suitable substitutions, computed in the course of the work of the algorithm. Before that, we present two other existing algorithms that can be used for computing first-order equivalents of modal formulae.

4.1 First-order definability as second-order quantifier elimination

Recall that the local validity of a modal formula $\phi = \phi(p_0, \dots, p_n)$ in a pointed Kripke frame $(\mathfrak{F}, w)$ is expressed as

$$\mathfrak{F}, w \models \phi \text{ iff } \mathfrak{F}, w \models \forall P_0 \dots \forall P_n \mathrm{ST}(\phi)(w/x),$$

where $\mathrm{ST}(\phi)(x)$ is the standard translation of ϕ over the free variable x.

Respectively, the global validity is expressed as

$$\mathfrak{F} \models \phi \text{ iff } \mathfrak{F} \models \forall P_0 \dots \forall P_n \forall x \mathrm{ST}(\phi)(x).$$

Thus, the search for a local or global first-order equivalent of ϕ can be thought of as an attempt to *eliminate the universally quantified second-order variables* $P_0, \dots, P_n$ and obtain a first-order formula equivalent to $\forall P_0 \dots \forall P_n \forall x \mathrm{ST}(\phi)$. Sometimes it is more convenient to eliminate existentially quantified second-order variables. Then, the negation $\neg\forall P_0 \dots \forall P_n \forall x \mathrm{ST}(\phi)$ is taken, and the resulting first-order formula is negated again.

Currently, there are two developed algorithms for second-order quantifier elimination: SCAN and DLS. They are both implemented and available online, and can be used to compute first-order equivalents of modal formulae.

SCAN

SCAN was developed in 1992 by Gabbay and Ohlbach [17]. Its current implementation, available online at http://www.mpi-sb.mpg.de/units/ag2/projects/SCAN/index.html, is based on the theorem prover OTTER. SCAN works on skolemized and clausified existentially quantified second-order formulae, and attempts to reduce them to equivalent first-order ones by generating sufficiently many logical consequences, and eventually keeping from the resulting set of formulae only those in which no second-order variables occur. As an input SCAN takes second-order formulae of the form $\exists Q_1 \ldots \exists Q_k\ \psi$, where Q_i are predicate variables and ψ is a first-order formula. The algorithm involves three stages:

(i) transformation to clausal form and Skolemization;

(ii) a special kind of *constraint resolution (C-resolution)*, involving a *purity deletion rule* allowing one to delete 'used up' clauses.

(iii) reverse Skolemization (unskolemization), if possible.

SCAN can fail to produce a first-order equivalent of an input formula for one of two reasons: either (i) the C-resolution stage fails to terminate due to looping, or (ii) the C-resolution terminates, yielding a set of clauses in which the specified second-order variables are eliminated, but for which the Skolemization cannot be reversed.

SCAN can be used to compute the first-order equivalent of a modal formula by running it on the negation of its standard translation.

THEOREM 17 ([27]). SCAN *is successful on all Sahlqvist formulae.*[7]

That result can be supplemented with the following.

THEOREM 18 ([12]). SCAN *is successful on all polyadic inductive formulae.*

The latter does not formally subsume the former, as the proof that SCAN succeeds on a conjunction of Sahlqvist formulae is technically involved, while this step is avoided in the case of inductive formulae.

We conjecture that all modal formulae on which SCAN succeeds are canonical. This conjecture can be proved under some idealizing assumptions about SCAN, consistent with its specification in [14]. The difficulty in proving it for the actual implementation of the algorithm is that it does not match precisely the specification. We venture an even stronger conjecture, viz. that all modal formulae on which SCAN succeeds are locally equivalent to inductive formulae. A currently open question is if the class of modal formulae on which SCAN succeeds is decidable. Our conjecture is 'no'.

---

[7]This result holds under the assumption that SCAN uses inner Skolemization. In fact, the current implementation of SCAN does not always unskolemize successfully when run on Sahlqvist formulae, because it does not always employ inner Skolemization.

DLS

DLS was originally introduced by A. Szalas in 1993 and further developed by Doherty, Lukaszewics, and Szalas [13]. Its original implementation is available online at http://www.ida.liu.se/labs/kplab/projects/dls/, and a new one is currently being tested. DLS works on existentially quantified second-order formulae and always terminates, by either producing a first-order equivalent, or reporting failure. It is based on applying, after suitable preprocessing including Skolemization, the following lemma due to W. Ackermann:

LEMMA 19 ([1]). *For any first-order formula A not containing the predicate P and a first-order formula B, the following hold:*

$$\exists P(\forall \overline{x}(A(\overline{x}) \rightarrow P(\overline{x})) \wedge B(P)) \;\equiv\; B(A/P) \text{ (Downwards-Ackermann)},$$

if B is negative in P, and respectively,

$$\exists P(\forall \overline{x}(P(\overline{x}) \rightarrow A(\overline{x})) \wedge B(P)) \;\equiv\; B(A/P) \text{ (Upwards-Ackermann)},$$

if B is positive in P, where $B(A/P)$ *is the result of uniform substitution of all occurrences of P in B by* $A(\overline{x})$*, with the arguments of each particular occurrence of P each time substituted for* $\overline{x}$ *in* $A(\overline{x})$*.*

We note that the lemma can be strengthened by replacing 'positive' and 'negative', by 'upward monotone' and 'downward monotone' respectively, as these semantic properties are, in fact, used in the proof.

THEOREM 20 ([12]). DLS *is successful on all conjunctions of polyadic inductive formulae. In particular,* DLS *is successful on all Sahlqvist formulae.*

In fact, it turns out that DLS does not have to Skolemize on the translations of inductive formulae. We furthermore conjecture that for every modal formula on which DLS succeeds, it can succeed without skolemization.

We also claim that all modal formulae on which DLS succeeds are canonical, but, as with SCAN, the difficulty in proving such claim lies in the fact that the available specification of DLS in [29] is only partial, and the actual implementation does not match it precisely[8].

Finally, we conjecture, as for SCAN, that all modal formulae on which DLS succeeds are locally equivalent to inductive formulae.

Comparing SCAN *and* DLS

The constraint resolution rule of SCAN is based on a particular case of Ackermann's lemma. However, DLS does not subsume SCAN because it

[8]In fact, we have discovered a bug in the original implementation of DLS, which consists in reports of success ('true') in some cases where the algorithm should not succeed, and the formula to which it is applied is not equivalent to 'true'.

does not apply Ackermann's lemma repeatedly on the same variable, and does not use a purity deletion rule. Moreover, the C-resolution rule is not equivalence preserving.

From our practical experience with both algorithms, we find that SCAN is generally more flexible and syntax-tolerant (but easier to fool into looping) as it works on a low level, with formulae decomposed into a simple (clausal) form, and with simple rules (constraint resolution and factorization) applied repeatedly. On the other hand, DLS is more rigid and syntax-dependent, as it works on a high level, with only one, 'macro' rule (Ackermann's lemma).

In particular, neither of the implemented algorithms subsumes the other, but it seems that SCAN is generally more successful on modal formulae. For instance, it succeeds on the formula

$$\Box(p \leftrightarrow q) \rightarrow q,$$

on which DLS fails.

On the other hand, SCAN loops on the formula

$$\Box(p \vee \Box\neg p) \rightarrow \Diamond(p \wedge \Diamond\neg p),$$

(an example due to Szalas) on which, theoretically, DLS succeeds.[9]

We currently have no examples of *modal formulae* in the basic modal language, on which SCAN *loops* while DLS succeeds, but such examples can be constructed if the universal modality with its standard semantics is added to the language.

Ackermann's lemma and the method of substitutions.

We have obtained (see [10]) the following *modal version of Ackermann's lemma:*

LEMMA 21 (Ackermann, modal version). *For any modal formula A not containing p and a modal formula B, the following hold:*[10]

$$\exists p([\mathbf{u}](A \rightarrow p) \wedge B(p)) \;\equiv\; B(A/p) \;\textit{(Modal Downward-Ackermann)},$$

if B is negative (or stronger, downward monotone) in p, and respectively

$$\exists p([\mathbf{u}](p \rightarrow A) \wedge B(p)) \equiv B(A/p), \;\textit{(Modal Upward-Ackermann)},$$

if B is positive (or stronger, upward monotone) in p, where B(A/p) is the result of uniform substitution of all occurrences of p in B by A, and ≡ denotes local equivalence.

[9] But, the current implementation does not.

[10] These equivalences can be interpreted as follows: the right-hand side gives a condition for existence of a solution in p of the 'modal equation' on the left-hand side.

Indeed, under the conditions above, e.g. for the modal upwards-Ackermann lemma, the following holds: for every Kripke model M and $w \in M$, $M, w \models B(A/p)$ iff there is a model M' possibly differing from M only at the valuation of p, such that $M', w \models [\mathbf{u}](p \to A) \wedge B(p)$.

Note the contrapositive form of the downward Ackermann lemma, after replacing $\neg B$ with B:

$$\forall p([\mathbf{u}](A \to p) \to B(p)) \equiv B(A/p),$$

for any modal formula A not containing p, and a modal formula B which is upward monotone in p. This equivalence can be interpreted as follows: $[\mathbf{u}](A \to p) \to B(p)$ is valid in a given frame iff $B(P)$ is true for the '*minimal*' valuation satisfying the antecedent, viz. A. *This is precisely the technical idea at the heart of the substitution method of Sahlqvist and van Benthem!*

4.2 SQEMA: a new algorithm for computing elementary canonical formulae

In [10] we have introduced SQEMA: an algorithm for **S**econd-order **Q**uantifier **E**limination in **M**odal formulae, using **A**ckermann's lemma. It has the following basic features:

- Combines ideas from both DLS and SCAN and uses the modal version of Ackermann's lemma to eliminate the existentially quantified propositional variables.
- Works directly on (negated) modal formulae and decomposes them into sets of modal implications, called 'equations'.
- Does not introduce Skolem functions, but only Skolem constants, as nominals.
- Preserves formulae up to local frame equivalence.
- When successful, eventually produces a pure modal formula in a language, possibly extending the original one with nominals and inverse (reversive) modalities. The standard translation of this formula produces the corresponding first-order condition of the original formula.

The core algorithm.

Here we will present the algorithm on languages with unary modalities only. For the general case, see [12] and [11].

The input of SQEMA is a modal formula ϕ.

Step 1 Negate ϕ, eliminate $\rightarrow$ and $\leftrightarrow$, and rewrite in negation normal form. Then distribute diamonds and conjunctions over disjunctions as much as possible. The algorithm now proceeds on each disjunct ψ, separately, as follows:

Step 2 Rewrite as $i \rightarrow \psi$, where i is a fixed nominal, reserved to name the initial state. This is the only initial equation.

Step 3 Eliminate every variable p in which the system is monotone (upwards or downwards), by replacing it with $\top$ or $\bot$.

Step 4 If there are propositional variables remaining in equations of the system, choose to eliminate one, say p, the elimination of which has not been attempted yet.

If all remaining variables have been attempted and Step 5 has failed, backtrack and attempt another order of elimination.

If all orders of elimination and all remaining variables have been attempted and step 5 has failed, report failure.

If all propositional variables have been eliminated from the system, proceed to Step 6.

Step 5 The goal now is, by applying the transformation rules listed below, to rewrite the system of equations so that the Ackermann-rule becomes applicable with respect to the chosen variable p in order to eliminate it. Thus, the current goal is to transform the system into one in which every equation is either negative in p, or of the form $\alpha \rightarrow p$, with p not occurring in α, i.e. to 'extract' p and 'solve' for it.

If this fails, backtrack, change the polarity of p by substituting $\neg p$ for it everywhere, and attempt again to prepare for the Ackermann-rule.

If this fails again, or after the completion of this step, return to Step 4.

Step 6 If this step is reached it means that all propositional variables have been successfully eliminated from all systems resulting from the input formula. What remains now is to return the desired first-order equivalent. In each system, take the conjunction of all equations to obtain a formula pure, and form the formula $\forall\overline{y}\exists x_0 \mathrm{ST}(\neg\mathsf{pure})$, where $\overline{y}$ is the tuple of all occurring variables corresponding to nominals, but with y_i (corresponding to the designated current state nominal i) left free if the local correspondent is to be computed. Then take the conjunction of these translations over the systems on all disjunctive branches. For motivation of the correctness of this translation the reader is referred to the examples in the following subsection as well as the correctness proof in [10].

Return the result, which is the (local) first-order condition corresponding to the input formula.

The transformation rules

I. Rules for the logical connectives:

$\wedge$-rule: $\dfrac{\beta \to \gamma \wedge \delta}{\beta \to \gamma, \beta \to \delta}$	$\Diamond$-rule: $\dfrac{j \to \Diamond\gamma}{j \to \Diamond k,\ k \to \gamma}$ where k is a new nominal.
Left-shift $\vee$-rule: $\dfrac{\beta \to \gamma \vee \delta}{(\beta \wedge \neg\gamma) \to \delta}$	Right-shift $\vee$-rule: $\dfrac{(\beta \wedge \neg\gamma) \to \delta}{\beta \to \gamma \vee \delta}$
Left-shift $\Box$-rule: $\dfrac{\gamma \to \Box\delta}{\Diamond^{-1}\gamma \to \delta}$	Right-shift $\Box$-rule: $\dfrac{\Diamond^{-1}\gamma \to \delta}{\gamma \to \Box\delta}$

We will write Rjk as an abbreviation of $j \to \Diamond k$.

II. Auxiliary propositional rules:

1. Commutativity and associativity of $\wedge$ and $\vee$ (tacitly used).
2. Replace $\gamma \vee \neg\gamma$ with $\top$, and $\gamma \wedge \neg\gamma$ with $\bot$.
3. Replace $\gamma \vee \top$ with $\top$, and $\gamma \vee \bot$ with γ.
4. Replace $\gamma \wedge \top$ with γ, and $\gamma \wedge \bot$ with $\bot$.
5. Replace $\gamma \to \bot$ with $\neg\gamma$ and $\gamma \to \top$ with $\top$.
6. Replace $\bot \to \gamma$ with $\top$ and $\top \to \gamma$ with γ.

III. Polarity switching rule: Switch the polarity of every occurrence of a chosen variable p within the current system.

IV. Ackermann rule:

$$\left\| \begin{array}{l} \alpha_1 \to p \\ \dots \\ \alpha_n \to p, \\ \beta_1(p), \\ \dots \\ \beta_m(p), \end{array} \right. \quad \Rightarrow \quad \left\| \begin{array}{l} \beta_1[(\alpha_1 \vee \ldots \vee \alpha_n)/p], \\ \dots \\ \beta_m[(\alpha_1 \vee \ldots \vee \alpha_n)/p]. \end{array} \right.$$

where p does not occur in $\alpha_1, \dots, \alpha_n$ and each β_i is negative in p.[11]

4.3 Examples

The best way to get a feel of the workings of the algorithm is perhaps to consider an example or two. (For more, see [10].)

EXAMPLE 22. We take as input the formula $\Diamond\Box p \rightarrow \Box\Diamond p$.

The initial system of equations is

$$\| i \rightarrow (\Diamond\Box p \wedge \Diamond\Box\neg p)$$

Applying the $\wedge$-rule gives

$$\left\| \begin{array}{l} i \rightarrow \Diamond\Box p \\ i \rightarrow \Diamond\Box\neg p \end{array} \right.$$

Applying the $\Diamond$-rule to the first equation yields:

$$\left\| \begin{array}{l} Rij \\ j \rightarrow \Box p \\ i \rightarrow \Diamond\Box\neg p \end{array} \right.$$

and then applying the Left-shift $\Box$-rule:

$$\left\| \begin{array}{l} Rij \\ \Diamond^{-1} j \rightarrow p \\ i \rightarrow \Diamond\Box\neg p \end{array} \right.$$

The Ackermann rule is now applicable, yielding the system

$$\left\| \begin{array}{l} Rij \\ i \rightarrow \Diamond\Box\neg(\Diamond^{-1} j) \end{array} \right.$$

Taking the conjunction of the equations gives

$$Rij \wedge (i \rightarrow \Diamond\Box\neg(\Diamond^{-1} j)).$$

Negating we obtain

$$Rij \rightarrow (i \wedge \Box\Diamond\Diamond^{-1} j),$$

which, translated, becomes

$$\forall y_j \exists x_0 [Ry_i y_j \rightarrow (x_0 = y_i) \wedge \forall y (Rx_0 y \rightarrow \exists u (Ryu \wedge \exists v (Rvu \wedge v = y_j)))],$$

[11] As already discussed, this rule can be strengthened by replacing 'negative' with 'downwards monotone', but this brings a higher complexity price.

and simplifies to

$$\forall y_j[Ry_iy_j \rightarrow \forall y(Ry_iy \rightarrow \exists u(Ryu \wedge Ry_ju))]$$

defining the Church-Rosser property, as expected.

EXAMPLE 23. We take as input the (non-inductive) formula

$$\Box(\Box p \leftrightarrow q) \rightarrow p$$

on which both SCAN and DLS fail.

This yields the initial equation

$$\left\| \; i \rightarrow \Box((\Diamond\neg p \vee q) \wedge (\neg q \vee \Box p)) \wedge \neg p \right.$$

Choose q to eliminate first. Applying the $\wedge$-rule and the Left-shift $\Box$-rule:

$$\left\| \begin{array}{l} \Diamond^{-1}i \rightarrow (\Diamond\neg p \vee q) \\ \Diamond^{-1}i \rightarrow (\neg q \vee \Box p) \\ i \rightarrow \neg p \end{array} \right.$$

Applying the Left Shift $\vee$-rule to the first equation yields

$$\left\| \begin{array}{l} (\Diamond^{-1}i \wedge \Box p) \rightarrow q \\ \Diamond^{-1}i \rightarrow (\neg q \vee \Box p) \\ i \rightarrow \neg p \end{array} \right.$$

to which the Ackermann-rule is applicable with respect to q. This gives

$$\left\| \begin{array}{l} \Diamond^{-1}i \rightarrow (\neg\Diamond^{-1}i \vee \neg\Box p \vee \Box p) \\ i \rightarrow \neg p \end{array} \right. .$$

The first equation is now a tautology and may be removed, yielding the system

$$\left\| \; i \rightarrow \neg p \right.$$

in which p may be replaced by $\bot$ since it occurs only negatively, resulting in the system

$$\left\| \; \top \right. .$$

Negating we obtain $\bot$.

Some results and comments on SQEMA.

THEOREM 24 ([10]).

1. SQEMA *is sound: if successful, it produces a first-order formula locally frame equivalent to the input modal formula.*
2. SQEMA *is successful on all conjunctions of inductive formulae. In particular,* SQEMA *is successful on all Sahlqvist formulae.*
3. *All modal formulae on which* SQEMA *succeeds are canonical.*

Note that the original Sahlqvist's theorem and its extension to inductive formulae now follow from the results above.

Again, we conjecture that all modal formulae on which SQEMA succeeds are locally equivalent to inductive formulae.

How does SQEMA compare to SCAN and DLS on modal formulae? We believe that it is stronger than both, but this claim, if correct, can only be proved if precise descriptions of the specifications and implementations of SCAN and DLS are available.

Finally, we note that SQEMA is amenable to various extensions, e.g. with a recursive version of the Ackermann rule, which enables computation of correspondents of modal formulae in FO+LFP, see [11],[26].

4.4 The power and limits of the algorithmic approach

The algorithmic approach is certainly more powerful as a generator of elementary canonical formulae than the syntactic approach. It produces effectively enumerable classes of elementary canonical formulae, which in general need not be decidable.

The different algorithms discussed here: method of substitutions, Simmons' algorithm, SCAN, DLS, and SQEMA, have different computing powers and scope of applicability on modal formulae. Yet, we believe that the algorithmic approach, if developed to its full potential, will generate a natural class of *algorithmically elementary canonical formulae.* The major challenge of this research area is to develop such an optimal algorithm.

5 Model-theoretic aspects of elementary canonical formulae

In this section we briefly discuss semantic characterizations of elementary canonical formulae. The main model-theoretic tool we use for such characterizations is *persistence.* While canonicity *is defined* in terms of persistence, first-order definability can only be approximated in such a way. The approximation which we discuss here, due to van Benthem [3], defines a large and natural class of elementary modal formulae.

5.1 Sahlqvist formulae and ample-persistence

How can one prove that a given elementary canonical formula is *not* equivalent to a Sahlqvist formula in any reasonable sense? Here is a method, introduced in [25], based on a special kind of persistence.

DEFINITION 25. A general frame $\langle W, R, \mathbb{W}\rangle$ is **ample** if for every $w \in W$ and $n \in \mathbf{N}$, $R^n(w) = \{u \mid wR^n u\} \in \mathbb{W}$.

Note that every ample general frame is discrete, for $R^0(w) = \{w\}$.

DEFINITION 26. A modal formula A is **locally a-persistent** if it is locally persistent with respect to every ample general frame, i.e. for every such frame $\mathfrak{F} = \langle F, \mathbb{W}\rangle$, where $F = \langle W, R\rangle$, and $w \in W$,

$$\mathfrak{F}, w \models A \text{ iff } F, w \models A.$$

The following can be proved by inspection of the minimal valuations corresponding to Sahlqvist formulae.

LEMMA 27. *Every Sahlqvist formula in the basic modal language is locally* a*-persistent.*

PROPOSITION 28 ([25]). *The inductive formula*

$$D = p \wedge \Box(\Diamond p \to \Box q) \to \Diamond\Box\Box q$$

is not (even globally) a*-persistent.*

COROLLARY 29. *The formula D is not frame equivalent to any Sahlqvist formula in the basic modal language.*

5.2 van Benthem formulae and the limits of the substitutions method

Let FO be the first-order language for Kripke frames, and $\beta(\overline{x})$ be a FO-formula with unary predicates $P_1, \ldots, P_n$, such that the variables $\overline{x}$ do not occur bound in β and the variables $z_1, \ldots, z_k, y$ do not occur in β at all.

A **universally parameterized** FO**-substitution instance** of β is any FO-formula

$$\forall z_1 \ldots \forall z_k \beta[\sigma_1/P_1, \ldots, \sigma_n/P_n]$$

obtained from β by selecting FO-formulae $\sigma_i = \sigma_i(\overline{x}, z_1, \ldots, z_k, y)$ for $i = 1, \ldots, n$, substituting $\sigma_i[x/y]$ for every occurrence of $P_i x$, and then universally quantifying over $z_1, \ldots, z_k$.

Let $\Theta(\beta)$ be the set of all universally parametrised FO-substitution instances of β.

DEFINITION 30. A modal formula $\phi = \phi(p_1, \ldots, p_n)$ is a **van Benthem formula** if

$$\Theta(\mathrm{ST}(\phi; x_0)) \models \forall P_1 \ldots \forall P_n \mathrm{ST}(\phi; x_0).$$

We let VB denote the class of van Benthem formulae (defined slightly differently by van Benthem himself in [3], as the class M_1^{sub}).

THEOREM 31 ([28]. Essentially first proved in [3]). *A modal formula is locally $\mathcal{E}$-persistent iff it is a van Benthem formula.*

Since, by compactness, all van Benthem formulae are locally first-order definable, we obtain the following.

COROLLARY 32. *Every locally $\mathcal{E}$-persistent modal formula is locally elementary.*

Some burning questions arise now:

- Is every van Benthem formula canonical ($\mathcal{D}$-persistent)?

  Sadly, no: van Benthem's incomplete formula $\mathbf{vB} = \Box\Diamond\top \to \Box(\Box(\Box p \to p) \to p)$ (see [3] or [2]) is a counter-example.

- Is every elementary canonical formula a van Benthem formula?

  Even more sadly, no: $(\Box p \to \Box\Box p) \wedge \Box(\Box p \to \Box\Box p) \wedge (\Box\Diamond p \to \Diamond\Box p)$ is a locally elementary canonical formula, but not locally $\mathcal{E}$-persistent [3].

Still, van Benthem formulae (in his own words) 'neatly delimit the range of the method of substitutions', and provide a natural and important upper bound for the class of elementary canonical formulae.

THEOREM 33 ([4]). *The set VB is recursively enumerable.*

Thus, there is an algorithm generating all van Benthem formulae, and essentially based on the method of substitutions. It is a natural challenge to develop a practical one.

An even more challenging question is whether the set of canonical van Benthem formulae is recursively enumerable, and if so, to construct a generating algorithm for it.

5.3 Elementary canonical formulae and persistence in reversive languages with nominals

The leading problem of our model-theoretic approach to elementary canonical formulae is to characterize them in terms of a natural persistence property. We do not have (yet) a solution to this problem for the basic modal language, but we do for '*rich enough*' languages, viz. reversive languages with nominals. Recall again, that the natural notion of canonicity in languages with nominals is 'discrete canonicity', i.e. $\mathcal{DI}$-persistence.

THEOREM 34 ([25]). *For every modal formula A in a reversive language with nominals, the following are equivalent:*

1. *A is locally $\mathcal{DI}$-persistent.*

2. *A is locally equivalent to an inductive formula.*

3. *A is locally equivalent in the class of discrete frames to a pure formula.*

5.4 Topological perspective on elementary canonical formulae

Following topological ideas going back to Sambin and Vaccaro [41], we show in [25] how first-order definability and canonicity of inductive formulae can be established in a uniform way. In a similar way we give a simultaneous proof of the correctness and the canonicity of the algorithm SQEMA in [10]. Here are the key points of these arguments. The formulae for which first-order definability and canonicity (persistence) is to be established, are transformed into 'simple' ones, for which these properties are immediate. Typically, these are pure formulae in a reversive hybrid extension of the original language. Such transformation can be semantic (as is the case of inductive formulae in [25]), deductive (inductive formulae in hybrid languages, in [24]), or algorithmic (the formulae on which SQEMA succeeds, in [10]), but in any case they preserve the desired properties. For first-order definability, such preservation is proved by a direct semantic argument on Kripke frames, but for the general frames over which the persistence is to be proved (e.g. descriptive frames), the argument involves suitable topological closure properties of the modal operators in the extended language, considered as operators in the topologies on the general frames of the original language. These closure properties guarantee that the formulae under consideration (Sahlqvist, inductive, SQEMA) allow the semantic argument proving preservation on Kripke frames to be simulated for them on e.g. descriptive frames, thus implying $\mathcal{D}$-persistence. Proving the desired topological behavior in all cases we have studied crucially depends on the effective (syntactic, or algorithmic) nature of these formulae.

This topological approach is still open to further development, and the main aim of that approach is to find a sufficiently general argument which applies to all elementary canonical formulae, regardless of their syntactic features.

6 Concluding remarks: closing about elementary canonical formulae

Each of the syntactic, algorithmic, and model-theoretic approaches provides a hierarchy of approximations of the class of elementary canonical formulae, but none of them seems to yield both a practical and precise characterization yet. It is currently unknown if such characterization, better than the definition itself, exists at all. In particular, we do not know the complexity of the class of elementary canonical formulae, nor that of the class of first-order formulae definable by elementary canonical modal formulae.

We note that there are interesting and important cases of canonical modal formulae that are not elementary, see e.g. [19], [46]. Moreover, it has recently been established in [21] that a canonical modal logic need not be complete with respect to any elementary class of frames. Thus, first-order definability and canonicity are not as closely related as it was been conjectured by Fine in the 1970's. It is therefore natural to extend this study along each of these properties separately. In that respect, we should also mention the *algebraic* approach to canonicity, developed by Jónsson who gave an algebraic proof of Sahlqvist's theorem in [30].

Finally, we should mention an important family of modal formulae and logics axiomatized with such formulae, for which elementariness and canonicity coincide. These are the *subframe formulae and logics* introduced by Fine [15] and further extended to *cofinal subframe formulae and logics* by Zakharyaschev [48, 6][12].

With each finite transitive general frame $\mathfrak{F}$ one can associate (see [48, 6]) a *subframe formula* $\alpha(\mathfrak{F}, \emptyset)$, and a *cofinal subframe formula* $\alpha(\mathfrak{F}, \emptyset, \bot)$, such that any transitive general frame $\mathfrak{G}$ refutes $\alpha(\mathfrak{F}, \emptyset)$ (resp. $\alpha(\mathfrak{F}, \emptyset, \bot)$) if and only if $\mathfrak{G}$ is subreducible (resp. cofinally subreducible) to $\mathfrak{F}$ by way of a bounded morphism. A normal extension of $K4$ is a *subframe logic* if it can be axiomatized over $K4$ by a set of subframe formulae $\{\alpha(\mathfrak{F}_i) : i \in I\}$, for some family of transitive general frames $\{\mathfrak{F}_i : i \in I\}$. *Cofinal subframe logics* are defined similarly.

It turns out that, on transitive frames, a cofinal subframe formula is elementary iff it is $\mathcal{D}$-persistent; for subframe formulas these are equivalent to $\mathcal{R}$-persistence, as well. The same equivalences apply to (cofinal) subframe logics.

Similar results were established by Wolter [47] for modal formulae preserved in subframes, and normal modal logics L characterized by classes of (general) frames closed under taking subframes. For these, elementariness, $\mathcal{D}$-persistence, and $\mathcal{R}$-persistence coincide.

[12] Note that the term 'canonical formulae' in [48] has different meaning from the commonly used one in the present paper.

Acknowledgments

V. Goranko acknowledges the financial support provided by the National Research Foundation of South Africa and the Faculty of Science at Rand Afrikaans University. Part of this work was completed during the visit of Goranko and Conradie to the Department of Computer Science at the University of Manchester, the financial support for which was provided by a research grant from the British Research Council for Science and Engineering. We are grateful to Renate Schmidt for obtaining that grant and organizing our visits, and for some useful comments on SQEMA. D. Vakarelov was partially supported by the EU COST project 274 TARSKI and the project RILA-12 sponsored by the Bulgarian Ministry of Science and Education. We thank an anonymous reader for some useful comments and references.

BIBLIOGRAPHY

[1] Ackermann, W., Untersuchung über das Eliminationsproblem der mathematischen Logic. *Mathematische Annalen*, 110:390-413, 1935.

[2] Blackburn, P., M. de Rijke, Y. Venema, **Modal Logic**, Cambridge Tracts in Theoretical Computer Science, Cambridge University Press, 2001.

[3] van Benthem J.F.A.K., **Modal Logic and Classical Logic**, Bibleapolis, 1985.

[4] van Benthem J.F.A.K., **Modal Correspondence Theory**, Ph.D. Thesis, Mathematisch Instituut & Instituut voor Grondslagenonderzoek, Univ. of Amsterdam, 1976.

[5] van Benthem J.F.A.K., Minimal predicates, fixed points, and definability, J. Symb. Logic, to appear.

[6] Chagrov, A. and M Zakharyaschev, **Modal Logic**, Clarendon Press, Oxford, 1997.

[7] Chagrov, A. and M. Zakharyaschev, Sahlqvist formulae are not so elementary, **Logic Colloquium'92**, L. Csirmaz, D. Gabbay and M. de Rijke (eds.), CSLI Publications, Stanford, 1995, 61-73.

[8] Chagrov, A., F. Wolter, and M. Zakharyaschev, Advanced modal Logic, in: Handbook of Philosophical Logic, 2nd edition, vol 3, Kluwer, 2001, 83–266.

[9] Conradie, W., Decidable Equivalences of Modal Formulae, 2005, in preparation.

[10] Conradie, W., V. Goranko, and D. Vakarelov: Algorithmic correspondence and completeness in modal logic. I. The algorithm SQEMA, 2004, submitted.

[11] Conradie, W., V. Goranko, and D. Vakarelov: Algorithmic correspondence and completeness in modal logic. II. Extensions of the algorithm SQEMA, 2005, in preparation.

[12] Conradie, W. and V. Goranko: Algorithmic Classes of Elementary Canonical Formulae, 2005, in preparation.

[13] Doherty, P, W. Lukaszewics, and A. Szalas. Computing circumscription revisited: A reduction algorithm, *Journal of Automated Reasoning*, 18(3):297–336, 1997.

[14] Engel, T. Quantifier Elimination in Second-Order Predicate Logic, Diploma Thesis, Univ. of Saarbruecken, 1996.

[15] Fine, K., Logics containing K4, part II. *Journal of Symbolic Logic, 50:619-651, 1985.*

[16] Gabbay, D., I. Hodkinson, and M Reynolds, ***Temporal Logic: Mathematical Foundations and Computational Aspects****, Vol. 1, Clarendon Press, Oxford, 1994.*

[17] Gabbay, D. and H.-J. Ohlbach, Quantifier elimination in second-order predicate logic, South African Computer Journal, vol. 7, 1992, 35-43.

[18] Gargov, G. and V. Goranko, Modal Logic with Names, Journal of Philosophical Logic, 22/6 (1993), pp. 607-636.

[19] Ghilardi, S. and G. Meloni, Constructive canonicity in non-classical logics, Annals of Pure and Applied Logic, 86 (1997), pp. 1-32.

[20] Goldblatt, R., First-order definability in modal logic, *Journal of Symbolic Logic*, 40(1975), 35-40.

[21] Goldblatt, R., I. Hodkinson, and Y. Venema, Erdös graphs resolve Fine's canonicity problem, Bull. of Symb. Logic, 10 (2), 2004, 186-208.

[22] Goranko, V., Axiomatizations with Context Rules of Inference in Modal Logic, *Studia Logica*, 61(2), 1998, 179-197.

[23] Goranko, V. and D. Vakarelov, Sahlqvist formulae Unleashed in Polyadic Modal Languages, **Advances in Modal Logic, vol. 3**, World Scientific, Singapore, 2002, pp. 221-240.

[24] Goranko, V. and D. Vakarelov, Sahlqvist formulae in Hybrid Polyadic Modal Languages, *J. of Logic and Computation*, vol. 11(5), 2001, 737-754.

[25] Goranko, V. and D. Vakarelov, Elementary Canonical Formulae: Extending Sahlqvist's Theorem, 2003, to appear in Annals of Pure and Applied Logic.

[26] Conradie, W., Goranko, V. and D. Vakarelov, Computing equivalents of modal formulae in FO(LFP), in preparation, 2005.

[27] Goranko, V., U. Hustadt, R. Schmidt, and D. Vakarelov. SCAN is complete for all Sahlqvist formulae, in: Relational and Kleene-Algebraic Methods in Computer Science (Proc. of RelMiCS 7). LNCS 3051, Springer, 2004, 149-162.

[28] Goranko, V., and M. Otto, Model Theory of Modal Logic, in: Handbook of Modal Logic, Kluwer, 2005, to appear.

[29] Gustafsson, J., *An Implementation and Optimization of an Algorithm for Reducing Formulae in Second-Order Logic*. Technical Report LiTH-MAT-R-96-04. Dept. of Mathematics, Linkoping University, Sweden, 1996.

[30] Jónsson, B., On the Canonicity of Sahlqvist Identities, *Studia Logica*, 53 (1995), pp. 473-491.

[31] Jónsson, B. and A. Tarski, Boolean Algebras with Operators, Part 1, *American J. of Mathematics* , 73 (1952), 891-939.

[32] Kracht, M., How Completeness and Correspondence Theory Got Married, in: **Diamonds and Defaults**, M. de Rijke (ed.), Kluwer, Synthese Library, 1993, 175-214.

[33] Kracht, M., **Tools and Techniques in modal Logic**, Elsevier, Amsterdam, 1999.

[34] Nonnengart, A., H. J. Ohlbach, and A. Szalas. Quantifier elimination for second-order predicate logic, in: **Logic, Language and Reasoning: Essays in honour of Dov Gabbay**, Part I, H. J. Ohlbach and U. Reyle (eds.), Kluwer, 1997.

[35] Nonnengart, A. and A. Szalas. A fixpoint approach to second-order quantifier elimination with applications to correspondence theory, in: **Logic at Work, Essays dedicated to the memory of Helena Rasiowa**, E. Orlowska (ed.), Springer Physica-Verlag, 1998, pp. 89-108.

[36] de Rijke, M., How not to generalize Sahlqvist's Theorem, Technical Note, ILLC, 1992.

[37] de Rijke, M., **Extending Modal Logic**, Ph.D. thesis, ILLC, University of Amsterdam, ILLC Dissertation Series 1993-4, 1993.

[38] de Rijke, M. and Y. Venema, Sahlqvist's Theorem For Boolean Algebras with Operators with an Application to Cylindric Algebras, *Studia Logica*, 54 (1995), 61-78.

[39] Sahlqvist, H., Correspondence and completeness in the first and second-order semantics for modal logic, **Proc. of the 3rd Scandinavial Logic Symposium, Uppsala 1973**, S. Kanger (ed.), North-Holland, Amsterdam, 1975, 110-143.

[40] Sambin, G. and V. Vaccaro, Topology and Duality in Modal Logic, *Annals of Pure and Applied Logic*, 37(1988), 249-296.

[41] Sambin, G. and V. Vaccaro, A New Proof of Sahlqvist's Theorem on Modal Definability and Completeness, *Journal of Symbolic Logic*, 54(1989), 992-999.

[42] Szałas, A., On the correspondence between modal and classical logic: An automated approach, 3(6):605–620, 1993.

[43] Simmons, H., The Monotonous Elimination of Predicate Variables. *Journal of Logic and Computation*, 4(1):23–63, 1994.

[44] Vakarelov, D., *Modal Definability in Languages with a Finite Number of Propositional Variables and a New extension of the Sahlqvist's Class*, Advances in Modal logic, vol. 4, 499-518.
[45] Venema, Y., *Derivation rules as anti-axioms in modal logic*, Journal of Symbolic Logic, **58** (1993), 1003– 1034.
[46] Venema, Y., *Canonical pseudo-correspondence*, in: **Advances in Modal Logic, vol. 2**, M. Kracht, M. de Rijke, H. Wansing, and M. Zakharyaschev (eds.), CSLI Publications, Stanford, 2000, 421-430.
[47] Wolter, F., *The structure of lattices of subframe logics*, Annals of Pure and Applied Logic, 86 (1997), 47-100.
[48] Zakharyaschev, M. V., *Canonical formulas for K4. Part II: Cofinal subframe logics*, Journal of Symbolic Logic, 61:421-449, 1996.

Willem Conradie
Department of Mathematics and Statistics, University of Johannesburg
PO Box 524, Auckland Park 2006, Johannesburg, South Africa
wec@rau.ac.za

Valentin Goranko
Department of Mathematics and Statistics, University of Johannesburg
PO Box 524, Auckland Park 2006, Johannesburg, South Africa
vfg@rau.ac.za

Dimiter Vakarelov
Department of Mathematical Logic with Laboratory for Applied Logic,
Faculty of Mathematics and Computer Science, Sofia University
blvd James Bouchier 5, 1126 Sofia, Bulgaria
dvak@fmi.uni-sofia.bg

Axioms for Logics of Knowledge and Past Time: Synchrony and Unique Initial States

TIM FRENCH, RON VAN DER MEYDEN, MARK REYNOLDS

ABSTRACT. Sound and complete axiomatizations are provided for two different logics involving modalities for knowledge and both past and future time modalities. The logics considered allow for multiple agents with unique initial state and synchrony. The synchrony restriction gives every agent access to a system clock. Such semantic restrictions are of particular interest in the context of past time modalities since both synchrony and unique initial state restrictions are not expressible using future time modalities.

1 Introduction

There has been significant interest in multi-modal logics combining operators for knowledge and time in recent years [1, 3, 4, 7]. With only a few exceptions [3], this literature deals with *future time* temporal operators. In this paper we consider the effect of adding past time operators to such logics.

There are some compelling reasons to consider this extension. One of the topics of interest in the literature has been the interaction between knowledge and time when a variety of semantic properties are assumed, such as uniqueness of initial states, synchrony, perfect recall and no learning (a dual of perfect recall) [8]. These properties lead to interaction axioms, which involve both epistemic and temporal operators. Halpern, van der Meyden and Vardi [7] provide complete axiomatizations of logics of knowledge and linear future time for all the axiomatizable cases arising out of combinations of these assumptions. However, their results indicate that in some cases, some of the properties have no impact on the axiomatization. For example, [7] obtains identical complete axiomatizations for the cases of no assumptions, synchrony alone, unique initial states alone, and for both synchrony and unique initial states. This indicates that the logic with future time axioms is too weak to fully express the unique initial states and synchrony properties. Temporal logics with past operators are known to be at least

Advances in Modal Logic, Volume 5. © *2005, Tim French, Ron van der Meyden, Mark Reynolds.*

exponentially more succinct than temporal logics without past operators [12]. The past operators thus allow for a neater representation of many properties. For example, past operators allow for a much cleaner axiom for perfect recall in the asynchronous setting [14].

Another reason to consider knowledge in combination with past time operators is that *knowledge-based programs* [2] are better behaved with past-time operators than with future time operators. A knowledge-based program is like a standard program with formulas expressing the knowledge of the agent allowed to occur as conditions in conditional statements. A concrete implementation of such a program replaces the knowledge conditions by concrete conditions of the agent's local state. Knowledge-based programs behave somewhat like specifications, and in general, may have zero, one, or many different implementations. However, it is possible to provide conditions under which there is guaranteed to be a unique implementation [2]. One of these conditions is when the system is synchronous, and all knowledge tests involve only past time operators.

We would like to have an interaction axiom for each of the properties mentioned above, such that combinations of properties can be handled by combining their corresponding axiom. In this paper, we take a step in this direction by providing axioms which individually characterize the properties of unique initial states and synchrony. (We will deal with combinations in future work.) As already remarked, a past time axiom for perfect recall is already given in [14]. The property of no learning is best captured by the future time axiom in [7].

The synchrony restriction is particularly interesting since the axiomatization appears to require a complex automaton-based rule. We sketch the rather interesting completeness proof here, based on completeness proofs given in [7]. We show we can construct a canonical model for any consistent formula by induction over the nestings of knowledge operators. To enforce the synchrony constraint we introduce transducers to represent sufficiently detailed information about the time. This transducer can be encoded in a characteristic formula, and we use the new rule, SYNC, to show that the synchrony constraint is maintained.

2 Syntax and Semantics

The language is given by the abstract syntax:

$$\alpha = x \mid \neg\alpha \mid \alpha \wedge \alpha' \mid \bigcirc\alpha \mid \alpha\,\mathcal{U}\,\alpha' \mid \textcircled{w}\alpha \mid \alpha\,\mathcal{S}\,\alpha' \mid K_i\alpha$$

where $x \in \mathcal{V}$ is some propositional atom, and $1 \leq i \leq k$ is the index of an agent. The operators are respectively *not, and, tomorrow, until, weak yesterday, since and i-knows*, and have their usual meaning. Along with the usual propositional abbreviations (*true*, *false*, $\vee$, $\rightarrow$) we will also use the temporal abbreviations: $\ominus\alpha = \neg\text{ⓦ}\neg\alpha$; $\Diamond\alpha = true\,\mathcal{U}\,\alpha$; $\diamondminus\alpha = true\,\mathcal{S}\,\alpha$; $\Box\alpha = \neg\Diamond\neg\alpha$; and $\boxminus\alpha = \neg\diamondminus\neg\alpha$, and the epistemic operator, $L_i\alpha = \neg K_i\neg\alpha$.

For the semantics, we suppose a model is given by a set of runs, and each formula is evaluated with respect to some time in some run. Given sets $\mathcal{L}_i$ representing the possible local states of agent i, for $i = 1\ldots k$, we define a *run* to be an element of the set

(1) $\mathcal{R} = \{r \mid r : \omega \longrightarrow \wp(\mathcal{V}) \times \mathcal{L}_1 \times \ldots \times \mathcal{L}_k\}$.

We define a model to be a subset of $\mathcal{R}$.

Given, $M \subseteq \mathcal{R}$ we give the semantic interpretation of formulas with respect to one run $r \in M$ and one moment of time, $n \in \omega$. We inductively define $M, r, n \models \alpha$ as follows:

$$\begin{array}{rcl}
M,r,n \models x & \Leftrightarrow & x \in r(n)_0 \\
M,r,n \models \neg\alpha & \Leftrightarrow & M,r,n \not\models \alpha \\
M,r,n \models \alpha \wedge \alpha' & \Leftrightarrow & M,r,n \models \alpha \text{ and } M,r,n \models \alpha' \\
M,r,n \models \bigcirc\alpha & \Leftrightarrow & M,r,n+1 \models \alpha \\
M,r,n \models \alpha\,\mathcal{U}\,\alpha' & \Leftrightarrow & \exists m \geq n,\ M,r,m \models \alpha' \text{ and } n \leq j < m \Rightarrow M,r,j \models \alpha \\
M,r,n \models \text{ⓦ}\alpha & \Leftrightarrow & n = 0 \text{ or } M,r,n-1 \models \alpha \\
M,r,n \models \alpha\,\mathcal{S}\,\alpha' & \Leftrightarrow & \exists m \leq n,\ M,r,m \models \alpha' \text{ and } m < j \leq n \Rightarrow M,r,j \models \alpha \\
M,r,n \models K_i\alpha & \Leftrightarrow & \forall r' \in M, \forall m \in \omega,\ r(n)_i = r'(m)_i \Rightarrow M,r',m \models \alpha
\end{array}$$

for each agent i.

This gives the most general description of a language that describes knowledge and past time. However there are several useful restrictions we will consider:

- We say a model has *unique initial states* if for all runs $r, r' \in M$, for all $i \in \{1,\ldots,k\}$, we have $r(0)_i = r'(0)_i$;
- We say a model is *synchronous* if for all runs $r, r' \in M$, for all $n, m \in \omega$, for all $i \in \{1,\ldots,k\}$, we have $r(n)_i = r'(m)_i \Longrightarrow n = m$;

There are several other semantic restrictions that can be applied to combinations of temporal and modal logic, including *perfect recall* and *no learning* [1]. We have chosen to focus on the synchrony and unique initial state restrictions in this paper as they are especially relevant to temporal logics with past. The synchrony and unique initial state restrictions have little effect in logics without past operators, as these restrictions do not alter the set of valid formulas.

The unique initial state restriction requires that no agent can initially distinguish between the possible initial states of the system. This restriction can have significant consequences for the language. In [8] it was shown that when the unique initial state restriction is combined with the no learning restriction the resulting language is highly undecidable. We also note that in the presence of past operators the unique initial state restriction allows us to express a *universal modality* [13, 6]. Specifically, for any formula α, we can define the formula $\diamondminus(\text{ⓦ}\mathit{false} \wedge L_i \diamondsuit \alpha)$ which is satisfied by a model M with the unique initial state restriction if and only if there is some run $r \in M$ and some n such that $M, r, n \models \alpha$.

Once past operators are added to the language, the synchrony restriction has a dramatic affect on the set of valid formulas. Since every agent *knows* the time, an axiomatization must allow reasoning about which formulas can be true at which times. For example, if there is some formula, α, that is true at only even times, then, if at some time an agent even suspects that α might be true, then that agent should know that every formula that is true at only odd times must be false. This situation is captured in the following formula, which is a validity in the synchronous semantics.

(2) $\quad L_i(x \wedge \boxminus(x \leftrightarrow \text{ⓦ}\neg x)) \rightarrow K_i(\boxminus(y \leftrightarrow \text{ⓦ}\neg y) \rightarrow y)$

Here, $L_i(x \wedge \boxminus(x \leftrightarrow \text{ⓦ}\neg x))$ means that agent i *considers it possible* that x is currently true, and up to (and including) now, x has only been true at even moments of time. Since agent i knows the time, agent i knows that the time must be even. Hence agent i *knows* that if up to and including now y has only been true at even moments of time, then y must currently be true, or $K_i(\boxminus(y \leftrightarrow \text{ⓦ}\neg y) \rightarrow y)$.

3 Axioms

In this section, we describe the axioms and inference rules that we need for reasoning about knowledge and time for various classes of systems, and state the completeness results.

For reasoning about knowledge alone, the following system, with axioms K1–K5 and rules of inference R1–R2, is well known to be sound and complete [1, 9]:

K1. All propositional tautologies
K2. $K_i\varphi \wedge K_i(\varphi \rightarrow \psi) \rightarrow K_i\psi, \ i = 1, \ldots, k$
K3. $K_i\varphi \rightarrow \varphi, \ i = 1, \ldots, k$
K4. $K_i\varphi \rightarrow K_iK_i\varphi, \ i = 1, \ldots, k$
K5. $\neg K_i\varphi \rightarrow K_i\neg K_i\varphi, \ i = 1, \ldots, k$
R1. From φ and $\varphi \rightarrow \psi$ infer ψ
R2. From φ infer $K_i\varphi, \ i = 1, \ldots, k$

This axiom system is known as $S5_m$.

For reasoning about pure temporal formulas (or formulas not containing knowledge operators), the following axioms and rules (together with K1 and R1), can be shown to be sound and complete [10]:

F1. $\bigcirc(\varphi \to \psi) \to \bigcirc\varphi \to \bigcirc\psi$
F2. $\bigcirc(\neg\varphi) \leftrightarrow \neg\bigcirc\varphi$
F3. $\varphi\,\mathcal{U}\,\psi \leftrightarrow \psi \vee (\varphi \wedge \bigcirc(\varphi\,\mathcal{U}\,\psi))$
P1. $ⓦ(\varphi \to \psi) \to ⓦ\varphi \to ⓦ\psi$
P2. $\ominus\neg\varphi \to \neg\ominus\varphi$
P3. $\varphi\,\mathcal{S}\,\psi \leftrightarrow \psi \vee (\varphi \wedge \ominus(\varphi\,\mathcal{S}\,\psi))$
P4. $\mathit{true}\,\mathcal{S}\,ⓦ\mathit{false}$
FP. $\varphi \to \bigcirc\ominus\varphi$
PF. $\varphi \to ⓦ\bigcirc\varphi$

RT1. From φ infer $\bigcirc\varphi$
RT2. From $\varphi' \to \neg\psi \wedge \bigcirc\varphi'$ infer $\varphi' \to \neg(\varphi\,\mathcal{U}\,\psi)$
RP1. From φ infer $ⓦ\varphi$.
RP2. From $\varphi' \to \neg\psi \wedge ⓦ\varphi'$ infer $\varphi' \to \neg(\varphi\,\mathcal{S}\,\psi)$

This set of axioms gives a sufficient axiomatization of knowledge with past time. To allow for the unique initial states restriction, we add the following axiom:

UIS. $\boxminus(ⓦ\mathit{false} \to K_i\alpha) \to K_j\boxminus(ⓦ\mathit{false} \to \alpha),\ i,j = 1,\ldots k.$

For the rules required for synchrony, we must first define a characteristic formula for a *transducer*. A transducer is deterministic finite automaton over a one letter alphabet, and can be described by the tuple (Q, q_0, δ) where Q is a finite set of states, $q_0 \in Q$ is the initial state, and $\delta : Q \longrightarrow Q$ is the transition function.

DEFINITION 1. A *characteristic formula* is a formula of the form

$$\Diamond\left(ⓦ\mathit{false} \wedge \overline{a_0} \wedge \Box \bigwedge_{a\subseteq X} (\overline{a} \to \bigcirc\overline{\delta(a)})\right).$$

where: X is an arbitrary finite set of propositional atoms; $a_0 \subseteq X$; for each $a \subseteq X$, $\overline{a}$ is the formula $\bigwedge_{x\in a} x \wedge \neg\bigvee_{x\in X\setminus a} x$; and $\delta : \wp(X) \longrightarrow \wp(X)$ is some function.

A characteristic formula describes the operation of a transducer in linear temporal logic. Specifically we can take $\wp(X)$ to be the set of states, a_0 to be the initial state and δ to be the transition function. It should be clear that a characteristic formula is always satisfiable. It simply declares which atoms should be true at which times in a deterministic manner.

In the case of synchronous systems we require two new rules:

AUT. From $\chi \rightarrow \beta$ infer β, where $var(\chi) \cap var(\beta) = \emptyset$ and χ is a characteristic formula.

SYNC. From $\alpha \rightarrow \beta$ infer $\alpha \rightarrow K_i\beta$, where $var(\alpha) \cap var(\beta) = \emptyset$ and $i = 1, \ldots, k$.

where given a formula α, we let $var(\alpha)$ be the set of propositional atoms appearing in α.

We require the rule, AUT, to add extra propositions into a proof when the propositions already in the proof do not yield sufficient information about the system clock. It should be clear that any transducer, (Q, q_0, δ) will generate a unique sequence of states, $s_0, s_1, \ldots$, where $s_0 = q_0$ and $s_{i+1} = \delta(s_i)$. Furthermore for every transducer there is a unique $k \geq 0$ and a unique $n \geq 1$ such that

1. $\forall i < k$, $\forall j \neq i$, $s_i \neq s_j$, and
2. $\forall i \geq k$, $\forall j > i$ $s_i = s_j$ if and only if n divides $j - i$.

In this respect a transducer is simply a clock which tells the time up to k, and then reports the time modulo n. The rule, AUT, is interesting in that it does not use any knowledge operators. Such a rule is valid in temporal logics with past but has rarely been used in proof systems (although it is similar to the AA rule of [11]).

Just as the AUT rule does not use knowledge operators, the SYNC rule does not explicitly use any temporal operators. In fact the complete axiomatization for synchronous systems has no axiom or rule that uses both temporal and epistemic operators. This may be surprising since the language clearly has a strong interaction between time and knowledge. However the SYNC rule allows an implicit interaction between time and knowledge. Suppose that $\vdash \alpha \rightarrow \beta$ and $var(\alpha) \cap var(\beta) = \emptyset$. Since α and β do not share any propositional atom, we can only infer β from α if α describes some *structural property* (i.e., some property that is independent of propositions, like "this is an initial state"). By examining the language we can see that the only structural information that can be expressed are conditions on the time, such as $x \wedge \boxminus(x \leftrightarrow \text{ⓦ}\neg x)$ ("the time is even"). Thus we can only infer β from α if for every time that α could be true in some model, β must be true in every model. Since agents know the time, and are logically omniscient, if α is true then any agent will be able to deduce β. Thus the SYNC rule captures the interaction between knowledge and time in a synchronous system.

For an example of the effectiveness of the SYNC rule, consider the formula (2). Since it is clear that

$$\vdash (x \wedge \boxminus(x \leftrightarrow @\neg x)) \rightarrow (\boxminus(y \leftrightarrow @\neg y) \rightarrow y),$$

by the completeness of the temporal rules and axioms, the provability of (2) follows directly from the SYNC rule and S5 reasoning.

4 Soundness for unique initial states

Suppose the axiom, UIS, was not sound. Then there would be some model M with unique initial states, such that for some $r \in M$ and some j,

$$M, r, j \models \boxminus(@\mathit{false} \rightarrow K_i\alpha) \wedge \neg K_i \boxminus(@\mathit{false} \rightarrow \alpha).$$

Therefore there must be some $r' \in M$ such that $r(j)_i = r'(j)_i$ such that $M, r', j \models \neg \boxminus(@\mathit{false} \rightarrow \alpha)$. Thus $M, r', 0 \models \neg\alpha$, and $M, r, 0 \models K_i\alpha$ contradicting the unique initial states requirement of the model.

5 Completeness for unique initial states

To prove the axiom system augmented with UIS is complete we use a standard Henkin-style construction with finite sets of formulas. Given a consistent formula, ψ, we show that ψ has a model generated from the maximal consistent subsets of some closure set (see, for example [5]). We define the closure set in two stages. Given ψ, let $\Gamma_\psi = \{\alpha, \neg\alpha, @\mathit{false} | \alpha \subseteq \psi\}$, where $\alpha \subseteq \psi$ if and only if α is a subformula of ψ. As usual we let Σ be the set of maximally consistent sets of formulas, and $S_\psi = \{\Delta \cap \Gamma_\psi \mid \Delta \in \Sigma\}$. We let $S^0_\psi = \{s \in S_\psi \mid @\mathit{false} \in s\}$.

For the next stage, we let $\Gamma = \Gamma_\psi \cup \{\diamondminus \hat{s} \mid s \in S^0_\psi\}$ where $\widehat{s}$ is the conjunction of the formulas in s. We define $S = \{\Delta \cap \Gamma \mid \Delta \in \Sigma\}$ and define the relations $\rightsquigarrow, \sim_i \subseteq S \times S$ as:

- $s \rightsquigarrow t$ if and only if there exists $\Delta, \Delta' \in \Sigma$ such that $s = \Delta \cap \Gamma$, $t = \Delta' \cap \Gamma$ and for all $\alpha \in \Delta'$, $\ominus\alpha \in \Delta$;
- $s \sim_i t$ if and only if there exists $\Delta, \Delta' \in \Sigma$ such that $s = \Delta \cap \Gamma$, $t = \Delta' \cap \Gamma$ and for all $K_i\alpha \in \Delta$, $K_i\alpha \in \Delta'$.

It can be seen using the S5 axioms that $\sim_i$ is an equivalence relation. The following lemma gives an alternative characterization of the relation $\sim_i$.

LEMMA 2. *For all s and t in S, $s \sim_i t$ if and only if $\widehat{s} \wedge L_i\widehat{t}$ is consistent.*

Proof. If $s \sim_i t$ then there exists maximally consistent sets, Δ and Δ' such that $s \subset \Delta$, $t \subset \Delta'$ and $K_i\alpha \in \Delta$ if and only if $K_i\alpha \in \Delta'$. Since $t \subset \Delta'$ it follows that $K_iL_i\widehat{t} \in \Delta'$ so $\widehat{s} \wedge L_i\widehat{t} \in \Delta$, hence $\widehat{s} \wedge L_i\widehat{t}$ is consistent.

Conversely, if $\widehat{t} \wedge L_i\widehat{s}$ is consistent then it is contained in some maximally consistent set, Δ. We define $\Lambda = \{\alpha \mid K_i(\widehat{t} \rightarrow \alpha) \in \Delta\}$. It should be clear that Λ is consistent, and $t \subset \Lambda$. Furthermore, $K_i\alpha \in \Lambda$ if and only if $K_i\alpha \in \Delta$ and $L_i\alpha \in \Lambda$ if and only if $L_i\alpha \in \Delta$. Since Λ is consistent it can be extended to a maximally consistent set which is sufficient to show $s \sim_i t$. ■

For all $s \in S$, we let $[s]_i$ be the corresponding equivalence class of $\sim_i$ and let R be the set of functions $r : \omega \longrightarrow S$ such that:

1. $@\mathit{false} \in r(0)$;

2. for all n, $r(n) \rightsquigarrow r(n+1)$; and

3. for all n, if $\alpha\,\mathcal{U}\,\beta \in r(n)$ then there exists $m \geq n$ such that $\beta \in r(m)$.

¿From R we can derive a model $M = \{\pi_r : \omega \rightarrow \wp(\mathcal{V}) \times \mathcal{L}_1 \times \ldots \times \mathcal{L}_k \mid r \in R\}$ where $\pi_r(j)_0 = r(j) \cap \mathcal{V}$, and $\pi_r(j)_i = [r(j)]_i$. Finally, for every $r \in R$ we let $M_r \subset M$ be defined to be the smallest set such that $\pi_r \in M_r$, and for every $\pi_t \in M_r$, and $j \in \omega$, $\{\pi_u \in M \mid \exists i, j' \text{ s.t. } t(j) \sim_i u(j')\} \subseteq M_r$.

The standard approach here is to extend ψ to a maximal consistent set and use this to find a run r with a state containing ψ. We then prove a truth lemma on M_r, i.e. for every j we show $\alpha \in r(j)$ if and only if $M, \pi_r, j \models \alpha$. Therefore to complete the proof all we have to do is show that the resulting model satisfies the unique initial states constraint. We use the following tautology:

LEMMA 3. $\vdash @\mathit{false} \rightarrow (\varphi \rightarrow \Box(K_i \boxminus(@\mathit{false} \rightarrow \neg K_j\neg\varphi)))$

Proof. We define the formula γ as

(3) $\gamma = @\mathit{false} \wedge (\varphi \wedge \Diamond(L_i \diamondminus(@\mathit{false} \wedge K_j\neg\varphi)))$.

By taking the contrapositive of UIS we can derive the tautology

(4) $\vdash \neg K_i \neg \diamondminus(@\mathit{false} \wedge \neg\alpha) \rightarrow \diamondminus(@\mathit{false} \wedge \neg K_j\alpha)$.

Let $\neg K_j\neg\varphi$. Applied to γ we have

$$\begin{array}{lll} UIS & \vdash \gamma \rightarrow (\textcircled{w}false \wedge \wp \wedge \Diamond \diamondminus (\textcircled{w}false \wedge \neg K_j \neg K_j \neg \varphi)) & (5) \\ K5 & \vdash \neg K_j \neg \varphi \rightarrow K_j \neg K_j \neg \varphi & (6) \\ K1 & \vdash \neg K_j \neg K_j \neg \varphi \rightarrow K_j \neg \varphi & (7) \\ LTL & \vdash \gamma \rightarrow (\textcircled{w}false \wedge \wp \wedge \Diamond \diamondminus (\textcircled{w}false \wedge K_j \neg \varphi)) & (8) \\ K1 & \vdash \gamma \rightarrow \varphi \wedge \neg \varphi & (9) \\ K1 & \vdash \neg \gamma & (10) \end{array}$$

Since $\neg\gamma$ is equivalent to $\textcircled{w}false \rightarrow (\varphi \rightarrow \Box(K_i \boxminus(\textcircled{w}false \rightarrow \neg K_j\neg\varphi)))$, the proof is complete. ■

We note that this lemma is the only place where we are required to use the UIS axiom, so this tautology could be used instead of the UIS axiom.

COROLLARY 4. *The model M_r satisfies the unique initial states constraint.*

Proof. If this were not true there would be some runs with non-unique initial states. Thus there would be some $s(0), t(0), s(u), t(v) \in S$ (where $s, t \in R$) such that $s(u) \sim_i t(v)$, but $s(0) \not\sim_j t(0)$. Since $\textcircled{w}false \in s(0)$ we can use the Lemma 3 to derive

(11) $\vdash \widehat{s(0)} \rightarrow \Box K_i \boxminus(\textcircled{w}false \rightarrow L_j\widehat{s(0)})$.

Let $a = s(0) \cap \Gamma_\psi$. Given the definition of the closure, Γ, it follows that $\vdash \widehat{s(u)} \rightarrow \diamondminus(\textcircled{w}false \wedge \widehat{a})$. Furthermore, it is clear that $\widehat{s(0)}$ is equivalent to $\widehat{a}$, since $s(0) = a \cup \{\diamondminus\widehat{a}\}$. Combining this with (11) we derive $\vdash \widehat{s(u)} \rightarrow K_i \boxminus(\textcircled{w}false \rightarrow L_j\widehat{s(0)})$, and since $s(u) \sim_i t(v)$ it follows using Lemma 2 and S5 reasoning that $\widehat{t(v)} \wedge \boxminus(\textcircled{w}false \rightarrow L_j\widehat{s(0)})$ is consistent. It follows that $\widehat{t(0)} \wedge L_j\widehat{s(0)}$ must be consistent, contradicting the assumption $s(0) \not\sim_j t(0)$, again by Lemma 2. ■

6 Soundness for synchronous systems

The soundness of the rule AUT is straightforward, and is left to the reader. To show SYNC is sound, suppose that α and β do not share propositional atoms, $\alpha \rightarrow \beta$ is a validity, but $\alpha \wedge L_i\neg\beta$ has some model, M. Therefore there are runs $r_\alpha, r_\beta \in M$ and some j such that $M, r_\alpha, j \models \alpha$, and $M, r_\beta, j \models \neg\beta$, and $r_\alpha(j)_i = r_\beta(j)_i$. Note that the interpretation of α, and the interpretation of β can only depend on the propositional atoms that appear in α or β (this can be seen by the recursive definition of the $\models$ relation).

Now let M^+ be a new model defined by $M^+ = \{r \cdot s | r, s \in M\}$, where the run $r \cdot s$ is defined by $r \cdot s(u) = (a, l_1, \ldots, l_k)$ where

- $a = (r(u)_0 \cap var(\alpha)) \cup (s(u)_0 \cap var(\beta))$
- $l_m = (r(u)_m, s(u)_m)$

Note that M^+ is synchronous if M is synchronous.

We can show that $M, r, j \models \alpha$ if and only if $M^+, r \cdot s, j \models \alpha$ for all runs s of M, and $M, s, j \models \beta$ if and only if $M^+, r \cdots, j \models \beta$ for all runs r of M. (This is done by induction over the complexity of formulas, using the semantic descriptions given, and is left to the reader). If we let $r = r_\alpha$ and $s = r_\beta$, it follows that $M^+, r_a^b, j \models \alpha \wedge \neg\beta$, contradicting the fact that $\alpha \rightarrow \beta$ is a validity.

7 Completeness for synchronous systems

We use the strategy used in [7] to construct the model as a series of levels, where each level defines the depth of nestings of knowledge operators in a formula. Given any consistent formula, ψ, we will create a model (a set of runs) by taking sequences of maximally consistent subsets of a closure set of ψ. We will then show that any formula that appears in a maximal consistent subset will be true at the corresponding state in the model. To create such a model we need to find a sequence of maximally consistent sets, where one of the sets contains ψ. We then need to provide additional runs to ensure that if $L_i\gamma$ appears in some set, γ appears in some other run. However these additional runs can be defined over a smaller closure set since we are only interested in the formulas that appear in the scope of a knowledge operator. We can apply this process recursively until we only have to add runs defined over a closure containing no knowledge operators.

When we add additional runs, we have to ensure the knowledge relations conform to the synchrony constraint as well as the normal rules for epistemic logic. To do this, at each level of the construction we include the characteristic formula of a transducer. The state of the transducer at a given level provides sufficient information about the time for us to be able to deduce which of the sets in a lower level will be inconsistent with the current time. The SYNC rule allows us to use this information to ensure the model satisfies the synchrony constraint. The following definitions contribute to this construction.

Each level in the model is represented by a string of agent indexes, (that is, an element of $\{1, \ldots, k\}^*$). We call such strings *knowledge sequences*, and use the following notation:

- we let λ refer to the empty string;

- we let τi be the string τ, concatenated with the index i and let $(\tau i)^- = \tau$;
- we let $\tau\backslash i$ be the largest string μ such that μi is a prefix of τ, or λ if such a string does not exist, and
- we define $\tau \leq \sigma$ ($\tau < \sigma$) to mean τ is a (proper) prefix of σ.

To construct a model of a consistent formula we will use the following hierarchy of languages. We let $\mathcal{L}$ be the language defined above (for k agents), and define the hierarchy over knowledge sequences.

1. $\mathcal{L}_\lambda = \{\alpha \in \mathcal{L} \mid \forall \beta \in \mathcal{L}, \forall i\ K_i\beta \not\subseteq \alpha\}$.
2. $\mathcal{L}_{\tau i} = \{\alpha \in \mathcal{L} \mid K_j\beta \subseteq \alpha \Rightarrow$ either $j = i$ and $\beta \in \mathcal{L}_\tau$ or $K_j\beta \in \mathcal{L}_\tau\}$.

We can see that $\mathcal{L}_\lambda$ is the set of all pure temporal formulas, and let σ be the smallest string such that $\psi \in \mathcal{L}_\sigma$.

We will now define the *closure* of a formula, ψ.

DEFINITION 5. Given a formula ψ, we let Γ_ψ be the *closure* of ψ, defined recursively by:

- $\psi \in \Gamma_\psi$.
- $\alpha \subseteq \varphi$ implies $\alpha, \neg\alpha \in \Gamma_\psi$
- $\alpha \in \Gamma_\psi$ implies $\neg K_i\alpha \in \Gamma_\psi$ and $K_i\alpha \in \Gamma_\psi$ for $i = 1, \ldots, k$.

Given a knowledge sequence, τ, we define the τ-*closure* of ψ to be $\Gamma_\psi^\tau = \Gamma_\psi \cap \mathcal{L}_\tau$.

To be able to create a model we require that maximal consistent subsets of the closure contain sufficient information about the time. We do this as follows. Let Σ be the set of *maximally consistent sets* of formulas taken from the language (with respect to the axioms given and the two rules for synchrony), and given a set X of formulas, we let S_X be the set of maximally consistent subsets of X, (ie $S_X = \{\Delta \cap X \mid \Delta \in \Sigma\}$).

We define the temporal relation $\rightsquigarrow \subseteq S_X \times S_X$ by $s \rightsquigarrow t$ if and only if $\bigwedge_{\alpha\in s} \alpha \wedge \bigcirc \bigwedge_{\alpha\in t} \alpha$ is consistent.

The knowledge relations are quite complex, and will be constructed using the following definitions and lemmas. These constructions are given so that if we are considering formulas in $\mathcal{L}_{\tau i}$, then the closure includes an additional formula, χ_τ, that describes a transducer, A_τ. A run of this transducer associates a state with each moment of time and this state describes the set of maximal consistent subsets of $\mathcal{L}_\tau$ which are consistent with the given time. We do this by induction, where the base case is

$$X_\lambda = \Gamma_\psi^\lambda \cup \Gamma_{\diamondsuit \textcircled{w} false}^\lambda.$$

Given X_τ, for any τ, we can then define S_τ (the maximally consistent subsets of X_τ), A_τ (a transducer showing which subsets are consistent with which times), χ_τ (the characteristic formula of the transducer), and $X_{\tau i}$ (the inductive step). This is done as follows:

- $S_\tau = S_{X_\tau}$.
- For all τ, given S_τ and $\leadsto$ (defined above) we let A_τ be a transducer given by the tuple $(Q_\tau, p_\tau, \delta_\tau)$ where:
  - $Q_\tau = \wp(S_\tau)$ is the set of states;
  - $p_\tau = \{s \in S_\tau \mid \textcircled{w} false \in s\}$
  - $\delta_\tau : Q_\tau \to Q_\tau$ is the transition function defined by $\delta_\tau(q) = \{t \mid \exists s \in q,\ s \leadsto t\}$.

  This transducer is defined to identify states which are reachable in the constructed model at a given time. The run of A_τ is the sequence from Q_τ, $(p_\tau, \delta_\tau(p_\tau), \delta_\tau^2(p_\tau), \ldots)$.
- χ_τ is the characteristic formula of A_τ. To define χ_τ, for each $s \in S_\tau$, let x_s be a propositional atom not appearing in Γ_τ, and for all $q \in Q_\tau$, let $\overline{q} = \bigwedge_{s \in q} x_s \wedge \neg \bigvee_{s \in S_\tau \setminus q} x_s$. Then

$$\chi_\tau = \diamondsuit \left(\textcircled{w} false \wedge \overline{p_\tau} \wedge \Box \bigwedge_{q \in Q_\tau} (\overline{q} \to \bigcirc \overline{\delta_\tau(q)}) \right).$$

- $X_{\tau i} = \Gamma_\psi^{\tau i} \cup \Gamma_{\chi_\tau}^\lambda$.

The proof of completeness will follow from the following lemmas. The first three are technical lemmas which contribute to the proof of the fourth.

LEMMA 6. *For all τ, and $q \in Q_\tau$, we have* $\vdash \chi_\tau \wedge \overline{q} \to K_i \bigvee_{s \in q} \hat{s}$.

Proof. By the construction the transducer A_τ the initial state p_τ is the set of all maximal consistent subsets which are consistent with $\textcircled{w}$*false*, so the following is a validity

(12) $\vdash \textcircled{w} false \wedge \chi_\tau \to \bigvee_{s \in p_\tau} \hat{s}.$

Since $\vdash \chi_\tau \to \diamondsuit(\text{ⓦ}\mathit{false} \wedge \chi_\tau)$, we can derive

$$(13)\ \vdash \chi_\tau \wedge \overline{q} \to \diamondsuit\left(\chi_\tau \wedge \bigvee_{q\in Q_\tau}\left(\overline{q} \wedge \bigvee_{s\in q}\hat{s}\right)\right).$$

The transition function δ_τ is defined to map a set, q, of maximal consistent subsets to the set of all maximal consistent subsets that are consistent with $\ominus \bigvee_{s\in q}\hat{s}$. Since $\vdash \chi_\tau \wedge \overline{q} \to \bigcirc\overline{\delta(q)}$ and $\vdash \bigvee_{s\in q}\hat{s} \to \bigcirc\bigvee_{s\in\delta(q)}\hat{s}$ are tautologies we can derive the following:

$$(14)\ \vdash \chi_\tau \wedge \bigvee_{q\in Q_\tau}\left(\overline{q} \wedge \bigvee_{s\in q}\hat{s}\right) \to \bigcirc\left(\chi_\tau \wedge \bigvee_{q\in Q_\tau}\left(\overline{q} \wedge \bigvee_{s\in q}\hat{s}\right)\right).$$

Applying the rule, RT2, we can show

$$(15)\ \vdash \chi_\tau \wedge \bigvee_{q\in Q_\tau}\left(\overline{q} \wedge \bigvee_{s\in q}\hat{s}\right) \to \boxminus\left(\chi_\tau \wedge \bigvee_{q\in Q_\tau}\left(\overline{q} \wedge \bigvee_{s\in q}\hat{s}\right)\right),$$

and the following tautology follows from (15) and (13):

$$(16)\ \vdash \chi_\tau \wedge \overline{q} \to \left(\chi_\tau \wedge \bigvee_{q\in Q_\tau}\left(\overline{q} \wedge \bigvee_{s\in q}\hat{s}\right)\right).$$

Since by definition, $\vdash \overline{q} \to \bigwedge_{r\neq q}\neg\overline{r}$, we have

$$(17)\ \vdash \chi_\tau \wedge \overline{q} \to \bigvee_{s\in q}\hat{s}.$$

Since the propositional atoms in χ_τ are defined to be disjoint from those in $\bigvee_{s\in q}\hat{s}$ the result follows from the SYNC rule. ■

This lemma shows that for any elements of S_τ which are not consistent with $\chi_\tau \wedge \overline{q}$, given $\chi_\tau \wedge \overline{q}$ we can prove an agent knows that those elements of S_τ are not true. The construction we will use requires us to also prove an agent knows which elements of $S_{\tau\backslash i}$ are not consistent with $\chi_\tau \wedge \overline{q}$. To do this we use the following lemma.

LEMMA 7. *For all knowledge sequences τ, for all $j \in \omega$ and for all $\mu \leq \tau$, $\vdash \chi_\tau \wedge \overline{\delta^j_\tau(p_\tau)} \to \left(\chi_\mu \to \overline{\delta^j_\mu(p_\mu)}\right)$, where the propositional atoms in χ_τ are disjoint from the propositional atoms in χ_μ*

Proof. We will prove this by induction. The base case is the tautology

$$(18)\ \vdash \chi_\tau \wedge \overline{\delta_\tau^j(p_\tau)} \rightarrow \left(\chi'_\tau \rightarrow \overline{\delta_\tau^j(p_\tau)'}\right)$$

where the sets of propositional atoms in χ'_τ and χ_τ are disjoint. The base case is a temporal validity (since transducers are deterministic) so we can assume this is provable from the temporal axioms and rules.

For the inductive step we will first show

$$(19)\ \vdash \chi_\tau \wedge \overline{\delta_\tau^j(p_\tau)} \rightarrow \left(\chi_{\tau^-} \rightarrow \overline{\delta_{\tau^-}^j(p_{\tau^-})}\right)$$

where the sets of propositional atoms in χ_τ and χ_{τ^-} are disjoint. Since (19) is a pure temporal formula, it is sufficient for us to show it is valid using semantic reasoning. Since χ_τ is always satisfiable, if $p = \delta_\tau^n(p_\tau)$ then there must be some $s \in p$ and some $q \in Q_{\tau^-}$ such that $\{\chi_{\tau^-}, \overline{q}\} \subset s$. Furthermore, given such an s for all $t \in p$ if $\chi_{\tau^-} \in t$, then $\overline{q} \in t$ (since by definition $s, t \in p$ if and only if s and t can be reached in n steps from an initial set, and for every transducer there is always a unique state that can be reached in n steps from the initial state). Therefore (19) is valid.

To complete the induction, suppose that for some $\mu \leq \tau$

$$(20)\ \vdash \chi_\tau \wedge \overline{\delta_\tau^j(p_\tau)} \rightarrow \left(\chi_\mu \rightarrow \overline{\delta_\mu^j(p_\mu)}\right).$$

Combining this with (19) we can derive

$$(21)\ \vdash \chi_\mu \rightarrow \left(\chi_\tau \wedge \overline{\delta_\tau^j(p_\tau)} \rightarrow \left(\chi_{\mu^-} \rightarrow \overline{\delta_{\mu^-}^j(p_{\mu^-})}\right)\right).$$

Since we can assume that the sets of propositional atoms in χ_τ, χ_μ and χ_{μ^-} are all disjoint, the AUT rule gives us

$$(22)\ \vdash \chi_\tau \wedge \overline{\delta_\tau^j(p_\tau)} \rightarrow \left(\chi_{\mu^-} \rightarrow \overline{\delta_{\mu^-}^j(p_{\mu^-})}\right)$$

and the lemma follows by induction. ■

We will now restrict our attention to sets $s \in S_{\tau i}$, such that $\chi_\tau \in s$. We define $T_\lambda = S_\lambda$ and let $T_{\tau i} = \{s \in S_{\tau i} \mid \chi_\tau \in s\}$. By the construction of the set $S_{\tau i}$ and the rule, AUT, every consistent formula in $\Gamma_\psi^{\tau i}$ must be an element of some set in $T_{\tau i}$. Given any set, t, of formulas we let $t^i = \{\alpha \mid K_i\alpha \in t\}$. We require the following definition to allow us to compare maximal consistent subsets at different levels.

DEFINITION 8. For all $\tau \neq \lambda$, we define the relation $\prec_i \subset T_{\tau\backslash i} \times T_\tau$ and say $t \prec_i s$ (t *i-supports* s) if

- For some $q \in Q_{\tau^-}$, $\overline{q} \in s$ and there is some $a \in q$ such that $t \subseteq a$.

- $t^i \subseteq s^i \subseteq t$.

This definition is constructed such that for all $s \in T_\tau$, and for all $t \in T_{\tau\backslash i}$, $t \prec_i s$ if and only if $\widehat{s} \wedge L_i\widehat{t}$ is consistent.

LEMMA 9. *For all $s \in T_\tau$, for all $t \in T_{\tau\backslash i}$, if $t \not\prec_i s$ then $\vdash \hat{t} \rightarrow K_i\neg\hat{s}$.*

Proof. If $t \not\prec_i s$ then there are three cases we must consider:

1. for all $q \in Q_{\tau^-}$, if $\overline{q} \in s$ then for all $a \in q$, $t \not\subseteq a$. From Lemma 6 we know $\vdash \chi_{\tau^-} \wedge \overline{q} \rightarrow \bigvee_{a\in q} \hat{a}$, so it follows $\vdash \hat{t} \rightarrow \neg(\chi_{\tau^-} \wedge \overline{q})$, and thus by the SYNC rule, $\vdash \hat{t} \rightarrow K_i\neg(\chi_{\tau^-} \wedge \overline{q})$. As $\vdash \neg(\chi_{\tau^-} \wedge \overline{q}) \rightarrow \neg\hat{s}$ the proof follows from K2 and R1.

2. $s^i \not\subseteq t$. In this case there must be some $K_i\gamma \in s$ such that $\gamma \notin t$. Since $\gamma \in X_{\tau\backslash i}$ we must have $\vdash \hat{t} \rightarrow \neg\gamma$. Applying the epistemic axioms K3, K5 we can show $\vdash \hat{t} \rightarrow K_iL_i\neg\gamma$ and by the rule R2 we can show $\vdash K_i(L_i\neg\gamma \rightarrow \neg\hat{s})$. The result follows from K2.

3. $t^i \not\subseteq s^i$. In this case there must be some $K_i\gamma \in t$ such that $K_i\gamma \notin s$. In this case the result follows trivially from R1 and K2.

■

We note that the converse of this lemma is a consequence of Lemma 15. The following lemma allows us create a synchronous model from a set of runs corresponding to different knowledge sequences.

LEMMA 10. *For all τ, for all j, for all $s \in \delta^j_\tau(p_\tau) \cap T_\tau$, if $L_i\gamma \in s$, then there is some $t \in \delta^j_{\tau\backslash i}(p_{\tau\backslash i}) \cap T_{\tau\backslash i}$ such that $\gamma \in t$ and $t \prec_i s$.*

Proof. We use the Lemma 6 and Lemma 7 to prove this lemma as follows. Suppose for contradiction that there exists some knowledge sequence, τ, some $j \in \omega$, some $s \in \delta^j_\tau(p_\tau) \cap T_\tau$ and some $L_i\gamma \in s$ such that for all $t \in \delta^j_{\tau\backslash i}(p_{\tau\backslash i}) \cap T_{\tau\backslash i}$, if $t \prec_i s$, then $\gamma \notin t$. We can convert this statement into a formula and use the proof theory to derive a contradiction.

For all $t \in \delta^j_{\tau\backslash i}(p_{\tau\backslash i})$, either $t \not\prec_i s$, or $\gamma \notin t$, or $t \notin T_{\tau\backslash i}$. Thus, by Lemma 9, the following would be a tautology:

$$(23) \vdash \bigwedge_{t\in\delta^j_{\tau\backslash i}(p_{\tau\backslash i})} \left(\hat{t} \rightarrow (K_i\neg\hat{s} \vee \neg\gamma \vee \neg\chi_{\tau\backslash i^-})\right).$$

Note in the case that $\tau\backslash i = \lambda$ we can consider $\chi_{\tau\backslash i^-}$ to be the formula, *true*.

Since $s \in \delta^j_\tau(p_\tau) \cap T_\tau$ it follows that $\chi_{\tau^-}, \overline{\delta^j_{\tau^-}(p_{\tau^-})} \in s$. By the Lemma 6, Lemma 7 and the AUT rule we can deduce

(24) $\vdash \widehat{s} \rightarrow K_i \bigvee_{t\in\delta^j_{\tau\backslash i}(p_{\tau\backslash i})} \widehat{t}$

Putting this together with (23) we can show

(25) $\vdash \widehat{s} \rightarrow K_i(K_i \neg\widehat{s} \vee \neg\gamma \vee \neg\chi_{\tau\backslash i^-})$.

We can then apply basic epistemic reasoning and the AUT rule to show $\vdash \neg\widehat{s}$, contradicting the fact that s is consistent. ■

Lemma 10 gives us the sufficient machinery to complete the proof. If ψ is consistent, then for some knowledge sequence, σ, ψ must belong to Γ^{σ}_{ψ} and ψ must be consistent with χ_τ, for all τ. Therefore we can find some $s \in T_\sigma$ such that $\psi \in s$. It is clear that the relation $\rightsquigarrow$ can be restricted to T_σ for all σ, so we can use this to create a σ-history (an infinite $\rightsquigarrow$-sequence in T_σ) where all eventualities are satisfied. For every set in this history we can then satisfy any knowledge formulas using Lemma 10.

The construction we will use here is given as follows: A *ranked set of height* σ is a disjoint union $R = \bigcup_{\tau\leq\sigma} R_\tau$, where for each $\tau \leq \sigma$, R_τ is a set.

For each r in R_τ we associate a τ-history via a *labeling*, described in the following definition:

DEFINITION 11. A *labeling*, ℓ, of a ranked set R of height σ is a collection of functions $\ell_\tau : R_\tau \times \omega \longrightarrow T_\tau$ for $\tau \leq \sigma$ where:

1. for all $r \in R_\tau$, $\ell_\tau(r, 0) \in p_\tau$;
2. for all $r \in R_\tau$, for all $j \in \omega$, $\ell_\tau(r, j) \rightsquigarrow \ell_\tau(r, j+1)$;
3. for all $r \in R_\tau$, for all $j \in \omega$ for all $\alpha\mathcal{U}\beta \in \ell_\tau(r, j)$ there is some $i \geq j$ such that $\beta \in \ell_\tau(r, i)$.

Hence, for any labeling ℓ, for any $r \in R_\tau$, $\ell_\tau(r,0)\ell_\tau(r,1)\ell_\tau(r,2)\ldots$ will be a τ-history. The construction must also satisfy all the knowledge formulas. To do this we use the observation that if $L_i\gamma$ appears at some level (say, $L_i\gamma \in \ell_\tau(r, j)$ where $r \in R_\tau$), then a history labeled by an element of $R_{\tau\backslash i}$ is all that is required to satisfy this formula. In this case we need to find a *witness* for $L_i\gamma$ in $R_{\tau\backslash i}$ that is consistent with both $L_i\widehat{\ell\tau(r, j)}$ and the time j. By Lemma 10 we know this is always possible. To allocate these witnesses we use the following definition:

DEFINITION 12. A *system of support*, ρ, for a ranked set R of height σ equipped with a labeling ℓ consists of, for all $\tau < \sigma$, for all agents i, a partial function $\rho^i_\tau : R_{\tau\backslash i} \hookrightarrow R_\tau \times \omega$, such that

1. for all $r \in R_\tau$, for all $j \in \omega$, if $\rho^i_\tau(t) = (r, j)$, then $\ell_{\tau\backslash i}(t, j) \prec_i \ell_\tau(r, j)$.
2. for all $r \in R_\tau$ for all $j \in N$ if $L_i\gamma \in \ell_\tau(r, j)$ then exactly one of the following holds:
   - there is some $t \in R_{\tau\backslash i}$ such that $\rho^i_\tau(t) = (r, j)$ and $\gamma \in \ell_\tau(t, j)$.
   - there is some $r \in R_\mu$ such that $\mu\backslash i = \tau$, and $\rho^i_\mu(r) = (t, j)$.

We note that in some cases the system of support does not directly allocate a witness for some formula $L_i\gamma$. This occurs if $L_i\gamma \in \ell_\tau(r, j)$, and for some μ, $\rho^i_\mu(r) = (t, j)$. In this case $L_i\gamma$ appears in a set that is itself a witness for a set, $\ell_\mu(t, j)$, at a higher level. In this case we must have $L_i\gamma \in \ell_\mu(t, j)$, and the witness for $L_i\gamma$ in $\ell_\mu(t, j)$ is sufficient to also witness $L_i\gamma$ in $\ell_\tau(r, j)$. This gives us enough to define the basic structure.

DEFINITION 13. Let $\varphi \in \Gamma_n$ be a formula, and σ an index such that $\psi \in \mathcal{L}_\sigma$. A *$\psi$-frame* is a triple (R, ℓ, ρ) where R is a ranked set of height σ, ℓ is a labeling of R and ρ is a system of support for R and ℓ, such that for some $r \in R_\sigma$, and some $j \in \omega$, we have $\psi \in \ell_\sigma(r, j)$.

LEMMA 14. *Given any consistent formula, ψ, there exists a ψ-frame.*

This is left to the reader. The existence of i-τ-supports follows from lemma 10, and the existence of the τ-labellings follows from the usual reachability arguments.

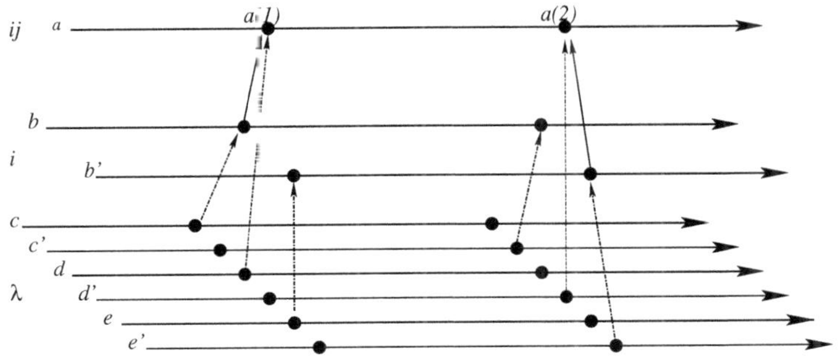

Figure 1. A basic φ-frame

A basic φ-frame is depicted in Figure 1. In this example $\sigma = ij$ and the levels of knowledge (λ, i, or ij) increase from bottom to top. Formulas containing K_j will only appear in the highest level (ij), whilst no formulas containing only epistemic operators will appear in the lowest level (λ). Time flows from left to right, a labels a run R_{ij}, b, b' label runs in R_i and $c, \ldots e'$

label runs in R_λ. The solid arrows between levels correspond to the partial function ρ^j_{ij} (so $\rho^j_{ij}(b) = (a,1)$), and the dashed arrows correspond to ρ^i_i (so $\rho^i_i(d') = (a,2)$).

Given a ψ-frame F, we can now construct a model, $M_F \subseteq \mathcal{R}$ (see (1)) as follows.

- We let the set of local states $\mathcal{L}_i$ be $R \times \omega$.
- For all $\tau \leq \sigma$, for each $r \in R_\tau$ we define a function $\pi_r : \omega \longrightarrow \wp(\mathcal{V}) \times \mathcal{L}_1 \times \ldots \times \mathcal{L}_d$ by $\pi_r(j) = (a, l_1, \ldots, l_d)$ where:
  - $a = \ell_\tau(r,j) \cap \mathcal{V}$;
  - for each $i = 1, \ldots d$ if $\rho^i_\tau(r) = (t,j)$, then $l_i = \pi_t(j)_i$, and otherwise $l_i = (r,j)$.

It is clear that the model M_F is synchronous.

LEMMA 15. *For all $\tau \leq \sigma$, $r \in R_\tau$, for all $j \in \omega$, and for all $\varphi \in \Gamma^\tau_\psi$*

$$M_F, \pi_r, j \models \varphi \iff \varphi \in \ell_\tau(r,j).$$

Proof. This is shown in the usual way, by induction over the complexity of formulas. We note that the definition ensures that for all propositional atoms x, $M_F, \pi_r, j \models x$ if and only if $x \in \ell_\tau(r,j)$. Given a formula α, the inductive hypothesis is $M_F, \pi_r, j \models \alpha \iff \alpha \in \ell_\tau(r,j)$. Assuming α and α' satisfy the inductive hypothesis it can be shown $\alpha \wedge \alpha'$, $\neg\alpha$, $\bigcirc\alpha$, ⓦα, $\alpha\,\mathcal{U}\,\alpha'$ and $\alpha\,\mathcal{S}\,\alpha'$ also satisfy the inductive hypothesis. This is relatively simple and is left to the reader (note that the $\mathcal{U}$ case relies on the last part of Definition 11).

The only interesting case is the inductive step for the knowledge operator, where $\varphi = K_i\alpha$. We assume by the inductive hypothesis that for all $\tau \leq \sigma$, for all $r \in R_\tau$, and for all $j \in \omega$, $M_F, \pi_r, j \models \alpha \iff \alpha \in \ell_\tau(r,j)$.
Suppose that $M_F, \pi_r, j \models \varphi$. In this case $M_F, \pi_r, j \models \alpha$ and for all t such that $\pi_r(j)_i = \pi_t(j)_i$ we have $M_F, \pi_t, j \models \alpha$. By the construction of M_F we have two possibilities:

1. For all $t \neq r$, such that $\pi_r(j)_i = \pi_t(j)_i$, $t \in R_{\tau\setminus i}$ and $\rho^i_\tau(t) = (r,j)$. By the induction hypothesis, for all such t where $\rho^i_\tau(t) = (r,j)$, $\alpha \in \ell_{\tau\setminus i}(t,j)$. Suppose for contradiction $\varphi \notin \ell_\tau(r,j)$. Then $L_i\neg\alpha \in \ell_\tau(r,j)$ and by the definition of ρ^i_τ there would be some t such that $\rho^i_\tau(t) = (r,j)$ and $\alpha \notin \ell_{\tau\setminus i}(t,j)$, giving us the required contradiction. Thus $\varphi \in \ell_\tau(r,j)$.

2. There is some μ and some $s \in T_\mu$ such that $\mu\backslash i = \tau$ and $\rho_\tau(j) = s$. For all $t \neq s$, such that
   $\pi_r(j)_i = \pi_t(j)_i$, we have $\rho^i_\mu(t) = (s,j)$ and hence $\alpha \in \ell_\tau(t,j)$. Since $\varphi \in \Gamma^{\mu\backslash i}_\psi$ we must have $K_i\varphi \in \Gamma^\mu_\psi$. If $L_i\neg\varphi \in \ell_\mu(s,j)$, by K4 we have $L_i\neg\alpha \in \ell_\mu(s,j)$ and by the definition of ρ^i_μ there must be some $t \in R_\tau$ such that $\rho^i_\mu(t) = (s,j)$ and $\alpha \notin \ell_\tau(t,j)$, contradicting the induction hypothesis. Therefore we must have $K_i\varphi \in \ell_\mu(s,j)$ and hence $\varphi \in \ell_\mu(s,j)^i$. By the definition of $\prec_i$ it follows that $\varphi \in \ell_\tau(r,j)$.

For the converse, suppose $\varphi \in \ell_\tau(r,j)$. Again we consider two possibilities:

1. For all $t \neq r$ such that $\pi_r(j)_i = \pi_t(j)_i$, $t \in R_{\tau\backslash i}$ and $\rho^i_\tau(t) = (r,j)$. In this case by the definition of ρ^i_τ and $\prec_i$, we have $\ell_\tau(r,j)^i \subseteq \ell_{\tau\backslash i}(t,j)$. Consequently $\alpha \in \ell_{\tau\backslash i}(t,j)^i$, and by K3, $\alpha \in \ell_\tau(r,j)$. By the inductive hypothesis, for all t such that $\pi_r(j)_i = \pi_t(j)_i$, we have $M_F, \pi_t, j \models \alpha$, so $M_F, \pi_r, j \models K_i\alpha$.

2. There is some μ, and $j \in \omega$ and some $s \in T_\mu$ such that $\mu\backslash i = \tau$ and $\rho(r) = (s,j)$. For all $t \neq s$ such that $\pi_r(j)_i = \pi_t(j)_i$, we have $\rho^i_\mu(t) = (s,j)$. Since $\ell_\tau(r,j) \prec_i \ell_\mu(s,j)$, $K_i\alpha \in \ell_\tau(r,j)$ implies $K_i\alpha \in \ell_\mu(s,j)$. For all $t \neq s$ such that $\pi_r(j)_i = \pi_t(j)_i$, we have $\rho^i_\mu(t) = (s,j)$ and thus $\ell_\tau(t,j) \prec_i \ell_\mu(s,j)$. By the definition of $\prec^i$, $K_i\alpha \in \ell_\mu(s,j)$ implies $\alpha \in \ell_\tau(t,j)$ for all such t. Since we must also have $\alpha \in \ell_\mu(s,j)$ (by K3) the results follows from the induction hypothesis.

■

8 Conclusion

In this paper we have presented sound and complete axiomatizations for logics of knowledge and past time with the synchrony and unique initial states constraints. While the proof of completeness for the unique initial states restriction is relatively straightforward, the proof of completeness for the synchrony restriction is surprisingly complicated. The axiomatization for synchrony relies on complex automata based rules, and finding a simpler axiomatization (and proof) would be of some interest.

For future work we will be investigating combinations of knowledge and past time given the semantic restrictions of *perfect recall* (where an agent retains the knowledge of previous times), and *no learning* (where an agent's knowledge can not increase over time) [1]. Along with synchrony and unique initial state restriction, this gives us sixteen different combinations of restrictions to consider. We will investigate axiomatizations for the resulting languages and extending these axiomatizations to include axioms for common knowledge.

Acknowledgements

Work supported by an Australian Research Council large grant. National ICT Australia is funded through the Australian Government's *Backing Australia's Ability* initiative, in part through the Australian Research Council.

BIBLIOGRAPHY

[1] R. Fagin, J. Halpern, Y. Moses, and M. Vardi. *Reasoning about knowledge.* MIT Press, 1995.

[2] R. Fagin, J. Halpern, Y. Moses, and M. Vardi. Knowledge-based programs. *Distributed Computing*, 10(4):199–225, 1997.

[3] M. Fisher, M. Wooldridge, and C. Dixon. A resolution-based proof method for temporal logics of knowledge and belief. In *Proceedings of the International Conference on Formal and Applied Practical Reasoning (FAPR)*, 1996.

[4] D. Gabbay, A. Kurucz, F. Wolter, and M. Zakharayashev. *Many Dimensional Modal Logics: Theory and Applications.* Elsevier, 2003.

[5] D. M. Gabbay, A. Pnueli, S. Shelah, and J. Stavi. On the temporal analysis of fairness. In *7th ACM Symposium on Principles of Programming Languages, Las Vegas*, pages 163–173, 1980.

[6] V. Goranko and S. Passy. Using the universal modality: gains and questions. *Journal of Logic and Computation*, 2:5–20, 1992.

[7] J. Halpern, R. van der Meyden, and M. Vardi. Complete axiomatizations for reasoning about knowledge and time. *SIAM Journal on Computing*, 33(3):674–703, 2004.

[8] J. Halpern and M. Vardi. The complexity of reasoning about knowledge and time, I:lower bounds. *Journal of Computer and System Science*, 38(1):195–237, 1989.

[9] J. Hintikka. *Knowledge and Belief.* Cornell University Press, 1962.

[10] O. Lichtenstein, A. Pnueli, and L. Zuck. The glory of the past. *Lecture Notes in Computer Science*, 193:196–218, 1985.

[11] M. Reynolds. An axiomatization of full computation tree logic. *Journal of Symbolic Logic*, 63(3):1011–1057, 2001.

[12] P. Schoebelen. The complexity of temporal logic model-checking. In *Advances in Modal Logic*, volume 4, pages 393–436, 2002.

[13] E. Spaan. *Complexity of Modal Logics.* PhD thesis, Universiteit van Amsterdam, 1993.

[14] R. van der Meyden. Axioms for knowledge and time in distributed systems with perfect recall. In *IEEE Symposium on Logic in Computer Science*, pages 448–457, 1994.

Tim French, Mark Reynolds
School of Computer Science & Software Engineering
University of Western Australia, Perth 6009, Australia
tim@csse.uwa.edu.au, mark@csse.uwa.edu.au

Ron van der Meyden
School of Computer Science and Engineering
University of New South Wales, and
National ICT Australia, Sydney 2052, Australia
meyden@cse.unsw.edu.au

A Two-sorted Hybrid Logic Including Guarded Jumps

Bernhard Heinemann

ABSTRACT. In this paper we present a two-sorted hybrid logic where formulas are interpreted in spaces of sets. The logical language we consider contains names for both points and sets. Usual nominals are then realized as pairs of names, consisting of one name from each category actually. Accompanying satisfaction operators are mimicked with the aid of the global modality. And we have a family of additional hybrid operators associated with names of sets in a natural way. These connectives make jumps possible that are 'guarded' in a sense, which yields in fact more expressiveness than admitting only unrestricted jumps (by means of the global modality) and the strictly controlled ones from basic hybrid logic. Below we take the first steps towards a hybrid logic comprising the features just indicated, by proving a corresponding completeness as well as a decidability result. Moreover, we discuss an application of the new system to reasoning about knowledge.

1 Introduction

Classical modal languages play an important part in modelling and reasoning about sytems. Some applications, however, call for more expressive power. *Hybrid logic* is a useful tool for extending 'faithfully' the expressive means of usual modal logic; cf [1] for a summary regarding this. But it is not always clear *how* to hybridize a modal language. Already on the elementary stage of defining appropriate satisfaction operators there are several possibilities, depending on the source language. In the present paper, we want to illustrate this by means of a sample language we are particularly interested in because it is related to reasoning about knowledge. (As to the latter notion, cf [7] or [15].)

In [16] (and more detailedly in [6]), a bi-modal system called *topologic* was developed. The two modalities K and $\Box$ appearing there, represent *knowledge* and *effort (in the course of time)*, respectively. Formulas are interpreted in *set spaces* $(X, \mathcal{O}, V)$ consisting of a non-empty set X of *states*, a set $\mathcal{O}$ of distinguished subsets of X, which can be taken as *knowledge states* of an agent, and a valuation V determining the states where the atomic propositions are true. The operator K quantifies then across any knowledge state, whereas $\Box$ quantifies 'downward' across $\mathcal{O}$. That is, descending within $\mathcal{O}$ with respect to the set inclusion relation means acquiring knowledge, and just this is modelled by $\Box$.

It is less noticed that evolving knowledge has also a *spatial* component besides the obvious temporal one. In fact, knowledge can be viewed as *closeness* and knowledge acquisition as *getting a closer proximity* in set spaces. Thus *topology* enters the context of knowledge (justifying the name 'topologic').

Unfortunately, it turned out that several topologically interesting classes of set spaces cannot be dealt with in the framework of *topologic*. However, hybridizing the underlying language can help a lot here. Already the naïve, 'unsorted' hybrid logic of set spaces allows us to characterize neatly the classes arising from linear flows of time; cf [10] and [13].[1] And following the *sorting strategy* from [1] makes the hybrid approach much more flexible, even without having any satisfaction operators at one's disposal; cf [11].

The hybrid logic considered in this paper is a development of the one from [11]. Satisfaction operators are now integrated. Actually, we have two kinds of such operators: usual ones which can be realized with the aid of the global modality, and new 'guarded jump'–operators coming along with names of sets. For technical reasons we treat here both types of hybrid connectives simultaneously. But in general, the intended application will tell one which language to prefer. The 'guarded jump'–operators seem to be particularly suited to a certain class of intersection closed set spaces we revisit repeatedly in this paper; cf [17].

In the next section we introduce the new hybrid language of set spaces. Afterwards we present an axiomatization, add suitable proof rules and prove completeness. Finally, we argue that the new logic is decidable. All this is done for general set spaces as well as so-called *directed* ones.

2 The language

In this section we define first the syntax and semantics of the two-sorted hybrid language for set spaces indicated above. Then, we give an example proving the expressive power of this language.

We extend the basic bi-modal language for set spaces by two sets of nominals on the one hand and a family of unary modalities on the other hand. The denotation of every nominal is either a unique state or a distinguished set of states, whereas the new modalities represent the global one and the 'guarded jump'–operators belonging to set names, respectively.

[1]We let nominals denote *neighbourhood situations* there; see Definition 1 below.

Let PROP $= \{p, q, \ldots\}$, $\mathrm{N}_{stat} = \{i, j, \ldots\}$ and $\mathrm{N}_{sets} = \{A, B, \ldots\}$ be three mutually disjoint denumerable sets of symbols called *proposition letters, names of states,* and *names of sets,* respectively. Then we define the set WFF of *well-formed formulas* over PROP $\cup \mathrm{N}_{stat} \cup \mathrm{N}_{sets}$ by the rule

$$\alpha ::= p \mid i \mid A \mid \neg\alpha \mid \alpha \wedge \beta \mid K\alpha \mid \Box\alpha \mid \mathsf{A}\alpha \mid [\epsilon_A]\alpha.$$

The modality $[\epsilon_A]$ is to be read 'in (the set denoted by) A (at the actual state)'. The missing boolean connectives $\top, \bot, \vee, \rightarrow, \leftrightarrow$ are treated as abbreviations, as needed. The duals of K, $\Box$, A and $[\epsilon_A]$ are denoted L, $\Diamond$, E and $\langle\epsilon_A\rangle$, respectively.

We give next meaning to formulas. To begin with, we define the domains where formulas will be interpreted in. We let $\mathcal{P}(X)$ designate the powerset of a given set X.

DEFINITION 1 (Set frames; set spaces with names).

1. A pair $(X, \mathcal{O})$ consisting of a non-empty set X and a set $\mathcal{O} \subseteq \mathcal{P}(X)$ of subsets of X such that $X, \emptyset \in \mathcal{O}$ is called a *set frame.*

2. Let $\mathcal{S} := (X, \mathcal{O})$ be a set frame. The set of *neighbourhood situations* of $\mathcal{S}$ is the set $\mathcal{N}_{\mathcal{S}} := \{x, U \mid x \in U \text{ and } U \in \mathcal{O}\}$.

3. Let $\mathcal{S}$ be a set frame as above. An $\mathcal{S}$*–valuation* is a mapping

$$V : \mathrm{PROP} \cup \mathrm{N}_{stat} \cup \mathrm{N}_{sets} \longrightarrow \mathcal{P}(X)$$

   such that

   (a) $V(i)$ is either $\emptyset$ or a singleton subset of X for every $i \in \mathrm{N}_{stat}$, and

   (b) $V(A) \in \mathcal{O}$ for every $A \in \mathrm{N}_{sets}$.

4. A *set space with names* (or, in short, an *SSN*) is a triple $(X, \mathcal{O}, V)$, where $\mathcal{S} = (X, \mathcal{O})$ is a set frame and V an $\mathcal{S}$–valuation. One says then that $\mathcal{M} := (X, \mathcal{O}, V)$ is *based on* $\mathcal{S}$.

The requirement '$X \in \mathcal{O}$' from item 1 of the above definition shrinks the class of frames we are interested in not very much, but facilitates matters here and there. The condition '$\emptyset \in \mathcal{O}$' takes into account that nominals may have an empty denotation, according to item 3. This is appropriate for our purposes, but not common in standard hybrid logic.[2]

[2] Because of the presence of the global modality we could force the usual interpretation of nominals by the formulas $\mathsf{E}\,i$ and $\mathsf{E}\,A$, respectively. This would make the proceeding a bit more difficult later on, so we prefer the way just taken.

For a given SSN $\mathcal{M}$ we define now the relation of satisfaction, $\models_{\mathcal{M}}$, between neighbourhood situations of the underlying frame and formulas in WFF.

DEFINITION 2 (Satisfaction and validity). Let $\mathcal{M} = (X, \mathcal{O}, V)$ be an SSN based on the set frame $\mathcal{S} = (X, \mathcal{O})$, and let x, U a neighbourhood situation of $\mathcal{S}$. Then

$$\begin{array}{lcl}
x, U \models_{\mathcal{M}} p & :\iff & x \in V(p) \\
x, U \models_{\mathcal{M}} i & :\iff & x \in V(i) \\
x, U \models_{\mathcal{M}} A & :\iff & V(A) = U \\
x, U \models_{\mathcal{M}} \neg\alpha & :\iff & x, U \not\models_{\mathcal{M}} \alpha \\
x, U \models_{\mathcal{M}} \alpha \wedge \beta & :\iff & x, U \models_{\mathcal{M}} \alpha \text{ and } x, U \models_{\mathcal{M}} \beta \\
x, U \models_{\mathcal{M}} K\alpha & :\iff & y, U \models_{\mathcal{M}} \alpha \text{ for all } y \in U \\
x, U \models_{\mathcal{M}} \Box\alpha & :\iff & \forall U' \in \mathcal{O} : (x \in U' \subseteq U \Rightarrow x, U' \models_{\mathcal{M}} \alpha) \\
x, U \models_{\mathcal{M}} \mathsf{A}\alpha & :\iff & y, U' \models_{\mathcal{M}} \alpha \text{ for all } y, U' \in \mathcal{N}_{\mathcal{S}} \\
x, U \models_{\mathcal{M}} [\epsilon_A]\alpha & :\iff & \text{if } x \in V(A), \text{ then } x, V(A) \models_{\mathcal{M}} \alpha,
\end{array}$$

for all $p \in$ PROP, $i \in \mathrm{N}_{stat}$, $A \in \mathrm{N}_{sets}$, and $\alpha, \beta \in$ WFF. In case $x, U \models_{\mathcal{M}} \alpha$ is true we say that α *holds in* $\mathcal{M}$ *at* the neighbourhood situation x, U.

A formula α is called *valid in* $\mathcal{M}$ ('$\mathcal{M} \models \alpha$') iff it holds in $\mathcal{M}$ at every neighbourhood situation.

The following remarks seem appropriate to elucidate this definition in some respects.

REMARK 3 (Peculiarities of the just defined language).

1. The meaning of both proposition letters and names of states is independent of neighbourhoods by definition, thus 'stable' with respect to $\Box$. This fact is reflected in two special axioms below (Axioms 6 and 11).

2. The formulas of the form $i \wedge A$, where $i \in \mathrm{N}_{stat}$ and $A \in \mathrm{N}_{sets}$, can be taken as names for elements of $\mathcal{N}_{\mathcal{S}}$. The satisfaction operator associated with such a name reads then $\mathsf{E}(i \wedge A \wedge \dots)$. I.e., pairs (i, A) can act like 'proper' nominals in set spaces with names; cf [10].

3. The last clause of Definition 2 explains what a guarded jump means for set spaces: it is allowed to jump from the neighbourhood U to the neighbourhood $V(A)$ for evaluating α there, provided that the 'guard' x, which is not affected by this jump, is also contained in the latter set. Note that $x, U \models_{\mathcal{M}} \langle\epsilon_A\rangle\alpha$ implies, in particular, that x is contained in the denotation of A.

4. The reader might wonder why a 'guarded jump'–operator belonging to a name $i \in \mathrm{N}_{stat}$ does not appear here. Such an operator is missing for the simple reason that it can be defined, in fact, by $K(i \to \dots)$.

Now, the question comes up naturally whether $[\epsilon_A]$ can be defined as well. But this is not the case for the following reason. One can 'leave' the actual neighbourhood U with the aid of the connective $[\epsilon_A]$ in a special, controlled way. However, leaving U is not at all possible by means of K or $\Box$, and not in that special way by means of A. Thus we have really got more expressiveness.

The example covered by the next definition and proposition points to an important application of the new language.

DEFINITION 4 (Directed frames and spaces). A set frame $\mathcal{S} = (X, \mathcal{O})$ is called *directed,* iff

$$\forall U_1, U_2 \in \mathcal{O}, \forall x \in X : (x \in U_1 \cap U_2 \Rightarrow \exists U \in \mathcal{O} : x \in U \subseteq U_1 \cap U_2).$$

An SSN $\mathcal{M}$ is called *directed,* iff it is based on a directed frame.

PROPOSITION 5 (Expressing directedness). *Let $\mathcal{S} = (X, \mathcal{O})$ be a set frame. Then $\mathcal{S}$ is directed, iff for all SSNs $\mathcal{M}$ based on $\mathcal{S}$ we have that*

$$\mathcal{M} \models \Diamond A \wedge \Diamond B \to \Diamond K(\langle \epsilon_A \rangle \top \wedge \langle \epsilon_B \rangle \top) \text{ for all } A, B \in \mathrm{N}_{sets}.$$

Proof. We prove only that the condition is sufficient for directedness. (The proof of its necessity is even easier.) So, let $\mathcal{S}$ be not directed. Then there are $U_1, U_2 \in \mathcal{O}$ and $x \in X$ such that $x \in U_1 \cap U_2$, but $x \notin U$ or $U \not\subseteq U_1 \cap U_2$ for all $U \in \mathcal{O}$. Take any $\mathcal{S}$–valuation V satisfying $V(A) = U_1$ and $V(B) = U_2$, and let $\mathcal{M} := (X, \mathcal{O}, V)$. Then $x, X \models_{\mathcal{M}} \Diamond A \wedge \Diamond B$. However, we have that $x, X \models_{\mathcal{M}} \Box L([\epsilon_A]\bot \vee [\epsilon_B]\bot)$ because we can find a point witnessing $U \not\subseteq V(A) \cap V(B)$ in every $U \in \mathcal{O}$ containing x. In this way we have found an SSN $\mathcal{M}$ based on $\mathcal{S}$ and an instance of the above formula schema which is not valid in $\mathcal{M}$, as desired. ∎

Directed frames correspond to the topological notion of *filter base;* cf [5], Ch. I, § 6.3. Filters are crucial to the general idea of *convergence;* cf [5], Ch. I, § 7. As set spaces represent a model for the development of knowledge, directed spaces are, therefore, associated with 'converging' knowledge acquisition procedures. The *modal* logic of knowledge arising from that was studied in the paper [17]. We will come back to this system later on, at the end of the next section (and in Section 4 as well).[3]

[3] Originally, a good portion of work was devoted to the axiomatizability problem for the logic of *intersection closed spaces;* cf [6]. Later on, in [17], it was announced that this logic and the logic of directed spaces coincide.

Concluding this section, we comment on the relevance of names and accompanying hybrid operators to the context of knowledge. We confine ourselves to names of sets here since the general usefulness of names of states has been sufficiently demonstrated elsewhere; cf, above all, [1]. As mentioned in the introduction already, the elements of $\mathcal{O}$ can be viewed as knowledge states of an agent, for any correspondingly given set frame $(X, \mathcal{O})$. Thus the new language supplies one with *names of knowledge states.* And the hybrid operators $[\epsilon_A]$, where $A \in \mathrm{N}_{sets}$, allow to switch over to named knowledge states as long as the point of evaluation is fixed. This additional means of expression is, therefore, perfectly in accord with the common 'external' view of knowledge in multi-agent systems, where knowledge is 'ascribed' to the agents; cf [7], Ch. 4.

3 Completeness

In this section we study a hybrid logic for the class of all set spaces with names. To start with, we propose an appropriate system of axioms. Then we formulate a couple of Gabbay–style proof rules. These and the usual ones of modal and hybrid logic lead to a logical system called GJ (indicating **g**uarded **j**umps). The completeness proof for GJ is contained in the remaining part of this section. It turns out that even *extended* completeness (in the sense of hybrid logic, cf [2], Sec. 7.3) is yielded.

The axiom schemata are arranged in four groups. First, the modal axioms for arbitrary set spaces are listed.

1. All instances of tautologies.
2. $K(\alpha \to \beta) \to (K\alpha \to K\beta)$
3. $K\alpha \to \alpha$
4. $K\alpha \to KK\alpha$
5. $L\alpha \to KL\alpha$
6. $(p \to \Box p) \wedge (\Diamond p \to p)$
7. $\Box(\alpha \to \beta) \to (\Box\alpha \to \Box\beta)$
8. $\Box\alpha \to \alpha$
9. $\Box\alpha \to \Box\Box\alpha$
10. $K\Box\alpha \to \Box K\alpha$,

where $p \in \mathrm{PROP}$ and $\alpha, \beta \in \mathrm{WFF}$. Note that the schemata 2 – 5 represent the usual S5 axioms of knowledge, 6 captures the aforementioned stability of the proposition letters with respect to $\Box$, 7 – 9 say that $\Box$ satisfies all the S4 laws, and 10 is the *Cross Axiom* from [6];[4] see also the comments on Definition 13 below.

The second group of axioms concerns names.

11. $(i \to \Box i) \wedge (\Diamond i \to i)$
12. $i \wedge \alpha \to K(i \to \alpha)$
13. $A \to KA$
14. $\mathsf{A}(A \wedge L\alpha \to L\beta) \vee \mathsf{A}(A \wedge L\beta \to L\alpha)$
15. $K(\Diamond B \to \Diamond A) \wedge L\Diamond B \to \Box(A \to L\Diamond B)$
16. $K\Diamond A \to A$,

[4]This schema plays a subtle part in multi-agent systems where agents have *perfect recall,* i.e., do not forget anything of their own history in the course of time; see [7], notes to Ch. 8, for a more detailed discussion.

where $i \in \mathrm{N}_{stat}$, $A, B \in \mathrm{N}_{sets}$ and $\alpha, \beta \in$ WFF. We mention that Axioms 11 and 12 are needed to force the right interpretation of names of states in the model constructed during the proof of completeness later on, and Axioms 13 and 14 serve the same purpose regarding names of sets. The remaining two axioms ensure that really a set frame structure can be defined for that model.[5] — Actually, a further axiom can be included in this group.

$$17.\ i \wedge A \wedge \mathsf{E}(j \wedge B) \rightarrow \mathsf{E}\left(\diamond(i \wedge A) \wedge L\diamond(j \wedge B)\right),$$

where $i, j \in \mathrm{N}_{stat}$ and $A, B \in \mathrm{N}_{sets}$. This axiom corresponds to a certain *left-directedness* property; see Lemma 9 below. Moreover, it reflects that the set frames $(X, \mathcal{O})$ we consider are *generated* in a sense. In fact, every neighbourhood situation of the form x, X (where $x \in X$) can be viewed as a generating element, according to both the clause '$X \in \mathcal{O}$' from item 1 of Definition 1 and the semantics of K and $\Box$; see also item 5 of Definition 13 below.[6]

Each axiom of the next group deals with the global modality.

18. $\mathsf{A}(\alpha \rightarrow \beta) \rightarrow (\mathsf{A}\alpha \rightarrow \mathsf{A}\beta)$	20. $\mathsf{A}\alpha \rightarrow \mathsf{AA}\alpha$	22. $\mathsf{A}\alpha \rightarrow K\Box\alpha$
19. $\mathsf{A}\alpha \rightarrow \alpha$	21. $\alpha \rightarrow \mathsf{AE}\alpha$,	

where $\alpha, \beta \in$ WFF. Here Axioms 18 – 21 mean that A is an S5 modality (like K), and Axiom 22 describes the obvious interplay between A and $K\Box$.

By the last group, the 'guarded jump'–operators are handled.

23. $[\epsilon_A](\alpha \rightarrow \beta) \rightarrow ([\epsilon_A]\alpha \rightarrow [\epsilon_A]\beta)$	24. $\diamond(A \wedge \alpha) \rightarrow \Box\langle\epsilon_A\rangle\alpha$	25. $[\epsilon_A]A$
26. $\diamond\langle\epsilon_A\rangle\alpha \wedge L\diamond A \rightarrow \diamond\alpha$	27. $\mathsf{A}\alpha \rightarrow [\epsilon_A]\alpha$,	

where $A \in \mathrm{N}_{sets}$ and $\alpha, \beta \in$ WFF. As to the properties corresponding to Axioms 24 and 25, see item 8 of Definition 13. Axiom 26 requires an extra investigation later on. Finally, Axiom 27 expresses inclusion (like Axiom 22). – Now we define the desired logical system GJ.

DEFINITION 6 (The logic). Let GJ be the smallest set of formulas containing all of the above axiom schemata and closed under application of the following rules:

$$(\text{MODUS PONENS})\ \frac{\alpha \rightarrow \beta, \alpha}{\beta} \qquad (\Delta\text{–NECESSITATION})\ \frac{\alpha}{\Delta\alpha}$$

$$(\text{NAME}_{stat})\ \frac{i \rightarrow \beta}{\beta} \qquad (\text{NAME}_{sets})\ \frac{B \rightarrow \beta}{\beta}$$

[5]Note that, apart from Axiom 14 (which had to be reformulated since the global modality belongs now to the language) the axioms of this group were also used for proving completeness of the satisfaction operator-free fragment of our logic in the paper [11].

[6]I am grateful to Valentin Goranko who pointed an error concerning generation in the preliminary proceedings version of this paper out to me.

$$(\Delta\text{-ENRICHMENT}) \quad \frac{\mathsf{E}\,(i \wedge A \wedge \nabla(j \wedge B \wedge \alpha)) \rightarrow \beta}{\mathsf{E}(i \wedge A \wedge \nabla\alpha) \rightarrow \beta},$$

where $\alpha, \beta \in$ WFF, $i, j \in \mathrm{N}_{stat}$, $A, B \in \mathrm{N}_{sets}$, $\Delta \in \{K, \Box, \mathsf{A}\} \cup \{[\epsilon_A] \mid A \in \mathrm{N}_{sets}\}$, $\nabla \in \{L, \Diamond, \mathsf{E}\} \cup \{\langle\epsilon_A\rangle \mid A \in \mathrm{N}_{sets}\}$, and j, B are *new* each time (i.e., do not occur in all the other syntactic building blocks of the respective rule).

The reader can easily check that GJ is sound with respect to the class of all SSNs.

In order to obtain completeness we proceed via the canonical model of GJ. To begin with, we call a maximal consistent set s of formulas

- *named* iff s contains some $i \in \mathrm{N}_{stat}$ and some $A \in \mathrm{N}_{sets}$, and
- *enriched* iff for every $\nabla \in \{L, \Diamond, \mathsf{E}\} \cup \{\langle\epsilon_A\rangle \mid A \in \mathrm{N}_{sets}\}$ we have that

  $$\mathsf{E}(i \wedge A \wedge \nabla\alpha) \in s \text{ implies } \mathsf{E}\,(i \wedge A \wedge \nabla(j \wedge B \wedge \alpha)) \in s$$

  for some $j \in \mathrm{N}_{stat}$ and $B \in \mathrm{N}_{sets}$.

Now we take a maximal consistent set Γ containing the negation γ of a given non-derivable formula, and we extend Γ to a named and enriched maximal consistent set Γ' in the language increased by enough new constants. This can be done by means of a suitable *Extended Lindenbaum Lemma* in the same way as in usual hybrid logic; cf [2], 7.25. (So, we may abstain from the details concerning this.)

Furthermore, we let $\widetilde{\mathcal{M}}$ be the submodel of the structure *generated* by Γ' of which the domain D consists of all points that are named. The accessibility relations induced on $\widetilde{\mathcal{M}}$ are denoted $\stackrel{\nabla}{\longrightarrow}$, where $\nabla \in \{L, \Diamond, \mathsf{E}\} \cup \{\langle\epsilon_A\rangle \mid A \in \mathrm{N}_{sets}\}$. Then, we get the following *Existence Lemma;* cf [2], 7.27.

LEMMA 7. *Assume that $s \in D$ contains the formula $\nabla\alpha$ (where ∇ is as above). Then there exists some $t \in D$ satisfying $s \stackrel{\nabla}{\longrightarrow} t$ and $\alpha \in t$.*

Proof. First, one realizes that, because of Axioms 22 and 27, the domain of the structure generated by Γ' coincides with the $\stackrel{\mathsf{E}}{\longrightarrow}$-equivalence class of Γ'. Second, Axioms 14 and 12 imply that, given $A \in \mathrm{N}_{sets}$ and $i \in \mathrm{N}_{stat}$, then there is at most one point in D containing the formula $i \wedge A$. (The arguments for that are similar to those from the proof of [11], Lemma 2. So, we may be brief here.) Now the assertion of the lemma follows easily from the fact that Γ' is enriched. ■

We have to provide for a set frame structure next. It turns out that we cannot guarantee that $\widetilde{\mathcal{M}}$ is based on such a frame, but we will clearly exploit the properties of the accessibility relations of this model. For every $s \in D$ we let $[s] := \{t \in D \mid s \stackrel{L}{\longrightarrow} t\}$ be the $\stackrel{L}{\longrightarrow}$-equivalence class of s (see Axioms 3 – 5), and we define a binary relation $\preccurlyeq$ on the set $\mathcal{Q} := \{[s] \mid s \in D\}$ of all such classes by

$$[s] \preccurlyeq [t] :\Longleftrightarrow \text{there are } s' \in [s],\, t' \in [t] \text{ satisfying } s' \stackrel{\Diamond}{\longrightarrow} t',$$

for all $s, t \in D$. Then we have the following lemma.

LEMMA 8 (Properties of $\preccurlyeq$).

1. *The relation $\preccurlyeq$ is reflexive and transitive.*
2. *If $[s] \preccurlyeq [t]$, then every $s' \in [s]$ has at most one $\stackrel{\Diamond}{\longrightarrow}$-successor in $[t]$.*
3. *If $[s] \preccurlyeq [t]$, then no two distinct points $s', s'' \in [s]$ have a common $\stackrel{\Diamond}{\longrightarrow}$-successor in $[t]$.*
4. *The relation $\preccurlyeq$ is antisymmetrical.*

Proof. This is most of Lemma 3.6 from [13]. The proofs of item 1 are the same, and the proofs of item 4 are very similar here and there. Actually, Axiom 16 is applied now instead of Axiom 21 from the paper just referred to. As to the proof of items 2 and 3, see [11], Lemma 3. Note that more axioms from the first and the second group above are needed to establish the lemma. In particular, Axiom 10 and the so-called *Cross Property* derived from it,

$$s \stackrel{\Diamond}{\longrightarrow} t \stackrel{L}{\longrightarrow} u \Rightarrow \exists v : s \stackrel{L}{\longrightarrow} v \stackrel{\Diamond}{\longrightarrow} u,$$

play a crucial part.[7] We omit further details. ■

The lemma says, in particular, that $(\mathcal{Q}, \preccurlyeq)$ is a partial order. There is an additional property of $\preccurlyeq$ which is important for our purposes.

LEMMA 9. *The relation $\preccurlyeq$ is* left-directed, *i.e., for all $[s], [t] \in \mathcal{Q}$ there exists $[u] \in \mathcal{Q}$ such that $[u] \preccurlyeq [s]$ and $[u] \preccurlyeq [t]$.*

Proof. This is an easy consequence of Axiom 17, due to the fact that we are working in a named model. ■

[7](A variant of) this property is used decisively in the proof of completeness of many logics related to *topologic*.

Now suppose that $[s] \preccurlyeq [t]$. Lemma 8.2 implies that $\stackrel{\Diamond}{\longrightarrow}$ induces an injective partial function $h^{[s]}_{[t]} : [s] \longrightarrow [t]$. With the aid of all such functions we construct now a model realizing γ. To this end, let X be the set of all partial functions $f : \mathcal{Q} \longrightarrow D$ of which the domain $\mathrm{dom}(f)$ is a maximal subset of $\mathcal{Q}$ satisfying the following conditions:

1. $f([s]) \in [s]$ for all $[s] \in \mathrm{dom}(f)$, and
2. for all $[s], [t] \in \mathrm{dom}(f)$ such that $[s] \preccurlyeq [t]$ it holds that $f([t]) = h^{[s]}_{[t]}([s])$.

We write $f_s := f([s])$ in case f([s]) exists. And we define

- $U_{[s]} := \{f \in X \mid f_s \text{ exists}\}$, for all $s \in D$,
- $\mathcal{O} := \{U_{[s]} \mid s \in D\} \cup \{X, \emptyset\}$, and
- $V : \mathrm{PROP} \cup \mathrm{N}_{stat} \cup \mathrm{N}_{sets} \longrightarrow \mathcal{P}(X)$ by

$$f \in V(c) \iff c \in f_s \text{ for some } s \in D \text{ such that } f_s \text{ exists},$$

  for all $c \in \mathrm{PROP} \cup \mathrm{N}_{stat} \cup \mathrm{N}_{sets}$.

With that we obtain the following *Truth Lemma.*

LEMMA 10. *The structure $\mathcal{M} := (X, \mathcal{O}, V)$ is an SSN. Moreover, for all formulas α, functions $f \in X$, and points $s \in D$ such that $f \in U_{[s]}$, we have that*

$$(*) \quad f, U_{[s]} \models_{\mathcal{M}} \alpha \iff \alpha \in f_s.$$

Proof. First note that $X \neq \emptyset$ holds because of Zorn's Lemma. Second, V is in fact well-defined. This is due to Axioms 6, 11, 13, and Lemma 9 (among other things). Third, the property

$$U_{[t]} \subseteq U_{[s]} \text{ implies } [s] \preccurlyeq [t]$$

can be proved by means of Axiom 15 and Lemma 9; cf [11], Proposition 4. This property and Lemma 8.4 are needed for the proof of the assertion $(*)$. In fact, $(*)$ can be established now for the basic cases and all the inductive steps apart from $\alpha = [\epsilon_A]\beta$. As the arguments are similar to those from [11], proof of Lemma 5, the details are omitted here.

Finally, we treat the remaining case. We assume first that $f, U_{[s]} \models_{\mathcal{M}} [\epsilon_A]\beta$. This means that whenever $f \in V(A)$, then $f, V(A) \models_{\mathcal{M}} \beta$. We must prove that $[\epsilon_A]\beta \in f_s$ holds in this case. So, take any $t \in D$ such that $f_s \stackrel{\langle \epsilon_A \rangle}{\longrightarrow} t$. It suffices to show that $\beta \in t$. To this end we state first that we have $A \in t$, because of Axiom 25. Next, there are $u, v \in D$ such that $u \xrightarrow{\Diamond} f_s$ and $u \xrightarrow{L} v \xrightarrow{\Diamond} t$. This is true because of Lemma 9. Now Axiom 26 gives us $u \xrightarrow{\Diamond} t$. That is, f_t exists, or, in other words, $f \in V(A)$. Consequently, $f, V(A) \models_{\mathcal{M}} \beta$. Because of $V(A) = U_{[t]}$, the induction hypothesis tells us that $\beta \in f_t = t$. This is what we wanted to show.

Conversely, assume that $f, U_{[s]} \not\models_{\mathcal{M}} [\epsilon_A]\beta$. From this we conclude that

$$f \in V(A) \text{ and } f, V(A) \not\models_{\mathcal{M}} \beta.$$

Let $f_{V(A)}$ be the (uniquely determined) component of f containing A. Due to the induction hypothesis, $\beta \notin f_{V(A)}$. Thus if we can prove that $f_s \xrightarrow{\langle \epsilon_A \rangle} f_{V(A)}$, then we are done. But, according to the definition of (the elements of) X there is some $u \in D$ such that $u \xrightarrow{\Diamond} f_s$ and $u \xrightarrow{\Diamond} f_{V(A)}$. Now axiom 24 can be applied, yielding $f_s \xrightarrow{\langle \epsilon_A \rangle} f_{V(A)}$, as desired. ■

¿From the *Truth Lemma* we get the desired Completeness Theorem in the standard manner.

THEOREM 11 (Completeness). *Every formula* $\alpha \in$ WFF *that is valid in all SSNs is contained in the logic* GJ.

And we obtain even more. The fact that we are dealing with *named* objects enables us to extend the completeness proof to other classes of set spaces, by adding to the above list of axioms a suitable defining correspondent. For example, we can prove the following theorem in this way.

THEOREM 12 (Extended completeness). *Let* D *be the formula schema from Proposition 5.*[8] *Then the system* GJ + D *is sound and complete with respect to the class of all directed SSNs.*

Proof. We construct the model $\mathcal{M} = (X, \mathcal{O}, V)$ as in the proof of the previous theorem, but for the extended logic. Our aim is to show that $\mathcal{M}$ is directed. For every $f \in X$ and $U \in \mathcal{O}$ containing f we let f_U denote the unique value of f contained in U. Now assume that $f \in U_1 \cap U_2$, and let $U_1 = V(A)$ and $U_2 = V(B)$. Then, $f, X \models_{\mathcal{M}} \Diamond A \wedge \Diamond B$. Consequently, $f, X \models_{\mathcal{M}} \Diamond K(\langle \epsilon_A \rangle \top \wedge \langle \epsilon_A \rangle \top)$, according to the validity of the schema D. It follows that there is some $U \in \mathcal{O}$ such that $f, U \models_{\mathcal{M}} K(\langle \epsilon_A \rangle \top \wedge \langle \epsilon_A \rangle \top)$; note that, in particular, $f \in U$ holds. Now take any $g \in U$. Then, $g, U \models_{\mathcal{M}} \langle \epsilon_A \rangle \top \wedge \langle \epsilon_A \rangle \top$. This means that there are $u, v \in D$ such that $g_U \xrightarrow{\langle \epsilon_A \rangle} u$ and $g_U \xrightarrow{\langle \epsilon_B \rangle} v$. As $g_X \xrightarrow{L} f_X$ and $f \in V(A)$ (and $\in V(B)$, respectively), Axiom

[8] The reader will not confuse this with the schema D from standard modal logic.

26 implies $g_X \stackrel{\Diamond}{\longrightarrow} u$ (and $g_X \stackrel{\Diamond}{\longrightarrow} v$, respectively). Thus $g \in V(A) = U_1$ and $g \in V(B) = U_2$. This proves $U \subseteq U_1 \cap U_2$, i.e., $\mathcal{M}$ is directed. ∎

A corresponding result holds for quite a lot of different classes of set spaces.

In view of the issues of the paper [17], Theorem 12 is somewhat surprising at first glance. It was proved there that the modal theory of directed set spaces is *not* finitely axiomatizable. As we have just shown it *is* in the hybrid setting, actually;[9] see also the end of the following section.

4 Decidability

Subsequently it is shown that GJ is a decidable set of formulas. In a preparatory step, we consider a certain class of auxiliary Kripke structures which are a bit involved. These models are supposed to realize as many of the above axioms as possible.

DEFINITION 13 (GJ–models). A tupel

$$\mathfrak{M} := \left(W, \stackrel{L}{\longrightarrow}, \stackrel{\Diamond}{\longrightarrow}, \stackrel{\mathsf{E}}{\longrightarrow}, \{\stackrel{\langle \epsilon_A \rangle}{\longrightarrow}\}_{A \in \mathrm{N}_{sets}}, V\right)$$

is called a GJ–*model*, iff the following conditions are satisfied:

1. W is a non-empty set,
2. the relation $\stackrel{L}{\longrightarrow} \subseteq W \times W$ (belonging to K) is an equivalence,
3. the relation $\stackrel{\Diamond}{\longrightarrow} \subseteq W \times W$ (belonging to $\Box$) is reflexive and transitive,
4. for all $u, v, w \in W$ such that $u \stackrel{\Diamond}{\longrightarrow} v \stackrel{L}{\longrightarrow} w$ there exists $t \in W$ such that $u \stackrel{L}{\longrightarrow} t \stackrel{\Diamond}{\longrightarrow} w$,
5. there is some $u_0 \in W$ such that $W = \{v \mid (u_0, v) \in (\stackrel{L}{\longrightarrow} \cup \stackrel{\Diamond}{\longrightarrow})^*\}$ (i.e., in particular, W is *generated* by $\stackrel{L}{\longrightarrow} \cup \stackrel{\Diamond}{\longrightarrow}$),
6. the relation $\stackrel{\mathsf{E}}{\longrightarrow} \subseteq W \times W$ (belonging to A) is universal,
7. $V : \mathrm{PROP} \cup \mathrm{N}_{stat} \cup \mathrm{N}_{sets} \longrightarrow \mathcal{P}(W)$ is a mapping satisfying

[9] The investigation of the part that unorthodox proof rules like the above ENRICHMENT rule play for hybrid completeness and, in particular, to what extent such rules can be taken as substitutes for infinite axiomatizations, is an actual field of research. Regarding this, cf, eg, [4].

(a) for all $c \in \mathrm{PROP} \cup \mathrm{N}_{stat}$ and $u, v \in W$: if $u \stackrel{\diamond}{\longrightarrow} v$, then $u \in V(c) \iff v \in V(c)$,

(b) for all $i \in \mathrm{N}_{stat}$ and $u \in W$ there is at most one $v \in W$ such that $u \stackrel{L}{\longrightarrow} v$ and $v \in V(i)$,

(c) for all $A \in \mathrm{N}_{sets}$, the set $V(A)$ equals either $\emptyset$ or a unique $\stackrel{L}{\longrightarrow}$–equivalence class,

8. for every $A \in \mathrm{N}_{sets}$, the relation $\stackrel{\langle \epsilon_A \rangle}{\longrightarrow}$ (belonging to $[\epsilon_A]$) satisfies

(a) for all $u, v \in W$: if $u \stackrel{\langle \epsilon_A \rangle}{\longrightarrow} v$, then $v \in V(A)$, and

(b) for all $u \in W$: if there exists some $v \in V(A)$ such that $u \stackrel{\diamond}{\longrightarrow} v$, then $\forall w \in W : (u \stackrel{\diamond}{\longrightarrow} w \Rightarrow w \stackrel{\langle \epsilon_A \rangle}{\longrightarrow} v)$.

We give a couple of comments on this definition. First, note that the fourth item above is the aforementioned Cross Property (see the proof of Lemma 8 above), which can really be taken as a certain diagram property of the relations $\stackrel{\diamond}{\longrightarrow}$ and $\stackrel{L}{\longrightarrow}$. Second, item 7 suits the notion of $\mathcal{S}$–valuation from Definition 1 to GJ–models. And finally, a certain diagram property is imposed on the relations $\stackrel{\langle \epsilon_A \rangle}{\longrightarrow}$ and $\stackrel{\diamond}{\longrightarrow}$, too, by (b) of the last item. As this property is occasionally referred to below, we call it the *Property (b).*

GJ–models are closely related to SSNs. In fact, by taking neighbourhood situations as points and defining then the accessibility relations and the valuation in a way suggesting itself (in accordance with the proceeding for *topologic,* cf [6], Section 2.3), every SSN induces a GJ–model. Therefore, it is natural to try GJ–models in order to establish the finite model property for GJ.[10] Unfortunately, not every GJ–model validates necessarily all of the above axioms. But, for a start, we get at least the following partial result.

PROPOSITION 14. *Apart from, possibly, Axioms 15,16, and 26, each of the above schemata is valid in every* GJ*–model.*

Proof. The proof is rather straightforward; see [12], proof of Proposition 2, for a few details. ∎

We introduce two ad hoc notations, by calling a GJ–model $\mathfrak{M}$ *(almost) faithful,* iff *all* the axioms (*except 26*) are valid in $\mathfrak{M}$. We want to show now that GJ satisfies the *Strong Finite Model Property* (cf [2], Def. 6.6) with

[10]Note that already *topologic* does not satisfy the fmp with respect to the class of all *set spaces;* cf [6]. Thus we cannot use much of our work from Section 3 here.

respect to the class of all faithful GJ–models. Then, the desired decidability of GJ will easily arise.

To establish this finite model property we use the method of filtration, followed by an appropriate model surgery. So, let $\alpha \in \mathrm{WFF}$ be a consistent formula for which we want to find a model of size at most $f(|\alpha|)$, where f is some computable function and $|\alpha|$ denotes the length of α. Moreover, let $\mathrm{sf}(\alpha)$ be the set of all subformulas of α. We construct a suitable filter set Σ via the following sets of formulas. We first let

$$\Sigma_0 := \begin{cases} \mathrm{sf}(\alpha) \cup \{\Box\neg A \mid A \in \mathrm{N}_{sets} \text{ occurs in } \alpha\} \\ \cup\{[\epsilon_A]A, [\epsilon_A]\neg A \mid A \in \mathrm{N}_{sets} \text{ occurs in } \alpha\} \\ \cup\{\Box\neg(A \wedge \neg\beta) \mid [\epsilon_A]\beta \in \mathrm{sf}(\alpha)\} \\ \cup\{\Box\neg[\epsilon_A]\neg\beta \mid A \text{ occurs in } \alpha \text{ and } \beta \in \mathrm{sf}(\alpha)\} \\ \cup\{[\epsilon_A]\neg\beta \mid [\epsilon_A]\beta \in \mathrm{sf}(\alpha) \text{ or } (A \text{ occurs in } \alpha \text{ and } \beta \in \mathrm{sf}(\alpha))\}, \end{cases}$$

and secondly $\Sigma^{\neg} := \{\neg\beta \mid \beta \in \Sigma_0\}$.[11] Then we take the set Σ' of all finite conjunctions of pairwise distinct elements of $\Sigma_0 \cup \Sigma^{\neg}$. Afterwards we close Σ' under single applications of the operator L and take, finally, the set of all subformulas of the resulting set. Let Σ be the union of all these intermediate sets of formulas. Then Σ is subformula closed, and $2^{c\cdot|\alpha|}$ is an upper bound of the cardinality of Σ (for some constant c).

Let $\mathcal{C}$ be the submodel of the canonical model of GJ generated by a maximal consistent set realizing α. Moreover, let the Kripke model

$$\mathfrak{M} := \left(W, \xrightarrow{L}, \xrightarrow{\Diamond}, \xrightarrow{\mathsf{E}}, \{\xrightarrow{\langle\epsilon_A\rangle}\}_{A\in\mathrm{N}_{sets}}, V\right)$$

be obtained from $\mathcal{C}$ as follows:

- W is the filtration of the carrier set of $\mathcal{C}$ with respect to Σ,
- $\xrightarrow{L}$, $\xrightarrow{\Diamond}$ and $\xrightarrow{\mathsf{E}}$ are the *smallest* filtrations, and $\xrightarrow{\langle\epsilon_A\rangle}$ is the *largest* filtration of the accessibility relations of $\mathcal{C}$ belonging to the respective modalities (where $A \in \mathrm{N}_{sets}$), and
- V is induced by the canonical valuation.

Then, exploiting the structure of the filter set Σ and the fixings for the filtrations of the relations we get two crucial propositions, of which the first one reads as follows.

PROPOSITION 15. *By modifying some of the relations* $\xrightarrow{\langle\epsilon_A\rangle}$ *and the valuation* V *suitably, the just defined model* $\mathfrak{M}$ *can be turned into an almost*

[11]The part the individual sets forming Σ_0 play, becomes apparent later on.

faithful GJ–*model* $\mathfrak{M}_1$ *which is semantically equivalent to* $\mathfrak{M}$ *with respect to* α.

Proof. We let $\mathfrak{M}_1$ be the model differing from $\mathfrak{M}$ in that the denotation of every element of PROP $\cup$ N_{stat} $\cup$ N_{sets} not occurring in α is the empty set and the corresponding relations $\stackrel{\langle\epsilon_A\rangle}{\longrightarrow}$ are empty as well. In this way, a structure results which satisfies items 1 – 7 of Definition 13 as well as Axioms 15 and 16. This was detailedly proved in the paper [12] (proof of Proposition 3). Note that the first subset of Σ_0 is relevant to that.

We prove now that the conditions from item 8 of Definition 13 are satisfied for $\mathfrak{M}_1$. To this end, we may assume that A actually occurs in α, for otherwise the desired properties hold trivially.

(a) Let $u, v \in W$ satisfy $u \stackrel{\langle\epsilon_A\rangle}{\longrightarrow} v$, and let $\hat{u}, \hat{v}$ be any representatives of u, v, respectively, contained in $\mathcal{C}$. Then we have that the implication $\mathcal{C}, \hat{u} \models [\epsilon_A]\beta \Rightarrow \mathcal{C}, \hat{v} \models \beta$ holds for all formulas $[\epsilon_A]\beta \in \Sigma$, due to the definition of filtration. Now, $[\epsilon_A]A$ is in Σ, and $\mathcal{C}, \hat{u} \models [\epsilon_A]A$ is in fact true because of Axiom 25. It follows that $\mathcal{C}, \hat{v} \models A$. This implies $\mathfrak{M}_1, v \models A$, i.e., $v \in V(A)$.

(b) Let $u, v, \hat{u}, \hat{v}$ be as above, and let $v \in V(A)$ and $u \stackrel{\Diamond}{\longrightarrow} v$ be satisfied. Furthermore, take any $w \in W$ such that $u \stackrel{\Diamond}{\longrightarrow} w$. We must show that $w \stackrel{\langle\epsilon_A\rangle}{\longrightarrow} v$. For that it suffices to prove that $\mathcal{C}, \hat{w} \models [\epsilon_A]\beta \Rightarrow \mathcal{C}, \hat{v} \models \beta$ for all formulas $[\epsilon_A]\beta \in \Sigma$, where $\hat{w}$ is any representative of w; this is true, actually, because of the fact that we are dealing with the largest filtration. So, let $\mathcal{C}, \hat{w} \models [\epsilon_A]\beta$ be valid. Then, also $\mathfrak{M}_1, w \models [\epsilon_A]\beta$ holds. Since $\stackrel{\Diamond}{\longrightarrow}$ is a smallest filtration and $[\epsilon_A]\beta \in \Sigma$, there are representatives $\check{u}, \check{w}$ of u, w, respectively, such that both $\check{w}$ is accessible from $\check{u}$ with respect to the modality $\Box$ and $\Diamond[\epsilon_A]\beta \in \check{u}$. With the aid of Axiom 24 we conclude $\Box\neg(A\wedge\neg\beta) \in \check{u}$ from that. As this formula is contained in Σ, too, we get that $\mathfrak{M}_1, u \models \Box\neg(A\wedge\neg\beta)$. Now $\mathfrak{M}_1, v \models \beta$ follows, which implies $\mathcal{C}, \hat{v} \models \beta$, as desired.

Thus we have shown that $\mathfrak{M}_1$ is an almost faithful GJ–model. The second assertion of the proposition is obvious since we have only changed the components of $\mathfrak{M}$ by which the semantics of α is certainly not affected. ∎

Another diagram property, corresponding to a weak version of Axiom 26, is established for the relations $\stackrel{\langle\epsilon_A\rangle}{\longrightarrow}$ and $\stackrel{\Diamond}{\longrightarrow}$ by the next lemma.

LEMMA 16. *For all* $u, v, w, x, y \in W$ *satisfying*

$$u \stackrel{\Diamond}{\longrightarrow} v \stackrel{\langle\epsilon_A\rangle}{\longrightarrow} w,\ u \stackrel{\Box}{\longrightarrow} x \stackrel{\Diamond}{\longrightarrow} y,\ \text{and } y \in V(A),$$

there exists some $z \in V(A)$ *such that* $u \stackrel{\Diamond}{\longrightarrow} z$.

Proof. Let $u, v, w, x, y \in W$ satisfy $u \stackrel{\Diamond}{\longrightarrow} v \stackrel{\langle \epsilon_A \rangle}{\longrightarrow} w$, $u \stackrel{L}{\longrightarrow} x \stackrel{\Diamond}{\longrightarrow} y$ and $y \in V(A)$. Again, we may assume that A occurs in α. In order to prove that $\exists z \in V(A) : u \stackrel{\Diamond}{\longrightarrow} z$ holds, it suffices to show that $\Diamond A \in \hat{u}$ for some representative $\hat{u}$ of u since $\Diamond A \in \Sigma$. For that, we will use Axiom 26 with $\alpha = A$. First, we infer $\Diamond A \in \check{x}$, where $\check{x}$ is an arbitrary representative of x, from $y \in V(A)$ and $\Diamond A \in \Sigma$. Second, it is a peculiarity of our filtration that for *every* representative $\check{u}$ of u there is some representative of x which is accessible from $\check{u}$ with respect to K.[12] Thus we obtain $L\Diamond A \in \check{u}$ for all $\check{u}$. Third, we have that $\mathfrak{M}_1, w \models A$ because $\mathfrak{M}_1, v \models [\epsilon_A]A$. Thus even $\mathfrak{M}_1, v \models \langle \epsilon_A \rangle A$ is valid. Since this formula is contained in Σ, too, we infer (similarly as above) that $\Diamond \langle \epsilon_A \rangle A \in \hat{u}$ holds for some representative $\hat{u}$ of u. Now Axiom 26 can be applied, yielding $\Diamond A \in \hat{u}$, as desired. ∎

The following very special property of the model $\mathfrak{M}$ (and $\mathfrak{M}_1$, respectively) turns out to be useful for later purposes.

LEMMA 17. *Let* $A \in \mathrm{N}_{sets}$ *occur in* α, *and let* $u, v, w, x, y \in W$ *satisfy* $u \stackrel{\Diamond}{\longrightarrow} w$, $u \stackrel{\Diamond}{\longrightarrow} x$, $v \stackrel{\Diamond}{\longrightarrow} w$, $v \stackrel{\Diamond}{\longrightarrow} y$, $w \stackrel{\langle \epsilon_A \rangle}{\longrightarrow} x$, $w \stackrel{\langle \epsilon_A \rangle}{\longrightarrow} y$, *and* $x, y \in V(A)$. *Then,* x *and* y *cannot be distinguished by some subformula of* α.

Proof. Let $\beta \in \mathrm{sf}(\alpha)$ and assume that $\mathfrak{M}, x \models \beta$. Then, $\mathfrak{M}, w \models \langle \epsilon_A \rangle \beta$. Furthermore, let $\hat{v}, \check{v}, \hat{y}, \hat{w}$ be representatives of v, y, w, respectively, such that $\hat{v} \stackrel{\Diamond}{\longrightarrow} \hat{w}$ and $\check{v} \stackrel{\Diamond}{\longrightarrow} \hat{y}$. Then $\Diamond \langle \epsilon_A \rangle \beta$ is contained in $\hat{v}$ because $\langle \epsilon_A \rangle \beta$ is contained in $\hat{w}$ (see the definition of Σ). Since $y \in V(A)$, we have also that $\Diamond A \in \check{v}$.

In the following we apply a couple of semantic arguments. Since

$$\Diamond \langle \epsilon_A \rangle \beta \wedge \Diamond A \rightarrow \Diamond (A \wedge \beta)$$

represents a validity, we conclude from Theorem 11 that this formula, and hence its right-hand side, too, is contained in $\check{v}$. According to Axiom 24 it follows that $\Box \langle \epsilon_A \rangle \beta \in \check{v}$. As this formula is also contained in Σ (see the definition of Σ_0), we infer $\langle \epsilon_A \rangle \beta \in \hat{y}$ from that. Now, $\langle \epsilon_A \rangle \beta \wedge A \rightarrow \beta$ is a validity, too.[13] Thus we obtain that β is contained in $\hat{y}$. That is, $\mathfrak{M}, y \models \beta$.

We may clearly reverse the roles of x and y (and of v and u, respectively). Thus the assertion of the lemma follows for reasons of symmetry. ∎

Now we are in a position to prove the second of our crucial propositions.

---

[12] This property was proved, eg, in [6], Sec. 2.3.

[13] Note that this is the *elimination* formula of usual hybrid logic; see [2], p 438.

PROPOSITION 18. *The model* $\mathfrak{M}_1$ *can be turned into a faithful* GJ*–model* $\mathfrak{M}'$ *which is semantically equivalent to* $\mathfrak{M}_1$ *with respect to* α.

Proof. We subject $\mathfrak{M}_1$ to two changes. First, all the points that are both contained in $V(A)$ and indistinguishable by α in the sense of Lemma 17, are identified; moreover, all the arrows that already exist are kept (and no new ones are introduced). The reader can easily convince himself or herself that an almost faithful GJ–model $\mathfrak{M}_2$ results from that which is semantically equivalent to $\mathfrak{M}_1$ with respect to α.

Second, we leave out some of the $\stackrel{\langle\epsilon_A\rangle}{\longrightarrow}$–connections. We explain which ones. Let u_0 be a generating element of $\mathfrak{M}_2$. For every u contained in the $\stackrel{L}{\longrightarrow}$–equivalence class $[u_0]$ of u_0 and satisfying $u \stackrel{\Diamond}{\longrightarrow} v$ for some $v \in V(A)$, we do the following: For all w such that $u \stackrel{\Diamond}{\longrightarrow} w$ we take the $\stackrel{\langle\epsilon_A\rangle}{\longrightarrow}$–connection from w to v. Afterwards we cross out all the other $\stackrel{\langle\epsilon_A\rangle}{\longrightarrow}$–connections. Let $\mathfrak{M}'$ be the resulting model.

We have to convince the reader that $\mathfrak{M}'$ is correctly defined. That is, an arrow $\stackrel{\langle\epsilon_A\rangle}{\longrightarrow}$ from w to v really exists (w and v as above). But this is clear from the Property (b) established in the proof of Proposition 15.

The decisive property we have just achieved for the relation $\stackrel{\langle\epsilon_A\rangle}{\longrightarrow}$ is *partial functionality,* which is a consequence of Lemma 17. In fact, Lemma 17 implies that two points having both a $\stackrel{\Diamond}{\longrightarrow}$–successor in $V(A)$ and a common $\stackrel{\Diamond}{\longrightarrow}$–successor, do have a *unique* $\stackrel{\Diamond}{\longrightarrow}$–successor in $V(A)$ (with regard to $\mathfrak{M}'$). Therefore, exactly one point in $V(A)$ is assigned via $\stackrel{\langle\epsilon_A\rangle}{\longrightarrow}$ to every w having a $\stackrel{\Diamond}{\longrightarrow}$–predecessor u in $[u_0]$ which points into $V(A)$ by means of $\stackrel{\Diamond}{\longrightarrow}$.

But we must be careful, for necessary $\stackrel{\langle\epsilon_A\rangle}{\longrightarrow}$–connections could have been deleted above. We give now reasons for that this is not the case. So, let v be any point and assume that $v \stackrel{\langle\epsilon_A\rangle}{\longrightarrow} w$ holds in the model $\mathfrak{M}_2$, for some point w. Take some $u \in [u_0]$ such that $u \stackrel{\Diamond}{\longrightarrow} v$. Then there exists some $z \in V(A)$ such that $u \stackrel{\Diamond}{\longrightarrow} z$, because of Lemma 16. Due to the Property (b), there is also an arrow $\stackrel{\langle\epsilon_A\rangle}{\longrightarrow}$ from v to z, i.e., to a previously chosen element of $V(A)$. This means that not every arrow $\stackrel{\langle\epsilon_A\rangle}{\longrightarrow}$ leaving v was deleted.

By an inductive argument we can show now that any subformula of α holds in the model $\mathfrak{M}'$ at some point u, iff it holds in $\mathfrak{M}_2$ at u. As to the proof of this assertion, we confine ourselves to the case $[\epsilon_A]\beta$ of the induction. However, regarding this we remark only that the following property is required for that:

$$(*) \quad \text{if } \exists v : u \xrightarrow{\langle \epsilon_A \rangle} v, \text{ then } (\mathfrak{M}, u \models \langle \epsilon_A \rangle \beta \iff \mathfrak{M}, u \models [\epsilon_A]\beta)\,.$$

In order to establish $(*)$, one has to apply a semantic argument once again.[14] Actually, $\langle \epsilon_A \rangle \beta \rightarrow [\epsilon_A]\beta$ is a valid formula, thus a theorem of GJ. Now the last set constituting Σ_0 (see above) plays its part, yielding the critical left-to-right direction of $(*)$.

On the basis of the properties of $\mathfrak{M}'$ we have just proved, Axiom 26 is easily seen to be valid there. And $\mathfrak{M}'$ is clearly an almost faithful GJ–model. This finishes the proof of the proposition. ■

The previous results imply that α is satisfiable in a finite model of the axioms, of which the size is in $O\left(2^{2^{c \cdot |\alpha|}}\right)$ (where $c \in \mathbb{N}$ is some constant). This gives us the first main issue of this section.

THEOREM 19 (Decidability). *The logic* GJ *is decidable.*

Now we turn to the system GJ + D. By changing the filter set Σ_0 slightly and performing the same filtration-and-model-surgery procedure as above, we obtain the following theorem.

THEOREM 20 (Extended decidability). *The logic* GJ + D *is decidable.*

Proof. We have to guarantee that the schema D remains valid by passing to the accordingly defined filtrated structure. Again, only the nominals occurring in the consistent formula α for which we want to construct a finite model have to be taken into account in the process. So, let

$$\Sigma_0^{\mathrm{D}} := \Sigma_0 \cup \{K\left(\langle \epsilon_A \rangle \top \wedge \langle \epsilon_B \rangle \top\right) \mid A, B \in \mathrm{N}_{sets} \text{ occur in } \alpha\},$$

and let Σ^{D} be derived from Σ_0^{D} as Σ was derived from Σ_0 above. Then, the proof of the validity of D in the final model (i.e., the counterpart of $\mathfrak{M}'$) can easily be done in the same manner as, eg, the proof of Property (b) in Proposition 15. We may, therefore, omit further details. ■

As an immediate consequence of Theorem 20 we get that the modal logic of directed spaces is decidable, too.

COROLLARY 21. *The modal logic of directed spaces is decidable.*

This result constitutes a striking application of hybrid logic to reasoning about knowledge. In fact, it solves the decidability problem raised in [17], Sec. 6.

[14]Note that $(*)$ resembles the *Fun–Lemma* used in the completeness proof for linear time temporal logic; cf [9], 9.9.

Concluding this section we comment on the complexity of the satisfiability problem of GJ. Unfortunately, we can say only little about this. The above decidability proof yields obviously a very large upper bound. On the other hand, Theorem 4.5 from [3] gives us good reason for suspecting that EXPTIME is a lower bound.[15]

5 Summary, comparison, and outlook for further options

In the present paper we developed the fundamental matters of a two-sorted hybrid logic, GJ, for set spaces with names. A family of new 'guarded jump'–operators assigned to names of sets was considered, in particular. We obtained a Completeness as well as a Decidability Theorem for GJ, and we could extend these results to the logic of directed spaces (which are significant to reasoning 'topologically' about knowledge).

It should be remarked that hybrid logics for spaces of sets received little attention in the literature up to now. At least, a (usual) hybrid extension of the *classical* topological semantics of modal logic (going back to [14]) was briefly considered in [8] (with regard to expressiveness).

Which other classes of set spaces should be studied from the point of view of hybrid logic? Concluding this paper, we touch on the class of frames $(X, \mathcal{O})$ where $\mathcal{O}$ is *closed under complementation.* This class is possibly of some interest to various applications. Addressing the complement of the actual neighbourhood at a situation x, U means, in particluar, jumping completely outside U. Thus a 'complementation operator' appears, which goes beyond all the previous abilities. Now, it is not too hard to write down some formulas describing certain properties of such an operator and its interaction with K and $\Box$. In order to obtain nice meta-results, however, it seems more promising to connect complementation operators with names. In a forthcoming paper we will develop a system like that.

BIBLIOGRAPHY

[1] Patrick Blackburn. Representation, reasoning, and relational structures: a hybrid logic manifesto. *Logic Journal of the IGPL*, 8:339–365, 2000.

[2] Patrick Blackburn, Maarten de Rijke, and Yde Venema. *Modal Logic*, volume 53 of *Cambridge Tracts in Theoretical Computer Science.* Cambridge University Press, Cambridge, 2001.

[3] Patrick Blackburn and Edith Spaan. A modal perspective on the computational complexity of attribute value grammar. *Journal of Logic, Language and Information*, 2(2):129–169, 1993.

[4] Patrick Blackburn and Balder ten Cate. Pure extensions, proof rules, and hybrid axiomatics. In R. Schmidt, I. Pratt-Hartmann, M. Reynolds, and H. Wansing, editors,

[15] Concerning this, note once again that $[\epsilon_A]$ is a partially functional modality, for every $A \in \mathrm{N}_{sets}$.

AiML-2004: Advances in Modal Logic, volume UMCS-04-9-1 of *Technical Report Series*, pages 16–29, Manchester, UK, 2004. Department of Computer Science of the University of Manchester.
[5] Nicolas Bourbaki. *General Topology, Part 1.* Hermann, Paris, 1966.
[6] Andrew Dabrowski, Lawrence S. Moss, and Rohit Parikh. Topological reasoning and the logic of knowledge. *Annals of Pure and Applied Logic*, 78:73–110, 1996.
[7] Ronald Fagin, Joseph Y. Halpern, Yoram Moses, and Moshe Y. Vardi. *Reasoning about Knowledge.* MIT Press, Cambridge, MA, 1995.
[8] David Gabelaia. Modal definability in topology. Master's thesis, ILLC, Universiteit van Amsterdam, 2001.
[9] Robert Goldblatt. *Logics of Time and Computation*, volume 7 of *CSLI Lecture Notes.* Center for the Study of Language and Information, Stanford, CA, second edition, 1992.
[10] Bernhard Heinemann. Axiomatizing modal theories of subset spaces (an example of the power of hybrid logic). In *HyLo@LICS, Proceedings*, pages 69–83, Copenhagen, Denmark, July 2002.
[11] Bernhard Heinemann. Extended canonicity of certain topological properties of set spaces. In M. Vardi and A. Voronkov, editors, *Logic for Programming, Artificial Intelligence, and Reasoning*, volume 2850 of *Lecture Notes in Artificial Intelligence*, pages 135–149, Berlin, 2003. Springer.
[12] Bernhard Heinemann. A hybrid logic of knowledge supporting topological reasoning. In C. Rattray, S. Maharaj, and C. Shankland, editors, *Algebraic Methodology and Software Technology*, volume 3116 of *Lecture Notes in Computer Science*, pages 181–195, Berlin, 2004. Springer.
[13] Bernhard Heinemann. The hybrid logic of linear set spaces. *Logic Journal of the IGPL*, 12(3):181–198, 2004.
[14] J. C. C. McKinsey. A solution to the decision problem for the Lewis systems S2 and S4, with an application to topology. *Journal of Symbolic Logic*, 6(3):117–141, 1941.
[15] J.-J. Ch. Meyer and W. van der Hoek. *Epistemic Logic for AI and Computer Science*, volume 41 of *Cambridge Tracts in Theoretical Computer Science.* Cambridge University Press, Cambridge, 1995.
[16] Lawrence S. Moss and Rohit Parikh. Topological reasoning and the logic of knowledge. In Y. Moses, editor, *Theoretical Aspects of Reasoning about Knowledge (TARK 1992)*, pages 95–105, San Francisco, CA, 1992. Morgan Kaufmann.
[17] M. Angela Weiss and Rohit Parikh. Completeness of certain bimodal logics for subset spaces. *Studia Logica*, 71:1–30, 2002.

Bernhard Heinemann
Fachbereich Informatik
FernUniversität in Hagen
58084 Hagen, Germany
E-mail: Bernhard.Heinemann@fernuni-hagen.de

On the Modularity of Theories

ANDREAS HERZIG AND IVAN VARZINCZAK

ABSTRACT. In this paper we give the notion of modularity of a theory and analyze some of its properties, especially for the case of action theories in reasoning about actions. We propose algorithms to check whether a given action theory is modular and that also make it modular, if needed. Completeness, correctness and termination results are demonstrated.

1 Introduction

In many cases knowledge is represented by logical theories containing multiple modalities $\alpha_1, \alpha_2, \ldots$ Then it is often the case that we have modularity, in the sense that our theory $\mathcal{T}$ can be partitioned into a union of theories

$$\mathcal{T} = \mathcal{T}^{\emptyset} \cup \mathcal{T}^{\alpha_1} \cup \mathcal{T}^{\alpha_2} \cup \ldots$$

such that

- $\mathcal{T}^{\emptyset}$ contains no modal operators, and
- the only modality of $\mathcal{T}^{\alpha_i}$ is α_i.

We call these subtheories *modules* (some modules might by empty). Examples of such theories can be found in reasoning about actions, where each $\mathcal{T}^{\alpha_i}$ contains descriptions of the atomic action α_i in terms of preconditions and effects, and $\mathcal{T}^{\emptyset}$ is the set of static laws (alias domain constraints, alias integrity constraints), i.e., those formulas that hold in every possible state of a dynamic system, and are thus global axioms.

For example, consider the following theory:

$$\mathcal{T}^{marry} = \left\{ \begin{array}{c} \neg Married \rightarrow \langle marry \rangle \top, \\ [marry]Married \end{array} \right\}$$

$$\mathcal{T}^{\emptyset} = \{\neg(Married \wedge Bachelor)\}$$

Such a theory is composed of two subtheories, one for expressing the dynamic part of the theory, $\mathcal{T}^{marry}$, and one to formalize the constraints of

Advances in Modal Logic, Volume 5. © *2005, Andreas Herzig and Ivan Varzinczak.*

the domain, $\mathcal{T}^{\emptyset}$. $\mathcal{T}^{marry}$ formalizes the behavior of the action of getting married, in this case the precondition for executing *marry* (viz. $\neg$*Married*) and the effect that obtains after its execution (viz. *Married*). $\mathcal{T}^{\emptyset}$ formalizes the domain constraint according to which it is not possible to be married and bachelor at the same time.

Another example is when mental attitudes such as knowledge, beliefs or goals of several independent agents are represented: then each $\mathcal{T}^{\alpha_i}$ contains the respective mental attitude of agent α_i.[1]

Let the underlying multimodal logic be independently axiomatized (i.e., the logic is a fusion and there is no interaction between the modal operators), and suppose we want to know whether $\mathcal{T} \models \varphi$, i.e., whether a formula φ follows from the theory $\mathcal{T}$. Then it is natural to expect that we only have to consider those elements of $\mathcal{T}$ which concern the modal operators occurring in φ. For instance the proof of some consequences of action α_1 should not involve laws for other actions α_2; querying the belief base of agent α_1 should not require bothering with that of agent α_2. Moreover, intensional information in any $\mathcal{T}^{\alpha_i}$ should not influence information about the laws of the world (encoded in $\mathcal{T}^{\emptyset}$). Note that this is not the case if the logic is not independently axiomatized, and there are interaction axioms such as $[\alpha_1]\varphi \rightarrow [\alpha_2]\varphi$.

Similar modularity principles can also be found in structural and object-oriented programming: a commonly used guideline in software development is to divide the software into modules, based on their functionality or on the similarity of the information they handle. This means that instead of having a "jack of all trades" program, it is preferable to split it up into specialized subprograms. For instance, a program made of a module for querying a database and a module for checking its integrity is more modular than a single module that does these two tasks at the same time.

The major benefits of modular systems are reusability, scalability and better management of complexity.[2] Among the criteria commonly used for evaluating how modular a piece of software is are the notions of *cohesion* and *coupling* [15, 17]. Roughly, cohesion is about how well defined a module is, while coupling is about how modules are interdependent. A common sense maxim in object-oriented design is maximize cohesion of modules and diminish their coupling, and this paradigm can also be applied to reasoning about actions [1, 4, 5].

---

[1] Here we should assume more generally that $[\alpha_i]$ is the only outermost modal operator of $\mathcal{T}^{\alpha_i}$; we think that this case could be analyzed in a way that is similar to ours.

[2] Observe that this is closely related to the concept of *elaboration tolerance* [12] in reasoning about actions.

In this work we pursue the following plan: after some logical preliminaries (Section 2) we formalize the concepts of modularity to be used throughout the paper (Sections 3 and 4). In Section 5 we focus then in a particular kind of theories that are commonly used in reasoning about actions and discuss how to decide (Section 6) and guarantee (Section 7) its modular property. We finish by addressing related work in the field and making some concluding remarks.

2 Preliminaries

Let $MOD = \{\alpha_1, \alpha_2, \ldots\}$ be the set of modal operators. Formulas are constructed in the standard way from these and the set of atomic formulas *ATM*. They are denoted by $\varphi, \psi, \ldots$ Formulas without modal operators (propositional formulas) are denoted by $PFOR = \{A, B, C, \ldots\}$.

Let $mod(\varphi)$ return the set of modal operators occurring in formula φ, and let $mod(\mathcal{T}) = \bigcup_{\psi \in \mathcal{T}} mod(\psi)$. For instance $mod([\alpha_1](p \rightarrow [\alpha_2]q)) = \{\alpha_1, \alpha_2\}$. If $\mathfrak{M} \subseteq MOD$ is a nonempty set of modalities, then we define

$$\mathcal{T}^{\mathfrak{M}} = \{\varphi \in \mathcal{T} : mod(\varphi) \cap \mathfrak{M} \neq \emptyset\}$$

For $\mathfrak{M} = \emptyset$, we define

$$\mathcal{T}^{\emptyset} = \{\varphi \in \mathcal{T} : mod(\varphi) = \emptyset\}$$

For example, if

$$\mathcal{T} = \left\{ \begin{array}{c} \neg(\mathit{Married} \wedge \mathit{Bachelor}), \\ \neg\mathit{Married} \rightarrow \langle \mathit{marry} \rangle \top, [\mathit{marry}]\mathit{Married}, \\ \mathit{Married} \rightarrow \langle \mathit{divorce} \rangle \top, [\mathit{divorce}]\neg\mathit{Married} \end{array} \right\}$$

then

$$\mathcal{T}^{\{divorce\}} = \{\mathit{Married} \rightarrow \langle \mathit{divorce} \rangle \top, [\mathit{divorce}]\neg\mathit{Married}\}$$

and

$$\mathcal{T}^{\emptyset} = \{\neg(\mathit{Married} \wedge \mathit{Bachelor})\}$$

We write $\mathcal{T}^{\alpha}$ instead of $\mathcal{T}^{\{\alpha\}}$.

We suppose from now on that $\mathcal{T}$ is *partitioned*, in the sense that $\{\mathcal{T}^{\emptyset}\} \cup \{\mathcal{T}^{\alpha_i} : \alpha_i \in MOD\}$ is a partition of $\mathcal{T}$. We thus exclude $\mathcal{T}^{\alpha_i}$ containing more than one modal operator.

Models of the logic under concern are of the form $M = \langle W, R, V \rangle$, where W is a set of possible worlds, $R : MOD \longrightarrow W \times W$ associates an accessibility relation to every modality, and $V : W \longrightarrow 2^{ATM}$ associates a valuation to every possible world.

Satisfaction of a formula φ in world w of model M ($M, w \models \varphi$) and truth of a formula φ in M (denoted $M \models \varphi$) are defined as usual. Truth of a set of formulas $\mathcal{T}$ in M (denoted $M \models \mathcal{T}$) is defined by: $M \models \mathcal{T}$ if and only if $M \models \psi$ for every $\psi \in \mathcal{T}$. $\mathcal{T}$ has global consequence φ (denoted $\mathcal{T} \models \varphi$) if and only if ($M \models \mathcal{T}$ implies $M \models \varphi$). Note that the underlying logic is an extension of classical propositional logic: if A is a logical consequence of $\mathcal{T}$ in classical propositional logic, then $\mathcal{T} \models A$.

We suppose that the logic under consideration is *compact*.

Given these fundamental concepts, we are able to formally define modularity of a theory.

3 Modularity

We make the following hypothesis:

$$\{\mathcal{T}^{\emptyset}\} \cup \{\mathcal{T}^{\alpha_i} : \alpha_i \in MOD\} \text{ partitions } \mathcal{T}^{3} \qquad \text{(H)}$$

We are interested in the following principle of modularity:

DEFINITION 1 A theory $\mathcal{T}$ is *modular* if and only if for every formula φ,

$$\mathcal{T} \models \varphi \text{ implies } \mathcal{T}^{mod(\varphi)} \cup \mathcal{T}^{\emptyset} \models \varphi$$

Modularity means that when investigating whether φ is a consequence of $\mathcal{T}$, the only formulas of $\mathcal{T}$ that are relevant are those whose modal operators occur in φ and the classical formulas in $\mathcal{T}^{\emptyset}$.

This is reminiscent of interpolation, which more or less[4] says:

DEFINITION 2 A theory $\mathcal{T}$ has the *interpolation property* if and only if for every formula φ, if $\mathcal{T} \models \varphi$, then there is a theory $\mathcal{T}_\varphi$ such that

- $mod(\mathcal{T}_\varphi) \subseteq mod(\mathcal{T}) \cap mod(\varphi)$
- $\mathcal{T} \models \psi$ for every $\psi \in \mathcal{T}_\varphi$
- $\mathcal{T}_\varphi \models \varphi$

Our definition of modularity is a strengthening of interpolation because it requires $\mathcal{T}_\varphi$ to be a subset of $\mathcal{T}$.

[3] $\{\mathcal{T}^{\emptyset}\} \cup \{\mathcal{T}^{\alpha_i} : \alpha_i \in MOD\}$ partitions $\mathcal{T}$ if and only if $\mathcal{T} = \mathcal{T}^{\emptyset} \cup \bigcup_{\alpha_i \in MOD} \mathcal{T}^{\alpha_i}$, and $\mathcal{T}^{\emptyset} \cap \mathcal{T}^{\alpha_i} = \emptyset$, and $\mathcal{T}^{\alpha_i} \cap \mathcal{T}^{\alpha_j} = \emptyset$, if $\alpha_i \neq \alpha_j$. Note that $\mathcal{T}^{\emptyset}$ and $\mathcal{T}^{\alpha_i}$ might be empty.

[4] We here present a version in terms of global consequence, as opposed to local consequence or material implication versions that can be found in the literature [6, 7]. We were unable to find such global versions in the literature.

Contrary to interpolation, modularity does not generally hold. Clearly if (H) is not satisfied, then modularity fails. To witness, consider

$$\mathcal{T} = \{p \rightarrow [\alpha][\beta]q, [\alpha][\beta]q \rightarrow r\}$$

Then $\mathcal{T} \models p \rightarrow r$, but $\mathcal{T}^{\emptyset} \not\models p \rightarrow r$.

Nevertheless even under our hypothesis modularity may fail to hold. For example, let

$$\mathcal{T} = \{p \vee [\alpha]\bot, p \vee \neg[\alpha]\bot\}$$

Then $\mathcal{T}^{\emptyset} = \emptyset$, and $\mathcal{T}^{\alpha} = \mathcal{T}$. Now $\mathcal{T} \models p$, but clearly $\mathcal{T}^{\emptyset} \not\models p$.

Being modular is a useful feature of theories: beyond being a reasonable principle of design that helps avoiding mistakes, it clearly restricts the search space, and thus makes reasoning easier. To witness, if $\mathcal{T}$ is modular then consistency of $\mathcal{T}$ amounts to consistency (in classical logic) of its propositional part $\mathcal{T}^{\emptyset}$. This is what we address in the following section.

4 Propositional modularity

How can we know whether a given theory $\mathcal{T}$ is modular? The following criterion is simpler:

DEFINITION 3 A theory $\mathcal{T}$ is *propositionally modular* if and only if for every propositional formula A,

$$\mathcal{T} \models A \text{ implies } \mathcal{T}^{\emptyset} \models A$$

And it will suffice to guarantee modularity:

THEOREM 4 Let the underlying logic be a fusion, and let $\mathcal{T}$ be a partitioned theory. If $\mathcal{T}$ is propositionally modular, then $\mathcal{T}$ is modular.

Proof. Let $\mathcal{T}$ be propositionally modular. Suppose $\mathcal{T}^{mod(\varphi)} \cup \mathcal{T}^{\emptyset} \not\models \varphi$. Hence there is a model $M = \langle W, R, V \rangle$ such that $M \models \mathcal{T}^{mod(\varphi)} \cup \mathcal{T}^{\emptyset}$, and there is some w in M such that $M, w \not\models \varphi$. We prove that $\mathcal{T} \not\models \varphi$ by constructing from M a model M' such that $M' \models \mathcal{T}$ and $M', w \not\models \varphi$.

First, as we have supposed that our logic is an extension of classical propositional logic and that it is compact, propositional modularity implies that for every propositional valuation $val \subseteq 2^{ATM}$ which is a model of $\mathcal{T}^{\emptyset}$ there is a possible worlds model $M_{val} = \langle W_{val}, R_{val}, V_{val} \rangle$ such that $M_{val} \models \mathcal{T}$, and there is some w in M_{val} such that $V_{val}(w) = val$. In other words, for every propositional model of $\mathcal{T}^{\emptyset}$ there is a model of $\mathcal{T}$ containing that propositional model.

Second, taking the disjoint union of all these models we obtain a 'big model' M_{big} such that $M_{big} \models \mathcal{T}$, and for every propositional model $val \subseteq 2^{ATM}$ of $\mathcal{T}^{\emptyset}$ there is a possible world w in M_{big} such that $V(w) = val$.

Now we can use the big model to adjust those accessibility relations $R(\alpha)$ of M whose α does not appear in φ, in a way such that the resulting model satisfies the rest of the theory $\mathcal{T} \setminus \mathcal{T}^{mod(\varphi)}$: let $M' = \langle W', R', V' \rangle$ be such that

- $W' = \{u_v : u \in W, v \in W_{big}, \text{ and } V(u) = V_{big}(v)\}$
- if $\alpha \in mod(\varphi)$, then $u_v R'(\alpha) u'_{v'}$ if and only if $u R(\alpha) u'$
- if $\alpha \notin mod(\varphi)$, then $u_v R'(\alpha) u'_{v'}$ if and only if $v R(\alpha) v'$
- $V'(u_v) = V(u) = V_{big}(v)$

W' is nonempty because $M \models \mathcal{T}^{\emptyset}$. M' is a model of the underlying logic because the latter is a fusion. Then for the sublanguage constructed from $mod(\varphi)$ it can be proved by structural induction that for every formula ψ of the sublanguage and every $u \in W$ and $v \in W_{big}$, $M, u \models \psi$ if and only if $M', u_v \models \psi$. The same can be proved for the sublanguage constructed from $MOD \setminus mod(\varphi)$. As, by hypothesis, $\mathcal{T}$ is partitioned, $\mathcal{T}^{\emptyset}$ and each of our modules $\mathcal{T}^{\alpha}$ are in at least one of these sublanguages, thus we have proved that $M' \models \mathcal{T}$, and $M', w_v \not\models \varphi$ for every v. ■

In the rest of the paper we investigate how it can be automatically checked whether a given theory $\mathcal{T}$ is modular or not, and how to make it modular, if needed. We do this for a particular kind of theories commonly used in reasoning about actions. First of all we say what an action theory is.

5 Action theories

We suppose that the underlying logic is multimodal K. (Note that this is a fusion and that it is compact.)

Every formalization of a dynamic domain contains a representation of action effects. We call *effect laws* formulas relating an action to its effects. *Executability laws* in turn stipulate the context where an action is guaranteed to be executable. Finally, *static laws* are formulas that do not mention actions and express constraints that must hold in every possible state. These are our four ingredients that we introduce more formally in the sequel.

Static laws Frameworks which allow for indirect effects make use of logical formulas that link invariant propositions about the world. Such formulas characterize the set of possible states. They do not refer to actions, and we suppose they are formulas of classical propositional logic $A, B, \ldots \in PFOR$.

A *static law*[5] is a formula $A \in PFOR$ that is consistent. An example is *Walking* $\rightarrow$ *Alive*, saying that if a turkey is walking, then it must be alive [19]. In our action theories $\mathcal{T}$, static laws correspond to $\mathcal{T}^{\emptyset}$.

Effect laws To speak about action effects we use the syntax of propositional dynamic logic (PDL) [3]. The formula $[\alpha]A$ expresses that A is true after every possible execution of α.

An *effect law*[6] *for* α is of the form $A \rightarrow [\alpha]C$, where $A, C \in PFOR$. The consequent C is the effect which obtains when α is executed in a state where the antecedent A holds. An example is $Loaded \rightarrow [shoot]\neg Alive$, saying that whenever the gun is loaded, after shooting the turkey is dead. Another one is $[tease]Walking$: in every circumstance, the result of teasing is that the turkey starts walking.

A particular case of effect laws are *inexecutability laws* of the form $A \rightarrow [\alpha]\bot$. For example $\neg HasGun \rightarrow [shoot]\bot$ expresses that *shoot* cannot be executed if the agent has no gun.

Executability laws With only static and effect laws one cannot guarantee that *shoot* is executable if the agent has a gun.[7] In dynamic logic the dual $\langle\alpha\rangle A$, defined as $\neg[\alpha]\neg A$, can be used to express executability. $\langle\alpha\rangle\top$ thus reads "the execution of action α is possible".

An *executability law*[8] *for* α is of the form $A \rightarrow \langle\alpha\rangle\top$, where $A \in PFOR$. For instance $HasGun \rightarrow \langle shoot\rangle\top$ says that shooting can be executed whenever the agent has a gun, and $\langle tease\rangle\top$ says that the turkey can always be teased.

Action theories $\mathcal{S} \subseteq PFOR$ denotes the set of all static laws of a domain. For a given action $\alpha \in MOD$, $\mathcal{E}_\alpha$ is the set of its effect laws, and $\mathcal{X}_\alpha$ is the set of its executability laws. We define $\mathcal{E} = \bigcup_{\alpha \in MOD} \mathcal{E}_\alpha$, and $\mathcal{X} = \bigcup_{\alpha \in MOD} \mathcal{X}_\alpha$. An *action theory* is a tuple of the form $\langle \mathcal{S}, \mathcal{E}, \mathcal{X} \rangle$. We suppose that $\mathcal{S}, \mathcal{E}$ and $\mathcal{X}$ are finite.

---

[5] Static laws are often called *domain constraints*, but the different laws for actions that we shall introduce in the sequel could in principle also be called like that.

[6] Effect laws are often called *action laws*, but we prefer not to use that term here because it would also apply to executability laws that are to be introduced in the sequel.

[7] Some authors [16, 2, 11, 19] more or less tacitly consider that executability laws should not be made explicit, but rather inferred by the reasoning mechanism. Others [10, 21] have executability laws as first class objects one can reason about. It seems a matter of debate whether one can always do without, but we think that in several domains one wants to explicitly state under which conditions a given action is guaranteed to be executable, e.g., that a robot should never get stuck and should always be able to execute a move action. In any case, allowing for executability laws gives us more flexibility and expressive power.

[8] Some approaches (most prominently Reiter's) use biconditionals $A \leftrightarrow \langle\alpha\rangle\top$, called precondition axioms. This is equivalent to $\neg A \leftrightarrow [\alpha]\bot$, such laws thus merge information about inexecutability with information about executability.

EXAMPLE 5 Consider the following formalization of a transaction domain:

$$\mathcal{S} = \{\neg Adult \rightarrow \neg OblgPay\}$$

$$\mathcal{E} = \left\{ \begin{array}{c} [order]OblgPay, \\ \neg Adult \rightarrow [order]\neg Adult \end{array} \right\}$$

$$\mathcal{X} = \{\langle order \rangle \top\}$$

Observe that by the fact that $\mathcal{S}, \mathcal{E}, \mathcal{X} \models \neg Adult \rightarrow [order]\bot$ we have $\mathcal{S}, \mathcal{E}, \mathcal{X} \models Adult$. But $\mathcal{S} \not\models Adult$, hence $\langle \mathcal{S}, \mathcal{E}, \mathcal{X} \rangle$ is also an example of an action theory that is not modular.

Our central hypothesis here is that the different types of laws in an action theory should be neatly separated and only interfere in one sense: static laws together with action laws for α may have consequences that do not follow from the action laws for α alone. The other way round, action laws should not allow to infer new static laws. That is what modularity of action theories establishes and we develop it in the sequel.

6 Deciding modularity

How can we check whether a given action theory $\mathcal{T} = \langle \mathcal{S}, \mathcal{E}, \mathcal{X} \rangle$ is modular? Following Theorem 4, it is enough to check for propositional modularity.

DEFINITION 6 $A \in PFOR$ is an *implicit static law* of an action theory $\langle \mathcal{S}, \mathcal{E}, \mathcal{X} \rangle$ if and only if $\mathcal{S}, \mathcal{E}, \mathcal{X} \models A$ and $\mathcal{S} \not\models A$.

In Example 5, *Adult* is an example of an implicit static law.

Theorem 4 tells us that an action theory is modular if and only if it has no implicit static law. Hence, checking the existence of such laws provides us a way to decide modularity of a given action theory. Assuming $\mathcal{T}$ is finite, the algorithm below does the job:

ALGORITHM 7 (Finding some implicit static laws)

input: $\langle \mathcal{S}, \mathcal{E}, \mathcal{X} \rangle$
output: a set of implicit static laws $\mathcal{S}^I$

$\mathcal{S}^I := \emptyset$
for all $\alpha \in mod(\mathcal{E}) \cap mod(\mathcal{X})$ **do**
  for all $B \rightarrow \langle \alpha \rangle \top \in \mathcal{X}$ **do**
    for all $\{A_1 \rightarrow [\alpha]C_1, \ldots, A_n \rightarrow [\alpha]C_n\} \subseteq \mathcal{E}_\alpha$ **do**
      if $\mathcal{S} \cup \{C_1, \ldots, C_n\} \vdash \bot$ **and** $\mathcal{S} \cup \{B, A_1, \ldots, A_n\} \not\vdash \bot$ **then**
        $\mathcal{S}^I := \mathcal{S}^I \cup \{\neg(B \wedge A_1 \wedge \ldots \wedge A_n)\}$

THEOREM 8 Algorithm 7 terminates.

Proof. Straightforward from finiteness of $\mathcal{X}$ and $\mathcal{E}$. ■

LEMMA 9 For every $A \in PFOR$ such that $\mathcal{S}, \mathcal{E}, \mathcal{X} \models A$, if $A \in \mathcal{S}^I$, then A is an implicit static law of $\langle \mathcal{S}, \mathcal{E}, \mathcal{X} \rangle$.

Proof. Let $A \in PFOR$ be such that $A \in \mathcal{S}^I$ and $\mathcal{S}, \mathcal{E}, \mathcal{X} \models A$. A is of the form $\neg(B \wedge A_1 \wedge \ldots \wedge A_n)$, for some $B, A_1, \ldots, A_n$, and $\mathcal{S} \wedge \neg(B \wedge A_1 \wedge \ldots \wedge A_n) \nvdash \bot$ is the case. Hence, $\mathcal{S} \wedge \neg A \nvdash \bot$, which means that $\mathcal{S} \not\models A$. Therefore A is an implicit static law. ■

REMARK 10 The converse of Lemma 9 does not hold: consider the quite simple action theory

$$\langle \mathcal{S}, \mathcal{E}, \mathcal{X} \rangle = \left\langle \begin{array}{c} \{\neg p_n\}, \\ \{p_{i-1} \to [\alpha]p_i,\ 1 \leq i \leq n\}, \\ \{\langle \alpha \rangle \top\} \end{array} \right\rangle$$

Thus, $\langle \mathcal{S}, \mathcal{E}, \mathcal{X} \rangle \models \neg p_i$, for $0 \leq i \leq n$, but running Algorithm 7 returns only $\mathcal{S}^I = \{\neg p_{n-1}\}$. This suggests that it is necessary to iterate the algorithm in order to find all implicit static laws. We shall do this in the next section, and now just observe that:

THEOREM 11 An action theory $\langle \mathcal{S}, \mathcal{E}, \mathcal{X} \rangle$ is modular if and only if $\mathcal{S}^I = \emptyset$.

Proof. The left-to-right direction is straightforward, by Lemma 9.

For the right-to-left direction, suppose $\mathcal{S}^I = \emptyset$. Therefore for all subsets $\{A_1 \to [\alpha]C_1, \ldots, A_n \to [\alpha]C_n\}$ of $\mathcal{E}_\alpha$ and all $B \to \langle \alpha \rangle \top \in \mathcal{X}$ we have that

(1) if $\mathcal{S} \cup \{B, A_1, \ldots, A_n\} \nvdash \bot$, then $\mathcal{S} \cup \{C_1, \ldots, C_n\} \nvdash \bot$.

By Theorem 4, then, it suffices to prove that then $\langle \mathcal{S}, \mathcal{E}, \mathcal{X} \rangle$ is propositionally modular. Therefore, suppose $\mathcal{S} \not\models A$ for some propositional A. Let W be the set of all propositional valuations satisfying $\mathcal{S}$ that falsify A. As $\mathcal{S} \not\models A$, $\mathcal{S} \cup \{\neg A\}$ is satisfiable, hence W must be nonempty. For every $w \in W$ let

$$\mathcal{E}_\alpha(w) = \{A : A \to [\alpha]C \in \mathcal{E}_\alpha \text{ and } w \models A\}.$$

We define $R(\alpha)$ such that $wR(\alpha)w'$ if and only if

- $w \models B$ for some $B \to \langle \alpha \rangle \top \in \mathcal{X}$, and
- $w' \models C$ for every $A \to [\alpha]C \in \mathcal{E}_\alpha$ such that $A \in \mathcal{E}_\alpha(w)$.

Taking the obvious definition of V we obtain a model $M = \langle W, R, V \rangle$. We have that $M \models \mathcal{S} \wedge \mathcal{E} \wedge \mathcal{X}$, because:

- $M \models \mathcal{S}$: by definition of W;
- $M \models \mathcal{E}$: for every world w, every α and every $A \rightarrow [\alpha]C \in \mathcal{E}_\alpha$ if $w \models A$, then, by the definition of $R(\alpha)$, $w' \models C$ for all $w' \in W$ such that $wR(\alpha)w'$;
- $M \models \mathcal{X}$: for every world w, every α and every $B \rightarrow \langle\alpha\rangle\top \in \mathcal{X}$, if $w \models B$, then from (1) and the definition of $R(\alpha)$, there exists at least one w' such that $wR(\alpha)w'$.

Clearly $M \not\models A$, by the definition of W. Hence $\mathcal{S}, \mathcal{E}, \mathcal{X} \not\models A$. ∎

7 Making action theories modular

Considering the action theory in Remark 10, we can see that running Algorithm 7 on $\langle \mathcal{S} \cup \{\neg p_{n-1}\}, \mathcal{E}, \mathcal{X}\rangle$ will give us $\mathcal{S}^I = \{\neg p_{n-2}\}$. This means that some of the implicit static laws of an action theory may be needed in order to derive others. Hence, Algorithm 7 must be iterated to get $\langle \mathcal{S}, \mathcal{E}, \mathcal{X}\rangle$ modular. This is achieved with the following algorithm, which iteratively feeds the set of static laws considered into the **if**-test of Algorithm 7.

ALGORITHM 12 (Finding all implicit static laws)
input: $\langle \mathcal{S}, \mathcal{E}, \mathcal{X}\rangle$
output: $\mathcal{S}^I_{\text{total}}$, the set of all implicit static laws of $\langle \mathcal{S}, \mathcal{E}, \mathcal{X}\rangle$
  $\mathcal{S}_{\text{new}} := \mathcal{S}$
  $\mathcal{S}^I_{\text{total}} := \emptyset$
  repeat
    $\mathcal{S}^I :=$ *find_imp_stat*$(\langle \mathcal{S}_{\text{new}}, \mathcal{E}, \mathcal{X}\rangle)$ /* a call to Algorithm 7 */
    $\mathcal{S}_{\text{new}} := \mathcal{S}_{\text{new}} \cup \mathcal{S}^I$
    $\mathcal{S}^I_{\text{total}} := \mathcal{S}^I_{\text{total}} \cup \mathcal{S}^I$
  until $\mathcal{S}^I = \emptyset$

THEOREM 13 Algorithm 12 terminates.

Proof. First, for given α the set of candidates to be an implicit static law is

$$\{\neg(B \wedge \bigwedge_{A_i \rightarrow [\alpha]C_i \in \mathcal{E}'_\alpha} A_i) : B \rightarrow \langle\alpha\rangle\top \in \mathcal{X} \text{ and } \mathcal{E}'_\alpha \subseteq \mathcal{E}_\alpha\}$$

This set is finite.

In each step either the algorithm ends because $\mathcal{S}^I = \emptyset$, or at least one of the candidates is put into $\mathcal{S}^I$ (by a call to Algorithm 7, which terminates). Such a candidate is not going to be put into $\mathcal{S}^I$ in future steps, because once added to $\mathcal{S}_{\text{new}}$, it will be in the set of laws of all subsequent calls to

Algorithm 7, falsifying its respective **if**-test for such a candidate. Hence the **repeat**-loop is bounded by the number of candidates, and therefore Algorithm 12 terminates. ■

THEOREM 14 Let $\mathcal{S}^I_{\text{total}}$ be the output of Algorithm 12 on input $\langle \mathcal{S}, \mathcal{E}, \mathcal{X} \rangle$. Then

1. $\langle \mathcal{S} \cup \mathcal{S}^I_{\text{total}}, \mathcal{E}, \mathcal{X} \rangle$ is modular.
2. $\mathcal{S}, \mathcal{E}, \mathcal{X} \models \bigwedge \mathcal{S}^I_{\text{total}}$.

Proof. Item 1. is straightforward from the termination of Algorithm 12 and Theorem 11. Item 2. follows from the fact that by the **if**-test in Algorithm 7, the only formulas that are put in $\mathcal{S}^I_{\text{total}}$ at each execution of the loop are exactly those that are implicit static laws of the original theory. ■

COROLLARY 15 For all $A \in \mathit{PFOR}$, $\mathcal{S}, \mathcal{E}, \mathcal{X} \models A$ if and only if $\mathcal{S} \cup \mathcal{S}^I_{\text{total}} \models A$.

Proof.
For the left-to-right direction, let $A \in \mathit{PFOR}$ be such that $\mathcal{S}, \mathcal{E}, \mathcal{X} \models A$. Then $\mathcal{S} \cup \mathcal{S}^I_{\text{total}}, \mathcal{E}, \mathcal{X} \models A$, by monotonicity. By Theorem 14-1., $\langle \mathcal{S} \cup \mathcal{S}^I_{\text{total}}, \mathcal{E}, \mathcal{X} \rangle$ is modular, hence $\mathcal{S} \cup \mathcal{S}^I_{\text{total}} \models A$.

The right-to-left direction is straightforward by Theorem 14-2. ■

This establishes that Algorithm 12 finds all implicit static laws of a given action theory $\langle \mathcal{S}, \mathcal{E}, \mathcal{X} \rangle$. Adding such laws to the original set of static laws $\mathcal{S}$ guarantees, hence, modularity of $\langle \mathcal{S}, \mathcal{E}, \mathcal{X} \rangle$.

In the next section we assess existing work on the field in the literature, emphasizing the points that make our approach a step further on a more fine-grained characterization of modularity.

8 Related work

Pirri and Reiter have investigated the metatheory of the Situation Calculus [14]. In a spirit similar to ours, they use executability laws and effect laws. Contrary to us, their executability laws are equivalences and are thus at the same time inexecutability laws. There are no static laws, i.e., $\mathcal{S} = \emptyset$. For this setting they give a syntactical condition on effect laws guaranteeing that they do not interact with the executability laws in the sense that they do not entail implicit static laws. Basically, the condition says that when there are effect laws $A \rightarrow [\alpha]C$ and $A' \rightarrow [\alpha]\neg C$, then A and A' are inconsistent (which essentially amounts to having in their theories a kind of "implicit static law schema" of the form $\neg(A \wedge A')$).

This then allows them to show that such theories are always consistent. Moreover they thus simplify the entailment problem for this calculus, and show for several problems such as consistency or regression that only some of the modules of an action theory are necessary.

Amir [1] focuses on design and maintainability of action descriptions applying many of the concepts of the object-oriented paradigm in the Situation Calculus. In that work, guidelines for a partitioned representation of a given theory are presented, with which the inference task can also be optimized, as it is restricted to the part of the theory that is really relevant to a given query. This is observed specially when different agents are involved: the design of an agent's theory can be done with no regard to others', and after the integration of multiple agents, queries about an agent's beliefs do not take into account the belief state of other agents.

In the above mentioned work, executabilities are as in [14] and the same condition on effect laws is assumed, which syntactically precludes the existence of implicit static laws.

Despite of using many of the object-oriented paradigm tools and techniques, no mention is made to the concepts of cohesion and coupling. In the approach presented in [1], even if modules are highly cohesive, they are not minimally coupled, due to the dependence between objects in the reasoning phase. We do not investigate this further here, but conjecture that this could be done there by, during the reasoning process defined for that approach, avoiding passing to a module a formula of a type different from those it contains.

The present work generalizes and extends Pirri and Reiter's result to the case where $\mathcal{S} \neq \emptyset$ and both Pirri and Reiter's and Amir's where the syntactical restriction on effect laws is not made. This gives us more expressive power, as we can reason about inexecutabilities, and a better modularity in the sense that we do not combine formulas that are conceptually different (viz. executabilities and inexecutabilities).

Zhang *et al.* [20] have also proposed an assessment of what a good action theory should look like. They develop the ideas in the framework of EPDL [21], an extended version of PDL which allows for propositions as modalities to represent causal connection between literals. We do not present the details of that, but concentrate on the main metatheoretical results.

Zhang *et al.* propose a normal form for describing action theories,[9] and

[9]But not as expressive as one might think: For instance, in modeling the non-

investigate three levels of consistency. Roughly speaking, an action theory $\mathcal{T}$ is *uniformly consistent* if it is globally consistent (i.e., $\mathcal{T} \not\models_{\mathsf{EPDL}} \bot$); a formula φ is *$\mathcal{T}$-consistent* if $\mathcal{T} \not\models_{\mathsf{EPDL}} \neg\varphi$, for $\mathcal{T}$ a uniformly consistent theory; $\mathcal{T}$ is *universally consistent* if (in our terms) every logically possible world is accessible.

Furthermore, two assumptions are made to preclude the existence of implicit qualifications. Satisfaction of such assumptions means the action theory under consideration is *safe*, i.e., it is uniformly consistent. Such a normal form justifies the two assumptions made and on whose validity relies their notion of good action theories.

Given these definitions, they propose algorithms to test the different versions of consistency for an action theory $\mathcal{T}$ that is in normal form. This test essentially amounts to checking whether $\mathcal{T}$ is *safe*, i.e., whether $\mathcal{T} \models_{\mathsf{EPDL}} \langle\alpha\rangle\top$, for every α. Success of this check should mean the action theory under analysis satisfies the consistency requirements.

Nevertheless, this is only a necessary condition: it is not hard to imagine action theories that are uniformly consistent but in which we can still have implicit laws that are not caught by the algorithm. Consider for instance a scenario with a lamp that can be turned on and off by a toggle action, and its EPDL representation given by:

$$\mathcal{T} = \left\{ \begin{array}{c} On \rightarrow [toggle]\neg On, \\ Off \rightarrow [toggle]\, On, \\ [On]\neg Off, \\ [\neg On]\, Off \end{array} \right\}$$

The causal statement $[On]\neg Off$ means that On causes $\neg Off$. Such an action theory satisfies each of the consistency requirements (in particular it is uniformly consistent, as $\mathcal{T} \not\models_{\mathsf{EPDL}} \bot$). Nevertheless, $\mathcal{T}$ is not safe because the static law $\neg(On \wedge Off)$ cannot be proved.[10]

Although they are concerned with the same kind of problems that have been discussed in this paper, they take an overall view of the subject, in

deterministic action of dropping a coin on a chessboard, we are not able to state $[drop](Black \vee White)$. Instead, we should write something like $[drop_{Black}]Black$, $[drop_{White}]White$, $[drop_{Black,White}]Black$ and $[drop_{Black,White}]White$, where $drop_{Black}$ is the action of dropping the coin on a black square (analogously for the others) and $drop = drop_{Black} \cup drop_{White} \cup drop_{Black,White}$, with "$\cup$" the nondeterministic composition of actions.

[10]A possible solution could be considering the set of static constraints explicitly in the action theory (viz. in the deductive system). For the running example, taking into account the constraint $On \leftrightarrow \neg Off$ (derived from the causal statements and the EPDL global axioms), we can conclude that $\mathcal{T}$ is safe. On the other hand, all the side effects such a modification could have on the whole theory has yet to be analyzed.

the sense that all problems are dealt with together. This means that in their approach no special attention (in our sense) is given to the different components of the action theory, and then every time something is wrong with it this is taken as a global problem inherent to the action theory as a whole. Whereas such a "systemic" view of action theories is not necessarily a drawback (we have just seen the strong interaction that exists between the different sets of laws composing an action theory), being modular in our sense allows us to circumscribe the "problematic" laws and take care of them. Moreover, the advantage of allowing to find the set of laws which must be modified in order to achieve the desired consistency is made evident by the algorithms we have proposed (while their results only allow to decide whether a given theory satisfies some consistency requirement).

Lang *et al.* [9] address consistency in the causal laws approach [11], focusing on the computational aspects. They suppose an abstract notion of completion of an action theory solving the frame problem. Given an action theory $\mathcal{T}^{\alpha}$ containing logical information about α's direct effects as well as the indirect effects that may follow, the completion of $\mathcal{T}^{\alpha}$ roughly speaking is the original theory $\mathcal{T}^{\alpha}$ amended of logical axioms stating the persistence of all non-affected (directly nor indirectly) literals.

Their EXECUTABILITY problem is to check whether α is executable in all possible initial states (Zhang *et al.*'s safety property). This amounts to testing whether every possible state w has a successor w' reachable by α such that w and w' both satisfy the completion of $\mathcal{T}^{\alpha}$. For instance, still considering the lamp scenario, the representation of the action theory for *toggle* is:

$$\mathcal{T}^{toggle} = \left\{ \begin{array}{l} On \stackrel{toggle}{\longrightarrow} \mathit{Off}, \\ \mathit{Off} \stackrel{toggle}{\longrightarrow} On, \\ \mathit{Off} \longrightarrow \neg On, \\ On \longrightarrow \neg \mathit{Off} \end{array} \right\}$$

where the first two formulas are conditional effect laws for *toggle*, and the latter two causal laws in McCain and Turner's sense. We will not dive in the technical details, and just note that the executability check will return "no" for this example as *toggle* cannot be executed in a state satisfying $On \wedge \mathit{Off}$.

In the mentioned work, the authors are more concerned with the complexity analysis of the problem of doing such a consistency test and no algorithm for performing it is given, however. In spite of the fact they have the same motivation as us, again what is presented is just a kind of "yes-no tool" which can help in doing a metatheoretical analysis of a given action theory, and many of the comments concerning Zhang and Chopra's approach could be repeated here.

9 Discussion and conclusion

In the perspective of independently axiomatized multimodal logics that are not serial we have investigated several criteria of modularity for simple theories. We have demonstrated the usefulness of modularity in reasoning about actions, where we have given an algorithmic checking for modularity of a given action theory.

We can have our criterion of modularity refined by taking into account polarity. Let $mod^{\pm}(\varphi)$ be the set of modalities of MOD occurring in φ together with their polarity. For instance $mod^{\pm}([\alpha_1]([\alpha_2]p \rightarrow q)) = \{+\alpha_1, -\alpha_2\}$. $mod^{\pm}(\mathcal{T})$ is defined accordingly. If $\mathcal{M}$ is a set of modalities with polarity then we define: $\mathcal{T}^{\mathcal{M}} = \{\varphi \in \mathcal{T} : mod^{\pm}(\varphi) \cap \mathcal{M} \neq \emptyset\}$.

DEFINITION 16 A theory $\mathcal{T}$ is $\pm$-modular if and only if for every formula φ,

$$\mathcal{T} \models \varphi \text{ implies } \mathcal{T}^{mod^{\pm}(\varphi)} \cup \mathcal{T}^{\emptyset} \models \varphi$$

There are other theories that are modular but not $\pm$-modular, e.g.,

$$\mathcal{T} = \{\neg[\alpha]p, [\alpha]p \vee [\alpha]\neg p\}$$

Indeed, $\mathcal{T} \models [\alpha]\neg p$, but $\mathcal{T}^{+\alpha} \cup \mathcal{T}^{\emptyset} \not\models [\alpha]\neg p$.

For the restricted case of action theories this has been proved in [4]. Moreover, a monotonic solution to the frame problem has been integrated there in such an algorithm. This makes the algorithm a bit more complex as it involves computing prime implicates. For the sake of simplicity this has not been done here.

With regard to the action theory in Example 5, it can be argued that unintuitive consequences in action theories are mainly due to badly written axioms and not to the lack of modularity. True enough, but what we have presented here is the case that making a domain description modular gives us a tool to detect at least some of such problems and correct it. (But note that we do not claim to correct badly written axioms automatically and once for all). Besides this, having separate entities in the ontology and controlling their interaction help us to localize where the problems are, which can be crucial for real world applications.

A topic for further investigations could be considering the notion of *coherence* defined in [8] as a guideline for "repairing" a given theory. Roughly, given an action theory $\mathcal{T}$ and an *unintuitive* implicit static law A, the formulas in $\mathcal{T}$ that are most likely to be revised are exactly those whose utility, in Kwok *et al.*'s sense, for deriving A are the highest.

Acknowledgments

Ivan Varzinczak has been supported by a fellowship from the government of the FEDERATIVE REPUBLIC OF BRAZIL. Grant: CAPES BEX 1389/01-7.

BIBLIOGRAPHY

[1] E. Amir. (De)composition of situation calculus theories. In *Proc. 17th Nat. Conf. on Artificial Intelligence (AAAI'2000)*, pages 456–463, Austin, 2000. AAAI Press/MIT Press.

[2] P. Doherty, W. Łukaszewicz, and A. Szałas. Explaining explanation closure. In *Proc. Int. Symposium on Methodologies for Intelligent Systems*, Zakopane, Poland, 1996.

[3] D. Harel. Dynamic logic. In D. M. Gabbay and F. Günthner, editors, *Handbook of Philosophical Logic*, volume II, pages 497–604. D. Reidel, Dordrecht, 1984.

[4] A. Herzig and I. Varzinczak. Domain descriptions should be modular. In R. López de Mántaras and L. Saitta, editors, *Proc. 16th Eur. Conf. on Artificial Intelligence (ECAI'04)*, pages 348–352, Valencia, 2004. IOS Press.

[5] A. Herzig and I. Varzinczak. On the cohesion and coupling of action theories. Technical report, Institut de recherche en informatique de Toulouse (IRIT), Université Paul Sabatier, March 2005. (Internal Report).

[6] M. Kracht and F. Wolter. Properties of independently axiomatizable bimodal logics. *J. of Symbolic Logic*, 56(4):1469–1485, 1991.

[7] M. Kracht and F. Wolter. Simulation and transfer results in modal logic: A survey. *Studia Logica*, 59:149–177, 1997.

[8] R. Kwok, N. Foo, and A. Nayak. Coherence of laws. In Sorge et al. [18], pages 1400–1401.

[9] J. Lang, F. Lin, and P Marquis. Causal theories of action – a computational core. In Sorge et al. [18], pages 1073–1078.

[10] F. Lin. Embracing causality in specifying the indirect effects of actions. In Mellish [13], pages 1985–1991.

[11] N. McCain and H. Turner. A causal theory of ramifications and qualifications. In Mellish [13], pages 1978–1984.

[12] J. McCarthy. *Mathematical logic in artificial intelligence.* Daedalus, 1988.

[13] C. Mellish, editor. *Proc. 14th Int. Joint Conf. on Artificial Intelligence (IJCAI'95)*, Montreal, 1995. Morgan Kaufmann Publishers.

[14] F. Pirri and R. Reiter. Some contributions to the metatheory of the situation calculus. *Journal of the ACM*, 46(3):325–361, 1999.

[15] R. S. Pressman. *Software Engineering: A Practitioner's Approach.* McGraw-Hill, 1992.

[16] L. K. Schubert. Monotonic solution of the frame problem in the situation calculus: an efficient method for worlds with fully specified actions. In H. E. Kyberg, R. P. Loui, and G. N. Carlson, editors, *Knowledge Representation and Defeasible Reasoning*, pages 23–67. Kluwer Academic Publishers, 1990.

[17] I. Sommerville. *Software Engineering.* Addison Wesley, 1985.

[18] V. Sorge, S. Colton, M. Fisher, and J. Gow, editors. *Proc. 18th Int. Joint Conf. on Artificial Intelligence (IJCAI'03)*, Acapulco, 2003. Morgan Kaufmann Publishers.

[19] M. Thielscher. Computing ramifications by postprocessing. In Mellish [13], pages 1994–2000.

[20] D. Zhang, S. Chopra, and N. Y. Foo. Consistency of action descriptions. In *PRICAI'02, Topics in Artificial Intelligence.* Springer-Verlag, 2002.

[21] D. Zhang and N. Y. Foo. EPDL: A logic for causal reasoning. In B. Nebel, editor, *Proc. 17th Int. Joint Conf. on Artificial Intelligence (IJCAI'01)*, pages 131–138, Seattle, 2001. Morgan Kaufmann Publishers.

Andreas Herzig
Institut de Recherche en Informatique de Toulouse (IRIT)
118 route de Narbonne
F-31062 Toulouse Cedex 4 (France)
e-mail: `herzig@irit.fr`
`http://www.irit.fr/recherches/LILAC`

Ivan Varzinczak
Institut de Recherche en Informatique de Toulouse (IRIT)
118 route de Narbonne
F-31062 Toulouse Cedex 4 (France)
e-mail: `ivan@irit.fr`
`http://www.irit.fr/~Ivan.Varzinczak`

Decidability of *IF* Modal Logic of Perfect Recall

TAPANI HYTTINEN AND TERO TULENHEIMO

ABSTRACT. IF Modal Logic of Perfect Recall is obtained from basic modal logic by allowing modal operators to be logically independent from certain syntactically superordinate dual operators. It is proven that the satisfiability and validity problems of this logic are decidable.

1 Plan of the Paper

The aim of this paper is to establish decidability of the satisfiability and validity problems of the logic (to be called 'IF Modal Logic of Perfect Recall') obtained from basic modal logic by allowing modal operators to be logically independent from certain syntactically superordinate dual operators. In *Section* 2, IF Modal Logic of Perfect Recall with k modality types (or, $\mathbf{IFML}^+[k]$) is defined in detail. Its semantics is defined in terms of evaluation games, which are simultaneously games of *imperfect information* and games of *perfect recall*. The former feature is used to implement modal operators' logical independence from syntactically superordinate modal operators. The latter feature follows from the syntactic restrictions regulating which operators may be marked as independent from which others. $\mathbf{IFML}^+[k]$ is a variant of the IF modal logic ($\mathbf{IFML}[k]$) of [16]. In *Section* 3 winning strategies of a player in evaluation games for $\mathbf{IFML}^+[k]$ are proven to have *history-free normal form*: if a player has a w.s., he has a w.s. whose value on a history only depends on the last 'known' position in the history. This result is shown not to be generalizable to $\mathbf{IFML}[k]$. In *Section* 4, $\mathbf{IFML}^+[k]$ is shown to have the strong finite model property, whence its satisfiablity problem is decidable. Our strategy for proving this result is hence the same as is typically used for proving the decidability of basic modal logic relative to the class of all modal structures (cf. [1], Th. 6.7). In *Section* 5 the validity and satisfiability problems of $\mathbf{IFML}^+[k]$ are observed not to be each others' duals. However, the validity problem of $\mathbf{IFML}^+[k]$ is proven to be decidable as well. *Section* 6 concludes the paper by pointing out related works, discussing the motivation behind IF modal logics, and mentioning open problems.

2 IF Modal Logic of Perfect Recall

Syntax. Let a class **prop** $:= \{p_i : i < \omega\}$ of propositional atoms be given. For all $k < \omega$, we define recursively a class $\mathcal{L}[k]$ of strings. Every string φ in $\mathcal{L}[k]$ is simultaneously associated with its *set of indices*, $|\varphi|$. The class of $\mathbf{IFML}^+[k]$-formulas will then be introduced as a class of strings from $\mathcal{L}[k]$ meeting certain syntactic conditions; semantically, k will be the number of accessibility relations available in the models in terms of which the semantics of the logic $\mathbf{IFML}^+[k]$ will be defined.

- If $p_i \in \mathbf{prop}$, then $p_i, \neg p_i \in \mathcal{L}[k]$; and $|p_i| = |\neg p_i| = \{0\}$.
- If $\varphi, \psi \in \mathcal{L}[k]$ and $n = max(|\varphi| \cup |\psi|) + 1$, then $(\varphi \vee_n \psi) \in \mathcal{L}[k]$ and $(\varphi \wedge_n \psi) \in \mathcal{L}[k]$; and $|(\varphi \vee_n \psi)| = |(\varphi \wedge_n \psi)| = \{n\} \cup |\varphi| \cup |\psi|$.
- If $\varphi \in \mathcal{L}[k]$, $n = max|\varphi| + 1$, $i < k$, $W_n \subseteq \,]n, \infty[$, and $O \in \{\Box, \Diamond\}$, then $(O_{i,n}/W_n)\varphi \in \mathcal{L}[k]$; and $|(O_{i,n}/W_n)\varphi| = \{n\} \cup |\varphi|$.

We will write $\Diamond_{i,n}$ for $(\Diamond_{i,n}/\varnothing)$ and $\Box_{i,n}$ for $(\Box_{i,n}/\varnothing)$. Further, in strings of $\mathcal{L}[1]$, we allow writing $(\Diamond_n/W_n)$ for $(\Diamond_{0,n}/W_n)$ and $(\Box_n/W_n)$ for $(\Box_{0,n}/W_n)$.

For every string φ in $\mathcal{L}[k]$ there corresponds in an obvious way a *syntactic tree*, whose nodes consist of modal operators, propositional connectives and (negated) propositional atoms. For instance, the syntactic tree of the string $\Box_4(\Box_1 p_0 \wedge_3 ((\neg p_1 \wedge_1 p_2) \vee_2 (\Diamond_1/\{4\})p_3))$ is the one depicted in Figure 1:

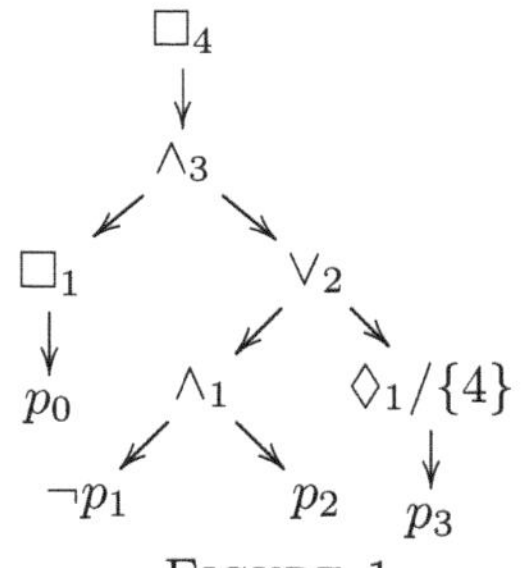

FIGURE 1

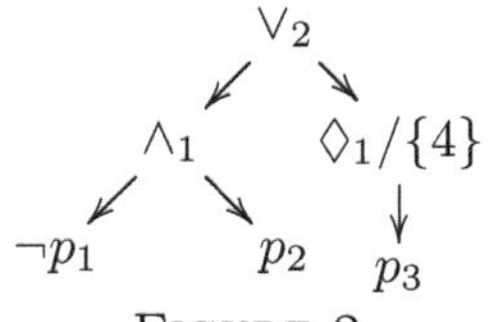

FIGURE 2

We call the ordering relation $(\prec_\varphi)$ of the syntactic tree of the string φ *(syntactic) superordination.* In Figure 1, $\Box_4$ for instance is superordinate to $\Diamond_1/\{4\}$; conversely $\Diamond_1/\{4\}$ is said to be *subordinate* to $\Box_4$. Whenever O is superordinate to O', there is a *path* from O to O', namely the sequence $\langle t_1, \ldots, t_n\rangle$, where $t_1 = O$, $t_n = O'$, and $t_i \prec_\varphi t_j$ iff $i < j$. For example, $(\wedge_3, \vee_2, \wedge_1, p_2)$ is a path from $\wedge_3$ to p_2 in the tree depicted in Figure 1. Operators appearing in a string and ordered according to the relation of superordinateness are said to be *nested.* If $(T, \prec_\varphi)$ is the syntactic tree of φ

and $t \in T$, the set $\{x : t = x \vee t \prec_\varphi x\}$ ordered by the relation $\prec_\varphi$ is a *full subtree* of $(T, \prec_\varphi)$. For instance the tree depicted in Figure 2 is a full subtree of the tree represented in Figure 1. The first index (i) of a modal operator $(O_{i,n}/W_n)$ appearing in string $\varphi \in \mathcal{L}[k]$ identifies a *modality type*, $i < k$; while the second (n) identifies the *occurrences* of nested modal operators.

The *length* of a string $\mathcal{L}[k]$, denoted $\sharp[\varphi]$, is the number of symbols appearing in φ. Hence if $p_i \in$ **prop**, then $\sharp[p_i] = 1 + d(i)$, where $d(i)$ is the number of digits of the numeral representing i in the binary system; $\sharp[(\varphi \vee \psi)] = \sharp[(\varphi \wedge \psi)] = \sharp[\varphi] + \sharp[\psi] + 3$; and $\sharp[(\Diamond_{i,n}/W_n)\varphi] = \sharp[(\Box_{i,n}/W_n)\varphi] = d(i)+d(n)+4+/W_n/+\sharp[\varphi]$, where $/W_n/ := 1$, if $W_n = \varnothing$, and $/W_n/ := d(n_1) + \ldots + d(n_m) + m + 1$, if $W_n = \{n_1, \ldots, n_m\} \neq \varnothing$.

The set $Sub[\varphi]$ of *substrings* of a string φ of $\mathcal{L}[k]$ is defined recursively in a straightforward way: $Sub[p_i] = \{p_i\}$ and $Sub[\neg p_i] = \{\neg p_i\}$; for $\circ \in \{\vee_n, \wedge_n\}$: $Sub[(\varphi \circ \psi)] = \{(\varphi \circ \psi)\} \cup Sub[\varphi] \cup Sub[\psi]$; and for $O \in \{\Box, \Diamond\}$: $Sub[(O_{i,n}/W_n)\varphi] = \{(O_{i,n}/W_n)\varphi\} \cup Sub[\varphi]$. If $\psi \in Sub[\varphi]$ and an operator O appears in φ, we say that O is *superordinate to the substring* ψ, if O is superordinate to all nodes of the full subtree of $(T, \prec_\varphi)$ determined by ψ.

The class of formulas of *IF Modal Logic of Perfect Recall with k modality types* (or the class of **IFML**$^+[k]$-formulas), is defined as the class of strings $\varphi \in \mathcal{L}[k]$ meeting the following conditions:

- **(1)** *Closedness:* If $O \in \{\Box, \Diamond\}$ and $(O_{i,n}/W_n)$ appears in φ, then there appear $(O_{i_1,n_1}/W_{n_1}), \ldots, (O_{i_m,n_m}/W_{n_m})$ in φ s.t. $W_n = \{n_1, \ldots, n_m\}$ and $(O_{i_1,n_1}/W_{n_1}) \prec_\varphi \ldots \prec_\varphi (O_{i_m,n_m}/W_{n_m}) \prec_\varphi (O_{i,n}/W_n)$.
- **(2.i)** *Diamonds independent only from boxes:* If $(\Diamond_{i,n}/W_n)$ appears in φ, $j \in W_n$ and $(O_{r,j}/W_j) \prec_\varphi (\Diamond_{i,n}/W_n)$, then $O_{r,j} = \Box_{r,j}$.
- **(2.ii)** *Boxes independent only from diamonds:* If $(\Box_{i,n}/W_n)$ appears in φ, $j \in W_n$, and $(O_{r,j}/W_j) \prec_\varphi (\Box_{i,n}/W_n)$, then $O_{r,j} = \Diamond_{r,j}$.
- **(3)** *Minimal requirement on inherited dependencies:*
  - **(i)** If $(\Box_{i,n}/W_n) \prec_\varphi \vee_m \prec_\varphi (\Diamond_{r,j}/W_j)$, and there is no operator $O \in \{\Box, \Diamond\}$ with $(\Box_{i,n}/W_n) \prec_\varphi O \prec_\varphi \vee_m$, then $n \notin W_j$.
  - **(ii)** If $(\Diamond_{i,n}/W_n) \prec_\varphi \wedge_m \prec_\varphi (\Box_{r,j}/W_j)$, and there is no operator $O \in \{\Box, \Diamond\}$ with $(\Diamond_{i,n}/W_n) \prec_\varphi O \prec_\varphi \wedge_m$, then $n \notin W_j$.

Observe that hence the *formulas* of **IFML**$^+[k]$ are a certain kind of *strings* of $\mathcal{L}[k]$, namely the strings satisfying the conditions (1), (2) and (3). The set of *subformulas* of an **IFML**$^+[k]$-formula φ is defined to be $Sub[\varphi]$, the set of substrings of φ. So in general subformulas of **IFML**$^+[k]$-formulas are *not* themselves *formulas* of **IFML**$^+[k]$, but merely strings of $\mathcal{L}[k]$.

Condition (1) rules out formulas with operators $(O_{i,n}/W_n)$, where some indices in W_n do not refer to any operators superordinate to $(O_{i,n}/W_n)$. Notice that the more subordinate operators an operator has, the greater its index; and that due to conjunctions and disjunctions, there may well be several operators with the same index. The semantic correlate of the restrictions (2.i,ii) will be that in the evaluation games for $\mathbf{IFML}^+[k]$, players recall their own past moves; the restrictions are motivated by the fact that while it certainly makes sense to speak of an operator being logically independent from its dual, it is not evident that there is such a properly logical phenomenon as the independence of a diamond, say, from another diamond.

Condition (3) imposes a specific restriction on possible independencies between dual modal operators. The semantic counterpart of condition (3.i) will be this: if the relevant player of an evaluation game knows, when making a move for a disjunction, the opponent's choice corresponding to a box – which in particular is the closest modal operator superordinate to the disjunction – then he will know the opponent's choice for the box also always later in the appropriate play of the game. This condition has an air of arbitrariness about it. The natural condition would be obtained by extending independence indications to disjunctions and conjunctions (allowing writing e.g. $\vee_2/\{5\}$ and $\wedge_1/\{9,11\}$), and requiring that the independencies of a formula do not violate the principle "a player never forgets a move he has once known." The formulation and study of such a more general syntax is left to another occasion; we content ourselves here with (3), which is the weakest condition that together with (1) and (2) results in a logic with the property that winning strategies in the associated evaluation games can be assumed to be 'history-free' (cf. Theorem 6 and Example 11). If a game is said to be of *perfect recall* whenever a player always recalls his own past moves, and does not forget what he has known, the evaluation games of $\mathbf{IFML}^+[k]$ are the most general 'approximation' of evaluation games of perfect recall that is possible without extending independence indications to $\vee$ and $\wedge$.

Semantics. The semantics is defined relative to k-ary modal structures $\mathcal{M} = (D, R_0, \ldots, R_{k-1}, \mathfrak{h})$, where $D \neq \varnothing$, $R_i \subseteq D^2$ for all $i < k$, and $\mathfrak{h} : \mathbf{prop} \longrightarrow Pow(D)$. If $\mathcal{M}$ is a k-ary modal structure and $w \in dom(\mathcal{M})$, we will call the pair $\langle \mathcal{M}, w \rangle$ a *k-ary model.* We associate with every formula φ and every k-ary model $\langle \mathcal{M}, w \rangle$ a game $G(\varphi, \mathcal{M}, w)$ between two players, $\forall$ and $\exists$. (The context will always take care that the names "$\forall$" and "$\exists$" of these players cannot be confused with universal *resp.* existential quantifier.) First we specify game rules for something a bit more complicated, namely the games $G(\varphi, \mathcal{M}, w, \pi_\varphi, \sigma_\varphi)$ proceeding from the extra information codified in the partial functions $\pi_\varphi : \omega + 1 \rightharpoonup dom(\mathcal{M})$ and $\sigma_\varphi : \omega \rightharpoonup \{0, 1\}$.

- If $\varphi \in \{p_i, \neg p_i\}$, no move is made. If $\varphi = p_i$ and $w \in \mathfrak{h}(p_i)$, or $\varphi = \neg p_i$

and $w \notin \mathfrak{h}(p_i)$, $\exists$ wins the play and $\forall$ loses it. Otherwise $\forall$ wins and $\exists$ loses.

- If $\varphi = \theta \wedge_n \psi$, $\forall$ chooses $\chi \in \{\theta, \psi\}$, and the play continues as $G(\chi, \mathcal{M}, w, \pi_\chi, \sigma_\chi)$, where $\pi_\chi := \pi_\varphi$ and $\sigma_\chi := \sigma_\varphi \cup \{(n, \flat_\chi)\}$, where $\flat_\chi = 0$ if $\chi = \theta$, and $\flat_\chi = 1$ otherwise.
- The case $\varphi = \theta \vee_n \psi$ is dual to the case for $\varphi = \theta \wedge_n \psi$.
- If $\varphi = (\Box_{i,n}/W_n)\psi$, $\forall$ chooses, if possible, $v \in dom(\mathcal{M})$ satisfying $R_i(w, v)$. The play continues as $G(\psi, \mathcal{M}, v, \pi_\psi, \sigma_\psi)$, where $\pi_\psi := \pi_\varphi \cup \{(n, v)\}$ and $\sigma_\psi := \sigma_\varphi$. If such a choice is not possible, the play ends in $\forall$'s choice FAIL, $\exists$ wins and $\forall$ loses. (No choice of v with $R_i(w, v)$ is possible iff w has no R_i-successor. If no other choice is possible, the player chooses by stipulation the object FAIL $\notin dom(\mathcal{M})$.)
- The case for $\varphi = (\Diamond_{i,n}/W_n)\psi$ is dual to the case for $\varphi = (\Box_{i,n}/W_n)\psi$.

STIPULATION 1. The game rules for $G(\varphi, \mathcal{M}, w)$ are the game rules of $G(\varphi, \mathcal{M}, w, \{(\omega, w)\}, \varnothing)$.

Evaluation of φ in $\langle \mathcal{M}, w\rangle$ begins in a situation where the point w is indexed by the ordinal ω and no conjunctive or disjunctive moves have been made: $\pi_\varphi = \{(\omega, w)\}$ and $\sigma_\varphi = \varnothing$. Thus ω will be in the domain of all functions π_χ corresponding to subformulas χ of φ. Plays of a game are also called its *histories*. An m-move play of $G(\varphi_0, \mathcal{M}, w)$ is a sequence $\langle(\pi_{\varphi_0}, \sigma_{\varphi_0}), \ldots, (\pi_{\varphi_m}, \sigma_{\varphi_m})\rangle$, satisfying $\pi_{\varphi_0} \subseteq \ldots \subseteq \pi_{\varphi_m}$ and $\sigma_{\varphi_0} \subseteq \ldots \subseteq \sigma_{\varphi_m}$. The play can be identified with the last member of the sequence, i.e. $(\pi_{\varphi_m}, \sigma_{\varphi_m})$. *Terminal plays* are plays ending with a (negated) propositional atom φ_m, or in the choice of FAIL by one of the players.

A token of a subformula χ of φ is uniquely determined by fixing the left/right choices for the conjunctions/disjunctions in φ to which χ is subordinate. Hence the tokens of χ may be identified with pairs (χ, σ_χ) s.t. for some π_χ, (π_χ, σ_χ) is a play. When no confusion may arise, we refer to subformula tokens equally as subformulas. A *strategy* of a player in $G(\varphi, \mathcal{M}, w)$ is a function specifying a move for every play at which it is this player's turn to move. Hence if f is $\exists$'s strategy and (π_χ, σ_χ) is a play, $f(\pi_\chi, \sigma_\chi) \in \{0, 1\}$, if $\chi = \psi \vee_n \theta$, and $f(\pi_\chi, \sigma_\chi) \in dom(\mathcal{M}) \cup \{\text{FAIL}\}$, if $\chi = (\Diamond_{i,n}/W_n)\psi$.

Given a subformula (χ, σ_χ), write V_{σ_χ} for the common domain of all functions π_χ s.t. (π_χ, σ_χ) is a play of $G(\varphi, \mathcal{M}, w)$. For each subformula (χ, σ_χ) with χ of the form $(\Diamond_{i,n}/W_n)\psi$, partition the set of plays $\{(\pi_\chi, \sigma_\chi) : \pi_\chi \text{ is a function } V_{\sigma_\chi} \longrightarrow dom(\mathcal{M})\}$ into equivalence classes under the following equivalence relation $\sim_\exists^{\sigma_\chi}$: $(\pi'_\chi, \sigma_\chi) \sim_\exists^{\sigma_\chi} (\pi''_\chi, \sigma_\chi)$, if $(\forall i \in V_{\sigma_\chi})(\pi'_\chi(i) \neq \pi''_\chi(i) \implies i \in W_n)$. This partition gathers together plays at which a given $\chi := (\Diamond_{i,n}/W_n)\psi$ is evaluated and which differ at most for moves interpreting an operator $(\Box_{j,m}/W_m)$ with $m \in W_n$. Let $\sim_\exists$ be the union of all relations $\sim_\exists^{\sigma_\chi}$, where σ_χ identifies a subformula (χ, σ_χ) with χ of the form $(\Diamond_{i,n}/W_n)\psi$. Relation $\sim_\forall$ is defined similarly. A strategy f of $\exists$ is a *winning strategy* (w.s.), if there is a set S of plays such that:

- S contains the play $(\{(\omega, w)\}, \varnothing)$.
- If $h \in S$, $f(h)$ is a move according to the game rules and $f(h) \in S$.
- S is closed under all moves of $\forall$ made according to the game rules.
- f is uniform, i.e. if $h \sim_\exists h'$, then $f(h) = f(h')$.
- Every terminal play in S is a win for $\exists$.

Winning strategies of $\forall$ are defined analogously. Semantics of a formula $\varphi \in \mathbf{IFML}^+[k]$ is defined as follows:

Truth: $\mathcal{M} \models^+ \varphi[w]$ **if** there exists a w.s. for $\exists$ in $G(\varphi, \mathcal{M}, w)$.
Falsity: $\mathcal{M} \models^- \varphi[w]$ **if** there exists a w.s. for $\forall$ in $G(\varphi, \mathcal{M}, w)$.
Indeterminacy: $\mathcal{M} \models^0 \varphi[w]$ **if** $\mathcal{M} \not\models^+ \varphi[w]$ and $\mathcal{M} \not\models^- \varphi[w]$.

When no confusion may arise, we will simply write $\models$ for the relation $\models^+$.

EXAMPLE 2. Think of the unary models $\langle \mathcal{M}, m_0 \rangle$ and $\langle \mathcal{N}, n_0 \rangle$:

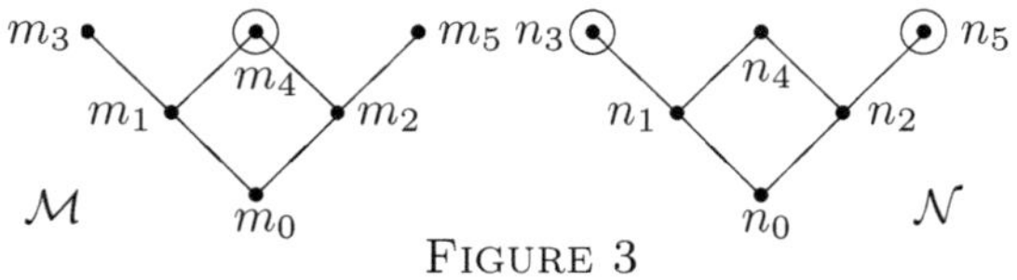

FIGURE 3

In Figure 3, the circles indicate the points making the atom p_0 true. Consider evaluating the formula $\varphi := \Box_2(\Diamond_1/\{2\})p_0$ relative to the two models.

(a) φ is true in $\langle \mathcal{M}, m_0 \rangle$. In fact, the function f defined by the condition $f(m_0, m_1) = f(m_0, m_2) = m_4$ is a w.s. for $\exists$ in $G(\varphi, \mathcal{M}, m_0)$.

(b) φ is indeterminate in $\langle \mathcal{N}, n_0 \rangle$. **(i)** Suppose for contradiction that f is a w.s. for $\exists$ in $G(\varphi, \mathcal{N}, n_0)$. Hence f is uniform. If the value $x = f(n_0, n_1) = f(n_0, n_2)$ is distinct from n_4, x constitutes an illegal move for one of the histories (n_0, n_1) and (n_0, n_2), and so f is not winning. If, again, $x = n_4$, f does not lead to a win for $\exists$ in any play of $G(\varphi, \mathcal{M}', n_0)$. **(ii)** Neither of the two possible strategies $g_1(n_0) = n_1$ and $g_1(n_0) = n_2$ of $\forall$ is winning: $\exists$ can extend both plays (n_0, n_1) and (n_0, n_2) with a choice yielding a win for her.

(c) The models $\langle \mathcal{M}, m_0 \rangle$ and $\langle \mathcal{N}, n_0 \rangle$ are bisimilar. Hence they are not distinguished by any formula of basic modal logic. ■

Known results of expressive power. If L and L' are modal logics and $\mathcal{K}$ is a class of modal structures on which the semantics of these logics is defined, L is *embeddable* in L' over $\mathcal{K}$ ($L \leq_{\mathcal{K}} L'$), if for each $\varphi \in L$ there is $\psi_\varphi \in L'$ s.t. for all $\langle \mathcal{M}, w \rangle$ with $\mathcal{M} \in \mathcal{K}$, $\mathcal{M} \models \varphi[w]$ iff $\mathcal{M} \models \psi_\varphi[w]$. *$L'$ has greater expressive power than L* over $\mathcal{K}$ ($L <_{\mathcal{K}} L'$), if $L \leq_{\mathcal{K}} L'$ but $L' \nleq_{\mathcal{K}} L$. *L and L' have the same expressive power* over $\mathcal{K}$ ($L =_{\mathcal{K}} L'$), if $L \leq_{\mathcal{K}} L'$ and $L' \leq_{\mathcal{K}} L$. (For abstract logics, the same terminology is found e.g. in [6].)

In [16] 'IF modal logic of k modality types' (or, $\mathbf{IFML}[k]$) was introduced. Its syntax can be obtained from that of $\mathbf{IFML}^+[k]$ by giving up the syntactic restrictions listed as (2) and (3) in *Section* 2, and requiring simply that modal operators may only be independent from other modal operators – whether these latter are duals of the former or not. The semantics of $\mathbf{IFML}[k]$ is completely analogous to the semantics of $\mathbf{IFML}^+[k]$. However, in the presence of diamonds indicated as independent from other diamonds, there are equivalence relations $\sim_\exists^{\sigma_X}$ that could not be induced in the case of $\mathbf{IFML}^+[k]$. The evaluation games of $\mathbf{IFML}[k]$ are *not* in general games of perfect recall: a player always knows the *strategy* he has followed up to any given stage in the game, but may fail to distinguish between two or more *individual moves* he himself has made in accordance to that strategy. $\mathbf{IFML}^+[k]$ is trivially embeddable in $\mathbf{IFML}[k]$. Basic modal logic of k modality types (or, $\mathbf{ML}[k]$) and basic tense logic of k temporal modality types (or, $\mathbf{TL}[k]$) are generated by closing the class of (negated) propositional atoms under $\wedge$, $\vee$ and applications of unary modal operators $\Box_i, \Diamond_i$ *resp.* $\Box_i$, $\Diamond_i$, $\Box_i^{-1}, \Diamond_i^{-1}$, with $i < k$; these modal operators have their usual semantics. Syntactically IF tense logic of perfect recall with k temporal modality types (or, $\mathbf{IFTL}^+[k]$) equals $\mathbf{IFML}^+[2k]$, and it is interpreted relative to modal structures with $2k$ accessibility relations of which the i-th and the $(i+k)$-th are each other's converses for all $i < k$, and which all are irreflexive and transitive ('temporal structures'). Let $\mathcal{C}_k$ be the class of all k-ary modal structures, and $\mathcal{T}_k$ the class of all k-ary temporal structures. Directly by (the proofs of) the corresponding theorems in [16], we have:

THEOREM 3. *Let $k \geq 1$.* **(a)** *Write $\mathbf{IFML}^+_{\mathbf{det}}[k]$ for the "determined fragment" of $\mathbf{IFML}^+[k]$, i.e. the class $\{\varphi \in \mathbf{IFML}^+[k]$: for all $\mathcal{M}, w$, either $\mathcal{M} \models^+ \varphi[w]$ or $\mathcal{M} \models^- \varphi[w]\}$. We have: $\mathbf{IFML}^+_{\mathbf{det}}[k] =_{\mathcal{C}_k} \mathbf{ML}[k]$.* **(b)** $\mathbf{ML}[k] <_{\mathcal{C}_k} \mathbf{IFML}^+[k]$. **(c)** $\mathbf{TL}[k] <_{\mathcal{T}_k} \mathbf{IFTL}^+[k]$.

Notice that Example 2 proves Theorem 3 (b). It is not difficult to show:

THEOREM 4. *For all $k \geq 1$,* $\mathbf{IFML}^+[k] <_{\mathcal{C}_k} \mathbf{IFML}[k]$.

Hence the IF modal logic of [16] is a proper extension of the IF modal logic of the present paper. However, the extra expressive power of $\mathbf{IFML}[k]$ is not needed for the main expressivity results concerning this logic: as shown by Theorem 3 (b,c), these results are obtained already within $\mathbf{IFML}^+[k]$.

3 History-Free Strategies

If φ is an $\mathbf{IFML}^+[k]$-formula and (χ, σ_χ) its subformula, write S_{σ_χ} for the set of all histories (plays) of game $G(\varphi, \mathcal{M}, w)$ that end with (χ, σ_χ). I.e., $S_{\sigma_\chi} = \{(\pi_\chi, \sigma_\chi) : (\pi_\chi, \sigma_\chi) \text{ is a play and } \pi_\chi \text{ is a function } V_{\sigma_\chi} \longrightarrow dom(\mathcal{M})\}$.

DEFINITION 5. (History-free strategy) A strategy f of $\exists$ is *history-free*, if for any subformula (χ, σ_χ) of φ with $\chi \in \{\theta \vee_n \psi, (\Diamond_{i,n}/W_n)\psi\}$, and any histories $(\pi'_\chi, \sigma_\chi), (\pi''_\chi, \sigma_\chi) \in S_{\sigma_\chi}$, we have:

$$\pi'_\chi(K) = \pi''_\chi(K) \Longrightarrow f(\pi'_\chi, \sigma_\chi) = f(\pi''_\chi, \sigma_\chi),$$

where $K := min\{x > n : x = \omega \text{ or } [O_x \prec_\varphi \vee_n \ \& \ O_x \text{ is of the form } \Diamond, \Box]\}$, if $\chi := \theta \vee_n \psi$; and $K := min\{x > n : x = \omega \text{ or } [O_x \prec_\varphi \Diamond_{i,n}/W_n \ \& \ O_x \text{ is of the form } \Diamond, \Box \ \& \ x \notin W_n]\}$, if $\chi := (\Diamond_{i,n}/W_n)\psi$.

Hence the value of a history-free strategy of $\exists$ for a history (π_χ, σ_χ) depends only on the last member of π_χ 'known' by $\exists$.

THEOREM 6. *Let $\varphi \in \mathbf{IFML}^+[k]$, $\mathcal{M}$ and w be arbitrary. If there is a w.s. for $\exists$ in $G(\varphi, \mathcal{M}, w)$, there is a history-free w.s. for her in this game.*

Proof. Assume $\exists$ has a w.s. f in $G(\varphi, \mathcal{M}, w)$. Given a subformula (χ, σ_χ) of φ and $y \in dom(\mathcal{M})$, consider the set $[w, y]^f_{\sigma_\chi}$ of all maps π_χ s.t. (π_χ, σ_χ) is a play of $G(\varphi, \mathcal{M}, w)$ in which $\exists$ has used f, and in which the last choice from $dom(\mathcal{M})$ is y. Whenever $[w, y]^f_{\sigma_\chi} \neq \varnothing$, choose a canonical element of this set, and denote it by $\iota([w, y]^f_{\sigma_\chi})$. Further, given a subformula (χ, σ_χ) of φ, and a play (π_ψ, σ_ψ) of $G(\varphi, \mathcal{M}, w)$ in which $\exists$ has applied f, and which satisfies $\sigma_\psi \subseteq \sigma_\chi$, think of the set $[\pi_\psi, \sigma_\psi]^f_{\sigma_\chi}$ of all maps π_χ s.t. (π_χ, σ_χ) satisfies $\pi_\psi \subseteq \pi_\chi$ and is a play of $G(\varphi, \mathcal{M}, w)$ in which $\exists$ has used f. Choose also for sets of this type a canonical representative, denoted by $\iota([\pi_\psi, \sigma_\psi]^f_{\sigma_\chi})$. Let us then define a strategy f' for $\exists$ in $G(\varphi, \mathcal{M}, w)$ as follows:

- If $\chi = \psi \vee \theta$, put $f'(\pi_\chi, \sigma_\chi) := f(\iota([w, y]^f_{\sigma_\chi}), \sigma_\chi)$, where $y := \pi_\chi(min(dom(\pi_\chi)))$.
- If $\chi = (\Diamond_{i,n}/W_n)\psi$, put $f'(\pi_\chi, \sigma_\chi) := f(\iota([\pi_\phi, \sigma_\phi]^f_{\sigma_\chi}), \sigma_\chi)$, the play (π_ϕ, σ_ϕ) being determined by the conditions (i) $\pi_\phi := \iota([w, \pi_\chi(K)]^f_{\sigma_\phi})$ and (ii) $\sigma_\phi := \{(x, \sigma_\chi(x)) : x \in dom(\sigma_\chi) \ \& \ x > K\}$, where $K := min\{x > n : x = \omega \text{ or } [O_x \prec_\varphi \Diamond_{i,n}/W_n \ \& \ O_x \in \{\Diamond, \Box\} \ \& \ x \notin W_n]\}$.

Observe that in the latter case (π_ϕ, σ_ϕ) can indeed be extended into a play (π_χ, σ_χ) so that $\exists$'s moves are given by the strategy f. This would not in general be possible if the syntactic restriction (3) of *Section* 2 did not hold:

then there might be operators $(\Box_{i',n'}/W_{n'}) \prec_\varphi \vee_m \prec_\varphi (\Diamond_{i,n}/W_n)$ with $n' \in W_n$ s.t. $\Box_{i',n'}$ would be interpreted in the course of the play (π_ϕ, σ_ϕ), and f would choose for $\vee_m$ a disjunct in which the operator $(\Diamond_{i,n}/W_n)$ does not appear. But when (3) is assumed, the set $[\pi_\phi, \sigma_\phi]^f_{\sigma_\chi}$ is non-empty.

CLAIM 7. f' is a w.s. for $\exists$ in $G(\varphi, \mathcal{M}, w)$.

Proof. **(1)** Let (χ, σ_χ) be any subformula of φ, and suppose $\exists$ has used f' in constructing (π_χ, σ_χ). We must show that f' gives legal extensions. For disjunctions this is trivial, so consider diamonds. The proof will require considering different plays of game $G(\varphi, \mathcal{M}, w)$; these will be schematically illustrated in Figure 4 below. (For the play (π_χ, σ_χ), cf. the sequence (w, w_1, u_2, w_3) in the figure.)

Write $\{\omega, n_1, \ldots, n_m\}$ for $dom(\pi_\chi)$, with $n_m < \ldots < n_1 < \omega$. If $\chi := (\Diamond_{i,n}/W_n)\psi$, let K be the smallest $x > n$ s.t. x is an index of a modal operator superordinate to $\Diamond_{i,n}/W_n$ with $x \notin W_n$, or if none exists, let $K := \omega$. Assume for contradiction that $\langle \pi_\chi(n_m), f'(\pi_\chi, \sigma_\chi)\rangle \notin R_i$. Write $\langle w, \pi'_\chi(n_1), \ldots, \pi'_\chi(n_{p-1}), \pi_\chi(K)\rangle$ for $\iota([w, \pi_\chi(K)]^f_{\sigma_\chi})$, $p \leq m$. Hence $h_1 := (\langle w, \pi'_\chi(n_1), \ldots, \pi'_\chi(n_{p-1}), \pi_\chi(K), \pi_\chi(n_{p+1}), \ldots, \pi_\chi(n_m)\rangle, \sigma_\chi)$ is a play. (For h_1, cf. the sequence (w, v_1, u_2, w_3) in Figure 4.) Moreover, by the syntax of $\mathbf{IFML}^+[k]$ and the definition of K, all moves $\pi_\chi(n_{p+1}), \ldots, \pi_\chi(n_m)$ are by $\forall$. So h_1 is not only a play, but in particular a play that can be realized when $\exists$ applies f. But also the play h_2 for which $f'(\pi_\chi, \sigma_\chi) = f(h_2)$, $h_2 := (\langle w, \pi'_\chi(n_1), \ldots, \pi'_\chi(n_{p-1}), \pi_\chi(K), \pi''_\chi(n_{p+1}), \ldots, \pi''_\chi(n_m)\rangle, \sigma_\chi)$, is constructible when $\exists$ applies f. (For h_2, cf. the sequence (w, v_1, u_2, v_3) in Figure 4.) Since f is a w.s. and $h_1 \sim_\exists h_2$, we have that $f(h_1) = f(h_2)$. By definition, on the other hand, $f'(\pi_\chi, \sigma_\chi) = f(h_2)$, whence $\langle \pi_\chi(n_m), f(h_1)\rangle \notin R_i$. This, however, is impossible, as f is a w.s. for $\exists$.

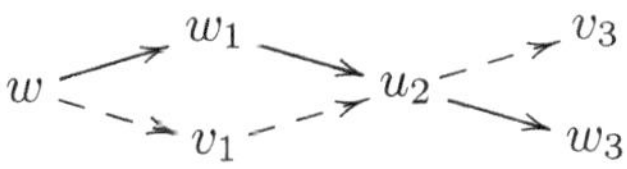

FIGURE 4

(2) f' agrees on equivalent histories: This is trivial, since the value of f' on equivalent histories is defined as the value of f on one and the same history.

(3) Clearly f' always leads to a win for $\exists$. ■

By definition f' is history-free. The statement of Theorem 6 follows. ■

Every $\mathbf{IFML}^+[k]$-formula φ has the following normal forms φ° and φ^+:

COROLLARY 8. *For all* $\varphi \in \mathbf{IFML}^+[k]$, $\mathcal{M}$, *and* w, *we have:*

$$\mathcal{M} \models \varphi^\circ[w] \iff \mathcal{M} \models \varphi[w] \iff \mathcal{M} \models \varphi^+[w],$$

φ° being obtained from φ by replacing any $(\Diamond_{i,n}/W_n)$ by $(\Diamond_{i,n}/V_n)$, and φ^+ from φ by replacing any $(\Diamond_{i,n}/W_n)$ by $(\Diamond_{i,n}/U_n)$, where V_n [resp. U_n] is the set of indices x of boxes superordinate to $(\Diamond_{i,n}/W_n)$ in φ such that $n < x < K_n$ [resp. $x \neq K_n$], for $K_n := min\{x > n : x = \omega$ or $[O_x \prec_\varphi \Diamond_{i,n}/W_n$ & $O_x \in \{\Diamond, \Box\}$ & $x \notin W_n]\}$. ■

E.g., if φ is the formula $\Box_7\Diamond_6\Box_5\Diamond_4\Box_3\Box_2(\Diamond_1/\{2,5\})p_0$, it is true in precisely the same models as $\varphi^\circ := \Box_7\Diamond_6\Box_5\Diamond_4\Box_3\Box_2(\Diamond_1/\{2\})p_0$, and also true in exactly the same models as $\varphi^+ := \Box_7\Diamond_6\Box_5(\Diamond_4/\{7\})\Box_3\Box_2(\Diamond_1/\{2,5,7\})p_0$. By Theorem 6, when evaluating any of the formulas φ, φ° and φ^+, the choice for $\Diamond_1$ can be made as a function of the last known move in the relevant history, i.e. depending only on the choice for $\Box_3$. Hence allowing the choice for $\Diamond_1$ to depend not only on the move for $\Box_3$ but also on all earlier moves by the opponent (φ°), or requiring this choice to be independent from all of these earlier moves (φ^+), simply results in the truth-condition of φ.

Theorem 6 cannot be generalized from **IFML**$^+[k]$ to **IFML**$[k]$:

EXAMPLE 9. Let $\varphi := \Box_4\Box_3(\Diamond_2/\{3\})(\Diamond_1/\{2\})\top \in$ **IFML**$[1]$. Let us evaluate φ relative to the unary model $\langle\mathcal{M}, a\rangle$ depicted in Figure 5:

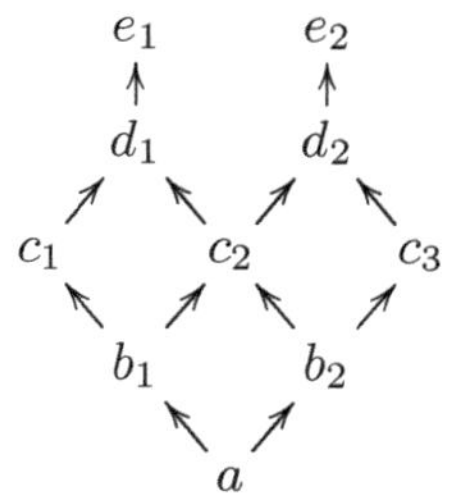

FIGURE 5

(a) To see that $\exists$ has a w.s. in $G(\varphi, \mathcal{M}, a)$, define a function f by setting: $f(a, b_1, c_1) = f(a, b_1, c_2) = d_1$, $f(a, b_2, c_2) = f(a, b_2, c_3) = d_2$, $f(a, b_1, c_1, d_1) = f(a, b_1, c_2, d_1) = e_1$, $f(a, b_2, c_2, d_2) = f(a, b_2, c_3, d_2) = e_2$. Let (a, x_1, x_2) and (a, x_1', x_2') be any plays of $G(\varphi, \mathcal{M}, a)$ s.t. $x_1 = x_1'$. So $f(a, x_1, x_2) = f(a, x_1', x_2')$. If R is the accessibility relation represented in the figure and $d = f(a, x_1, x_2)$, then $R(x_2, d)$. Further, if (a, x_1, x_2, x_3) and (a, x_1', x_2', x_3') are plays in which $\exists$ has used f and which satisfy $x_1 = x_1'$ and $x_2 = x_2'$, then $f(a, x_1, x_2, x_3) = f(a, x_1', x_2', x_3')$. And if $e = f(a, x_1, x_2, x_3)$, then $R(x_3, e)$. Hence f is a w.s. for $\exists$ in $G(\varphi, \mathcal{M}, a)$. Observe that f is *not* history-free: for both plays (a, b_1, c_2, d_1) and (a, b_2, c_2, d_2), the last move 'known' by $\exists$ is c_2, but f associates these plays with distinct choices.

(b) We claim that in fact there is *no* history-free w.s. for $\exists$ in $G(\varphi, \mathcal{M}, a)$. Suppose for contradiction that f^* is such a history-free winning strategy.

Hence in particular $f^*(a, b_1, c_2, d_1) = f^*(a, b_2, c_2, d_2)$. On the other hand, necessarily $f^*(a, b_1, c_1) = d_1$ and $f^*(a, b_2, c_3) = d_2$. But then, by the uniformity of f^*, we must further have $f^*(a, b_1, c_2) = d_1$ and $f^*(a, b_2, c_2) = d_2$. Thus (a, b_1, c_2, d_1) and (a, b_2, c_2, d_2) are both histories constructed by using f^* against a sequence of moves by $\forall$. However, the value $f^*(a, b_1, c_2, d_1) = f^*(a, b_2, c_2, d_2)$ cannot be accessed along R from both d_1 and d_2. This is a contradiction in view of f^* being a winning stratregy. We may conclude that there is no history-free w.s. for $\exists$ in game $G(\varphi, \mathcal{M}, a)$. ■

THEOREM 10. *For all $k \geq 1$, $\exists$'s winning strategies in evaluation games of* **IFML**$[k]$ *do not in general admit of history-free normal form.* ■

Finally, let $\mathbf{IFML}^+_{12}[k]$ be the logic that results from giving up condition (3) in the definition of the syntax of $\mathbf{IFML}^+[k]$. Let us consider the impact of this condition to the existence of history-free winning strategies.

EXAMPLE 11. Let $\varphi := \Box_{0,3}\Box_{0,2} \bigvee_{i\in\{0,1\}} (\Diamond_{i,1}/\{2\})\top \in \mathbf{IFML}^+_{12}[2]$, and think of evaluating φ relative to the binary model $\langle \mathcal{M}, w\rangle$ of Figure 6:

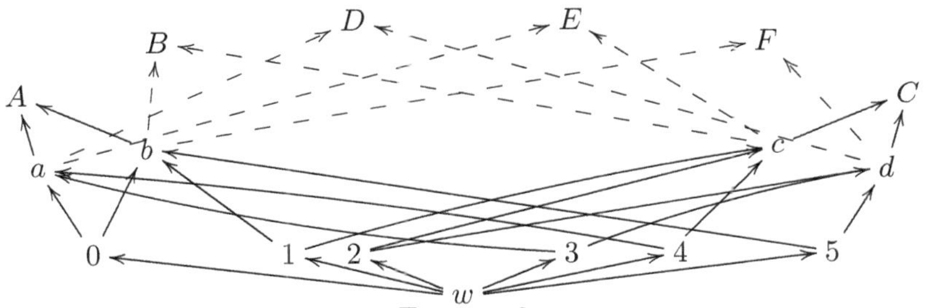

FIGURE 6

In Figure 6, the solid lines represent the accessibility relation R_0, while the dashed lines stand for the accessibility relation R_1. The domain of $\mathcal{M}$ consist of four levels: $G_0 = \{w\}$, $G_1 = \{0, 1, 2, 3, 4, 5\}$, $G_2 = \{a, b, c, d\}$ and $G_3 = \{A, B, C, D, E, F\}$. Write $Z = \{0, 2\}$.

(a) We claim that there is a w.s. for $\exists$ in $G(\varphi, \mathcal{M}, w)$. To see this, define a function f as follows: if $x \in G_1$ and $y \in G_2$ are any consecutive choices of $\forall$ in a play of $G(\varphi, \mathcal{M}, w)$, made according to the game rules, we set:

$$f(w, x, y) = 0 \text{ if } x \in Z;\text{ and } f(w, x, y) = 1 \text{ otherwise;}$$
$$f(w, 0, y, 0) = A;\ f(w, 2, y, 0) = C;$$
$$f(w, 1, y, 1) = B;\ f(w, 3, y, 1) = D;\ f(w, 4, y, 1) = E; f(w, 5, y, 1) = F.$$

Let (w, x, y, i) and (w, x', y', i') be any plays of $G(\varphi, \mathcal{M}, w)$ s.t. $x = x'$ and $i = i'$. So $f(w, x, y, i) = f(w, x', y', i')$. Write $X := f(w, x, y, i)$. If $x \in Z$ and hence $i = 0$, we have $R_0(y, X)$. And if $x \notin Z$ and thus $i = 1$, we have

$R_1(y, X)$. We may conclude that f is a w.s. for $\exists$ in $G(\varphi, \mathcal{M}, w)$. It is readily observed that f is *not* history-free: e.g. $0 = f(w, 0, b) \neq f(w, 1, b) = 1$, while for both plays $(w, 0, b)$ and $(w, 1, b)$, the last move 'known' by $\exists$ is b.

(b) We claim there is *no* history-free w.s. for $\exists$ in $G(\varphi, \mathcal{M}, w)$. For a contradiction, suppose f^* is such a strategy. If $x \in G_2$, denote by $f^*(x) \in \{0,1\}$ the value f^* gives, as a choice corresponding to $\vee_{i\in\{0,1\}}$, to plays ending with x. Let $x, y \in G_2$ and $z \in G_1$ be points s.t. $R_0(z, x)$, $R_0(z, y)$. If $z \in Z$, it is not possible that $f^*(x) = f^*(y) = 1$, because there is no $p \in G_3$ s.t. $R_1(x, p)$ and $R_1(y, p)$; if $z \notin Z$, it is not possible that $f^*(x) = f^*(y) = 0$, because then there is no $p \in G_3$ s.t. $R_0(x, p)$ and $R_0(y, p)$.

In fact, $f^*(x) \neq f^*(y)$. To see this, suppose for contradiction that $f^*(x) = f^*(y)$, and let x', y' be the elements in $G_2 \setminus \{x, y\}$. Consider the case $z \in Z$; the case of $z \notin Z$ is similar. We must have: $(*)$ $f^*(x') = f^*(y') \neq f^*(x) = f^*(y) = 0$. For otherwise there is $v \in \{x', y'\}$ s.t. $f^*(v) = 0$ and $z' \in G_1 \setminus Z$ satisfying $R_0(z', x)$, $R_0(z', v)$, which is not possible in view of what observed above. By $(*)$, $f^*(x') = f^*(y') = 1$. But there is no $p \in G_3$ s.t. $R_1(x', p)$ and $R_1(y', p)$, whence f^* is not a w.s. for $\exists$, a contradiction.

So either $f^*(a) = f^*(c) = 0$ and $f^*(b) = f^*(d) = 1$, or $f^*(a) = f^*(c) = 1$ and $f^*(b) = f^*(d) = 0$. In both cases we can find $x, y \in G_2$ and $z \in G_1 \setminus Z$ such that $R_0(x, z)$, $R_0(y, z)$ and $f^*(x) = f^*(y) = 0$. Again, there is no $p \in G_3$ such that $R_0(x, p)$ and $R_0(y, p)$, a contradiction. We may conclude that there is no history-free w.s. for $\exists$ in $G(\varphi, \mathcal{M}, w)$. ■

THEOREM 12. *For all $k \geq 2$, $\exists$'s winning strategies in evaluation games of* $\mathbf{IFML}^+_{12}[k]$ *do not in general admit of history-free normal form.* ■

4 Satisfiability Problem

Supposing $\varphi \in \mathbf{IFML}^+[k]$ is satisfied in $\langle \mathcal{M}, w\rangle$, we (1) construct a tree $\mathcal{T}$ out of $\langle \mathcal{M}, w\rangle$; (2) generate a labeled subtree $\mathcal{T}^f$ of $\mathcal{T}$; (3) produce a model $\langle \mathcal{M}^*, w\rangle$ of φ out of $\mathcal{T}^f$; (4) generate from $\langle \mathcal{M}^*, w\rangle$ a finite model $\langle \mathcal{N}, v\rangle$ of φ, whose size has a recursive upper bound w.r.t. $\sharp[\varphi]$. For the rest of the section, let φ be any satisfiable $\mathbf{IFML}^+[k]$-formula, and $\langle \mathcal{M}, w\rangle$ its model.

DEFINITION 13. The *modal depth* of an $\mathbf{IFML}^+[k]$-formula is defined as follows: $md(p_i) = 0 = md(\neg p_i)$; $md(\varphi \vee_n \psi) = max\{md(\varphi), md(\psi)\} = md(\varphi \wedge_n \psi)$; and $md((\Diamond_{i,n}/W_n)\varphi) = md(\varphi) + 1 = md((\Box_{i,n}/W_n)\varphi)$.

Towards a finite model: preparations. For each $n \leq md(\varphi)$, let T^n_w be the set of sequences $(w, i_1, s_1, \ldots, i_m, s_m)$ s.t. $R_{i_1}(w, s_1), \ldots, R_{i_m}(s_{m-1}, s_m)$ for some $m \leq n$ and some $i_1, \ldots, i_m \in \{0, \ldots, k-1\}$. Further, define relations R'_i $(i < k)$ on the set $\bigcup_{n \leq md(\varphi)} T^n_w$ as follows: $(w, i_1, s_1, \ldots, i_m, s_m)\ R'_i\ (w, i'_1, s'_1, \ldots, i'_{m'}, s'_{m'})$ if $m' = m+1$ and $(i_j, s_j) = (i'_j, s'_j)$ $[0 \leq j \leq m]$ and $i'_{m'} = i$. Finally, fix an assignment $\mathfrak{h}'$ by putting, for all $p_i \in$ **prop**: $(w\ i_1, s_1, \ldots, i_m, s_m) \in \mathfrak{h}'(p_i)$ if $s_m \in \mathfrak{h}(p_i)$. We have obtained a tree structure $\langle \mathcal{T}, w\rangle = \langle \bigcup_{n\leq md(\varphi)} T^n_w, R'_0, \ldots, R'_{k-1}, \mathfrak{h}', w\rangle$. In general, $\langle \mathcal{T}, w\rangle$ is *not* a model of φ.

DEFINITION 14. A *φ-label* is a set $\{\sigma_{\varphi_1}, \ldots, \sigma_{\varphi_n}\}$ of partial functions $\sigma_{\varphi_i} : \omega \rightharpoonup \{0,1\}$ such that φ_i is a subformula of φ, and $dom(\sigma_{\varphi_i})$ consists of indices of disjunctions and conjunctions that are superordinate to φ_i in φ. When no confusion may arise, we refer to φ-labels as labels.

We will write $\dot{x} := (w, s_0, \ldots, s_m)$, if $x = (w, i_0, s_0, \ldots, i_m, s_m)$. Let f be a w.s. of $\exists$ in game $G(\varphi, \mathcal{M}, w)$, and define a labeling on $\mathcal{T}$ as follows:

- First let all points in $\mathcal{T}$ be labeled with the empty label $\emptyset$. Then label the root w of $\mathcal{T}$ with $\{\emptyset\}$, and go on as follows.
- If a point x in $\mathcal{T}$ has the label $\Sigma = \{\sigma_{\varphi_1}, \ldots, \sigma_{\varphi_N}\}$ with $\varphi_m = (\Box_{i,n}/W_n)\psi$ $(1 \leq m \leq N)$, and if Σ' is the label of an R_i-successor of x, replace Σ' by $\Sigma' \cup \{\sigma_\psi\}$, where $\sigma_\psi := \sigma_{\varphi_m}$.
- If a point x in $\mathcal{T}$ has the label $\Sigma = \{\sigma_{\varphi_1}, \ldots, \sigma_{\varphi_N}\}$ with $\varphi_m = (\psi \wedge_j \theta)$ $(1 \leq m \leq N)$, replace the label Σ itself by $(\Sigma\backslash\{\sigma_{\varphi_m}\}) \cup \{\sigma_\psi, \sigma_\theta\}$, where $\sigma_\psi := \sigma_{\varphi_m} \cup \{(j,0)\}$ and $\sigma_\theta := \sigma_{\varphi_m} \cup \{(j,1)\}$.
- If a point x in $\mathcal{T}$ has the label $\Sigma = \{\sigma_{\varphi_1}, \ldots, \sigma_{\varphi_N}\}$ with $\varphi_m = (\Diamond_{i,n}/W_n)\psi$ $(1 \leq m \leq N)$, take the R_i-successor $y := f(\dot{x}, \sigma_{\varphi_m})$ of x, and if Σ' is the label of y, replace Σ' by $\Sigma' \cup \{\sigma_\psi\}$, where $\sigma_\psi := \sigma_{\varphi_m}$.
- If a point x in $\mathcal{T}$ has the label $\Sigma = \{\sigma_{\varphi_1}, \ldots, \sigma_{\varphi_N}\}$ with $\varphi_m = (\psi \vee_j \theta)$ $(1 \leq m \leq N)$, take the disjunct $\chi \in \{\psi, \theta\}$ given by $f(\dot{x}, \sigma_{\varphi_m}) \in \{0,1\}$, and replace the label Σ itself by $(\Sigma\backslash\{\sigma_{\varphi_m}\}) \cup \{\sigma_\chi\}$, where $\sigma_\chi := \sigma_{\varphi_m} \cup \{(j, f(\dot{x}, \sigma_{\varphi_m}))\}$.

We define $\mathcal{T}^f$ as the subtree of $\mathcal{T}$ containing precisely those of its elements that have a non-empty label under the above labeling.

Next we turn the labeled tree $\mathcal{T}^f$ into a modal structure $\mathcal{M}^*$ that will provide a model for φ. Specify $\mathcal{M}^* := \langle D^*, R^*_0, \ldots, R^*_{k-1}, \mathfrak{h}^*\rangle$ by setting:

$$D^* := \{w\} \cup \{(n,i,s) : \text{for some } (w, i_1, s_1, \ldots, i_m, s_m) \in T^f,\ n = m \ \&\ i = i_m \ \&\ s = s_m\}.$$
$$\forall j < k\text{: } (n,i,s)\ R^*_j\ (n',i',s') \ \textbf{ if } \ n' = n{+}1 \ \&\ j = i' \ \&\ R_{i'}(s,s').^1$$
$$\forall p_j \in \textbf{prop}\text{: } (n,i,s) \in \mathfrak{h}^*(p_j) \ \textbf{ if } \ s \in \mathfrak{h}(p_j).$$

CLAIM 15. $\mathcal{M}^* \models \varphi[w]$. ■

[1] Here $R_{i'}$ is the relation of type i' of the original model $\langle \mathcal{M}, w\rangle$.

Finite model $\langle \mathcal{N}, v\rangle$ of φ. We move on to build up a *finite* model $\langle \mathcal{N}, v\rangle$ of φ out of the model $\langle \mathcal{M}^*, w\rangle$.

OBSERVATION 16. **(a)** For a fixed φ, the number m_φ of different φ-labels is finite: if r is the number of subformulas of φ, and n is the maximal number of nested conjunctions and disjunctions in φ, then $m_\varphi \leq 2^{r\cdot 2^n}$. **(b)** The labeling of $\mathcal{T}^f$ induces a labeling of $\mathcal{M}^*$. The label of an element $(n, i, s) \in D^*$ is obtained as the union of the labels of elements of the set $T_{(n,i,s)} := \{(w, i_1, s_1, \ldots, i_m, s_m) \in T^f : m = n \;\&\; i_m = i \;\&\; s_m = s\}$. ■

To construct $\langle \mathcal{N}, v\rangle$, we partition $dom(\mathcal{M}^*)$ into $md(\varphi)$ 'levels', the level of an element $(N, i, s) \in D^*$ being given by its 'height', N. On each level we 'identify' as many elements as it goes without affecting the fact that the resulting structure is a model of φ. The identification yields a finite structure. Write $M := md(\varphi)$. We define structures $\mathcal{M}^N := (D^N, R_0^N, \ldots, R_{k-1}^N, \mathfrak{h}^N)$ $[N := 1, \ldots, M]$. First we define an equivalence relation $\equiv_M$ on D^*:

$$\begin{aligned}&(n, i, s) \equiv_M (n', i', s') \textbf{ if } (n, i, s) = (n', i', s') \;\vee\\ &n = n' = M \quad \& \quad i = i' \quad \& \quad \forall p_j \in \textbf{prop}[s \in \mathfrak{h}(p_j) \iff s' \in \mathfrak{h}(p_j)].\end{aligned}$$

The modal structure $\mathcal{M}^M$ is then specified by putting:

$$\begin{aligned}&D^M := D^*/\equiv_M.\\ &[\xi]_{\equiv_M} R_j^M [\zeta]_{\equiv_M} \quad \textbf{if} \quad \text{there is } \zeta' \text{ s.t. } \zeta \equiv_M \zeta' \quad \& \quad R_j^*(\xi, \zeta').\\ &\text{For all } p_i \in \textbf{prop}: \; [\xi]_{\equiv_M} \in \mathfrak{h}^M(p_i) \;\textbf{ if }\; \xi \in \mathfrak{h}^*(p_i).\end{aligned}$$

It can be checked that the definitions of $\mathfrak{h}^M$ and the R_j^M are independent of the choices of representatives of the equivalence classes involved.

OBSERVATION 17. The size of the set $S_M := \{\xi \in D^M : \xi = [(M, i, s)]_{\equiv_M}$ for some $i, s\}$ is finite. In fact, it has at most $k \cdot 2^{p_\varphi}$ elements, where p_φ is the number of propositional atoms actually appearing in φ. ■

CLAIM 18. There is a history-free w.s. for $\exists$ in $G(\varphi, \mathcal{M}^M, [w]_{\equiv_M})$. ■

Fix $N > 1$. Let $D^{M+1} := D^*$. Suppose that if $N \leq K \leq M$, we already have defined: **(a)** an equivalence relation $\equiv_K \subseteq (D^{K+1})^2$; **(b)** a modal structure $\mathcal{M}^K := (D^K, R_0^K, \ldots, R_{k-1}^K, \mathfrak{h}^K)$ with $D^K := D^{K+1}/\equiv_K$; and **(c)** a history-free w.s. f^K for $\exists$ in $G^K := G(\varphi, \mathcal{M}^K, [[w]_{\equiv_M} \ldots]_{\equiv_K})$. We go on to define an equivalence relation $\equiv_{N-1}$ on D^N by putting $\xi \equiv_{N-1} \xi'$ if and only if either $\xi = \xi'$, or else there are n, i, s, n', i', s' such that

$$n = n' = N{-}1 \;\&\; \xi = [[(n, i, s)]_{\equiv_M} \ldots]_{\equiv_N} \;\&\; \xi' = [[(n', i', s')]_{\equiv_M} \ldots]_{\equiv_N},$$

and the following conditions (i) to (v) hold:

(i) $i = i'$.

(ii) For all $p_j \in \mathbf{prop}: s \in \mathfrak{h}^N(p_j) \Longleftrightarrow s' \in \mathfrak{h}^N(p_j)$.

(iii) The points (n, i, s) and (n', i', s') have the same label in $\mathcal{M}^*$.

(iv) For all $i < k$ and all ζ: $R_i^N(\xi, \zeta) \Longleftrightarrow R_i^N(\xi', \zeta)$.

(v) For all subformulas (χ, σ_χ) of φ, and all plays (π'_χ, σ_χ) and $(\pi''_\chi, \sigma_\chi)$ of G^N in which $\exists$ has applied f^N, if the last choices from the domain 'known' to $\exists$ are, respectively, $\pi'_\chi(K) = \xi$ and $\pi''_\chi(K) = \xi'$, and for all $l < K$, $\pi'_\chi(l) = \pi''_\chi(l)$, then $f^N(\pi'_\chi, \sigma_\chi) = f^N(\pi''_\chi, \sigma_\chi)$.

For condition (v), notice that given any subformula (χ, σ_χ) of φ, and any plays (π'_χ, σ_χ) and $(\pi''_\chi, \sigma_\chi)$ of G^N, we necessarily have: $dom(\pi'_\chi) = dom(\pi''_\chi)$. Further, the index K of the last modal operator 'known' to $\exists$ is obtained as follows: $K := min\{x > n : x = \omega$ or $[O_x \prec_\varphi \vee_n$ & O_x is of the form $\Diamond, \Box]\}$, if $\chi := \psi \vee_n \theta$; and $K := min\{x > n : x = \omega$ or $[O_x \prec_\varphi \Diamond_{i,n}/W_n$ & O_x is of the form $\Diamond, \Box$ & $x \notin W_n]\}$, if $\chi := (\Diamond_{i,n}/W_n)\psi$. As f^N is history-free, its values on the plays (π'_χ, σ_χ) and $(\pi''_\chi, \sigma_\chi)$ do not depend on the values $\pi'_\chi(l)$ *resp.* $\pi''_\chi(l)$ for $l > K$. Define $\mathcal{M}^{N-1}$ by setting:

$D^{N-1} := D^N / \equiv_{N-1}$.
$[\xi]_{\equiv_{N-1}} R_j^{N-1} [\zeta]_{\equiv_{N-1}}$ **if** there is ζ' s.t. $\zeta \equiv_{N-1} \zeta'$ and $R_j^N(\xi, \zeta')$.
For all $p_i \in \mathbf{prop}$: $[\xi]_{\equiv_{N-1}} \in \mathfrak{h}^{N-1}(p_i)$ **if** $\xi \in \mathfrak{h}^N(p_i)$.

It can be checked that the definition of the modal structure $\mathcal{M}^{N-1}$ is independent of the choices for representatives of the equivalence classes involved.

CLAIM 19. Let $1 \leq K \leq M$. **(a)** The size of the set $S_K = \{\xi \in D^K : \xi = [[(K, i, s)]_{\equiv_M} \ldots]_{\equiv_K}$ for some $i, s\}$ has a recursive upper bound w.r.t. the length of φ. **(b)** There is a history-free w.s. for $\exists$ in G^K.

Proof. The inductive proof of **(b)** is straightforward. For **(a)**, let us write $N := k \cdot 2^{p_\varphi} \cdot m_\varphi$, where p_φ is the number of propositional atoms appearing in φ, and m_φ is the number of distinct φ-labels. Let $c(K)$ denote the size of the set S_K $(K := 1, \ldots, M)$. By Observation 17, $c(M) \leq N$. For $K := 1, \ldots, M-1$, we have by the construction of $\mathcal{M}^K$: $c(K) \leq N \cdot 2^{c(K+1)} \cdot \Pi_{i=K+1}^{M} c(i)$. Hence statement (a) follows. ■

To finish the construction, we put: $\mathcal{N} := \mathcal{M}^1$ and $v := [[w]_{\equiv_M} \cdots]_{\equiv_1}$.

THEOREM 20. *Let* $k \geq 1$. **(a)** $\mathbf{IFML}^+[k]$ *has the strong finite model property.* **(b)** *The satisfiability problem of* $\mathbf{IFML}^+[k]$ *is decidable.*

Proof. Statement **(b)** is immediate from **(a)**. For **(a)**, suppose $\mathcal{M} \models \varphi[w]$. Let $\langle \mathcal{N}, v \rangle$ be the model constructed out of $\langle \mathcal{M}, w \rangle$ as explained above. By Claim 19 (b), $\mathcal{N} \models \varphi[v]$. And by Claim 19 (a), the size of $\mathcal{N}$ has a recursive upper bound w.r.t. the length of φ. ■

5 Validity Problem

IFML$^+[k]$-formulas are in negation normal form: the symbol $\neg$ appears only as prefixed to propositional atoms. Closing **IFML**$^+[k]$ under $\neg$, and interpreting $\neg$ as initiating a 'role switch' between the players of evaluation games (from verifier to falsifier and *vice versa*), we have: $\exists$ [$\forall$] has a w.s. in $G(\neg\varphi)$ iff $\forall$ [*resp.* $\exists$] has a w.s. in $G(\varphi)$. The negation hence interpreted game-theoretically may be termed *dual negation*. As there are indeterminate **IFML**$^+[k]$-formulas (formulas s.t. neither $\exists$ nor $\forall$ has a w.s. in the associated evaluation game), $\neg$ cannot have the force of *contradictory negation* ($\sim$), whose semantics is simply: $\sim\varphi$ is true in $\langle\mathcal{M}, w\rangle$ iff φ is *not* true in $\langle\mathcal{M}, w\rangle$. So if φ is indeterminate in $\langle\mathcal{M}, w\rangle$, $\sim\varphi$ is true therein.

FACT 21. The satisfiability and validity problems of **IFML**$^+[k]$ are *not* each others' duals.

Proof. If they were duals, we would have: φ is valid iff $\neg\varphi$ is not satisfiable. Let $\psi \in$ **IFML**$^+[k]$ be any formula which is indeterminate in a model $\langle\mathcal{M}, w\rangle$. Then the formula $\neg(\psi \vee \neg\psi)$ is not satisfiable, but still the formula $(\psi \vee \neg\psi)$ is not valid, since it is indeterminate in $\langle\mathcal{M}, w\rangle$. ■

We prove that the validity problem of **IFML**$^+[k]$ is decidable by reducing it to the validity problem of **ML**$[k]$. By Fact 21, this result neither implies nor is implied by Theorem 20. We define a syntactic transformation $\varphi \mapsto \varphi^*$ among **IFML**$^+[k]$-formulas. Recall the definitions of syntactic tree and path in a syntactic tree from *Section* 2. If $\varphi \in$ **IFML**$^+[k]$ and $r \geq 0$, consider all paths of the form $(*)$ $\langle(\Box_{i',n'}/W_{n'}), \beta_1, \ldots, \beta_r, (\Diamond_{i'',n''}/W_{n''})\rangle$ in its syntactic tree, where $\beta_i \in \{\vee, \wedge\}$ and the diamond is independent from the box (i.e. $n' \in W_{n''}$). For each such path, replace the full subtree with the diamond at its root by the symbol $\bot$. The result of these replacements is denoted by φ^*.

CLAIM 22. For any $\varphi \in$ **IFML**$[k]^+$, φ is valid $\Leftrightarrow$ φ^* is valid.

Proof. Let $\varphi \in$ **IFML**$[k]^+$ be arbitrary. The validity of φ^* trivially implies the validity of φ. For the converse, suppose φ is valid. We assume that the syntactic tree of φ contains at least one path of the form $(*)$: otherwise there is nothing to prove. Let $\langle\mathcal{M}, w\rangle = \langle D, R_0, \ldots, R_{k-1}, \mathfrak{h}, w\rangle$ be an arbitrary model. We must show that $\mathcal{M} \models \varphi^*[w]$. Now let $N := 2^{2^{\sharp[\varphi]}}$. We construct a structure $\mathcal{M}^* = (D^*, R_0^*, \ldots, R_{k-1}^*, \mathfrak{h}^*)$ as follows:

$$D^* := \{(\eta, d) : d \in D \text{ and } \eta : n \to N \text{ for some } n < \omega\}$$
$$(\eta, d)\ R_i^*\ (\eta', d') \quad \textbf{if} \quad (d\ R_i\ d' \ \&\ \eta' = \eta^\frown\langle k\rangle \text{ for some } k < N)$$
$$(\eta, d) \in \mathfrak{h}^*(p) \quad \textbf{if} \quad d \in \mathfrak{h}(p).$$

Because φ is valid, we have in particular: $\mathcal{M}^* \models \varphi[(\emptyset, w)]$. Let f be a w.s. of $\exists$ in the corresponding evaluation game. We prove:

SUBCLAIM 23. For any $d_1, \ldots, d_n \in D$ there are numbers $m_1, \ldots, m_n < N$ s.t. we have: *if* a play of $G(\varphi, \mathcal{M}^*, (\emptyset, w))$ is played so that (i) $\exists$ has followed the strategy f, and (ii) $\forall$ has chosen for each box an element $(\eta^\frown\langle m_i\rangle, d_i)$, where i is the least $j \leq n$ with $m_j \notin dom(\eta)$, and the previous element chosen from D^* is (η, d) for some $d \in D$, *then* assuming that the play has gone according to the rules, there is no path of the form (*) in the syntactic tree of φ s.t. the play has proceeded to the diamond $(\Diamond_{i'',n''}/W_{n''})$.

Proof of the Subclaim. The proof is by induction on the number n of boxes reached in a play of game $G(\varphi, \mathcal{M}^*, (\emptyset, w))$. The base case of 0 boxes holds trivially. Suppose then that the claim holds for the case of n boxes. Let h be a play of game $G(\varphi, \mathcal{M}^*, (\emptyset, w))$ involving choices interpreting $n+1$ boxes. Suppose that it is $\forall$'s turn to move at h to interpret the $(n+1)$-th box. Assume that $\exists$ has followed f when playing h. Given that the first n moves from D^* by $\forall$ are $(\eta^\frown\langle m_i\rangle, d_i)$ $[i := 1, \ldots, n]$ and that they are made according to the game rules, by the induction hypothesis we may assume that the numbers m_i are chosen so that the play h has not proceeded along any path of the form (*) to a diamond $(\Diamond_{i'',n''}/W_{n''})$.

Let $d_{n+1} \in D$ be any element s.t. $d \; R_i \; d_{n+1}$, where $d \in D$ is given by the previous choice (η, d) from D^* in h. Consider the different elements $(\eta^\frown\langle p\rangle, d_{n+1})$ with $p < N$ that $\forall$ may choose, following the game rules, to interpret the $(n+1)$-th box $(\Box_{i',n'}/W_{n'})$. Suppose from $(\Box_{i',n'}/W_{n'})$ there begins a path of the form (*) in the syntactic tree of φ. The sequence $\beta_1, \ldots, \beta_r$ contains some finite number $m \leq r$ of conjunctions; so there are at most $2^{2^m} < 2^{2^{\sharp[\varphi]}}$ different ways in which the values of the strategy f can be chosen for the disjunctions (if any) among $\beta_1, \ldots, \beta_r$. Hence there are numbers $p_1, p_2 < N$ with $p_1 \neq p_2$, and plays h_1, h_2, s.t. up to the box $(\Box_{i',n'}/W_{n'})$, h_1 and h_2 both coincide with h; in h_1 the box is interpreted as $(\eta^\frown\langle p_1\rangle, d_{n-1})$ and in h_2 as $(\eta^\frown\langle p_2\rangle, d_{n+1})$; and proceeding from both points $(\eta^\frown\langle p_1\rangle, d_{n+1})$ and $(\eta^\frown\langle p_2\rangle, d_{n+1})$, the strategy f gives the same choices for the disjunctions (if any) in $\beta_1, \ldots, \beta_r$. Choose $m_{n+1} := min\{p_1, p_2\}$. For a contradiction, suppose there are choices for the conjunctions in $\beta_1, \ldots, \beta_r$, and replies to them by f, so that these choices $\bar{b} \in \{0,1\}^r$ extend both h_1 and h_2 in such a way that the diamond $(\Diamond_{i'',n''}/W_{n''})$, which is independent from the box $(\Box_{i',n'}/W_{n'})$, is reached. As f is uniform, $f(h_1{}^\frown\bar{b}) = f(h_2{}^\frown\bar{b})$. But then f is not a w.s. for $\exists$, since from distinct points in $\mathcal{M}^*$ only distinct points are accessible along R_i^*. ■

Given the truth of Subclaim 23, the strategy f clearly induces a w.s. for $\exists$ in game $G(\varphi, \mathcal{M}, w)$, and so $\mathcal{M} \models \varphi^*[w]$. Hence Claim 22 follows. ■

THEOREM 24. *For all $k \geq 1$, $\mathbf{IFML}^+[k]$-validity is decidable.*

Proof. Define a syntactic transformation $\varphi \mapsto \varphi^-$ from $\mathbf{IFML}[k]^+$-formulas to $\mathbf{ML}[k]$-formulas by letting φ^- be the result of replacing any operator of the form $(O_{i,n}/W_n)$ in the formula φ by $O_{i,n}$. By Theorem 6 it is immediate that for any $\varphi \in \mathbf{IFML}[k]^+$ and for any model $\langle \mathcal{M}, w \rangle$, $\mathcal{M} \models \varphi^*[w]$ iff $\mathcal{M} \models (\varphi^*)^-[w]$, where * stands for the above-defined transformation $\varphi \mapsto \varphi^*$ among $\mathbf{IFML}^+[k]$-formulas. The transformations $\varphi \mapsto \varphi^* \mapsto (\varphi^*)^-$ are effective; so to decide whether a given $\varphi \in \mathbf{IFML}^+[k]$ is valid, it suffices to apply an algorithm deciding $\mathbf{ML}[k]$-validity to the formula $(\varphi^*)^-$. ■

COROLLARY 25. *$\mathbf{IFML}^+[k]$-validity is decidable in PSPACE.*

Proof. The statement follows from the proof of Theorem 24 by the fact that the complexity of $\mathbf{ML}[k]$-validity is in PSPACE (cf. [8]). ■

6 Concluding Remarks

Related work. *Independence-Friendly (IF) first-order logic* was introduced by Hintikka and Sandu in [10]. Its basic properties were studied by Sandu in [13]. The idea behind this logic is to allow arbitrary relations of logical (in)dependence between quantifiers. One way of thinking how to accomplish this is by separating the syntactic relation of subordinateness and the semantic relation of logical dependence, so that a quantifier can be logically independent from some quantifiers in whose syntactic scope it nevertheless lies. IF first-order sentences coincide in expressive power with Σ^1_1-sentences, whence IF first-order logic is a proper extension of first-order logic. This logic has been discussed in connection with foundations of mathematics, e.g. in the monograph [9] of Hintikka, and in the papers [11] and [12] by Hyttinen and Sandu. Sandu and Väänänen showed in [15] how to extend the idea of informational independence from the case of quantifiers to propositional connectives. Bradfield was in [2] the first to define a version of IF modal logic; his logic was designed to be used when studying transition systems with concurrency. With Fröschle, he studied further properties of this logic in [5]. In [3] and [4], Bradfield has extended IF first-order logic into an IF fixpoint logic, and shown how to obtain an IF modal μ-calculus.

Motivation. Given the existence of IF first-order logic and the fact that its expressive power exceeds that of usual first-order logic, a systematic question arises: how to formulate an IF modal logic. A natural approach is to modify basic modal logic ($\mathbf{ML}[k]$) so as to allow modal operators to be logically independent from syntactically superordinate dual operators. The relevant notion of logical independence is implemented by the requirement of uniformity of players' winning strategies. In the corresponding evaluation games, a player may be ignorant of earlier moves of the opponent, but always remembers his or her own previous moves. Allowing this kind of independence in a modal logical setting typically allows expressing the existence of a *common* endpoint of several transitions: for instance the requirement that various computations end in the *same* state becomes expressible. More general motivation for studying IF modal logics is that they provide one possible, and arguably natural, way of producing relatively expressive modal languages. Such languages are interesting not only for modal logic itself, but in particular for finite model theory, in which modal logics are of interest in connection with different definability issues (cf., e.g., [7]).

Changing $\mathbf{ML}[k]$ by allowing more logical independencies results in powerful logics. As noted in the present paper, $\mathbf{IFML}^+[k]$ has greater expressive power than $\mathbf{ML}[k]$. On the other hand, it is possible to show that this logic is translatable into first-order logic. (In [17] a translation is provided to a fragment of $\mathbf{IFML}^+[k]$; the proof of the general result can be found in [14].) By contrast, in [17] it is shown that generalizing $\mathbf{IFML}^+[k]$ by allowing diamonds (boxes) to be independent from conjunctions (disjunctions) results in a modal logic which is no longer translatable into first-order logic, but has in fact some genuine *second-order* expressive power (i.e. can express properties of modal structures which can be expressed of the corresponding first-order structures only by genuine second-order formulas).

Open problems. In the present paper the 'perfect recall' fragment $\mathbf{IFML}^+[k]$ of the IF modal logic (**IFML**) of [16] was studied. Which of the two logics better deserves to be termed 'IF modal logic'? The answer depends on whether or not it makes sense to speak of a diamond (box) being *logically independent* from another diamond (*resp.* box). It is not obvious how to separate cases where the game-theoretical framework merely serves to model logical phenomena from cases where this tool adds to the picture something that does not really belong to the domain of logic.

The upper bound for the complexity of $\mathbf{IFML}^+[k]$-satisfiability given by the proof of Theorem 20 is far from feasible. (The obtained recursive bound of the size of the finite model of φ has the form of tower function w.r.t. the length of φ.) The complexity is at least PSPACE, because $\mathbf{ML}[k]$ is embeddable in $\mathbf{IFML}^+[k]$, and $\mathbf{ML}[k]$ has PSPACE-complete satisfiability problem w.r.t. $\mathcal{C}_k$ (cf. [8]). The exact complexity of $\mathbf{IFML}^-[k]$-satisfiability is an open question. It is also an open problem whether the satisfiability and validity problems of the logics $\mathbf{IFML}^+_{12}[k]$ and $\mathbf{IFML}[k]$ are decidable. However, if their satisfiability problems are decidable at all, by Theorems 12 and 10 this cannot be proven in the same way as Theorem 20, i.e. relying on the existence of history-free winning strategies. We pointed out in *Section* 2 that IF modal logic of perfect recall would be most naturally formulated as a logic whose semantics obeys the principle "a player never forgets a move he has once known." This would involve allowing disjunctions (conjunctions) independent from boxes (diamonds). Studying this general version of IF modal logic of perfect recall is left for future research.

Acknowledgements

We wish to thank the anonymous referees for their valuable comments.

BIBLIOGRAPHY

[1] P. Blackburn & al., 2002. *Modal Logic*, Cambridge University Press, Cambridge.
[2] J. Bradfield, 2000. "Independence: Logics and Concurrency," LNCS 1862.
[3] J. Bradfield, 2004a. "Parity of Imperfection" in *Proc. CSL 2003*, LNCS 2803, 72-85.
[4] J. Bradfield, 2004b. "On Independence-Friendly Fixpoint Logics," *Philosophia Scientiae* 8(2), 125-44.
[5] J. Bradfield & S. Fröschle, 2002. "Independence-Friendly Modal Logic and True Concurrency" in *Nordic Journal of Computing*, vol. 9, No. 2, 102-117.
[6] H.-D. Ebbinghaus & J. Flum, 1999. *Finite Model Theory*, Springer-Verlag.
[7] E. Grädel & al., 2004. *Finite Model Theory and Its Applications*, Springer-Verlag.
[8] J. Halpern & Y. Moses, 1992. "A guide to completeness and complexity for modal logics of knowledge and belief", *Artificial Intelligence* 54(3), 319-79.
[9] J. Hintikka, 1996. *The Principles of Mathematics Revisited*, Cambridge University Press, New York.
[10] J. Hintikka & G. Sandu, 1989. "Informational Independence as a Semantical Phenomenon," in Fenstad et al. (eds.), *Logic, Methodology and Philosophy of Science*, vol. 8, Amsterdam, North-Holland, 571-589.
[11] T. Hyttinen & G. Sandu, 2000. "Henkin Quantifiers and the Definability of Truth," *Journal of Philosophical Logic*, vol. 29, 507-527.
[12] T. Hyttinen & G. Sandu, 2001. "IF-logic and foundations of mathematics," *Synthese*, vol. 126, 37-47.
[13] G. Sandu, 1993. "On the Logic of Informational Independence and its Applications," *Journal of Philosophical Logic*, 22, 29-60.
[14] G. Sandu & T. Tulenheimo, 2005. *Logics of Imperfect Information*, manuscript.
[15] G. Sandu & J. Väänänen, 1992. "Partially ordered connectives," *Zeitschrift für Matematische Logik und Grundlagen der Mathematik*, 38, 361-372.
[16] T. Tulenheimo, 2003. "On IF Modal Logic and its Expressive Power" in Balbiani & al. (eds.): *Advances in Modal Logic*, vol. 4, King's College Publications, London, 475-98.
[17] T. Tulenheimo, 2004. *Independence-Friendly Modal Logic*, Philosophical Studies from the University of Helsinki vol. 4 (Doctoral dissertation).

Tapani Hyttinen
Department of Mathematics and Statistics
P.O. Box 68 (Gustaf Hällströmin katu 2b)
University of Helsinki
FIN-00014, Finland
E-mail: tapani.hyttinen@helsinki.fi

Tero Tulenheimo
Department of Philosophy
P.O. Box 9 (Siltavuorenpenger 20 A)
University of Helsinki
FIN-00014, Finland
E-mail: tero.tulenheimo@helsinki.fi

A Lower Complexity Bound for Propositional Dynamic Logic with Intersection

MARTIN LANGE

ABSTRACT. This paper shows that satisfiability for Propositional Dynamic Logic with Intersection is EXPSPACE-hard. The proof consists of a reduction from the word problem for alternating, exponential time bounded Turing Machines.

1 Introduction

Propositional Dynamic Logic, PDL, was defined in [5] to reason about program behaviour but is nowadays mainly interesting because of its connection to other logics like Description Logics [18] or Information Logics [15, 4] for example. It is an extension of multi-modal logic where modalities take as arguments elements of a Kleene Algebra with a test operator.

PDL enjoys nice algorithmic properties: its model checking problem is PTIME-complete and solvable in linear running time, its satisfiability problem is complete for EXPTIME [5, 16, 17]. PDL is embeddable into infinitary multi-modal logic and thus has the tree model property. It also has the finite model property [8]. It is finitely axiomatisable [19, 11]. However, it is rather weak in expressive power since it is strictly less expressive than the alternation-free fragment of Kozen's modal μ-calculus [10].

Several variants of PDL have been studied since, for example restrictions to deterministic atomic programs, etc. Most variants aim at extending the set of operators in the underlying Kleene Algebra in order to allow properties of more programs to be expressed. Examples of these are loop constructs, the converse operator [20], an interleaving operator [13], etc. In most cases axiomatisations and decision procedures for these extension can be obtained by extending PDL's axiom system and its decision procedures.

One variant for which this approach fails entirely is PDL with Intersection, IPDL [7]. Using the connection to Description Logics mentioned above, IPDL can be seen as a Description Logic which can express intersection of roles. Although intersection looks like yet another regular operation it cannot be defined using the operators of a Kleene Algebra. This is because it does not retain the tree model property. In addition, it also does not have the finite model property which makes it a good candidate for an undecidable logic.

Nevertheless, IPDL is only undecidable if atomic programs are required to be deterministic [7]. If they are allowed to be non-deterministic then its satisfiability problem is decidable [3]. Danecki showed that it can be decided in double exponential time. This is proved by constructing Büchi tree automata for IPDL formulas. They do not work on the formula's models directly but on trees describing these models. The size of an automaton obtained from a formula is doubly exponential in the formula's size, and the emptiness problem for non-deterministic Büchi automata is solvable in polynomial time.

These are the only complexity results about IPDL so far. It is not known whether there is a better decision procedure for IPDL or whether it is complete for double exponential time.

Complexity issues for modal logics containing the intersection operator have been addressed in [12] for instance. However, there modalities take elements of a Boolean algebra rather than a Kleene algebra like PDL does.

Fragments of IPDL have been studied in [6, 14, 1] for instance, mainly regarding the issue of axiomatisability. They also show that the presence of the intersection operator makes IPDL a "strange" logic compared to PDL. For instance, it loses both the tree model property and the finite model property.

In this paper we show that the satisfiability problem for IPDL is hard for exponential space. The proof consists of a reduction from the word problem for alternating exponential time bounded Turing Machines. This is inspired by [21] where satisfiability of the temporal logic CTL* is shown to be hard for double exponential time.

2 Preliminaries

2.1 Propositional Dynamic Logic with Intersection

Let $\mathcal{P} = \{p, q, \ldots\}$ be a finite set of *propositional constants* which includes $\mathtt{tt}$ and $\mathtt{ff}$. Let $\mathcal{A} = \{a, b, \ldots\}$ be a finite set of *atomic program names.* A *Kripke structure* is a triple $(\mathcal{S}, \{\xrightarrow{a} \mid a \in \mathcal{A}\}, L)$ with $\mathcal{S}$ being a set of *points,* $\xrightarrow{a}$ for every $a \in \mathcal{A}$ is a binary relation on points, and $L : \mathcal{S} \to 2^{\mathcal{P}}$ labels the points with propositions, s.t. for all $s \in \mathcal{S}$: $\mathtt{tt} \in L(s)$ and $\mathtt{ff} \notin L(s)$.

Formulas φ and *programs* α of IPDL are defined as follows:

$$\begin{array}{lcl} \varphi & ::= & q \mid \varphi \vee \varphi \mid \neg\varphi \mid \langle\alpha\rangle\varphi \\ \alpha & ::= & a \mid \alpha \cup \alpha \mid \alpha \cap \alpha \mid \alpha;\alpha \mid \alpha^* \mid \varphi? \end{array}$$

where q ranges over $\mathcal{P}$, and a ranges over $\mathcal{A}$.

We will use the standard abbreviations $\varphi \wedge \psi := \neg(\neg\varphi \vee \neg\psi)$, $\varphi \rightarrow \psi := \neg\varphi \vee \psi$, $[\alpha]\varphi := \neg\langle\alpha\rangle\neg\varphi$ and $\alpha^+ := \alpha; \alpha^*$.

IPDL formulas are interpreted over Kripke structures. Given a Kripke structure $\mathcal{T}$, the semantics of an IPDL formula is defined by simultaneous induction on the structure of programs and formulas:

$$
\begin{array}{lll}
s \xrightarrow{\alpha;\beta} t & \text{iff} & \exists u \in \mathcal{S} \text{ s.t. } s \xrightarrow{\alpha} u \text{ and } u \xrightarrow{\beta} t \\
s \xrightarrow{\alpha\cup\beta} t & \text{iff} & s \xrightarrow{\alpha} t \text{ or } s \xrightarrow{\beta} t \\
s \xrightarrow{\alpha\cap\beta} t & \text{iff} & s \xrightarrow{\alpha} t \text{ and } s \xrightarrow{\beta} t \\
s \xrightarrow{\alpha^*} t & \text{iff} & \exists n \in \mathbb{N}, s \xrightarrow{\alpha^n} t \text{ where} \\
 & & \forall s, t \in \mathcal{S} : s \xrightarrow{\alpha^0} s, \text{ and } s \xrightarrow{\alpha^{n+1}} t \text{ iff } s \xrightarrow{\alpha;\alpha^n} t \\
s \xrightarrow{\varphi?} s & \text{iff} & \mathcal{T}, s \models \varphi
\end{array}
$$

where

$$
\begin{array}{lll}
\mathcal{T}, s \models q & \text{iff} & q \in L(s) \\
\mathcal{T}, s \models \varphi \vee \psi & \text{iff} & \mathcal{T}, s \models \varphi \text{ or } \mathcal{T}, s \models \psi \\
\mathcal{T}, s \models \neg\varphi & \text{iff} & \mathcal{T}, s \not\models \varphi \\
\mathcal{T}, s \models \langle\alpha\rangle\varphi & \text{iff} & \exists t \in \mathcal{S} \text{ s.t. } s \xrightarrow{\alpha} t \text{ and } \mathcal{T}, t \models \varphi
\end{array}
$$

2.2 Alternating Turing Machines.

We use the following model of an alternating Turing Machine, which differs slightly from the standard model [2] in the way that it either moves its head or it writes a symbol and branches existentially or universally. It is not hard to see that this model is equivalent to the standard one w.r.t. the standard time and space complexity classes.

An alternating Turing Machine $\mathcal{M}$ is of the form $\mathcal{M} = (Q, \Sigma, q_0, q_{acc}, \delta)$, where Q is the set of states, Σ is the alphabet which contains a blank symbol $\square$, and $q_0, q_{acc} \in Q$ – the starting and the accepting state. W.l.o.g. we assume that $q_0 \neq q_{acc}$.

Q is partitioned into $Q_\exists \cup Q_\forall \cup Q_m \cup \{q_{acc}\}$, where we write Q_b for $Q_\exists \cup Q_\forall$, these are the *branching* states. We assume that $\mathcal{M}$, in order to accept an input, moves to the left end of the tape once its computation is finished and enters state q_{acc}.

The transition relation δ is of the form

$$\delta \subseteq \big(Q_b \times \Sigma \times Q \times \Sigma\big) \cup \big(Q_m \times \Sigma \times Q \times \{L, R\}\big) .$$

We also write $(q', b) \in \delta(q, a)$ to denote $(q, a, q', b) \in \delta$ for given q and a.

In a branching state $q \in Q_b$, the machine can overwrite the symbol under the head but not move the head. Furthermore, it can branch non-deterministically or universally. In a state $q \in Q_m$, the machine acts deterministically and moves its head, i.e., for each $a \in \Sigma$, there is exactly one transition $(q, a, q', D) \in \delta$, for $q' \in Q$ and $D \in \{L, R\}$, meaning that the head moves to the left (L) or right (R), and the machine enters state q'.

A configuration of a Turing Machine is a snapshot in time consisting of the actual state that the machine is in, the current content of the tape, and the position of the tape head. The initial configuration consists of q_0, the leftmost position of the tape, and the input word written on the tape followed by blank symbols.

$\mathcal{M}$ accepts the input w if its initial configuration is accepting. This is denoted $w \in L(\mathcal{M})$. The configurations' acceptance behaviour depends on the kind of state:

- If the state is q_{acc} then the configuration is accepting.
- If the state is in $Q_m \setminus \{q_{acc}\}$, then the configuration is accepting iff its unique successor is accepting.
- If the state is in $Q_\exists$, then the configuration is accepting iff at least one of its successors is accepting.
- If the state is in $Q_\forall$, then the configuration is accepting iff all of its successors are accepting.

The entire computation is accepting if the initial configuration is. Note that an accepting computation can be represented as a tree of configurations with the starting configuration as its root, and where every existential and deterministic configuration has exactly one successor whereas all possible successors of a universal configuration are present in the tree.[1]

2.3 Complexity classes

We will quickly recall the definitions of the complexity classes used in this paper. Let DTIME($f(n)$), resp. DSPACE($f(n)$), be the classes of problems that can be decided by a deterministic Turing Machine in time $f(n)$, resp. with space $f(n)$, where n is the size of the input to the machine. Equally, let ATIME($f(n)$) and ASPACE($f(n)$) be the classes of problems that can be decided by an alternating Turing Machine in time $f(n)$, resp. with space $f(n)$.

[1]If computations of alternating Turing Machines are regarded as a game then such a witness is nothing more than a winning strategy for the existential player.

The classes we refer to are those of exponential time, space and double exponential time. They are defined as

$$\begin{aligned} \text{EXPTIME} &:= \bigcup \text{DTIME}(2^{p(n)}) \\ \text{EXPSPACE} &:= \bigcup \text{DSPACE}(2^{p(n)}) \\ 2\text{-EXPTIME} &:= \bigcup \text{DTIME}(2^{2^{p(n)}}) \end{aligned}$$

where $p(n)$ ranges over all polynomials in n. Furthermore, the classes of problemss decidable in alternating exponential time, resp. space, are

$$\begin{aligned} \text{AEXPTIME} &:= \bigcup \text{ATIME}(2^{p(n)}) \\ \text{AEXPSPACE} &:= \bigcup \text{ASPACE}(2^{p(n)}) \end{aligned}$$

THEOREM 1. [2] *AEXPTIME = EXPSPACE, AEXPSPACE = 2-EXPTIME.*

3 The Reduction

Let $\mathcal{M} = (Q, \Sigma, q_0, q_{acc}, \delta)$ be an alternating and exponential time bounded Turing Machine. Assume that it has been tailored to obey the restrictions laid out in the previous section. Let $w = a_0 \dots a_{n-1} \in \Sigma^*$ be an input for $\mathcal{M}$. We assume the space used by $\mathcal{M}$ on w to be bounded by $2^{p(n)}$ for some polynomial p. Let $N := 2^{p(n)} - 1$ be the maximal index of a tape cell used in the computation, and $P := p(n) - 1$ the maximal index of a bit needed for a counter that can store values between 0 and N.

In the following we will construct an IPDL formula $\varphi_{\mathcal{M},w}$ over a singleton set $\mathcal{A} = \{x\}$ of atomic programs s.t. $w \in L(\mathcal{M})$ iff $\varphi_{\mathcal{M},w}$ is satisfiable. Informally, a witness for an accepting computation of $\mathcal{M}$ on w will serve as a model for $\varphi_{\mathcal{M},w}$.

The encoding of $\mathcal{M}$'s configurations is very similar to the ones presented in [21] or [9]. A configuration of the form

a_0	a_1	a_2	$\dots$	a_N

with the tape head on, say, the third cell from the left and the machine in state q is modeled by a sequence of states of the form

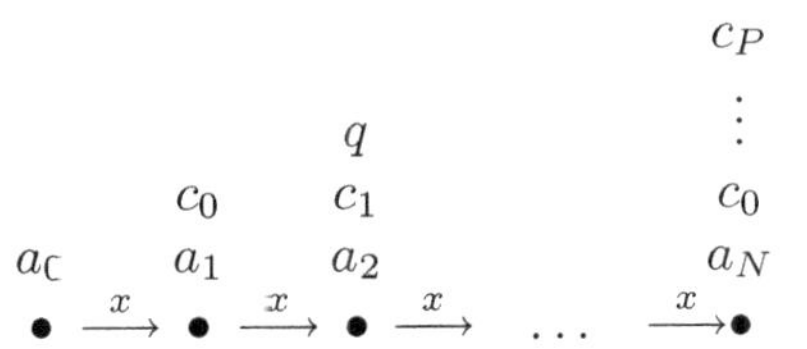

Here, the counter bits for D are left out to avoid clutter. Successive configurations are modeled by concatenating these sequences.

The following propositions are used.

$$\mathcal{P} = Q \cup \Sigma \cup \{c_0, \ldots, c_P\} \cup \{d_0, \ldots, d_P\}$$

- $q \in Q$ is true in a state of the model iff the head of $\mathcal{M}$ is on the corresponding tape cell in the corresponding configuration while the machine is in state q. The formula $h := \bigvee_{q \in Q} q$ says that the tape head is on the current cell.

- $a \in \Sigma$ is true iff a is the symbol on the corresponding tape cell in the corresponding configuration.

- $c_P, \ldots, c_0$ represent a counter C in binary coding. The counter value will be 0 in the leftmost and N in the rightmost tape cell for instance.

- $d_P, \ldots, d_0$ represent a counter D. The value of this counter will be the index of the actual configuration (i.e. the distance to the starting configuration in $\mathcal{M}$'s computation tree).

For every fixed $m \in \{0, \ldots, N\}$ we can write a formula $\chi_{C=m}$, resp. $\chi_{D=m}$, which says that the counter value of C or D is m in the current state, e.g.

$$\chi_{C=0} := \bigwedge_{i=0}^{P} \neg c_i \ , \ \chi_{C=1} := c_0 \wedge \bigwedge_{i=1}^{P} \neg c_i \text{ and } \chi_{C=N} := \bigwedge_{i=0}^{P} c_i$$

for the leftmost ($m = 0$), second ($m = 1$) and rightmost ($m = N$) position in a configuration.

For the next part we need auxiliary formulas $\varphi^C_{inc}, \varphi^D_{inc}$ that increase the value of C, resp. D, as long as they do not equal N.

$$\varphi^C_{inc} := \bigvee_{i=0}^{P} (\neg c_i \wedge [x]c_i \wedge \bigwedge_{j<i} (c_j \wedge [x]\neg c_j) \wedge \bigwedge_{j>i} (c_j \rightarrow [x]c_j) \wedge (\neg c_j \rightarrow [x]\neg c_j))$$

This is just the standard formula for incrementing a binary value: for some $i \in \{0, \ldots, P\}$ the i-th bit changes from 0 to 1, all higher bits remain the same and all lower bits change from 1 to 0. We also need to be able to say that the value of D remains unchanged.

$$\varphi^D_{remain} := \bigwedge_{i=0}^{P} (d_i \rightarrow [x]d_i) \wedge (\neg d_i \rightarrow [x]\neg d_i)$$

Then we can write down a formula which requires the counter C to be increased by one modulo $N+1$ in every move from a state to a successor. At the same time, counter D is incremented iff the value of C changes from N back to 0.

$$
\begin{aligned}
\varphi_{count} \;:=\; [x^*](\; (\; \chi_{C=N} \;\wedge\; [x]\chi_{C=0} \;\wedge\; \varphi^D_{inc} \;) \;\vee \\
(\; \neg\chi_{C=N} \;\wedge\; \varphi^C_{inc} \;\wedge\; \varphi^D_{remain} \;)\;)
\end{aligned}
$$

We form programs

- α_{rest} which goes from any state to the beginning of the next configuration, i.e. it traverses the rest of the current configuration,

$$\alpha_{rest} \;:=\; (\neg\chi_{C=N}?; x)^*; \chi_{C=N}?$$

- α_{2h}, α_{0h} which do an arbitrary amount of x-actions whilst seeing at least two, resp. no tape heads.

$$
\begin{aligned}
\alpha_{2h} \;&=\; x^*; h?; x^+; h?; x^* \\
\alpha_{0h} \;&:=\; (\neg h?; x)^*; \neg h?
\end{aligned}
$$

Then we can formalise the general requirements on a Turing Machine: every tape cell is marked with exactly one symbol from Σ and never with two different states; no configuration has more or less than one cell marked with the tape head.

$$
\begin{aligned}
\varphi_{gen} \;:=\; [x^*](\; (\bigvee_{a\in\Sigma} a\,) \;\wedge\; \bigwedge_{a,b\in\Sigma,b\neq a} \neg(a\wedge b) \;\wedge\; \bigwedge_{q,q'\in Q,q\neq q'} \neg(q\wedge q') \\
\wedge\; (\; \chi_{C=0} \;\rightarrow\; [\alpha_{rest} \cap (\alpha_{2h} \cup \alpha_{0h})]\mathtt{ff} \;)\;)
\end{aligned}
$$

At the beginning, the input word $w = a_0 \ldots a_{n-1}$ is written on the tape, followed by blank symbols $\Box$ until a state with counter value 0 is reached again.

$$
\begin{aligned}
\varphi_{start} \;:=\; & \chi_{C=0} \;\wedge\; q_0 \;\wedge\; a_0 \;\wedge \\
& [x](\; a_1 \;\wedge \\
& \quad [x](\; a_2 \;\wedge \\
& \qquad \ldots \;\wedge\; \ldots \\
& \qquad\quad [x](\; a_{n-1} \;\wedge \\
& \qquad\qquad [(x; \neg\chi_{C=0}?)^+]\Box\;)\ldots))
\end{aligned}
$$

Next we give a formula which expresses the fact that $\mathcal{M}$'s computation is accepting. Note that we assume $\mathcal{M}$ to move its head to the very left, go into state q_{acc}, and leave the head there once its computation is finished.

$$\varphi_{acc} \;:=\; [x^*](\chi_{C=0} \wedge \chi_{D=N} \rightarrow q_{acc})$$

Before we can encode $\mathcal{M}$'s transition function δ, we need to write programs that relate a tape cell in one configuration to itself or its neighbours in the next configuration. Program $\alpha_{0!}$ runs the atomic program x arbitrarily often whilst seeing the counter value $C = 0$ only once.

$$\alpha_{0!} \;:=\; (x; \neg\chi_{C=0}?)^*; x; \chi_{C=0}?; (x; \neg\chi_{C=0}?)^*$$

Using this program and the usual trick of incrementing, resp. decrementing a binary counter we can write a program α_{-1} that relates a tape cell with its left neighbour in the following configuration.

$$\begin{aligned}\alpha_{-1} \;:=\; \neg\chi_{C=0}?; \alpha_{0!} \;\cap\; \bigcup_{i=0}^{P} (\; c_i?; x^+; \neg c_i? \;\cap\; \bigcap_{j<i} \neg c_j?; x^+; c_j? \\ \cap\; \bigcap_{j>i} (\; c_j?; x^+; c_j? \cup \neg c_j?; x^+; \neg c_j? \;)\;)\end{aligned}$$

Equally, $\alpha_=$ and α_{+1} turn it into itself, resp. its right neighbour.

$$\begin{aligned}\alpha_= \;&:=\; \alpha_{0!} \;\cap\; \bigcap_{i=0}^{P} (\; c_i?; x^+; c_i? \;\cup\; \neg c_i?; x^+; \neg c_i? \;) \\ \alpha_{+1} \;&:=\; \neg\chi_{C=N}?; \alpha_=; x\end{aligned}$$

Now we are able to formalise δ's transitions in IPDL. Again, there are different requirements on the model depending on the nature of a machine's state.

In an existential state, there is one transition that determines all successors. In a universal state, all successors behave according to one of the possible transitions and every possible transition is present in the computation tree. In a moving state the machine acts deterministically, hence, all successor configurations must be the same. In all cases we need to require the existence of a successor configuration, though.

Finally, the label of any tape cell which is not under the tape head remains the same in the following configuration.

$$\begin{aligned}
\varphi_\delta \;:=\; & [x^*](\bigwedge_{q\in Q_\exists, a\in\Sigma} (q \wedge a \;\rightarrow\; \bigvee_{(p,b)\in\delta(q,a)} [\alpha_=](p\wedge b) \;\wedge\; \langle\alpha_=\rangle(p\wedge b) \,) \\
& \wedge \bigwedge_{q\in Q_\forall, a\in\Sigma} (q \wedge a \;\rightarrow\; (\, [\alpha_=] \bigvee_{(p,b)\in\delta(q,a)} (p\wedge b) \,) \;\wedge\; \bigwedge_{(p,b)\in\delta(q,a)} \langle\alpha_=\rangle(p\wedge b) \,)\,) \\
& \wedge \bigwedge_{(q,a,q',L)\in\delta} (\neg\chi_{C=0} \wedge q \wedge a \;\rightarrow\; \langle\alpha_{-1}\rangle q' \wedge [\alpha_{-1}] q' \,) \\
& \wedge \bigwedge_{(q,a,q',R)\in\delta} (\neg\chi_{C=N} \wedge q \wedge a \;\rightarrow\; \langle\alpha_{+1}\rangle q' \wedge [\alpha_{+1}] q' \,) \\
& \wedge \bigwedge_{a\in\Sigma} (\neg h \wedge a \;\rightarrow\; [\alpha_=] a \,)\,)
\end{aligned}$$

Altogether, the machine's behaviour is described by the formula

$$\varphi_{\mathcal{M},w} \;:=\; \varphi_{count} \wedge \varphi_{gen} \wedge \varphi_{start} \wedge \varphi_{acc} \wedge \varphi_\delta$$

LEMMA 2. *Let $\mathcal{M}$ be an alternating, exponential time bounded Turing machine, and $w \in \Sigma^*$. If $w \in L(\mathcal{M})$ then $\varphi_{\mathcal{M},w}$ is satisfiable.*

Proof. Suppose there is a successful computation of $\mathcal{M} = (Q, \Sigma, q_0, q_{acc}, \delta)$ on $w = a_0 \dots a_{n-1}$, and assume that $\mathcal{M}$ decides w in time $2^{p(n)}$ for some polynomial p. The computation can be represented by a tree of configurations with the starting configuration C_0 as the root, and successor configurations as sons of a configuration in the tree. Let $\mathcal{C}$ be the set of all configurations occurring in this computation. We write $C' \in Succ(C)$ to denote that C' is a successor of C in this tree.

Since $\mathcal{M}$'s computation time is bounded by $2^{p(n)}$, so is its space. Hence, the tree of configurations can be refined into a tree of tape cells with each cell being the son of its left neighbour. The leftmost tape cell is the son of the rightmost tape cell of the previous configuration.

Formally, let $V := \{(C, i) \mid C \in \mathcal{C}, i \in \{0, \dots, 2^{p(n)} - 1\}\}$. Define a binary relation $\xrightarrow{x}$ on V as

$$(C,i) \xrightarrow{x} (C',j) \text{ iff } \begin{cases} C = C', j = i+1, & \text{if } i < 2^{p(n)} - 1 \\ C' \in Succ(C), j = 0, & \text{if } i = 2^{p(n)} - 1 \end{cases}$$

which can be extended to programs over x in the usual way.

Finally, define a labeling function $L : V \rightarrow 2^{\mathcal{P}}$ satisfying the following and not more.

$a \in L(C,i)$	if the symbol on cell i in configuration C is a
$q \in L(C,i)$	if the head is on cell i in configuration C and the machine is in state q
$c_j \in L(C,i)$	if the j-th bit of i in binary coding is 1
$d_j \in L(C,i)$	if there are $n \leq 2^{p(n)} - 1$, and $C_1, \ldots, C_n \in \mathcal{C}$ s.t. $C = C_n$ and for all $k = 0, \ldots, n-1 : C_{k+1} \in Succ(C_k)$, and the j-th bit of n in binary coding is 1

Now let $T := (V, \{\xrightarrow{x}\}, L)$. We will show that $T, (C_0, 0) \models \varphi_{\mathcal{M},w}$. Remember that $\varphi_{\mathcal{M},w}$ consist of the following five conjuncts: (1) φ_{count} to identify tape cells, (2) φ_{start} to describe the starting configuration, (3) φ_{gen} to describe the general requirements on a Turing Machine, (4) φ_{acc} to describe the acceptance of the input, and (5) φ_{δ} for the correct evolving of the configurations.

(1) Note that φ_{count} formalises binary counting modulo $2^{p(n)}$ which is exactly what happens along each path of T by the definition of $\xrightarrow{x}$. Also, for every C and every i: $T, (C,i) \models \chi_{C=i}$. Now, if $(C,i) \xrightarrow{x} (C',j)$ then either $C = C'$, $i < 2^{p(n)} - 1$, and $j = i+1$, or $C' \in Succ(C)$, $i = 2^{p(n)} - 1$, and $j = 0$. Furthermore, if $C' \in Succ(C)$ then $T, (C,i) \models \chi_{D=k}$ for some k, and $T, (C',j) \models \chi_{D=k+1}$. Otherwise $T, (C',j) \models \chi_{D=k}$.

Thus, for all (C,i) we have $T, (C,i) \models \chi_{C=N} \wedge [x]\chi_{C=0} \wedge \varphi^D_{inc}$ or $T, (C,i) \models \neg\chi_{C=N} \wedge \varphi^C_{inc} \wedge \varphi^D_{remain}$, therefore

$$T, (C_0, 0) \models \varphi_{count}$$

(2) We have $T, (C_0, 0) \models \varphi_{start}$ because along each path of T, the first $2^{p(n)}$ tape cells form the starting configuration.

(3) We have $T, (C_0, 0) \models \varphi_{gen}$ because by the construction of L and $\xrightarrow{x}$,

- every (C,j) is labeled with exactly one tape symbol and no more than one machine state;

- from any state $(C, 0)$ one cannot reach a state $(C, 2^{p(n)} - 1)$ whilst seeing less or more than one state carrying a label q. Hence, for all (C,i) we have

$$T, (C,i) \models \chi_{C=0} \rightarrow [\alpha_{rest} \cap (\alpha_{2h} \cup \alpha_{0h})]\mathbf{ff}$$

(4) Every path through T ends in a sequence $(C, 0), \ldots, (C, 2^{p(n)} - 1)$ of states s.t. $d \in L(C, i)$ for all $i = 0, \ldots, 2^{p(n)} - 1$. By assumption, $\mathcal{M}$ has moved its head to the left end of the tape and entered state q_{acc}. Hence, for all (C, i) we have

$$T, (C, i) \models \chi_{C=0} \wedge \chi_{D=N} \rightarrow q_{acc}$$

(5) What remains to be seen is that $T, (C_0, 0) \models \varphi_\delta$. Take any $C \in \mathcal{C}$ and any $i \in \{0, \ldots, 2^{p(n)} - 1\}$. Suppose that $q \in L(C, i)$ for some $q \in Q$. Note that

- $(C, i) \xrightarrow{\alpha_=} (C', j)$ iff $C' \in Succ(C)$ and $i = j$;
- $(C, i) \xrightarrow{\alpha_{+1}} (C', j)$ iff $C' \in Succ(C)$, $i < 2^{p(n)} - 1$ and $j = i + 1$;
- $(C, i) \xrightarrow{\alpha_{-1}} (C', j)$ iff $C' \in Succ(C)$, $i > 0$ and $j = i - 1$.

Now, there are three possibilities.

(a) $q \in Q_\exists$. Then C has exactly one successor configuration C' in the computation tree, i.e. $Succ(C) = \{C'\}$. Furthermore, the symbol a at cell i is replaced by b for some $(p, b) \in \delta(q, a)$ and the tape head is not moved. Hence, $T, (C', i) \models p \wedge b$.

(b) $q \in Q_\forall$. This is similar to case (a). However, C has several successor configurations which are all modeled in T. Here, for any (C', i) with $C' \in Succ(C)$ we have $T, (C', i) \models p \wedge b$ for some $(p, b) \in \delta(q, a)$. Since T arose from the computation tree of $\mathcal{M}$, every $(p, b) \in \delta(q, a)$ is covered by some branch in T.

(c) $q \in Q_m$. Here, only the tape head is moved. C has one successor C'. If $(p, L) \in \delta(q, a)$ then $p \in L(C', i - 1)$ and accordingly for $(p, R) \in \delta(q, a)$. Finally, the remaining conjuncts in φ_δ are also satisfied because for any C, C' with $C' \in Succ(C)$ and any i we have: if $Q \cap L(C, i) = \emptyset$ then $T, (C, i) \models a$ implies $T, (C', i) \models a$ for any $a \in \Sigma$. ■

LEMMA 3. *Let $\mathcal{M}$ be an alternating exponential time bounded Turing Machine and $w \in \Sigma^*$. Then $\varphi_{\mathcal{M},w}$ has the tree model property.*

Proof. Note that PDL without the intersection operator has the tree model property. Also, $\varphi_{\mathcal{M},w}$ is built from one single atomic program x only. Hence, a sufficient condition for the tree model property of $\varphi_{\mathcal{M},w}$ is the fact that within each subformula of the form $\langle\alpha\rangle\psi$ in a positive position – or $[\alpha]\psi$ in a negative position – there is no subprogram β that cannot be realised by a sequence $s_0, \ldots, s_n$ of points s.t. $s_{i-1} \xrightarrow{x} s_i$ for all $i = 1, \ldots, n$.

Note that the only programs occurring in such a position are $\alpha_=$, α_{-1} and α_{+1}. Take $\alpha_=$ for instance, and a point s_0. Now $\alpha_=$ can be realised by the sequence $s_0, \ldots, s_n$ for any $n > 0$ if only the following hold:

- for all $k = 0, \ldots, 2^{p(n)} - 1$: $s_0 \models \chi_{C=k}$ iff $s_n \models \chi_{C=k}$,
- there is exactly one $i \in \{1, \ldots, n\}$ s.t. $s_i \models \chi_{C=0}$.

Obviously, these requirements are not unsatisfiable. Given that $\alpha_{+1} = \alpha_{=}; x$, the same holds for α_{+1}, too. For α_{-1} it can be shown in a similar way.

Then a standard model construction procedure for PDL can be employed in order to build a tree model for $\varphi_{\mathcal{M},w}$. ■

LEMMA 4. *Let $\mathcal{M}$ be an alternating, exponential time bounded Turing machine, and $w \in \Sigma^*$. If $w \notin L(\mathcal{M})$ then $\varphi_{\mathcal{M},w}$ is unsatisfiable.*

Proof. Suppose that $\mathcal{M} = (Q, \Sigma, q_0, \delta, q_{acc})$ and $w = a_0 \ldots a_{n-1}$. Let $\varphi' := \varphi_{count} \wedge \varphi_{gen} \wedge \varphi_{start} \wedge \varphi_\delta$. First we need to see something similar to the converse of Lemma 2: a model for φ' bears a computation (not necessarily successful) for $\mathcal{M}$ on w. Suppose there is an $M = (V, \xrightarrow{x}, L)$ with a $v \in V$ s.t. $M, v \models \varphi_{\mathcal{M},w}$. According to Lemma 3 we can assume M to be a tree.

Let v be the root of M. Since $M, v \models \varphi_{start}$ we have $L(v) = \{q, a_1\}$ for some $q \in Q$. But then either $q \in Q_\exists$, $q \in Q_\forall$, $q \in Q_m$, or $q = q_{acc}$. The last case is excluded which will become clear later. In any other case, since we also have $M, v \models \varphi_\delta$, there must be at least one v' s.t. $v \xrightarrow{\alpha} v'$ with $\alpha \in \{\alpha_=, \alpha_{-1}, \alpha_{+1}\}$. $M, v \models \varphi_{count}$ ensures that between v and v' there are $2^{p(n)}$, resp. $2^{p(n)} + 1$ atomic x-transitions, hence, there are states $v_0, \ldots, v_{2^{p(n)}-1}$ with $v = v_0$. Now, $M, v \models \varphi_{start}$ ensures that these states are labeled successively with the input word w followed by blank symbols. $M, v \models \varphi_{gen}$ ensures that no other labelings, including machine states can occur along this sequence. Then $v_0, \ldots, v_{2^{p(n)}-1}$ can be regarded as the starting configuration of $\mathcal{M}$.

Note that there is a $q' \in Q$ s.t. $M, v' \models q'$. The above argument can now be iterated forming further configurations of $\mathcal{M}$. It is important to note that a state v is required to have an α-successor – $\alpha \in \{\alpha_=, \alpha_{-1}, \alpha_{+1}\}$ – iff there is a $q \in Q$ with $M, v \models q$. Therefore, the building of configurations can be continued until a tree of configurations is obtained. Depending on the nature of each q, several different successors may have to be found in M.

Hence, we have $M, v_0 \models \varphi'$ and M can be regarded as a computation tree of $\mathcal{M}$ on w. But $w \notin L(\mathcal{M})$ by assumption. Therefore, there must be at least one path through the computation tree that does not contain an accepting configuration. In other words, $M, v_0 \models \varphi'$ implies $M, v_0 \models \neg\varphi_{acc}$. Thus, $\varphi_{\mathcal{M},w}$ is unsatisfiable. ■

LEMMA 5. *Let $\mathcal{M}$ be an alternating, exponential time bounded Turing machine. For any $w \in \Sigma^*$: $|\varphi_{\mathcal{M},w}|$ is polynomial in $|w|$ and $|\mathcal{M}|$.*

Proof. Let $\mathcal{M} = (Q, \Sigma, q_0, \delta, q_{acc})$ and $w \in \Sigma^*$ s.t. $\mathcal{M}$ decides each such w in time $2^{p(|w|)}$ for some polynomial p. Consider the five conjuncts that make up $\varphi_{\mathcal{M},w}$. We have

- $|\varphi_{count}| = O(p(|w|)^2)$,
- $|\varphi_{gen}| = O(|\Sigma|^2 + |Q|^2 + p(|w|))$,
- $|\varphi_{start}| = O(|w|)$,
- $|\varphi_{acc}| = O(p(|w|))$,
- $|\varphi_{\delta}| = O(|Q|^2 \cdot |\Sigma|^2 \cdot p(|w|) - |Q| \cdot |\Sigma|)$.

■

THEOREM 6. *Satisfiability of IPDL is EXPSPACE-hard*

Proof. According to Theorem 1 there is an alternating, exponential time bounded Turing Machine $\mathcal{M}$ whose word problem is EXPSPACE-hard. Let w be an input to $\mathcal{M}$, and assume that $\mathcal{M}$ decides every such w in time $2^{p(|w|)}$ for some polynomial p. The construction above yields an IPDL formula $\varphi_{\mathcal{M},w}$ whose size is polynomial in $|\mathcal{M}|$ and $|w|$ according to Lemma 5. Lemmas 2 and 4 show that that $\varphi_{\mathcal{M},w}$ is satisfiable iff $w \in L(\mathcal{M})$. ■

COROLLARY 7. *Satisfiability of IPDL over a singleton set of atomic programs and tests restricted to literals is already EXPSPACE-hard.*

Proof. Given $\mathcal{M}$ and w, it is possible to rewrite $\varphi_{\mathcal{M},w}$ s.t. the test operator is only applied to atomic propositions or their negations. Note that the only tests on non-literals occurring in $\varphi_{\mathcal{M},w}$ are $h?, \neg h?, \chi_{D=N}?, \chi_{C=k}?$, and $\neg\chi_{C=k}?$ for some $k \in \{0, 2^{p(n)} - 1\}$. They can be rewritten as programs over atomic tests in the following way:

- $h? := \bigcup_{q \in Q} q?$
- $\neg h? := \neg q_0?; \neg q_1?; \ldots; \neg q_m?$ if $Q = \{q_0, \ldots, q_m\}$
- $\chi_{C=0} := \neg c_0?; \ldots; \neg c_P?$
- $\neg\chi_{C=0} := \bigcup_{i=0}^{P} c_i?$

and similarly for $\chi_{C=N}?, \neg\chi_{C=N}?$ and $\chi_{D=N}?$. ■

COROLLARY 8. *Satisfiability of IPDL is 2-EXPTIME-hard under EXPTIME-reductions.*

Proof. According to Theorem 1 there is an alternating, exponential space bounded Turing Machine whose word problem is hard for double exponential time and polynomial time reductions. If the reduction is allowed to take exponential time then one can use exponentially many counter bits for D and make the reduction go through for an AEXPSPACE machine. Note that with exponentially many counter bits one can count up to $2^{2^{p(n)}}$ which – with the right choice of p – is the maximal number of different configurations an AEXPSPACE machine can have. ∎

COROLLARY 9. *Satisfiability for IPDL is already EXPSPACE-hard over the class of trees.*

Proof. Immediately from Lemma 3. ∎

4 Conclusions

This is the first step towards closing the complexity gap for satisfiability of IPDL. It remains to be seen whether this lower bound can be improved in order to achieve 2-EXPTIME-completeness or Danecki's upper bound can be improved in order to obtain EXPSPACE-completeness.

The other main question that remains open is the exact complexity (i.e. both upper and lower bound) for test-free IPDL. The proof of the lower bound presented here relies heavily on the presence of the test operator. A similar reduction might work for test-free IPDL, but then the encoding of a model would have to be altered.

Acknowledgments I would like to thank Carsten Lutz for valuable suggestions concerning the improvement of this paper. For example, he pointed out Corollary 9. He also has found a simpler reduction using non-deterministic, exponential space bounded Turing Machines. However, there is reason to doubt that non-deterministic machines can be used to show 2-EXPTIME-hardness of IPDL.

I would also like to thank the two anonymous referees as well as Jan Johannsen and Stéphane Demri for their comments and suggestions.

BIBLIOGRAPHY

[1] P. Balbiani and D. Vakarelov. Iteration-free PDL with intersection: a complete axiomatization. *Fundamenta Informaticae*, 45(3):173–194, February 2001.

[2] A. K. Chandra, D. C. Kozen, and L. J. Stockmeyer. Alternation. *Journal of the ACM*, 28(1):114–133, January 1981.

[3] S. Danecki. Nondeterministic propositional dynamic logic with intersection is decidable. In A. Skowron, editor, *Proc. 5th Symp. on Computation Theory*, volume 208 of *LNCS*, pages 34–53, Zaborów, Poland, December 1984. Springer.

[4] S. Demri and E. Orłowska. *Incomplete Information: Structure, Inference, Complexity.* EATCS Monographs. Springer, 2002.
[5] M. J. Fischer and R. E. Ladner. Propositional dynamic logic of regular programs. *Journal of Computer and System Sciences*, 18(2):194–211, April 1979.
[6] R. I. Goldblatt and S. K. Thomason. Axiomatic classes in propositional modal logic. In J. N. Crossley, editor, *Algebra and Logic: Papers 14th Summer Research Inst. of the Australian Math. Soc.*, volume 450 of *Lecture Notes in Mathematics*, pages 163–173. Springer, 1975.
[7] D. Harel. Recurring dominoes: Making the highly undecidable highly understandable. *Annals of Discrete Mathematics*, 24:51–72, 1985.
[8] D. Harel, D. Kozen, and J. Tiuryn. *Dynamic Logic.* MIT Press, 2000.
[9] J. Johannsen and M. Lange. CTL$^+$ is complete for double exponential time. In J. C. M. Baeten, J. K. Lenstra, J. Parrow, and G. J. Woeginger, editors, *Proc. 30th Int. Coll. on Automata, Logics and Programming, ICALP'03*, volume 2719 of *LNCS*, pages 767 – 775, Eindhoven, NL, June 2003. Springer.
[10] D. Kozen. Results on the propositional μ-calculus. *TCS*, 27:333–354, December 1983.
[11] D. Kozen and R. Parikh. An elementary proof of the completeness of PDL (note). *TCS*, 14:113 – 118, 1981.
[12] C. Lutz and U. Sattler. The complexity of reasoning with boolean modal logic. In *Advances in Modal Logic 2000 (AiML 2000)*, Leipzig, Germany, 2000. Final version appeared in Advances in Modal Logic Volume 3, 2001.
[13] A. J. Mayer and L. J. Stockmeyer. The complexity of PDL with interleaving. *TCS*, 161(1–2):109–122, 15 July 1996.
[14] S. Passy and T. Tinchev. An essay in combinatory dynamic logic. *Information and Computation*, 93(2):263–332, 1991.
[15] Z. Pawlak. Information systems theoretical foundations. *Information Systems*, 6(3):205–218, 1981.
[16] V. R. Pratt. A practical decision method for propositional dynamic logic. In *Proc. 10th Symp. on Theory of Computing STOC'78*, pages 326–337, San Diego, California, May 1978.
[17] V. R. Pratt. Models of program logics. In *Proc. 20th Symp. on Foundations of Computer Science, FOCS'79*, pages 115–122. IEEE, 1979.
[18] K. Schild. A correspondence theory for terminological logics: Preliminary report. In *Proc. 12th Int. Joint Conf. on Artificial Intelligence, IJCAI'91*, pages 466–471, Sydney, Australia, August 1991. Morgan Kaufmann.
[19] K. Segerberg. A completeness theorem in the modal logic of programs. *Notices of the AMS*, 24(6):A–552, October 1977.
[20] R. S. Streett. Propositional dynamic logic of looping and converse is elementarily decidable. *Information and Control*, 54(1/2):121–141, July 1982.
[21] M. Y. Vardi and L. Stockmeyer. Improved upper and lower bounds for modal logics of programs. In *Proc. 17th Symp. on Theory of Computing, STOC'85*, pages 240–251, Baltimore, USA, May 1985. ACM.

Martin Lange
University of Munich, Institut für Informatik
Oettingenstr. 67, D-80538 München, Germany
Martin.Lange@ifi.lmu.de

On Notions of Completeness Weaker than Kripke Completeness

TADEUSZ LITAK

ABSTRACT. We are going to show that the standard notion of Kripke completeness is the strongest one among many provably distinct algebraically motivated completeness properties, some of which seem to be of intrinsic interest. More specifically, we are going to investigate notions of completeness with respect to algebras which are either atomic, complete, completely additive or admit residuals (the last notion of completeness coincides with conservativity of minimal tense extensions); we will be also interested in combinations of these properties.

1 Motivation

It is known that Kripke frames correspond to complete, atomic and completely additive Boolean algebras with operators (BAO's). This fact became the basis of duality theory for Kripke frames, developed in the 1970's by Thomason [13], Goldblatt [4] and others. In this paper, we are going to investigate notions of completeness and consequence weaker than those associated with standard Kripke frames from an algebraic perspective. Our starting point is a simple question: can we still obtain incompleteness results if we drop at least one of the properties which hold in BAO's corresponding to Kripke frames? Is the phenomenon of Kripke incompleteness caused by any particular combination of these properties? It will be shown than in this way one obtains several provably distinct notions, whose mutual relationships seem to be of independent interest.

The structure of the present work is as follows. Section 2 introduces and systematizes those completeness notions and related consequence relations. Section 3 proves the existence of complete and completely additive BAO whose logic is inconsistent with respect to atomic BAO's;

Advances in Modal Logic, Volume 5. © *2005, Tadeusz Litak.*

that solves two open problems in the field (see Section 3 for details). Section 4 proves that non-finite tense logics of linear time are inconsistent with respect to complete BAO's; this is a strengthening of results obtained by Wolter [18] and Kowalski [6]. Finally, Section 5 discusses completeness with respect to BAO's which admit residuals, connection with van Benthem [14] and the notion of *weak second-order consequence* introduced there. The main new result of that section is an example of atomic and completely additive BAO whose logic is incomplete with respect to algebras admitting residuals.

The author hopes that his work provides new arguments for importance of algebraic methods and insights in modal logic.

2 Introduction

This section sums up some existing duality results and gives definitions crucial to what follows. We assume familiarity with basic (1) set-theoretical, (2) topological and, most importantly, (3) algebraic notions. Thus, the reader is supposed to be familiar with notions like: (1) cartesian product, powersets, cardinality of a set, (2) Euclidean topology (of the reals), open and closed intervals, regular open sets, (3) lattices, complete lattices, boolean algebras, products, homomorphic images, subalgebras, varieties, discriminator terms and so on. For arbitrary set X, let 2^X denote its powerset, X^* — the set of all finite sequences from X, X^2 — the cartesian product of X with itself and $[X]^2$ — the set of all dubletons (two-element sets) from X. $W - X$ denotes the set-theoretical difference; if the universe is understood from the context, we sometimes write $-X$. The converse of a relation R is denoted as R^{-1}.

DEFINITION 1 (Syntax). A *modal similarity type* is a finite set TYPE of unary operators; *the basic modal similarity type* is an arbitrary singleton. Formulas of propositional modal language are defined as

$$\varphi ::= \bot \mid \top \mid p \mid \neg\varphi \mid \psi \wedge \varphi \mid \psi \vee \varphi \mid \psi \rightarrow \varphi \mid \Diamond_\pi \varphi$$

for every $\pi \in$ TYPE. In addition, for every $\pi \in$ TYPE, we define the *dual operator* $\Box_\pi \varphi \leftrightharpoons \neg\Diamond_\pi\neg\varphi$. For the basic modal similarity type, we often drop the subscript.

A modal logic is any set of formulas closed under substitution, Modus Ponens, necessitation and containing all axioms of classical logic plus $\Box_\pi(p \to q) \to (\Box_\pi p \to \Box_\pi q)$ for every $\pi \in$ TYPE. Define also $\Diamond_\pi^+\varphi \leftrightharpoons \varphi \vee \Diamond_\pi\varphi$. $\Box_\pi^+$ is defined dually.

REMARK 2. As it may be seen from the above definition, we are interested only in *normal* modal logics. The definition entails also that the set of all logics in a given language is closed under arbitrary intersections. Thus, for every Γ, there exists the smallest logic containing Γ. We call it *the logic axiomatized by* Γ.

DEFINITION 3 (Frames). *A Kripke frame* is a structure $\langle W, \{R_\pi\}_{\pi\in\text{TYPE}}\rangle$, where $W \neq \emptyset$ and $R_\pi \subseteq W \times W$ for every $\pi \in$ TYPE; for every $\pi \in$ TYPE and every $X \subseteq W$, let $\Diamond_\pi X \leftrightharpoons \{y \in W \mid \exists x \in X\ yR_\pi x\}$ and $\Box_\pi \leftrightharpoons -\Diamond_\pi - X$. *A (normal) neighbourhood frame* is a structure $\langle W, \{\mathcal{N}_\pi\}_{\pi\in\text{TYPE}}\rangle$ where $W \neq \emptyset$ and every $\mathcal{N}_\pi$ is a function assigning a filter over W to every point. Recall that a filter is a non-empty family of elements of the powerset which is upward closed and closed under finite intersections.

DEFINITION 4 (BAO's). *A boolean algebra with (unary) operators* (BAO) is a structure $\langle \mathfrak{A}, \to, \wedge, \vee, \neg, \{\Diamond_\pi\}_{\pi\in\text{TYPE}}, \top, \bot\rangle$ s.t. $\langle \mathfrak{A}, \to, \wedge, \vee, \neg, \top, \bot\rangle$ is a boolean algebra and $\Diamond_\pi\bot = \bot$, $\Diamond_\pi(x \vee y) = \Diamond_\pi x \vee \Diamond_\pi y$ hold for every $\pi \in$ TYPE and every $x, y \in \mathfrak{A}$. *A trivial algebra* is any algebra whose universe is a singleton.

Thus, if not stated otherwise, we will make systematic confusion between (1) an algebra and its carrier set, (2) syntactic connectives (or derived terms) and corresponding algebraic operations; at least, if the underlying algebra is clear from the context. Definition of satisfaction and validity in frames and algebras are standard; the latter is done, as usual, by identifying modal formulas with algebraic terms. Just as we introduced the notion of the logic axiomatized by Γ, we may introduce the notion of the variety corresponding to Γ; it is the smallest variety s.t. all formulas from Γ are valid in all algebras from the variety. We write $\Gamma \models \varphi$ if φ is valid in the variety axiomatized by Γ. It is well-known that $\Gamma \models \varphi$ iff φ belongs to the logic axiomatized by Γ.

DEFINITION 5 (Basic properties).

- $\mathfrak{A}$ is a $\mathcal{C}$-BAO if it is non-trivial and lattice-complete, i.e, closed under arbitrary joins and meets.

- $\mathfrak{A}$ is a $\mathcal{A}$-BAO if it is non-trivial and atomic, i.e., below every element distinct from the bottom there is an *atom* — smallest element which is not equal to the bottom itself. The set of all atoms will be denoted by $At\mathfrak{A}$.

- $\mathfrak{A}$ is a $\mathcal{V}$-BAO if it is non-trivial and completely additive, i.e., for every $\pi \in$ TYPE and any family of elements $X \subseteq \mathfrak{A}$ s.t. whenever the join $\bigvee\limits_{x \in X} x$ exists, the join $\bigvee\limits_{x \in X} \Diamond_\pi x$ exists as well and is equal to $\Diamond_\pi \bigvee\limits_{x \in X} x$.

- $\mathfrak{A}$ is a $\mathcal{T}$-BAO if it is non-trivial and admits residuals, i.e., for any $\pi \in$ TYPE there exists a function $\mathbf{h}_\pi$ s.t. for each $x, y \in \mathfrak{A}$, $\Diamond_\pi x \leq y$ iff $x \leq \mathbf{h}_\pi y$.

The reader is asked to observe that we don't require residuals to be term-definable. This is the difference between our $\mathcal{T}$-BAO's and *residuated* BAO's of Jipsen [5]. Also, we have the following well-known

FACT 6. $\Diamond_\pi$ has a residual $\mathbf{h}_\pi$ iff f has a *conjugate* $\mathbf{p}_\pi$, i.e, for $x, y \in \mathfrak{A}$, $\Diamond_\pi x \wedge y = \bot$ iff $x \wedge \mathbf{p}_\pi y = \bot$. Both operations are then related to each other by equation $\mathbf{h}_\pi x = \neg \mathbf{p}_\pi \neg x$, i.e., $\mathbf{h}_\pi$ is the dual of $\mathbf{p}_\pi$.

DEFINITION 7 (Complex properties). Let PROPERTIES $\leftrightharpoons \{\mathcal{C}, \mathcal{A}, \mathcal{V}, \mathcal{T}\}$. For any $\mathcal{X}^* \in$ PROPERTIES*, we say that $\mathfrak{A}$ is a $\mathcal{X}^*$-BAO if $\mathfrak{A}$ is a $\mathcal{X}$-BAO for any $\mathcal{X}$ appearing in $\mathcal{X}^*$. E.g., a $\mathcal{AV}$-BAO is atomic and completely additive.

Some of these notions coincide. For example, we have the following

LEMMA 8. *Every $\mathcal{T}$-BAO is a $\mathcal{V}$-BAO. For $\mathcal{C}$-BAO's the converse also holds: $\mathfrak{A}$ is a $\mathcal{CV}$-BAO iff it is a $\mathcal{CT}$-BAO.*

Proof. The first statement is well-known. The second statement is justified by the observation that in a $\mathcal{CV}$-BAO $\mathfrak{A}$ for every $\pi \in$ TYPE

Dualities

	Kripke frames	neighbourhood frames
frames → algebras	$\mathcal{CAV}$-BAO: powerset algebra of W with operators $\lozenge X$ as in Def. 3 (*the complex algebra*)	$\mathcal{CA}$-BAO: powerset algebra of W with operators $\lozenge X \rightleftharpoons -\{x \in W \mid -X \in \mathcal{N}x\}$
algebras → frames	for a $\mathcal{CAV}$-BAO $\mathfrak{A}$: $\langle At\mathfrak{A}, R \rangle$ aRb iff $a \leq \lozenge b$ (*the atom structure*)	for a $\mathcal{CA}$-BAO $\mathfrak{A}$: $\langle At\mathfrak{A}, \mathcal{N} \rangle$ $\mathcal{N} : At\mathfrak{A} \ni a \longrightarrow \{X \subseteq At\mathfrak{A} \mid a \leq \Box \bigvee X\}$
morphisms	p-morphisms $\eta : \langle W, R \rangle \longrightarrow \langle V, S \rangle$ $\eta x S y$ iff $\exists z\ (xRz \& \eta z = y)$	neighbourhood morphisms $\eta : \langle W, \mathcal{M} \rangle \longrightarrow \langle V, \mathcal{N} \rangle$ $\eta^{-1}B \in \mathcal{N}a$ iff $B \in \mathcal{N}\eta a$
dual categories (by contravariant functors)	$\mathcal{CAV}$-BAO's with complete morphisms	$\mathcal{CA}$-BAO's with complete morphisms

Table 1. Summary of basic information. The part concerning neighbourhood frames based on Došen [3] . For simplicity, we work in the basic similarity type.

we may define the conjugate of $\lozenge_\pi$ as

$$\mathbf{p}_\pi a \rightleftharpoons \bigwedge \{x \mid a \leqslant \Box_\pi x\},$$

where $\Box_\pi$, as everywhere in this work, denotes the dual of $\lozenge_\pi$. ⊣

It is known that Kripke frames correspond to $\mathcal{CAV}$-BAO's and neighbourhood frames correspond to $\mathcal{CA}$-BAO. See Table 1 for the summary of known dualities.

REMARK 9. Lemma 8 implies that, in particular, $\mathcal{CAV}$-BAO's are also $\mathcal{CAT}$-BAO's. Thus, by Table 1 we may treat Kripke frames as $\mathcal{CAT}$-BAO's in disguise. The reason why we may choose to work with $\mathcal{T}$-BAO's instead of $\mathcal{V}$-BAO's is that the former are much more tractable. In particular, the property of being a $\mathcal{T}$-BAO in a finite type is expressible by a Σ_1^1-sentence (*there exists a function such that* ...) and hence preserved by ultraproducts. In general, however, it is not preserved by subalgebras, hence even the universal class generated by a given class of such BAO's contains algebras which are not $\mathcal{T}$-BAO's. Fortunately, the situation changes when residuals are term-definable, as it is often the case. In such a situation, all algebras in the variety must be $\mathcal{T}$-BAO's. Nothing like this can happen with $\mathcal{A}$-BAO's or $\mathcal{C}$-BAO's; non-degenerate varieties always contain a BAO which is neither complete nor atomic. Further discussion is postponed till Section 5.

DEFINITION 10 (Completeness notions). For $\mathcal{X}^* \in$ PROPERTIES*, Γ — arbitrary set of modal formulas in a fixed similarity type, φ — a formula in the same language, let

$$\Gamma \models_{\mathcal{X}^*} \varphi \quad \leftrightharpoons \quad \varphi \text{ holds in all } \mathcal{X}^*\text{-BAO's in the variety of BAO's corresponding to } \Gamma.$$

Thus, a logic axiomatized by Γ is $\mathcal{X}^*$-*incomplete* iff for some φ, $\Gamma \models_{\mathcal{X}^*} \varphi$ and yet $\Gamma \not\models \varphi$.

A particularly interesting notion arising in this way is

DEFINITION 11 ($\mathcal{X}^*$-inconsistency). A logic axiomatized by Γ is $\mathcal{X}^*$-inconsistent if it is consistent, i.e., distinct from the set of all formulas in a given similarity type and yet $\Gamma \models_{\mathcal{X}^*} \bot$, i.e., there is no $\mathcal{X}^*$-BAO in the corresponding variety.

REMARK 12. $\mathcal{X}^*$-inconsistency implies $\mathcal{X}^*$-incompleteness. For any $\mathcal{Y}^* \supseteq \mathcal{X}^*$, $\mathcal{X}^*$-inconsistency implies $\mathcal{Y}^*$-inconsistency and $\mathcal{X}^*$-incompleteness implies $\mathcal{Y}^*$-incompleteness. Thus, the construction of Thomason [12] may be viewed as the first example of a $\mathcal{CAT}$-inconsistent logic, or even a $\mathcal{CA}$-inconsistent one, as for tense logics there is no difference between those two notions; see also Section 5.

REMARK 13 (The basic similarity type). It is not possible to produce any example of $\mathcal{CAT}$-inconsistent logic in the basic modal similarity type. By Makinson's theorem, any nontrivial variety of BAO's with one unary operator contains at least one two-element algebra (cf., e.g., [2, theorem 8.67]). Nevertheless, the method of reduction invented by Thomason and developed by Kracht and Wolter (cf., e.g., [7, chapter 6]) shows how to transfer examples of $\mathcal{X}^*$-inconsistent logics into $\mathcal{X}^*$-incomplete ones in the basic similarity type.

Relationships between those notions of completeness (or consequence) are as shown by Figure 1 below. For $\models_{\mathcal{X}^*} \subseteq \models_{\mathcal{Y}^*}$, solid lines denote established proper inclusion $\models_{\mathcal{X}^*} \subsetneq \models_{\mathcal{Y}^*}$, i.e. the existence of $\mathcal{X}^*$-complete logics which are $\mathcal{Y}^*$-incomplete.

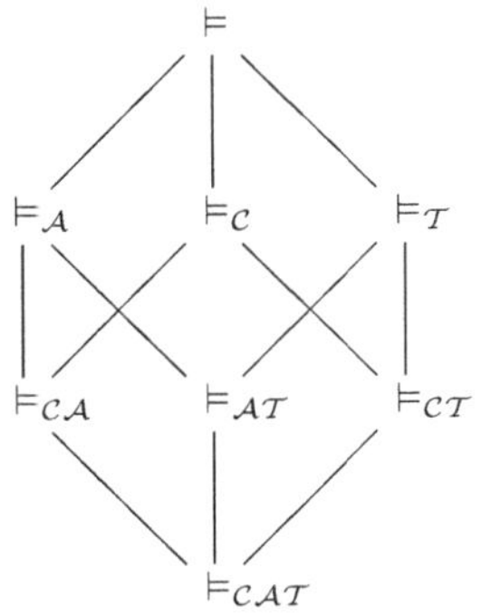

Figure 1. Proper inclusions between various algebraic consequence relations. This is not necessarily the Hasse diagram — neither joins nor meets have to be faithful.

As the drawing suggests, one may show that all inclusions are proper. The most economical way to achieve this goal is to prove three facts:

$$\models_{\mathcal{A}} \;\not\subseteq\; \models_{\mathcal{CT}}, \quad \models_{\mathcal{C}} \;\not\subseteq\; \models_{\mathcal{AT}}, \quad \models_{\mathcal{T}} \;\not\subseteq\; \models_{\mathcal{CA}}.$$

The first one will be proven in Section 3, the second one in Section 4, the last one will be discussed in Section 5. The reader certainly noted that $\mathcal{V}$-completeness is missing from the picture. Some reasons for that were already given by Remark 9. Further discussion and an updated diagram will be given in Section 5.

REMARK 14. The consequence relation symbol introduced above should not be confused with $\mathfrak{A} \models \varphi$ ($\mathfrak{A} \models_{\mathfrak{V}} \varphi$), which means, as usual, that the equation corresponding to φ holds in $\mathfrak{A}$ (under valuation $\mathfrak{V}$).

In forthcoming sections, it will be often useful to represent BAO's as *general frames*.

DEFINITION 15. *A general frame* is a pair $\langle \mathfrak{F}, \mathfrak{A} \rangle$ s.t. $\mathfrak{F}$ is a Kripke frame and $\mathfrak{A}$ is a subalgebra of the full complex algebra of $\mathfrak{F}$ (see Table 1 for the definition). An algebra thus represented will be called *a complex algebra*.

3 Varieties with no atomic algebras

Let us start with the proof that $\models_{\mathcal{A}} \not\subseteq \models_{\mathcal{CT}}$. We generalize here a result of Venema [16], who shows that there are $\mathcal{A}$-inconsistent logics. We are going to prove that there is $\mathcal{CV}$-BAO (hence a $\mathcal{CT}$-BAO) whose logic is $\mathcal{A}$-inconsistent. That solves at least two open problems in the field. The first one was posed to the present author by V. Shehtman: according to him, the question whether every $\mathcal{C}$-complete logic is neighbourhood complete (i.e., whether $\models_{\mathcal{CA}} = \models_{\mathcal{C}}$) was a folklore since V. Rybakov and L. Maksimova's work on the subject in the 1970's. Another one was posed by Y. Venema himself: can one produce a variety with no atomic members using a $\mathcal{T}$-BAO?

The original construction of Venema [16] is unsuited for the goal we have in mind. It is based on a particular representation of the atomless countable boolean algebra. Hence, lattice-incompleteness and lack of complete additivity seem irreparable. Nevertheless, we try to exhibit existing analogies between that construction and ours by the choice of notation and terminology. On the other hand, the earlier construction gives an example of an *atomless* variety, whereas ours — just a variety with no atomic members. It is enough for our purposes, though.

NOTATION 16. Let $\mathbf{I}$ be the unit interval $(0, 1)$, $\mathbf{P} = [\mathbf{I} \cup \{0, 1\}]^2$, $\mathfrak{P}$ — the boolean algebra whose universe is $2^{\mathbf{P}}$ and the operations coincide with standard set-theoretical ones, $\mathfrak{O}$ — the boolean algebra of regular open sets (i.e., interiors of closed sets) from $\mathbf{I}$. Note that $\vee$ and $\neg$ in $\mathfrak{O}$ differ from the set-theoretical operations. It is well-known that both $\mathfrak{P}$ and $\mathfrak{O}$ are complete lattices. For $0 \leq a < b \leq 1$, the open interval determined by a and b will be denoted as (a, b), whereas $\langle a, b\rangle$ will be simply an ordered pair.

THEOREM 17. *There exists a $\mathcal{CV}$-BAO $\mathfrak{A}$ s.t. the variety generated by $\mathfrak{A}$ — i.e., $HSP(\mathfrak{A})$ – contains no atomic members.*

Proof. We will proceed via a series of easily verifiable claims.

CLAIM 18. $B \in \mathfrak{O}$ only if there exists at most countable set J s.t. $B = \bigcup_{j \in J} (a_j, b_j)$, where for each $j \in J$, $0 \leq a_j < b_j \leq 1$ and for each $j \neq k$, either $b_j < a_k$ or $b_k < a_j$. The set $\{\{a_j, b_j\} \mid j \in J\}$ will be called *the canonical representation of* B and denoted as Z_B.

NOTATION 19. For any $z \in \mathbf{P}$, $z_< \leftrightharpoons inf(z)$, $z_> \leftrightharpoons sup(z)$ and $\llcorner z \lrcorner \leftrightharpoons (z_<, z_>)$.

Our main object of interest will be a BAO whose underlying boolean reduct is $\mathfrak{A}_- \leftrightharpoons \mathfrak{P} \times \mathfrak{O}$. As this is a product of two complete boolean algebras, we get

CLAIM 20. $\mathfrak{A}_-$ is a complete boolean algebra.

Now we are going to define five unary operations on the universe of our algebra. Subscripts will be chosen so as to emphasize analogies with Venema [16] construction.

First, we are going to define four auxiliary mappings:

$$\begin{aligned}
\eta_\lhd : \mathfrak{O} \ni B &\longrightarrow \{z \in \mathbf{P} \mid \exists y \in B\ y \in \llcorner z \lrcorner\} \in \mathfrak{P}, \\
\eta_\rhd : \mathfrak{P} \ni A &\longrightarrow \bigvee_{a \in A}{}^{\mathfrak{O}} \llcorner a \lrcorner \in \mathfrak{O}, \\
\eta_> : \mathfrak{P} \ni A &\longrightarrow \{z \in \mathbf{P} \mid \exists z^* \in A\ \llcorner z^* \lrcorner \supsetneq \llcorner z \lrcorner\} \in \mathfrak{P}, \\
\eta_L : \mathfrak{P} \ni A &\longrightarrow \{z \in \mathbf{P} \mid \exists z^* \in A\ z_< = z^*_<, z_> = z^*_< + \\
&\qquad (z^*_> - z^*_<)/2\} \in \mathfrak{P}.
\end{aligned}$$

Through this section, we set TYPE $\leftrightharpoons \{\lhd, \rhd, >, L, \mathbf{E}\}$.

DEFINITION 21. Define $\mathfrak{A} \leftrightharpoons \langle \mathfrak{A}_-, \{\Diamond_\pi\}_{\pi \in \mathrm{TYPE}}\rangle$, where $\Diamond_\lhd \langle A, B\rangle \leftrightharpoons \langle \eta_\lhd B, \emptyset\rangle$, $\Diamond_\rhd \langle A, B\rangle \leftrightharpoons \langle \emptyset, \eta_\rhd A\rangle$, $\Diamond_> \langle A, B\rangle \leftrightharpoons \langle \eta_> A, \emptyset\rangle$, $\Diamond_L \langle A, B\rangle \leftrightharpoons \langle \eta_L A, \emptyset\rangle$, and $\Diamond_\mathbf{E}$ will be *the unary boolean discriminator function* on $\mathfrak{A}_-$ (Jipsen [5]); i.e., $\Diamond_\mathbf{E} x = \bot$ iff $x = \bot$ and $\Diamond_\mathbf{E} x = \top$ otherwise.

We want to prove that these operations distribute over arbitrary joins. It is obvious for the unary discriminator, $\Diamond_>$ and $\Diamond_L$ as both $\Diamond_>$ and $\Diamond_L$ are easily seen to be operators given by certain relations on a Kripke frame whose universe is $\mathbf{P}$. Assume now we have a family of pairs $\{\langle A_j, B_j\rangle\}_{j \in J} \subseteq \mathfrak{A}$. Then

$$\bigvee_{j \in J} \Diamond_\lhd \langle A_j, B_j\rangle = \bigvee_{j \in J} \langle \eta_\lhd B_j, \emptyset\rangle = \langle \bigcup_{j \in J} \eta_< B_j, \emptyset\rangle$$

and, similarly,

$$\bigvee_{j\in J} \Diamond_{\triangleright}\langle A_j, B_j\rangle = \bigvee_{j\in J}\langle \emptyset, \eta_{\triangleright} A_j\rangle = \langle \emptyset, \bigvee_{j\in J} \eta_{\triangleright} A_j\rangle.$$

Hence, it is enough to establish

(a) $\bigcup_{j\in J} \eta_{\triangleleft} B_j = \eta_{\triangleleft} \bigvee_{j\in J} B_j$

and

(b) $\bigvee_{j\in J} \eta_{\triangleright} A_j = \eta_{\triangleright} \bigcup_{j\in J} A_j$.

Observe that in both cases, the $\subseteq$-direction is trivial. To establish the converse for a, assume there exists $i \in \mathbf{P}$ s.t.

$$(\exists y \in \bigvee_{j\in J} B_j \; y \in \llcorner i \lrcorner) \;\&\; (\forall j \in J \; \forall y \in B_j \; y \notin \llcorner i \lrcorner).$$

But then for each $j \in J$, $B_j \subseteq \neg\llcorner i \lrcorner$, hence $B_j = Int(B_j) \subseteq Int(\neg\llcorner i \lrcorner) \in \mathfrak{O}$. Thus, we get $\bigvee_{j\in J} B_j \subseteq Int(\neg\llcorner i \lrcorner) \subseteq \neg\llcorner i \lrcorner$, a contradiction.

To establish

$$\bigvee_{a\in \bigcup_{j\in J} A_j} \llcorner a \lrcorner \subseteq \bigvee_{j\in J} \bigvee_{a\in A_j} \llcorner a \lrcorner,$$

take any X s.t. for each $j \in J$, $\bigvee_{a\in A_j} \llcorner a \lrcorner \subseteq X$. Obviously, for each $a \in \bigcup_{j\in J} A_j$ there exists $j_a \in J$ s.t. $a \in A_{j_a}$. Hence for each $a \in \bigcup_{j\in J} A_j$, $\llcorner a \lrcorner \subseteq \bigvee_{a\in A_{j_a}} \llcorner a \lrcorner \subseteq X$ and thus we get $\bigvee_{a\in \bigcup_{j\in J} A_j} \llcorner a \lrcorner \subseteq X$, which gives us b. In fact, this a sort of law of infinite associativity which holds in every complete lattice; cf., e.g., [9]. Thus. we got the desired

CLAIM 22. $\Diamond_{\triangleleft}$, $\Diamond_{\triangleright}$, $\Diamond_{>}$, $\Diamond_L$ and $\Diamond_{\mathbf{E}}$ are completely additive operators.

Define $c \leftrightharpoons \Diamond_{\triangleright}\top$ and $Fx \leftrightharpoons \Diamond_{\triangleright}\Box_L(\Box_{\triangleleft}x \wedge \neg\Diamond_{>}\Box_{\triangleleft}x)$.

CLAIM 23. In $\mathfrak{A}$, c is equal to $\langle\emptyset, \mathbf{I}\rangle$. Hence, an element of $\mathfrak{A}$ is below c iff it is of the form $\langle\emptyset, B\rangle$ for some $B \in \mathfrak{O}$.

CLAIM 24. For any $B \in \mathfrak{O}$,

$$\Box_{\lhd}\langle\emptyset, B\rangle = \langle\{z \in \mathbf{P} \mid \llcorner z \lrcorner \subseteq B\}, \mathbf{I}\rangle,$$

$$\Box_{\lhd}\langle\emptyset, B\rangle \wedge -\Diamond_{>}\Box_{\lhd}\langle\emptyset, B\rangle = \langle Z_B, \mathbf{I}\rangle,$$

$$F\langle\emptyset, B\rangle = \langle\emptyset, \bigcup_{z \in Z_B} (z_{<}, z_{<} + (z_{>} - z_{<})/2)\rangle.$$

This gives us the following

CLAIM 25. There exists a constant term c and an unary term Fx in language determined by TYPE s.t. $\mathfrak{A} \models c > \bot$ and $\mathfrak{A} \models \forall x\ (\bot < x \leq c \rightarrow \bot < Fx < x)$. As $\mathfrak{A}$ is a discriminator algebra, i.e., $\Diamond_{\mathbf{E}}$ behaves like universal modality, those two facts may be reformulated as follows: $\mathfrak{A} \models \Diamond_{\mathbf{E}}c = \top$ and $\mathfrak{A} \models \Diamond_{\mathbf{E}}x \wedge \Box_{\mathbf{E}}(x \rightarrow c) \leq \Diamond_{\mathbf{E}}Fx \wedge \Diamond_{\mathbf{E}}(x \wedge \neg Fx) \wedge \Box_{\mathbf{E}}(Fx \rightarrow x)$.

The above claim implies Theorem 17. For similar arguments cf. Venema [16] or an earlier, undebugged attempt of such a construction by Kracht and Kowalski [8] . ⊣

4 Varieties with no complete algebras

This section establishes $\models_{\mathcal{C}} \not\subseteq \models_{\mathcal{AT}}$. Examples of $\mathcal{C}$-incomplete logics (above the unimodal logic **K4** and higher) have been already provided in a previous work of the author [11]. Nevertheless, we are now going to obtain a stronger result ($\mathcal{C}$-inconsistency) by means of surprisingly natural examples. We are going to work in the similarity type of tense logic in this section, and hence we set TYPE $\leftrightharpoons \{<, >\}$. In the 90's, Kowalski [6] and Wolter [18] proved independently that there are exactly countably many maximal consistent extensions of **Lin** — the tense logic of all linear time flows, i.e., frames which are transitive and connected $(\forall x \forall y(xRy \vee yRx \vee x = y)))$. We are going to show that every maximal **Lin** logic which is not *tabular*, i.e., which is not determined by a finite frame, is $\mathfrak{C}$-inconsistent. As far as the author is aware, this is a new result: up to now, it was known only that such logics do not have any Kripke frames.

DEFINITION 26. *A head-and-tail logic* is the tense logic determined by a general frame $\mathfrak{F}_{\mathbf{a}} \leftrightharpoons \langle W_{\mathbf{a}}, \{R^{\mathbf{a}}_{\pi}\}_{\pi \in \mathrm{TYPE}}, \mathfrak{A}_{\mathbf{a}} \rangle$, where

- $\mathbf{a} = \langle \kappa, r, \lambda \rangle$, $\kappa, \lambda \in \omega + 1$, $r \in \{0, 1\}$. If $\omega \in \{\kappa, \lambda\}$, then $r = 0$.
- $W_{\mathbf{a}} \leftrightharpoons \{n_< \mid n \in \kappa\} \cup \{o_* \mid o \in r\} \cup \{m_> \mid m \in \lambda\}$,
- $R^{\mathbf{a}}_< \leftrightharpoons \{\langle n_<, m_> \rangle \mid n \in \kappa, m \in \lambda\} \cup \{\langle n_<, m_< \rangle \mid n, m \in \kappa, n < m\} \cup \{\langle n_>, m_> \rangle \mid n, m \in \lambda, n > m\} \cup \{\langle n_<, o_* \rangle, \langle o_*, o_* \rangle, \langle o_*, m_> \rangle \mid n \in \kappa, m \in \lambda, o \in r\}$. $R^{\mathbf{a}}_> \leftrightharpoons R^{\mathbf{a}-1}_<$.
- $\mathfrak{A}_{\mathbf{a}}$ is the algebra of finite and co-finite subsets over $W_{\mathbf{a}}$.

Note that to make the definition more condensed, we followed von Neumann convention of representing ordinals; hence, e.g., $1 = \{0\}$. Finite irreflexive chain of length m may thus be represented as $\mathfrak{F}_{\langle k,0,l \rangle}$, for any k, l s.t. $k + l = m$.

FACT 27. On every $\mathfrak{A}_{\mathbf{a}}$, the following term (*universal modality*) defines the unary boolean discriminator:

$$\mathbf{E}x \leftrightharpoons \Diamond_> x \wedge x \wedge \Diamond_< x.$$

THEOREM 28. *A tense logic containing* **Lin** *is maximal consistent iff it is a head-and-tail logic.*

Proof. Cf. Kowalski [6] or Wolter [18]. ⊣

Thus, we will use the names *head-and-tail logic* and *maximal* **Lin** *logic* interchangeably. Those determined by a frame of the form $\mathfrak{F}_{\langle \kappa, r, \lambda \rangle}$ with either κ or λ (or both) equal to ω are called *infinite.* Those which are not infinite are called *tabular* (or *finite*).

¿From now on in this section, to fix the discourse we concentrate on head-and-tail logics with $\kappa = \omega$, but proofs are instantly adaptable for those with finite κ and $\lambda = \omega$ — it is enough to exchange modalities. As in the infinite case r is always equal to 0, instead of notation $\mathcal{F}_{\langle \omega, r, \lambda \rangle}$, we write $\mathfrak{F}_{\omega, \lambda}$. Define a sequence of variable-free formulas which will serve as names for points: $\underline{i} \leftrightharpoons \Diamond^i_> \top \wedge \Box^{i+1}_> \bot$ $(n \in \omega)$.

LEMMA 29. *The following variable-free formulas are true in every* $\mathfrak{F}_{\omega,\lambda}$*:*

(c) $\mathbf{E}\underline{i}$ $(i \in \omega)$,

(d) $\underline{i} \to \lozenge_< \underline{i+n}$ $(i \in \omega, n > 0)$.

In addition,

(e) $\underline{i} \to \neg \underline{j}$

is a theorem of **Lin** *for every* $i \neq j$.

Proof. The easy observation that *numeral* $\underline{i}$ is true exactly in one point (namely i) and Fact 27 entail statement c. To prove d it is enough to observe in addition that in the future of every i there is a point verifying $\underline{j}$ for every $j > i$. Statement e follows from the fact that the right conjunct of $\underline{i}$ is a negation of the left conjunct of $\underline{i+1}$ and that $\Box_< p \to \Box^2_< p$ is a theorem of **Lin**; the rest is proven by simple induction. ⊣

LEMMA 30. *Every complete (or even ω-complete)* BAO *verifying all variable-free formulas true in $\mathfrak{F}_{\omega,\lambda}$ refutes* the weak Grzegorczyk axiom *or* the law of weak foundation, *i.e., the formula*

$$\mathbf{wf} \leftrightharpoons \neg(p \wedge \Box^+_<(p \to \lozenge_<(\neg p \wedge \lozenge_< p))).$$

Proof. Take any complete BAO $\mathfrak{B}$ verifying all variable-free formulas true in $\mathfrak{F}_{\omega,\lambda}$. Define two sequences of variable-free formulas and corresponding elements of $\mathfrak{B}$: $a_n \leftrightharpoons \underline{2n}$ and $b_n \leftrightharpoons \underline{2n+1}$ $(n \in \omega)$.

By statement e of Lemma 29, any a and b intersect at $\bot$ and it follows that $\bigvee_{n\in\omega} a_n \leqslant \neg \bigvee_{n\in\omega} b_n$ and $\bigvee_{n\in\omega} b_n \leqslant \neg \bigvee_{n\in\omega} a_n$. By claim d, $a_n \leqslant \lozenge_< b_n \leqslant \lozenge_< \bigvee_{n\in\omega} b_n$ and it follows that $\bigvee_{n\in\omega} a_n \leqslant \lozenge_< \bigvee_{n\in\omega} b_n$; in a similar way, one proves $\bigvee_{n\in\omega} b_n \leqslant \lozenge_< \bigvee_{n\in\omega} a_n$. Hence, if we set $\mathfrak{V}(p) \leftrightharpoons \bigvee_{n\in\omega} a_n$, $\mathfrak{V}(\mathbf{wf}) = \mathfrak{V}(\neg p)$, as the value of the subformula of **wf** preceded by $\Box^+_<$ is $\top$. By statement c of Lemma 29, $\mathfrak{V}(p) \neq \bot$, hence $\mathfrak{V}(\neg p) \neq \top$. ⊣

THEOREM 31. *There is no complete* BAO *in the variety corresponding to an infinite maximal* **Lin** *logic.*

Proof. By Lemma 30 it is enough to prove that the law of weak foundation (both for $\Diamond_>$ and $\Diamond_<$) cannot be refuted in any frame of the form $\mathfrak{F}_{\omega,\lambda}$. Assume there is a point y_0 where formula **wf** is refuted under some valuation $\mathfrak{V}'$. In the future of this point, there must a point y_1 belonging to $\mathfrak{V}'(\neg p)$; in the future of y_1 there must be a point y_2 in $\mathfrak{V}'(p)$ and proceeding in this way we can prove there is an infinite set with infinite complement in $\mathfrak{A}_{\omega,\lambda}$, namely $\mathfrak{V}'(p)$, which is a contradiction. Hence, the law of weak foundation belongs to every maximal consistent **Lin** logic and yet it is refuted in any complete algebra verifying all variable-free formulas true in a given infinite maximal **Lin** logic. ⊣

REMARK 32. One may prove that all head-and-tail logics are decidable and their satisfiability problem is NP-complete. Thus, $\mathcal{C}$-inconsistency does not imply a high level of computational complexity. Moreover, all of them with the exception of the one determined by $\mathfrak{F}_{\omega,\omega}$ are finitely axiomatizable. These results follow from Wolter [18] and a forthcoming paper Litak, Wolter [10]. By use of methods from those papers one may actually prove that all extensions of **Lin** are $\mathcal{AT}$-complete. Thus, the situation in the lattice of all extensions of **Lin** is rather peculiar: although incompleteness is a common phenomenon for logics from that lattice, the sole perpetrator of all problems is lattice-completeness of BAO's corresponding to Kripke frames. Section 3 showed that in general it is not true even for logics with conjugated operators and universal modality. Proof of this result would require too much techniques outside the scope of the present work. Therefore, it will be provided in a separate paper together with a more in-depth analysis of incompleteness problems for tense logics of linear time.

5 Varieties which do not admit residuals

This section deals with $\mathcal{T}$-incompleteness and $\mathcal{AV}$-completeness. Let us formulate the following representation theorem, which, albeit simple, will be of some importance later on.

THEOREM 33. *An algebra is a $\mathcal{AV}$-BAO iff it can be represented as a complex algebra containing all singletons of the underlying frame.*

An algebra is a $\mathcal{AT}$-BAO iff it can be represented as a reduct of a complex algebra containing all singletons of a frame in which every relation has its converse — i.e., a tense frame.

Proof. Follows from Venema [15, Theorem 5.1]. The second part may be proven analogously. ⊣

OBSERVATION 34. A term P_π defines the conjugate of an operator $\Diamond_\pi$ on all algebras in the variety $\mathfrak{V}$ iff $\mathfrak{A} \models p \to \neg P_\pi \neg \Diamond_\pi p$ and $\mathfrak{V} \models p \to \Box_\pi P_\pi p$ hold for all $\mathfrak{A} \in \mathfrak{V}$. Hence, the most straightforward method to obtain a $\mathcal{T}$-complete extension of a given logic is to add for $\pi \in$ TYPE a new operator P_π and stipulate the above equalities. This procedure is known as forming *minimal tense extensions.* Thus we arrive at the most important reason for independent interest in $\mathcal{T}$-completeness; it boils down to conservativity of minimal tense extensions, as the latter means that logic is complete with respect to a class of algebras which are reducts of $\mathcal{T}$-BAO's.

For the time being, we are going to work in the basic modal similarity type. It is already known that logics with non-conservative minimal tense extensions indeed do exist; in fact, Zakharyaschev et al. [19, Theorem 124] generalizes the Blok Incompleteness Theorem for degrees of $\mathcal{T}$-incompleteness. For the purpose of this work, it is enough to observe that

LEMMA 35. $\Box\Diamond\top \to \Box(\Box(\Box p \to p) \to p) \models_{\mathcal{T}} \Diamond\Box\bot \vee \Box\bot$.

Proof. Assume $\mathfrak{A} \not\models \Diamond\Box\bot \vee \Box\bot = \top$ and $\Diamond$ has a conjugate $\mathbf{p}_\Diamond$ (not necessarily term-definable). It means that $a \leftrightharpoons \Box\Diamond\top \wedge \Diamond\top$ is distinct from the bottom element. Define $b \leftrightharpoons \mathbf{p}_\Diamond a$. $b \leq \Diamond\top$ because of the fact that $a \leq \Box\Diamond\top$. Our goal is to show that

(f) $a \leq \Diamond b$

and

(g) $b \leq \Diamond b$.

As, by assumption, $a \leq \Box\Diamond\top$, it will follow that

(h) $\bot < a \leq \Box\Diamond\top \wedge \Diamond(\Box(\Diamond b \vee \neg b) \wedge b)$, which implies our theorem. h follows from g and f, because g implies that $b \wedge \Box\neg b = \bot$ and hence $b = \Box(\Diamond b \vee \neg b) \wedge b$. f follows from the fact that for any $x \leq \Diamond\top$, $x \leq \Diamond \mathbf{p}_{\Diamond} x$. As $b \leq \Diamond\top$ as well and for every x, $x \leq \Box \mathbf{p}_{\Diamond} x$, we get that $a \leq \Box\Diamond b$, which gives us g. ⊣

Nevertheless, Wolter [17, Section 4.6] defined a $\mathcal{CA}$-BAO which is readily seen to separate $\Diamond\Box\bot \vee \Box\bot$ from $\Box\Diamond\top \rightarrow \Box(\Box(\Box p \rightarrow p) \rightarrow p)$. This completes the proof that $\models_{\mathcal{T}} \not\subseteq \models_{\mathcal{CA}}$ and that all inclusions in Figure 1 are proper. By adding the unary discriminator to the similarity type of the algebra defined by Wolter and following an argument analogous to the proof of Lemma 35, one may actually prove a $\mathcal{T}$-inconsistency theorem:

FACT 36. There exists a consistent bimodal logic that is $\mathcal{T}$-inconsistent and $\mathcal{CA}$-complete.

Observe that we may prove also

LEMMA 37. $\Box\Diamond\top \rightarrow \Box(\Box(\Box p \rightarrow p) \rightarrow p) \models_{\mathcal{AV}} \Diamond\Box\bot \vee \Box\bot$

Proof. This time, a is defined as an arbitrary atom below $\Box\Diamond\top \wedge \Diamond\top$ in an $\mathcal{AV}$-BAO $\mathfrak{D}$. As $a \leq \Diamond \bigvee\{x \mid x \in At\mathfrak{D}\}$ and $\mathfrak{D}$ is completely additive, there must exist an atom b s.t. $a \leq \Diamond b$ (we use here the fact that a is an atom). If $b \leq \Diamond b$, then we have both f and g and that gives us h in a manner similar to the proof of lemma 35. If g does not hold, then, as b is an atom, $b \leq \Box\neg b$ and that gives us h anyhow. ⊣

REMARK 38. We have proven above that $\models_{\mathcal{T}} \cap \models_{\mathcal{AV}} \not\subseteq \models_{\mathcal{CA}}$. Unfortunately, it is not at all clear how we could strengthen the above observations to a proof that $\models_{\mathcal{V}} \not\subseteq \models_{\mathcal{CA}}$ or at least to a proof of existence of a $\mathcal{V}$-incomplete logic. Atomicity in the proof of Lemma 37 or existence of residuals in the proof of Lemma 35 are used in a crucial way. This shows once again that complete additivity itself is not a very tractable property.

Van Benthem [14] formulated a monadic *weak second-order logic* complete with respect to Henkin models closed under first-order definability. Closure under first-order definability means that for any formula ψ without second-order quantifiers, any sequence of admissible subsets $\bar{X}$ and any sequence of elements $\bar{w}$, the set of elements x satisfying $\psi(x, \bar{w}, \bar{X})$ is admissible. Such a model is always a general frame, as $\Diamond X$ is definable in the standard way. This motivates definition of yet another property of BAO's and associated notion of consequence/completeness:

DEFINITION 39. $\mathfrak{A}$ is a $\mathcal{E}$-BAO if it can be depicted as a complex algebra containing all first-order definable subsets of some general frame. $\Gamma \vDash_{\mathcal{E}} \varphi$ and the notion of $\mathcal{E}$-completeness are defined in an analogous way to other properties.

THEOREM 40. *Every $\mathcal{E}$-BAO is a $\mathcal{AT}$-BAO.*

Proof. All singletons are definable by identity formulas with one parameter. $\mathbf{p}_{\Diamond}X$ may be defined as $\exists y(yRx \& y \in X)$. It is enough now to apply Theorem 33. ⊣

OBSERVATION 41. We have thus a following sequence of inclusions:

(i) $\vDash_{\mathcal{AV}} \ \subseteq \ \vDash_{\mathcal{AT}} \ \subseteq \ \vDash_{\mathcal{E}} \ \subsetneq \ \vDash_{\mathcal{CAT}}$.

As $\vDash_{\mathcal{AT}} \ \not\subseteq \ \vDash_{\mathcal{CA}}$, the notion of weak second-order consequence relation is too strong for analysis of consequence relation over neighbourhood frames. Nevertheless, the question now arises whether all of the inclusions in i are proper. Van Benthem [14] showed that

$$\Box\Diamond\top \to \Box(\Box(\Box p \to p) \to p) \vDash_{\mathcal{E}} \Diamond\Box\bot \vee \Box\bot.$$

By inequality i, this observation is weaker than either of Lemmas 35 and 37. It was, however, an open question whether Lemma 37 is an actual strengthening of van Benthem's result. Now we are going to show that it is indeed the case. Actually, the result we are going to prove is stronger: $\vDash_{\mathcal{T}} \ \not\subseteq \ \vDash_{\mathcal{AV}}$. This is the main new result of this section. Set TYPE $\leftrightharpoons \{<, >\}$.

THEOREM 42. *There exists a $\mathcal{AV}$-algebra which generates a $\mathcal{T}$-incomplete logic.*

Proof. Consider a frame $\mathfrak{G} \leftrightharpoons \langle W, \{R_\pi\}_{\pi\in\text{TYPE}}, \mathfrak{A}\rangle$. $W \leftrightharpoons \{\infty\} \cup \{a_n\}_{n\in\omega}$, $R_< \leftrightharpoons \{\langle a_i, a_j\rangle | i < j\} \cup \{\langle\infty, a_{2i}\rangle | i \in \omega\}$, $R_> \leftrightharpoons \{\langle a_i, a_j\rangle | i > j\}$ (i.e., $R_> = R_<^{-1} \cap (W - \{\infty\})^2$), $\mathfrak{A}$ is the algebra of finite and cofinite subsets of W. Observe that — just like in case of frames from Section 4 — all points are definable by variable-free formulas. For our purposes, we need only the following

CLAIM 43. Define $\underline{1} \leftrightharpoons \Diamond_> \top \wedge \Box^2_> \bot$, $\underline{\infty} \leftrightharpoons \Diamond_< \Box_> \bot$. Then $\underline{\infty}^{\mathfrak{A}} = \{\infty\}$ and $\underline{1}^{\mathfrak{A}} = \{a_1\}$.

LEMMA 44. *Let* $\blacksquare\varphi \leftrightharpoons \neg\varphi \wedge \Box_<\varphi$. *The following formulas hold in* $\mathfrak{G}$*:*

$$\begin{aligned}
\underline{\infty} &\rightarrow \Diamond^2_< \underline{1} \wedge \Box_< \neg\underline{1}, && \text{(j)}\\
\underline{\infty} \wedge \Diamond_<\Box_< x \wedge \Diamond^2_< \neg p &\rightarrow \Diamond_< \blacksquare p \vee \Diamond^2_< \blacksquare p, && \text{(k)}\\
\Diamond_< \blacksquare p &\rightarrow \Diamond_<(p \wedge \Box_< p \wedge \Diamond^2_> \blacksquare p \wedge \Box^3_> \neg\blacksquare p), && \text{(l)}\\
\underline{\infty} \wedge \neg\Diamond_< \blacksquare p &\rightarrow \Box_< \neg(p \wedge \Box_< p \wedge \Diamond^2_> \blacksquare p \wedge \Box^3_> \neg\blacksquare p), && \text{(m)}\\
\underline{\infty} \wedge \Box_< p &\rightarrow \Box_< \Diamond_< p, && \text{(n)}\\
\Box_< \Diamond_< p &\rightarrow \Diamond_< \Box_< p. && \text{(o)}
\end{aligned}$$

Conjunction of these formulas will be denoted as Γ.

Proof. Statement j follows directly from the definition of the frame and Claim 43. For k, assume $\{\infty\} \leq \mathfrak{V}(\Diamond_<\Box_< p \wedge \Diamond^2_< \neg p)$. It means that for some i, $\{a_i\} \leq \mathfrak{V}(\Box p)$ and yet $\mathfrak{V}(\neg p)$ is nonempty. Thus, there must exist a maximal point a_j in $\mathfrak{V}(\neg p)$ and ∞ $R_<$-sees a_j in one or two steps. For statement l, assume xRa_i and $\{a_i\} \leq \mathfrak{V}(\blacksquare p)$. But then $xR_< a_{i+2}$ and a_{i+2} is the one and only point $\mathfrak{V}$-satisfying the formula arising from the successor by erasing the initial $\Diamond_<$. Similar reasoning establishes m. Statement n is straightforward. For o, assume $\{w\} \leq \mathfrak{V}(\Box_< \Diamond_< p)$. It means that for every i, there exists $j > i$ s.t. $\{a_j\} \leq \mathfrak{V}(p)$, hence $\mathfrak{V}(p)$ is infinite. But then the complement of $\mathfrak{V}(p)$ must be finite and for some i, $a_i \leq \mathfrak{V}(\Box_< p)$. ⊣

LEMMA 45. $\Gamma \models_{\mathcal{T}} \neg\underline{\infty}$.

Proof. Assume that for some $\mathcal{T}$-BAO $\mathfrak{A}$ $\mathfrak{A} \not\models \underline{\infty} = \bot$ and there exists a conjugate $\mathbf{p}_<$ of $\Diamond_<$. We will show that $\underline{\infty} \leq \Box_< \Diamond_< \mathbf{p}_< \underline{\infty} \wedge$

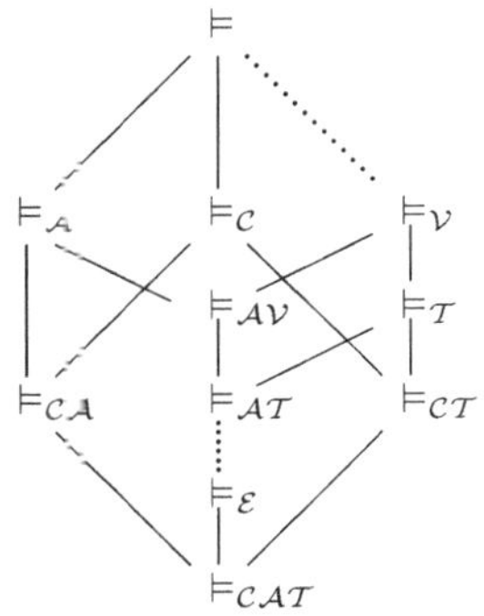

Figure 2. Dotted lines denote inclusions in whose case it is unclear whether they are proper or not.

$\Box_<\Diamond_<\neg\mathbf{p}_{<\underline{\infty}}$, thus contradicting the fact that $\mathfrak{A}$ validates McKinsey Axiom (statement o). That $\underline{\infty} \leq \Box_<\Diamond_<\mathbf{p}_{<\underline{\infty}}$ follows from n of the previous lemma. Assume now $\underline{\infty} \wedge \Diamond_<\Box_<\mathbf{p}_{<\underline{\infty}} \neq \bot$. By j, $\underline{\infty} \leq \Diamond^2_<\underline{1} \leq \Diamond^2_<\neg\mathbf{p}_{<\underline{\infty}}$. Thus, $\underline{\infty} \wedge \Diamond_<\Box_<\mathbf{p}_{<\underline{\infty}} \wedge \Diamond^2_<\neg\mathbf{p}_{<\underline{\infty}} \neq \bot$. By k, it means that

(p) $\Diamond_<\blacksquare\mathbf{p}_{<\underline{\infty}} \neq \bot$.

Define

$$c \leftrightharpoons \mathbf{p}_{<\underline{\infty}} \wedge \Box_<\mathbf{p}_{<\underline{\infty}} \wedge \Diamond^2_>\blacksquare\mathbf{p}_{<\underline{\infty}} \wedge \Box^3_>\neg\blacksquare\mathbf{p}_{<\underline{\infty}}.$$

By definition, $c \leq \mathbf{p}_{<\underline{\infty}}$. On the other hand, m and the fact that

$$\underline{\infty} \leq \Box_<\mathbf{p}_{<\underline{\infty}} \leq \Box_<(\mathbf{p}_{<\underline{\infty}} \vee \Diamond_<\neg\mathbf{p}_{<\underline{\infty}}) = \Box_<\neg\blacksquare\mathbf{p}_{<\infty}$$

imply $\mathbf{p}_{<\underline{\infty}} \leq \neg c$. Thus, $c = \bot$ but this contradicts l and p. ⊣

As $\mathfrak{A} \not\models \neg\underline{\infty}$, Theorem 42 follows. ⊣

Let us finish then with Figure 2 — a refined version of Figure 1. It is less symmetric, but perhaps more thought-provoking.

Acknowledgements. The author wishes to thank his supervisor Hiroakira Ono for his attention and criticism leading to rearrangement of the paper, Yde Venema and Tomasz Kowalski for stimulating discussions and convincing him the subject is worth writing a paper on, Valentin Shehtman for his correspondence, Johan van Benthem for all the inspiration he provided and anonymous referees for helpful comments. Thanks are also due to all members of the Ono Laboratory at JAIST for creating and maintaining a great environment to work.

BIBLIOGRAPHY

[1] P. Blackburn, M. de Rijke and Y. Venema, **Modal Logic**, Cambridge Tracts in Theoretical Computer Science 53, 2001.

[2] A.V. Chagrov and M.V. Zakharyaschev, **Modal Logic**, Clarendon Press, Oxford, 1997.

[3] K. Došen, *Duality between modal algebras and neighbourhood frames*, **Studia Logica**, 48:219–234, 1989.

[4] R.I. Goldblatt, **Mathematics of modality**, CSLI Lecture Notes 43, CSLI Publications, 1993.

[5] P. Jipsen, *Discriminator varieties of Boolean algebras with operators*, **Algebraic Logic**, Banach Center Publications vol. 28, Warsaw 1993.

[6] T. Kowalski, *Varieties of tense algebras*, **Reports on Mathematical Logic**, 32 (1998), pp. 53–95.

[7] M. Kracht, **Tools and Techniques in Modal Logic**, Studies in Logic and the Foundations of Mathematics vol. 142, Elsevier, 1999.

[8] M. Kracht, T. Kowalski, *Atomic incompleteness or how to kill one bird with two stones*, **Bulletin of the Section of Logic**, 30:71–78, 2001.

[9] K. Kuratowski, A. Mostowski, **Set Theory : with An Introduction to Descriptive Set Theory**, Studies in logic and the foundations of mathematics v. 86, Elsevier, 1976.

[10] T. Litak, F. Wolter, *All finitely axiomatizable tense logics of linear time flows are coNP-complete*, to appear in **Studia Logica**.

[11] T. Litak, *Modal incompleteness revisited*, **Studia Logica**, 76:329–342, 2004.

[12] S.K. Thomason, *Semantic analysis of tense logics*, **Journal of Symbolic Logic**, 37:150–158, 1972.

[13] S.K. Thomason, *Categories of frames for modal logic*, **Journal of Symbolic Logic**, 40:439–442, 1975.

[14] J.F.A.K. van Benthem, *Syntactic aspects of modal incompleteness theorems*, **Theoria**, 45:63–77, 1979.

[15] Y. Venema, *Atom structures*, **Advances in Modal Logic I**, California, CSLI, 1998, pp. 63–72.

[16] Y. Venema, *Atomless varieties*, **Journal of Symbolic Logic**, 68:607–614, 2003.

[17] F. Wolter, **Lattices of Modal Logics**, PhD Thesis, 1993.

[18] F. Wolter, *Tense logic without tense operators*, **Mathematical Logic Quarterly**, 42 (1996), 145–171.

[19] M. Zakharyaschev, F. Wolter and A. Chagrov, *Advanced modal logic*, [in:] **Handbook of Philosophical Logic, 2nd Edition**, vol. 3, Kluwer, 2001, pp. 83–266.

Tadeusz Litak
School of Information Science, JAIST
Asahidai 1-8, Nomi-shi, Ishikawa

923-1292 Japan
litak@jaist.ac.jp

Normal Modal Logics Containing KTB with some Finiteness onditions

YUTAKA MIYAZAKI

ABSTRACT. In this paper, we investigate normal modal logics over **KTB** with additional axioms which impose some finiteness conditions on Kripke frames for the logics. First we formulate the logics of reflexive and symmetric frames with finite *diameter*, and show some nice properties of such logics. We also prove of the existence of a continuum in the normal modal logics over $\mathbf{KTB} \oplus \Box^2 p \supset \Box^3 p$. Finally, we try to clarify the structure of the upper part of the lattice $Next(\mathbf{KTB})$.

1 Introduction

1.1 Background

The semantical study of propositional modal logics using Kripke type semantics has brought us a great success, especially to the investigation of the class of logics containing **K4**. This class of modal logics is characterized by the class of *transitive* frames. In order to analyze this class of logics, many algebraic and frame-theoretic techniques have been developed, and by using them, several general results have been established, which give us a perspective of the lattice of normal modal logics characterized by transitive frames (for example,[3],[6],[7],[17], [18]).

However, there is no effective way of clarifying the lattice structure of modal logics of non-transitive logics, and almost all techniques for logics of transitive frames are not valid, at least in their own form, for the logics of non-transitive frames. Therefore the study of logics of non-transitive frames has not yet grown to be so fruitful as that of logics of transitive frames.

Here we discuss the structure of the lattice of logics containing **KTB**, whose frames are reflexive and symmetric. Although the logic **KTB** has been known as *Brouwerian system*, there is very few results established about the structure of the lattice of extensions of **KTB**. In fact, various questions on the structure of this lattice remain open and untouched.

In this paper, we will focus on a little smaller class of the whole extensions of **KTB**, that is, a class of logics containing **KTB** that are characterized by

Advances in Modal Logic, Volume 5. ©

modal algebras, or general frames with some *finiteness* conditions. These conditions include '*diameter*' of frames and a weakened form of transitivity. These can be expressed by modal formulas, and they correspond to some nice first order conditions on frames. To be the most important, they enable us to employ the standard technique of propositional modal logic.

The construction of this paper is as follows. The rest of this section below is devoted to preliminary facts. We introduce the syntax of our logics, algebraic and frame semantics of our logics, and we explain the correspondence between these two types of semantics.

In section 2, we characterize subdirectly irreducible members and simple members of KTB- Kripke algebras, and give an example of KTB- Kripke algebras that is subdirectly irreducible but not simple. Here it is seen that every KTB- Kripke frame that corresponds to a subdirectly irreducible KTB- algebra can be regarded as a *connected* undirected graph in a graph-theoretical sense.

We introduce the notion of '*diameter*' of a KTB frame, and investigate some nice properties of KTB-logics of frames with finite diameter in section 3. In particular, we explain the splitting technique here, and we show that this standard technique is applicable if the class of KTB-logics are characterized by KTB- frames of finite diameter. In fact, it is proved here that every finite algebra for a KTB-logic of frames with finite diameter splits the lattice above that KTB-logic.

In section 4, we show that there exists a continuum of normal modal logics over $Next(\mathbf{KTB} \oplus \Box^2 p \supset \Box^3 p)$. The author proved in [14] of the existence of a continuum of normal modal logics over $Next(\mathbf{KTB})$, where he used an embedding of a continuum of orthologics into $Next(\mathbf{KTB})$. Here we make use of the standard technique of modal logic, that is, splitting and p-morphism of frames. This result is in stark contract to a well known fact that there are only countable many normal modal logics over $\mathbf{KTB} \oplus \Box p \supset \Box^2 p$ $(= \mathbf{S5})$.

In section 5, we investigate the structure of the upper part of the lattice $Next(\mathbf{KTB})$, and in particular, we show that the logic determined by a frame of two reflexive points jointed with a symmetric relation is the third greatest logic of all Kripke complete KTB-logics.

In the last section, we list up some open questions about the KTB-logics and the structure of the lattice $Next(\mathbf{KTB})$.

1.2 Preliminaries

We will follow notions and nomenclature of modal logics from [5]. But here, we want to recall the basic terminology of syntax and semantics of modal logics that we will use in this paper.

First of all, we fix our language and introduce propositional modal logics. Our language consists of: (1) a denumerable set of propositional variables $\{p_0, p_1, \ldots\}$, (2) a propositional constant $\bot$, (3) conjunction $\wedge$, (4) negation $\neg$, (5) box $\Box$, and (6) a pair of parentheses (,). The connectives $\vee$ (disjunction), $\supset$ (implication), and $\Diamond$ (diamond) can be treated as auxiliary connectives (abbreviations). The set Φ of formulas on our language is defined as usual.

$\mathbf{L} \subseteq \Phi$ is a *normal modal logic* (a logic for short) if: (1) $\mathbf{L}$ contains the classical tautologies, (2) $\Box(p \wedge q) \supset (\Box p \wedge \Box q) \in \mathbf{L}$, and (3) $\mathbf{L}$ is closed under modus ponens, substitution, and the rule of necessitation. The smallest normal modal logic is denoted by $\mathbf{K}$. For a set of formulas Γ, the smallest normal modal logic that includes both $\mathbf{K}$ and Γ is denoted by $\mathbf{K} \oplus \Gamma$. For a normal modal logic $\mathbf{L}$, the class of normal extensions of $\mathbf{L}$ is denoted by $Next(\mathbf{L})$.

In this paper, we concentrate on the class $Next(\mathbf{KTB})$, where $\mathbf{T} := \Box p \supset p$ and $\mathbf{B} := p \supset \Box\Diamond p$, and $\mathbf{KTB} = \mathbf{K} \oplus \mathbf{T} \oplus \mathbf{B}$. Each member of $Next(\mathbf{KTB})$ is called a *KTB-logic* for short.

Next we review semantics for propositional modal logics. The first one is the algebraic semantics. A structure $\mathfrak{A} = \langle \mathsf{A}, \cap, \cup, -, I, 0, 1 \rangle$ is a *modal algebra* if: (1) $\langle \mathsf{A}, \cap, \cup, -0, 1 \rangle$ is a Boolean algebra, and (2) I is a unary operator that satisfies $(i) I(1) = 1$ and $(ii) I(a \cap b) = I(a) \cap I(b)$ for $a, b \in \mathsf{A}$. Formulas are interpreted in a modal algebra by a valuation in the standard way and it is also defined in the standard way that a formula A is *valid* in a modal algebra $\mathfrak{A}$ ($\mathfrak{A} \models A$). For a class $\mathcal{K}$ of modal algebras, $\mathbf{L}(\mathcal{K}) := \{A \in \Phi \mid \mathfrak{A} \models A$ for any $\mathfrak{A} \in \mathcal{K}\}$ is called the normal modal logic determined by $\mathcal{K}$. A modal algebra $\mathfrak{A}$ is a *KTB-algebra* if it satisfies that $I(x) \leq x$ and that $x \leq I(-I(-x))$ for any $x \in \mathsf{A}$.

Another semantical tools are called *frames*. A structure $\mathcal{F} = \langle W, R, P \rangle$ is a (*general*) *frame* if: (1) W is not an empty set, (2) R is a binary relation on W, and (3) $P \subseteq \mathcal{P}(W)$ contains $\emptyset, W$ and closed under $\cap, -, I_R$, where $I_R(X) := \{x \in W \mid \forall y \in W(xRy$ implies $y \in X)\}$ for $X \in \mathcal{P}(W)$. A frame $\langle W, R \rangle := \langle W, R, \mathcal{P}(W) \rangle$ is called a *Kripke frame*. The way how we interpret each formula in a frame is in the standard way. A class $\mathcal{C}$ of frames also determines a normal modal logic as: $\mathbf{L}(\mathcal{C}) := \{A \in \Phi \mid \mathfrak{A} \models A$ for any $\mathfrak{A} \in \mathcal{C}\}$. As is well known, the axioms $\mathbf{T}$ and $\mathbf{B}$ correspond to *reflexivity* and *symmetry* of Kripke frames respectively, that is, the following holds: for any Kripke frame $\mathcal{F}$, $\mathcal{F} \models \mathbf{T}$ iff $\mathcal{F} \models \forall x(xRx)$, and $\mathcal{F} \models \mathbf{B}$ iff $\mathcal{F} \models \forall x, y(xRy$ implies $yRx)$. Here, the symbol $\models$ represents the relation that the frame $\mathcal{F}$ is a model of the first order condition on the right hand side. A reflexive and symmetric general frame is called a *KTB-frame*.

It is well established as *representation theory* that there exists a close relation between modal algebras and general frames. For any modal algebra $\mathfrak{A}$, we define $\mathfrak{A}_* = \langle W_{\mathfrak{A}}, R_{\mathfrak{A}}, P_{\mathfrak{A}} \rangle$ as follows: (1) $W_{\mathfrak{A}}$ is the set of all prime filters in A, (2) $R_{\mathfrak{A}}$ is the binary relation on $W_{\mathfrak{A}}$ defined as: for any $F, G \in W_{\mathfrak{A}}$, ${}_F R_{\mathfrak{A}G}$ iff for all $a \in \mathsf{A}$, $I(a) \in F$ implies $a \in G$, and (3) $P_{\mathfrak{A}} := \{\theta(a) | a \in \mathsf{A}\}$, where $\theta(a) := \{F \in W_{\mathfrak{A}} \mid a \in F\}$. Then, it is easily shown that $\mathfrak{A}_*$ is a general frame and that $\mathfrak{A}$ and $\mathfrak{A}_*$ validates the same set of formulas. Conversely, for any given general frame $\mathcal{F} = \langle W, R, P \rangle$, the corresponding modal algebra $\mathcal{F}^*$ can be defined as: $\mathcal{F}^* = \langle P, \cap, \cup, -, I_R, \emptyset, W \rangle$. Here, $\cap, \cup, -$ are the set theoretic operations, whereas I_R is the same one defined above. It is also easy to prove that both $\mathcal{F}$ and $\mathcal{F}^*$ validate the same set of formulas. Moreover, about two transformation $(\cdot)^*$ and $(\cdot)_*$, it is known that (1): $\mathfrak{A} \cong (\mathfrak{A}_*)^*$ for any modal algebra $\mathfrak{A}$ and that (2): $\mathcal{F} \cong (\mathcal{F}^*)_*$ iff the frame $\mathcal{F}$ is *descriptive*. We call an algebra $\mathfrak{A}$ a *Kripke algebra* if there exists a Kripke frame $\mathcal{F}$ such that $\mathfrak{A} \cong \mathcal{F}^*$.

2 Subdirectly irreducible and simple members of KTB-algebras

The first result in this paper is about subdirectly irreducible members, and simple members of KTB-algebras. An algebra $\mathfrak{A}$ is a *subdirect product* of an indexed family $\{\mathfrak{A}_i \mid i \in I\}$ of the same type if there exists a one-to-one homomorphism $f : \mathfrak{A} \to \prod_{i \in I} \mathfrak{A}_i$ such that for any $i \in I$, $\pi_i \circ f : \mathfrak{A} \to \mathfrak{A}_i$ is onto, where π_i is a projection map to i-th coordinate. Here the map f is called a *subdirect representation* of $\mathfrak{A}$. A non-trivial algebra $\mathfrak{A}$ is *subdirectly irreducible* (s.i. for short), if for any subdirect representation $f : \mathfrak{A} \to \prod_{i \in I} \mathfrak{A}_i$ of $\mathfrak{A}$, there exists $i \in I$ such that $\pi_i \circ f : \mathfrak{A} \to \mathfrak{A}_i$ is one-to-one.

The notion of subdirectly irreducible algebras is very important in universal algebra, since it is shown by G. Birkhoff ([1]) that every algebra is isomorphic to a subdirect product of subdirectly irreducible algebras of the same type. This famous theorem means that subdirectly irreducible algebras do form the building blocks of algebra. The notion is also important for logicians, because the following holds for logics determined by algebras.

PROPOSITION 1. *Let $\mathfrak{A}$ be a modal algebra that has a subdirect representation $f : \mathfrak{A} \to \prod_{i \in I} \mathfrak{A}_i$. Then $\mathbf{L}(\mathfrak{A}) = \bigcap_{i \in I} \mathbf{L}(\mathfrak{A}_i)$.* ■

Due to this proposition together with Birkhoff's theorem, when we consider a logic which is determined by a class of algebras, we have only to take into account the logics determined by a s.i. member of them, and focus on the intersection of such logics. The logics determined by subdirectly irreducible algebras do form the building blocks of algebraic logic. It is well known that a non-trivial $\mathfrak{A}$ is s.i. iff $\mathfrak{A}$ has the smallest non-trivial congruence. A non-trivial algebra $\mathfrak{A}$ is *simple* if the congruence lattice of $\mathfrak{A}$ consists of only two elements.

Subdirectly irreducible members, and simple members of modal algebras are well understood by the following characterization.

PROPOSITION 2. *Let* $\mathfrak{A} = \langle \mathsf{A}, \cap, \cup, -, I, 0, 1 \rangle$ *be a modal algebra.*

(1) *$\mathfrak{A}$ is subdirectly irreducible if and only if there exists $d \in \mathsf{A}$ ($d \neq 1$) such that for any $a \in \mathsf{A}$ ($a \neq 1$), there exists $n \in \omega$, $a \cap I(a) \cap \cdots \cap I^n(a) \leq d$.*

(2) *$\mathfrak{A}$ is simple if and only if for any $a \in \mathsf{A}$ ($a \neq 1$), there exists $n \in \omega$, $a \cap I(a) \cap \cdots \cap I^n(a) = 0$.* ■

On the above proposition, (1) is by Rautenberg ([15]). The element d in (1) is called an *opremum*. (2) is shown by the fact that for any non-empty subset $X \subseteq \mathsf{A}$, the open filter F generated by X is given by: $F = \{a \in \mathsf{A} \mid \exists x_1, \ldots, x_k \in X,$ and $\exists n_1, \ldots, n_k \in \omega, I^{n_1}(x_1) \cap \cdots \cap I^{n_k}(x_k) \leq a\ \}$. Here, in a modal algebra $\mathfrak{A}$, an *open filter* F is a filter which satisfies that $a \in F$ implies $I(a) \in F$ for any $a \in \mathsf{A}$. Note that if $\mathfrak{A}$ is a KT-algebra, that is $\mathfrak{A} \models \Box p \supset p$, then the above conditions can be slightly simplified as follows: (1): $\mathfrak{A}$ is s.i. if and only if there exists $d \in \mathsf{A}$ ($d \neq 1$) such that for any $a \in \mathsf{A}$ ($a \neq 1$), there exists $n \in \omega$, $I^n(a) \leq d$. (2): $\mathfrak{A}$ is simple if and only if for any $a \in \mathsf{A}$ ($a \neq 1$), there exists $n \in \omega$, $I^n(a) = 0$.

Now we prove the following characterization of s.i. and simple members of KTB- Kripke algebras. Below, R^n is defined for $n \in \omega$ as: xR^0y iff $x = y$, and $xR^{n+1}y$ iff there is $z \in W$ such that xR^nz and zRy.

THEOREM 3. *Let* $\mathcal{F} = \langle W, R \rangle$ *be a KTB- Kripke frame.*

(1) *$\mathcal{F}^*$ is subdirectly irreducible if and only if $\mathcal{F} \models \forall x, y, \exists n \in \omega\ (xR^ny)$.*

(2) *$\mathcal{F}^*$ is simple if and only if $\mathcal{F} \models \forall x, \exists n \in \omega, \forall y\ (xR^ny)$.*

Proof. (1): Suppose that $\mathcal{F}^*$ is s.i. Then, since $\mathcal{F}^*$ is a KT-algebra, we have that there exists $V \subsetneq W$ such that for any $U \subsetneq W$, for some $n \in \omega$, $I_R^n(U) \subseteq V$. Take arbitrary $x, y \in W$. Fix some element $w \in W - V$. First, put $U_0 := W - \{x\}$. Then by our condition, for this U_0, there exists $n \in \omega$ such that $I_R^n(U_0) \subseteq V$, which implies that $w \in -V \subseteq -I_R^n(U_0)$. Therefore we have in particular that wR^nx. Similarly we can show that wR^my for some $m \in \omega$. Thus we have $xR^{n+m}y$ because of symmetry of R. Conversely, suppose $\mathcal{F} \models \forall x, y, \exists n \in \omega(xR^ny)$. Here the following claim can be proved: for an arbitrary $x_0 \in W$, and for any $A \subset W$, there exists $n \in \omega$ such that

$I_R^n(A) \subseteq W - \{x_0\}$. In fact, for any $y \in W - A$, by our assumption there is some $n_0 \in \omega$ such that $x_0 R^{n_0} y$. Since $y \notin A$, this means $\{x_0\} \subseteq -I_R^{n_0}(A)$. Consider the open filter F in $\mathcal{F}^*$ that is generated from $W - \{x_0\}$. Then by the previous claim we can say that F is the smallest non-trivial open filter in $\mathcal{F}^*$, which corresponds to the smallest non-trivial congruence. Therefore $\mathcal{F}^*$ is subdirectly irreducible.
(2): Suppose that $\mathcal{F}$ is simple, which means that for any $U \subsetneq W$, there exists $n \in \omega$ such that $I_R^n(U) = \emptyset$. For any $x \in W$, put $U_0 := W - \{x\}$. Then for this U_0, there exists some $n_0 \in \omega$ such that $I_R^{n_0}(U_0) = \emptyset$. Therefore for any y, $y \in W = -I_R^n(U_0)$. Hence we have $xR^n y$. Conversely, suppose $\mathcal{F} \models \forall x, \exists n \in \omega, \forall y (xR^n y)$. Take an arbitrary non-trivial open filter F in $\mathcal{F}^*$. Then there exists $X \subsetneq W$ such that $X \in F$. Now, by our assumption, for any $x \in W - X$ there is some $n \in \omega$ such that for any $y \in W$ $xR^n y$. This means that $I_R^n(X) = \emptyset$. Since F is an open filter, $\emptyset \in F$, which implies that $F = \mathcal{P}(W)$. Therefore, there are only two open filters in $\mathcal{F}^*$. Hence $\mathcal{F}^*$ is simple. ■

In general, Kripke frames can be regarded as graphs, and reflexive and symmetry Kripke frames (KTB- Kripke frames) as undirected graphs. The above characterization shows that a KTB- Kripke frame which corresponds to a subdirectly irreducible algebra can be seen as a *connected* undirected graph. Here, the term *connected* is used in a graph-theoretical sense. As seen in an example below, any KTB- Kripke frame that corresponds to an s.i., but not simple algebra must be an infinite frame.

EXAMPLE 4. Let a Kripke frame $\mathcal{H} := \langle \omega, R \rangle$, where the relation R is defined as: mRn iff $|m-n| \leq 1$ for $m, n \in \omega$. Then, clearly this R is reflexive and symmetric, and so, $\mathcal{H}$ is a KTB- Kripke frame. By THEOREM 3, it is also easy to prove that $\mathcal{H}^*$ is s.i. but not simple.

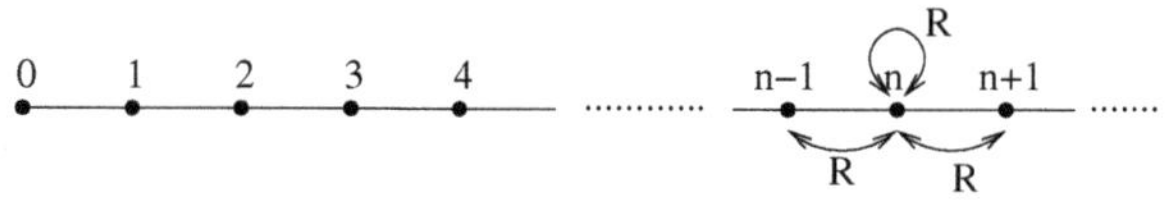

Figure 1. A frame $\mathcal{H}$ that corresponds to an s.i. but not simple algebra

To close this section, we make one more comment on the decomposition of a KTB- Kripke algebra into s.i. members. According to Blok's result ([2]), any Kripke algebra can be decomposed into s.i. Kripke algebras in a following way.

THEOREM 5. *Let $\mathcal{F} = \langle W, R \rangle$ be a Kripke frame. For each $x \in W$, define a Kripke frame $\mathcal{F}_x = \langle W_x, R_x \rangle$, where $W_x := \{a \in W \mid \exists n \in \omega\, xR^n a\}$, and $R_x = R \cap (W_x \times W_x)$. Then, for each $x \in W$, $\mathcal{F}_x^*$ is subdirectly irreducible, and $\mathcal{F}^*$ can be represented as a subdirect product of $\{\mathcal{F}_x^* \mid x \in W\}$, where the subdirect representation $f : \mathcal{F}^* \to \prod_{x \in W} \mathcal{F}_x^*$ is determined by $(f(X))_x := X \cap W_x$ for $X \in \mathcal{P}(W)$.* ■

For a KTB- Kripke frame $\mathcal{F} = \langle W, R \rangle$, define a binary relation $\approx$ on W as: $x \approx y$ if there exists $n \in \omega$ such that $xR^n y$. Then, because of symmetry of R, this $\approx$ turns out to be an equivalence relation, and so, $x \approx y$ iff $W_x = W_y$ for any $x, y \in W$. Therefore, in employing the above theorem to decompose a given KTB- Kripke algebra $\mathcal{F} = \langle W, R \rangle$, we factorize W by the relation $\approx$ first, and we collect one representative x_λ from every equivalence class W_λ, and then we can obtain our desirable subdirect representation of $\mathcal{F}$ from $\{\mathcal{F}_{x_\lambda}{}^* \mid x_\lambda \in W_\lambda\}$.

In this way, every KTB- Kripke algebra is decomposed into s.i. algebras. Therefore, when we consider a KTB-logic which is Kripke complete, we may assume that a class of KTB- Kripke frames, each member $\mathcal{F}$ of which satisfies $\mathcal{F} \models \forall x, y, \exists n \in \omega (xR^n y)$.

3 KTB-logics determined by frames with finite diameter

In this section, we will consider a finiteness condition on KTB- frames and KTB- logics determined by such frames. For transitive frames, a notion of '*depth*' of a frame is defined, by which locally tabular members of $Next(\mathbf{K4})$ can be characterized, that is, it is shown that a logic $\mathbf{L} \in Next(\mathbf{K4})$ is locally tabular if and only if it is characterized by a class of frames with finite depth ([11],[16]). However, in our case, the depth of a KTB- frame cannot be defined because of symmetry of our frame. Here we will introduce the notion of '*diameter*' of a frame, and present some nice properties of KTB logics characterized by a class of KTB frames with finite diameter.

Let $\mathcal{F} = \langle W, R, P \rangle$ be a general frame. A *path* (of length n) in $\mathcal{F}$ is a finite sequence $\{x_i\}_{i=0}^n \subseteq W$ of distinct points which satisfies $x_i R x_{i+1}$ for $i = 0, 1, \ldots, n-1$. The *diameter* of $\mathcal{F}$ is the maximum length of a path that $\mathcal{F}$ contains. It may happen that the diameter of a frame is infinite.

DEFINITION 6. For $n \geq 1$, a frame $\mathcal{F}$ is *n-bounded* if the diameter of $\mathcal{F}$ is at most $n-1$, specifically $\mathcal{F}$ is a model of the following first order sentence $\delta(n)$:

$$\delta(n) := \forall x_0, x_1, \ldots, x_n [x_0 R x_1 R x_2 \cdots x_{n-1} R x_n \text{ implies } \mho_{i \neq j}(x_i = x_j)]$$

$\mathcal{F}$ is *bounded* if it is n-bounded for some $n \geq 1$. ∎

In the above condition, $\mho$ is a logical connective of "big or". For each n, n-bounded Kripke frames can be characterized by a modal formula. For this purpose, first we define an exclusive sequence of formulas. An *exclusive sequence of formulas* is a sequence $\{F_i\}_{i=0}^{n}$ $(n \geq 1)$ of classical formulas (i.e. formulas on our language without $\Box, \Diamond$) such that (1) each F_i is satisfiable in one-point frame, or in the classical logic, and that (2) every pair of formulas F_i, F_j $(i \neq j)$, $F_i \wedge F_j$ is not satisfiable at any one point in a frame. Of course, we can easily construct an exclusive sequence of formulas for any $n \geq 1$. For example, if $n = 1$, then we can provide a sequence $\{F_i\}_{i=0}^{3}$, where $F_0 := p = 0 \wedge p = 1$, $F_1 := \neg p_0 \wedge p_1$, $F_2 := p_0 \wedge \neq p_1$, and $F_3 := \neg p_0 \wedge \neg p_1$. Of course, we can choose other combination of classical formulas to fulfill the conditions of an exclusive sequence. Now, we define a formula $\mathbf{D}_n$ by employing an exclusive sequence of formulas $\{F_i\}_{i=0}^{n}$.

$$\begin{aligned} \mathbf{D}_n &:= \neg\{F_0 \wedge \Diamond(F_1 \wedge \Diamond(\cdots \wedge \Diamond(F_{n+1} \wedge \Diamond F_n)\cdots))\} \\ &= F_0 \supset \Box\{F_1 \supset \Box(\cdots \supset \Box(F_{n-1} \supset \Box\neg F_n)\cdots)\} \end{aligned}$$

Then the following holds.

LEMMA 7. *Let $\mathcal{F} = \langle W, R\rangle$ be a Kripke frame and $n \geq 1$. Then $\mathcal{F} \models \mathbf{D}_n$ if and only if $\mathcal{F} \models \delta(n)$*

Proof. Suppose $\mathcal{F} \not\models \delta(n)$. Then there exists a sequence of distinct points $x_0, x_1, \ldots, x_n \in W$ such that $x_0 R x_1 R \cdots R x_n$. Since $\{F_i\}_{i=0}^{n}$ is exclusive, we can define a valuation V on $\mathcal{F}$ as: for each i, $x_i \models F_i$. Then the formula $\neg\mathbf{D}_n$ is satisfiable at x_0 in $\mathcal{F}$, which means $(\mathcal{F}, V) \not\models_{x_0} \mathbf{D}_n$. Thus we have $\mathcal{F} \not\models \mathbf{D}_n$. Conversely, suppose $\mathcal{F} \models \delta(n)$, and consider any valuation V on $\mathcal{F}$. Take arbitrary points $x_0, \ldots, x_n \in W$ such that $x_0 \models F_0$, $x_{i-1} R x_i$ and $x_i \models F_i$ for $1 \leq i \leq n-1$, and $x_{n-1} R x_n$. Then, because of $\delta(n)$, there exists at least one pair (x_i, x_j) such that $x_i = x_j$. But since $\{F_i\}_{i=0}^{n}$ is exclusive, we can see that for some i $(0 \leq i \leq n-1)$, $x_i = x_n$. Therefore, $x_i = x_n \models F_i$, which means that $x_n \not\models F_n$. Thus, we have $(\mathcal{F}, V) \models_{x_0} \mathbf{D}_n$, and so $\mathcal{F} \models \mathbf{D}_n$ ∎

By this lemma, it is obvious that the logic $\mathbf{K} \oplus \mathbf{D}_n$ does not depend on what its exclusive sequence of formulas $\{F_i\}_{i=0}^{n}$ really looks like. To be more important, this lemma also says that for $n \geq 1$, the logic $\mathbf{K} \oplus \mathbf{D}_n$ is elementary. This fact implies that these logics have the following nice property. Below, a logic $\mathbf{L}$ is *d-persistent* if for every descriptive frame $\mathcal{F} = \langle W, R, P\rangle$, $\mathcal{F} \models \mathbf{L}$ implies $\kappa\mathcal{F} \models \mathbf{L}$, where $\kappa\mathcal{F} = \langle W, R\rangle$ (underlying Kripke frame of $\mathcal{F}$).

THEOREM 8. *For every $n \geq 1$,*

(1) $\mathbf{K} \oplus \mathbf{D}_n$ *is d-persistent, in particular canonical.*

(2) $\mathbf{K} \oplus \mathbf{D}_n$ *is Kripke complete.*

(3) $\mathbf{K} \oplus \mathbf{D}_n$ *has the finite model property.* ■

For the proof of (3) above, we can appeal to the filtration method. Since the axioms $\mathbf{T}$ and $\mathbf{B}$ will not cause any trouble along the proof, we can apply the same argument to the logic $\mathbf{KTB} \oplus \mathbf{D}_n$ to show the following theorem on it.

THEOREM 9. *For every $n \geq 1$,*

(1) $\mathbf{KTB} \oplus \mathbf{D}_n$ *is d-persistent, in particular canonical.*

(2) $\mathbf{KTB} \oplus \mathbf{D}_n$ *is Kripke complete.*

(3) $\mathbf{KTB} \oplus \mathbf{D}_n$ *has the finite model property.* ■

Note that it is easily seen that every bounded KTB- Kripke frame corresponds to a simple algebra, and that we can build an infinite KTB- Kripke frame which gives us a simple algebra. It is still unknown whether the notion of diameter of a frame can characterize the locally tabular members of $Next(\mathbf{KTB})$.

4 Splittings in $Next(\mathbf{KTB})$

The notion of splitting was first introduced in lattice theory, and was used for classifying some subvarieties of the variety of lattices. In modal logic, this technique was used for investigating the structure of the lattice $Next(\mathbf{L})$ for some particular logics $\mathbf{L}$. Specifically, by splitting technique, the problem of the degree of Kripke incompleteness was solved by Blok ([2]) for the lattices $Next(\mathbf{K})$, $Next(\mathbf{KT})$, and $Next(\mathbf{KD})$, where the axiom $\mathbf{D}$ is $\Box p \supset \Diamond p$. It has not yet become clear whether there is any valuable splitting in the lattice $Next(\mathbf{KTB})$. Instead of the whole $Next(\mathbf{KTB})$, we investigate splittings of $Next(\mathbf{KTB} \oplus \mathbf{D}_n)$ and splittings of $Next(\mathbf{KTB} \oplus \Box^n p \supset \Box^{(n+1)} p)$ for $n \in \omega$. In consequence of the latter analysis, we show that there exists a continuum of normal modal logics over $\mathbf{KTB} \oplus \Box^2 p \supset \Box^3 p$.

4.1 Splittings in $Next(\mathbf{KTBD_n})$ and in $Next(\mathbf{KTB} \oplus \Box^n p \supset \Box^{(n+1)} p)$

First of all, we recall the notion of *splitting* of a lattice, that is often used for investigating the lattice of modal logics.

DEFINITION 10 (Splitting). Let $\mathcal{L} = \langle L, \wedge, \vee, 0, 1\rangle$ be a complete lattice and $a \in L$. Then a *splits* $\mathcal{L}$ if there exists $b \in L$ such that for any $x \in L$, either $x \leq a$ or $b \leq x$, but not both. Such a pair (a, b) is called a *splitting pair* of the lattice $\mathcal{L}$. In this case the element b is a *splitting* of $\mathcal{L}$.

When we think about splittings of a lattice of logics, if a logic $\mathbf{L}(\mathfrak{A})$ (or $\mathbf{L}(\mathcal{F})$) splits the lattice, then we say that the algebra $\mathfrak{A}$ (or the frame $\mathcal{F}$) splits it. Next we explain a general splitting theorem by Rautenberg ([15]).

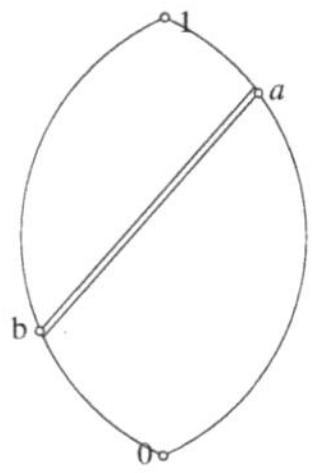

Figure 2. A splitting of a complete lattice $\mathcal{L}$

Consider the following formula: $t^m := (p \wedge \Box p \wedge \cdots \wedge \Box^m p) \supset \Box^{(m+1)} p$ for $m \in \omega$. This axiom corresponds to the following first order condition on Kripke frames. For a Kripke frame $\mathcal{F} = \langle W, R\rangle$,

$$\mathcal{F} \models t^m \quad \text{if and only if} \quad \mathcal{F} \models \forall x, y[xR^{(m+1)}y \text{ implies } \mho_{i=0}^{m}(xR^i y)]$$

The condition on the right means that if it is possible to go from x to y by $m+1$ steps, then that is possible by at most m steps. This condition is a weaker form of usual transitivity and is called *m-transitivity.*

THEOREM 11 (Rautenberg). *Let* $\mathbf{L}_0 \in Next(\mathbf{K} \oplus t^m)$ $(m \in \omega)$. *Let* $\mathfrak{A}$ *be a finite, subdirectly irreducible modal algebra such that* $\mathfrak{A} \models \mathbf{L}_0$. *Then* $\mathfrak{A}$ *splits* $Next(\mathbf{L}_0)$. ∎

Proof of this theorem proceeds in the following. For each element $a \in \mathsf{A}$, prepare propositional variable p_a, and let δ be the conjunction of all the following formulas. For any $a, b \in \mathsf{A}$,

$$p_a \wedge p_b \leftrightarrow p_{a\cap b}, \qquad p_a \vee p_b \leftrightarrow p_{a\cup b}, \qquad \neg p_a \leftrightarrow p_{-a}, \qquad \Box p_a \leftrightarrow p_{Ia}$$

Here $\alpha \leftrightarrow \beta := (\alpha \supset \beta) \wedge (\beta \supset \alpha)$. Put $\kappa_{\mathfrak{A}} := (\delta \wedge \Box\delta \wedge \cdots \wedge \Box^m \delta) \supset p_d$, where d is an opremum of $\mathfrak{A}$. This $\kappa_{\mathfrak{A}}$ is called the *characteristic formula* for $\mathfrak{A}$. Then it can be verified that for any algebra $\mathfrak{B}$ that validates t^m, the following three conditions are equivalent.

(1) $\mathfrak{A} \in SH(\mathfrak{B})$. (This means that $\mathfrak{A}$ is a subalgebra of a homomorphic image of $\mathfrak{B}$.)

(2) $\mathbf{L}(\mathfrak{B}) \subseteq \mathbf{L}(\mathfrak{A})$.

(3) $\kappa_{\mathfrak{A}} \notin \mathbf{L}(\mathfrak{B})$.

Now let $\mathbf{L}^{\sharp} := \mathbf{L}_0 \oplus \kappa_{\mathfrak{A}}$. For any $\mathbf{L} \in Next(\mathbf{L}_0)$, there is a modal algebra $\mathfrak{C}$ such that $\mathbf{L} = \mathbf{L}(\mathfrak{C})$ and $\mathfrak{C} \models t^m$. Then by the above equivalence, $\mathbf{L} = \mathbf{L}(\mathfrak{C}) \subseteq \mathbf{L}(\mathfrak{A})$ if and only if $\kappa_{\mathfrak{A}} \notin \mathbf{L}(\mathfrak{C})$ if and only if $\mathbf{L}_0 \oplus \kappa_{\mathfrak{A}} \nsubseteq \mathbf{L}(\mathfrak{C})$. Hence $(\mathbf{L}(\mathfrak{A}), \mathbf{L}^{\sharp})$ is a splitting pair.

It is rather difficult to find splittings in $Next(\mathbf{KTB})$. The reason seems that the variety of KTB-algebras does not have the *EDPC* (equationally definable principal congruence). So we discuss here splittings in $Next(\mathbf{KTB} \oplus \mathbf{D}_n)$ and in $Next(\mathbf{KTB} \oplus \Box^n p \supset \Box^{(n+1)} p)$. By the first order characterization of the formula $\mathbf{D}_n$, it is easy to show that for a modal algebra $\mathfrak{A}$ and for $n \in \omega$, if $\mathfrak{A} \models \mathbf{D}_n$ then $\mathfrak{A} \models t^n$. Therefore in considering splittings in $Next(\mathbf{KTB} \oplus \mathbf{D}_n)$, we can make use of Rautenberg's theorem. Because every finite KTB- algebra is a Kripke algebra, and so a finite KTB- algebra for $\mathbf{D}_n$ is simple by THEOREM 3, the following holds.

THEOREM 12. *Let $\mathfrak{A}$ be an arbitrary finite KTB-algebra such that $\mathfrak{A} \models \mathbf{D}_n$ $(n \in \omega)$. Then $\mathfrak{A}$ splits $Next(\mathbf{KTBD}_n)$.* ■

Here, the characteristic formula $\kappa_{\mathfrak{A}}$ for $\mathfrak{A}$ is $\Box^n \delta \supset p_0$, because we have the axiom $\mathbf{T}$ and in this case an opremum is 0. The obtained splitting pair is $(\mathbf{L}(\mathfrak{A}), \mathbf{KTBD}_n \oplus \kappa_{\mathfrak{A}})$

It is easily proved that for any finite KTB- algebra $\mathfrak{A}$, if $\mathfrak{A}_* \models \mathbf{D}_n$ for some $n \in \omega$, then $\mathfrak{A}_* \models \mathbf{D}_m$ for any $m (m > n)$. Therefore, THEOREM 12 says that any finite KTB-algebra splits the lattice $Next(\mathbf{KTBD}_\ell)$ for every sufficiently large $\ell \in \omega$. From the opposite point of view, this means that if there exists a KTB-logic that prevents some KTB-algebra from being a splitting algebra of $Next(\mathbf{KTB})$, it must be the logic determined by infinite s.i. algebras such as $\mathcal{H}^*$ in EXAMPLE 4, and that logic must be located in the bottom part of the lattice $Next(\mathbf{KTB})$.

Now we consider another splittings. Let $\mathbf{Tra}(n, n+1) := \Box^n p \supset \Box^{(n+1)} p$. Apparently this formula is a generalization of transitivity, indeed the following first order characterization holds. For a Kripke frame $\mathcal{F} = \langle W, R \rangle$,

$$\mathcal{F} \models \mathbf{Tra}(n, n+1) \ \text{ if and only if } \ \mathcal{F} \models \forall x, y (x R^{(n+1)} y \text{ implies } x R^n y)$$

The following can be proved by the similar argument as in the previous theorem.

THEOREM 13. *Let $\mathfrak{A}$ be an arbitrary finite KTB-algebra such that $\mathfrak{A} \models \mathbf{Tra}(n, n+1)$ $(n \in \omega)$. Then $\mathfrak{A}$ splits $Next(\mathbf{KTB} \oplus \mathbf{Tra}(n, n+1))$.* ■

Here, again the characteristic formula $\kappa_{\mathfrak{A}}$ for $\mathfrak{A}$ is $\Box^n\delta \supset p_0$. We will use this theorem for $n = 2$ in the next part.

4.2 The existence of a continuum of logics over $\mathbf{KTB} \oplus \Box^2 p \supset \Box^3 p$

First, we introduce the tool of *p-morphism* between two general frames.

DEFINITION 14. Let $\mathcal{F} = \langle W, R, P\rangle$ and $\mathcal{G} = \langle U, S, Q\rangle$ be general frames. A map $f : W \to U$ is a *p-morphism* from $\mathcal{F}$ to $\mathcal{G}$, if f satisfies the following:

(pi) f is onto.

(pii) For all $x, y \in W$, xRy implies ${}_{f(x)}S_{f(y)}$

(piii) For all $x \in W$ and for all $a \in U$, if ${}_{f(x)}S_a$, then there exists $y \in W$ such that xRy and $f(y) = a$.

(piv) For all $X \in Q$, $f^{-1}(X) \in P$.

There is a close connection between the fact that there is a p-morphism from one frame to another and the fact that one an algebra is a subalgebra of another. Specifically the following holds.

PROPOSITION 15. *Let $\mathcal{F} = \langle W, R, P\rangle$ and $\mathcal{G} = \langle U, S, Q\rangle$ be general frames, and suppose that there is a p-morphism from $\mathcal{F}$ to $\mathcal{G}$. Then $\mathcal{G}^*$ is a subalgebra of $\mathcal{F}^*$.* ■

We have also the converse proposition as follows.

PROPOSITION 16. *Let $\mathfrak{A}, \mathfrak{B}$ be modal algebras, and suppose that $\mathfrak{B}$ is a subalgebra of $\mathfrak{A}$. Then there exists a p-morphism from $\mathfrak{A}_*$ to $\mathfrak{B}_*$.* ■

We will use the following corollary derived from the above two propositions in the sequel.

COROLLARY 17. *Let $\mathcal{F}, \mathcal{G}$ be finite Kripke frames such that both $\mathcal{F}^*, \mathcal{G}^*$ are simple. Then $\mathbf{L}(\mathcal{F}) \subseteq \mathbf{L}(\mathcal{G})$ if and only if there exists a p-morphism from $\mathcal{F}$ to $\mathcal{G}$.*

Proof. For an algebra $\mathfrak{A}$, let $V(\mathfrak{A})$ denote the variety generated by $\mathfrak{A}$. For a class $\mathcal{C}$ of algebras, let $(\mathcal{C})_{SI}$ denote the class of all subdirectly irreducible members in $\mathcal{C}$, and $P_u(\mathcal{C})$ the class of *ultraproducts* of members in $\mathcal{C}$. Now $\mathbf{L}(\mathcal{F}) \subseteq \mathbf{L}(\mathcal{G})$ iff $\mathbf{L}(\mathcal{F}^*) \subseteq \mathbf{L}(\mathcal{G}^*)$ iff $V(\mathcal{G}^*) \subseteq V(\mathcal{F}^*)$ iff $\mathcal{G}^* \in (V(\mathcal{F}^*))_{SI} \subseteq HSP_u(\mathcal{F}^*)$ by Jónsson's lemma ([9]). Further, by the congruence extension property of modal algebras, finiteness and simplicity of $\mathcal{F}^*$, this is equivalent to $\mathcal{G}^* \in SH(\mathcal{F}^*) = S(\mathcal{F}^*)$. Hence, by the above propositions, this means equivalently that there exists a p-morphism from $(\mathcal{F}^*)_*$ to $(\mathcal{G}^*)_*$. These are isomorphic to $\mathcal{F}$ and $\mathcal{G}$ respectively. ■

Now we will prove the existence of a continuum of logics in $Next(\mathbf{KTB} \oplus \Box^2 p \supset \Box^3 p)$. Roughly speaking, we will construct an infinite series of finite frames $\{\mathcal{F}_n\}_{n\in\omega}$ and an infinite series of formulas $\{A_m\}_{m\in\omega}$ such that (1): for every $n \in \omega$, $\mathcal{F}_n \not\models A_n$ and (2): for every $m \in \omega (m \neq n)$, $\mathcal{F}_m \not\models A_n$. We employ the following frames.

DEFINITION 18 (Wheel frame). For $n \in \omega, n \geq 5$, the *wheel frame* $\mathcal{W}_n = \langle W, R\rangle$ of degree n consists of the following underlying set and binary relation: $W = \text{rim}(W) \cup \{h\}$, where $\text{rim}(W) := \{0, 1, \ldots, n-1\}$ and $h \notin \text{rim}(W)$. An element in $\text{rim}(W)$ is called a *rim element*, whereas the element h the *hub element*. The relation R is defined as: $R := \{(x,y) \in (\text{rim}(W))^2 \mid |x-y| \leqq 1 (\text{mod}\ n)\} \cup \{(h,h)\} \cup \{(h,x),(x,h) \mid x \in \text{rim}(W)\}$.

Note that $|x - y|$ means the absolute value of the difference, and for example $|(n-1) - 0| = |n-1| \equiv 1 (\text{mod}\ n)$. The following is the wheel frame $\mathcal{W}_7$. It is trivial that for $n \geq 5$, $\mathcal{W}_n \models \mathbf{B}, \mathbf{T}$ and $\mathcal{W}_n \models \mathbf{Tra}(2,3)$.

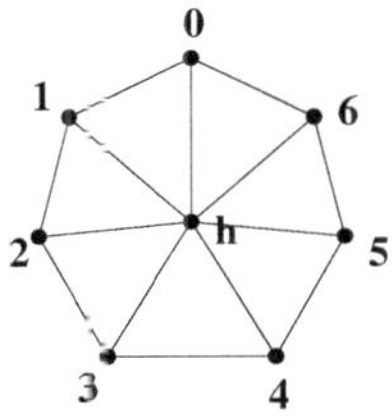

Figure 3. The wheel frame $\mathcal{W}_7$

The following proposition can be also verified.

PROPOSITION 19. *For $m > n \geq 5$, $\mathbf{L}(\mathcal{W}_n) \not\subseteq \mathbf{L}(\mathcal{W}_m)$.*

Proof. Because $m > n$, $\mathcal{W}_m \not\models \mathbf{D}_m$, but $\mathcal{W}_n \models \mathbf{D}_m$. Thus we have $\mathbf{D}_m \in \mathbf{L}(\mathcal{W}_n)$, but $\mathbf{D}_m \notin \mathbf{L}(\mathcal{W}_m)$. ■

The following lemma is crucial to our result in this section.

LEMMA 20. *For $m \geq n \geq 5$, suppose there is a p-morphism from $\mathcal{W}_m$ to $\mathcal{W}_n$. Then m is divisible by n.*

Proof. Let $\mathcal{W}_m := \langle W, R\rangle$, where $W = \text{rim}\ (W) \cup \{h\} = \{0, 1, \ldots m-1, h\}$ and $\mathcal{W}_n := \langle U, S\rangle$, where $U = \text{rim}\ (U) \cup \{\underline{h}\} = \{\underline{0}, \underline{1}, \ldots, \underline{n-1}, \underline{h}\}$. This means that we distinguish elements in $\mathcal{W}_n$ from elements in $\mathcal{W}_m$ by underlining. Let $f : W \to U$ be the p-morphism from $\mathcal{W}_M$ to $\mathcal{W}_n$. For the relation R, let $R(a) := \{x \in W \mid aRx\}$ for $a \in W$. The same notation is used

for the relation S. We consider hub elements and rim elements separately.
(i) On hub elements

We claim that $f(h) = \underline{h}$ and $f^{-1}(\underline{h}) = \{h\}$. Because, suppose there is $\underline{a} \in \text{rim}\,(U)$ such that $f(h) = \underline{h}$. For hub element h, we have hRx for all $x \in W$, so ${}_{f(h)}S_{f(x)}$ in $\mathcal{W}_n$ for all $x \in W$. But for the rim element $\underline{a} \in U$, card $(S(\underline{a})) = 4$. If $f(z) \in S(\underline{a})$ for all $z \in W$, then this contradicts to the condition that f is onto because $n \geq 5$. Thus we have $f(h) = \underline{h}$. Moreover, suppose there exists $b \in \text{rim}\,(W)$ such that $b \in f^{-1}(\underline{h})$. Then since card$(R(b)) = 4$ and $n \geq 5$, we can pick up an element $c \in W$ such that ${}_{f(b)}S_{f(c)}$ but there is no $z \in W$ that satisfies bRz and $f(z) = f(c)$. This is also a contradiction to (piv) of p-morphism. Thus we have $f^{-1}(\underline{h}) = \{h\}$.
(ii) On rim elements

By (i) above, we have that for any $x \in \text{rim}\,(W)$, $f(x) \in \text{rim}\,(U)$, and that for any $\underline{x} \in \text{rim}\,(U)$, $f^{-1}(\underline{x}) \subseteq \text{rim}\,(W)$. We claim here that for any distinct elements $x, y, z \in\in \text{rim}\,(W)$ such that $xRyRz$, $f(x), f(y), f(z)$ are all distinct and ${}_{f(x)}S_{f(y)}S_{f(z)}$ holds. Now we go on to case analysis.

(Case A): $f(x) = f(y) = \underline{a}$.
Let $\underline{b}, \underline{c} \in \text{rim}\,(U)$ be elements that are next to $\underline{a}$, that is, $\underline{b}S\underline{a}S\underline{c}$. Since $n \geq 5$, $\underline{b}, \underline{a}, \underline{c}$ can be all distinct. Then, by ${}_{f(y)}S_{\underline{b}}$ and $f(x) = \underline{a} \neq \underline{b}$, $z \in f^{-1}(\underline{b})$ holds. Also, by ${}_{f(y)}S_{\underline{c}}$ and $f(x) = \underline{a} \neq \underline{c}$, $z \in f^{-1}(\underline{c})$ holds. This is a contradiction. Therefore the case $f(x) = f(y)$ never happens. Similarly, the case $f(y) = f(z)$ never happens.

(Case B): $f(x) = f(z) = \underline{a}$.
Let $\underline{b}, \underline{c} \in \text{rim}(U)$ be elements that are next to $\underline{a}$, that is, $\underline{b}S\underline{a}S\underline{c}$. Of course, $\underline{b}, \underline{a}, \underline{c}$ can be all distinct

(subcase B-1): $f(y) = \underline{a}$.
We have ${}_{\underline{b}}S_{f(y)}$, but $f(x) = f(z) = \underline{a}$, which means that $x, z \notin f^{-1}(\underline{b})$. This contradicts to (piv) of conditions for p-morphism f.

(subcase B-2): $f(y) = \underline{b}$.
Let $\underline{d} \in \text{rim}\,(U)$ be the element next to $\underline{b}$ but not $\underline{a}$. Then we have ${}_{\underline{d}}S_{f(y)}$, but $x, z \notin f^{-1}(\underline{d})$. This also contradicts to (piv) of conditions for p-morphism f.

(subcase B-3): $f(y) = \underline{c}$.
Similar treatment can be applied to show that this case is impossible.

Our case analysis is completed and the claim has been justified. Now we will show that m is divisible by n. If $m = n$, then we have nothing else to prove. If $m > n$, then there exists $\underline{t} \in \text{rim}\,(U)$, and $x, y \in \text{rim}\,(W)(x \neq y)$ such that $x, y \in f^{-1}(\underline{t})$. Let $\underline{s}, \underline{u} \in \text{rim}\,(U)$ be elements that are next to

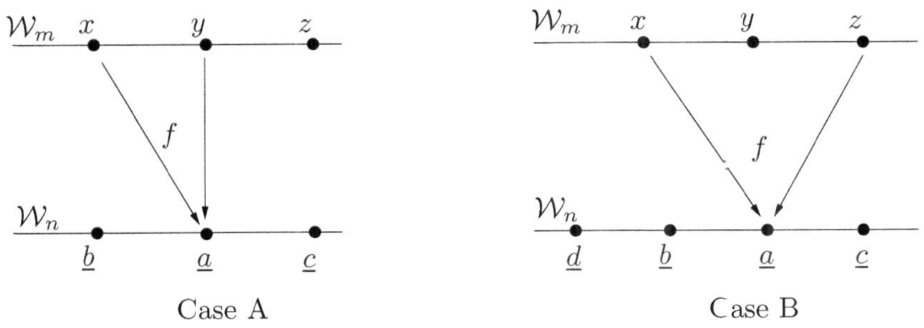

Figure 4. Case A and Case B

$\underline{t}$, that is, $\underline{s}S\underline{t}S\underline{u}$. Here $\underline{s}, \underline{t}, \underline{u}$ are all distinct. Then by our claim that we have shown above, we can find distinct elements $x', x'', y', y'' \in$ rim (W) such that $x', y' \in f^{-1}(\underline{s})$, $x'', y'' \in f^{-1}(\underline{u})$, $x'RxRx''$, and $y'RyRy''$ hold. Put $k := \max_{\underline{x} \in \text{rim}\,(U)} \text{card}\,(f^{-1}(\underline{x}))$. By the observation just above, if card $(f^{-1}(\underline{a})) = k$ for some $\underline{x} \in$ rim (U), then for elements $\underline{b}, \underline{c}$ that are next to $\underline{a}$, card $(f^{-1}(\underline{b}))$ = card $(f^{-1}(\underline{c})) = k$. Thus we can state that card $(f^{-1}(\underline{z})) = k$ for all $\underline{z} \in$ rim (U). Hence we have that $m = k \cdot n$. ■

Now we are in a position to prove our main result in this part. Let $Prim := \{n \in \omega \mid n \text{ is prime }, \ m \geq 5\}$. The following theorem holds.

THEOREM 21. *For any $p \in Prim$, there exists a formula A_p such that:*

(1) $\mathcal{W}_p \not\models A_p$.

(2) $\mathcal{W}_q \models A_p$ *for all* $q \in Prim$ $(p \neq q)$

Proof. As already mentioned in this section, $\mathcal{W}_p^*$ is a finite simple KTB-algebra and $\mathcal{W}_p^* \models \mathbf{Tra}(2,3)$. Then by THEOREM 11, $(\mathbf{L}(\mathcal{W}_p^*), \mathbf{KTB} \oplus \mathbf{Tra}(2,3) \oplus A_p)$ is a splitting pair of $Next(\mathbf{KTB} \oplus \mathbf{Tra}(2,3))$, where $A_p := \Box^2\delta \supset p_0$ is the characteristic formula for $\mathcal{W}_p^*$ defined just below THEOREM 11. Denote $\mathbf{L}^\sharp := \mathbf{KTB} \oplus \mathbf{Tra}(2,3) \oplus A_p$. Because of this splitting pair, we have in particular, $A_p \notin (\mathbf{L}(\mathcal{W}_p^*)$, which means that $\mathcal{W}_p \not\models A_p$.
Furthermore, take any $q \in Prim$ $(q \neq p)$. If $q > p$, then, since q is not divisible by p, there does never exist any p-morphism from $\mathcal{W}_q$ to $\mathcal{W}_p$ by LEMMA 20. This means by COROLLARY 17, that $\mathbf{L}(\mathcal{W}_q^*) \not\subseteq \mathbf{L}(\mathcal{W}_p^*)$, thus $\mathbf{L}^\sharp \subseteq \mathbf{L}(\mathcal{W}_p^*)$ by splitting. Therefore we have $\mathcal{W}_q \models A_p$. If $p < q$, then by PROPOSITION 19, $\mathbf{L}(\mathcal{W}_q^*) \not\subseteq \mathbf{L}(\mathcal{W}_p^*)$, thus also by splitting, $\mathbf{L}^\sharp \subseteq \mathbf{L}(\mathcal{W}_p^*)$. Therefore $\mathcal{W}_q \models A_p$. ■

COROLLARY 22. *There exists a continuum of normal modal logics over* $\mathbf{KTB} \oplus \mathbf{Tra}(2,3)$. ■

This proof is one of the quite standard ways to show an existence of a continuum of modal logics, by employing a series of finite algebras and using the splitting technique effectively. Aside from the fact that it turned out to be possible to apply such a standard tool of modal logic to KTB-logics, there is the following significance of this result.

The author showed already the existence of a continuum in $Next(\mathbf{KTB})$ in [14], where he first presented a continuum of orthologics , and then used an embedding from orthologics to KTB-logics. In that proof, he used a class of ortholattices to show the continuum of orthologics, where such ortholattices are not orthomodular ([13]). Therefore, the resulting continuum of KTB-logics is located not very near from the top of the lattice $Next(\mathbf{KTB})$. The continuum we obtained in this study, instead, is in $Next(\mathbf{KTB} \oplus \mathbf{Tra}(2,3))$, which means that this is located in quite upper part of $Next(\mathbf{KTB})$.

It is well known that there exists only countably many normal modal logics in $Next(\mathbf{S5})$, and that $\mathbf{S5}$ is pretabular. In context of this paper, it is helpful to rewrite $\mathbf{S5} = \mathbf{KTB} \oplus \mathbf{Tra}(1,2)$. Then we recognize that there lies a boundary of countability or uncountability of a class of normal modal logics between $\mathbf{KTB} \oplus \mathbf{Tra}(1,2)$ and $\mathbf{KTB} \oplus \mathbf{Tra}(2,3)$.

5 The upper part of the lattice $Next(\mathbf{KTB})$

As we mentioned above, the logic $\mathbf{L}(\bullet)$ is the second greatest element in $Next(\mathbf{KTB})$. Then what kind of logics follow $\mathbf{L}(\bullet)$? A plausible candidate is the logic $\mathbf{L}(\bullet\!\!-\!\!\bullet)$, where . $\bullet\!\!-\!\!\bullet$ is a frame of two reflexive points jointed with symmetric relation. In fact, by Jónsson's lemma ([9]) we can show that above it are only $\mathbf{L}(\bullet)$ and the inconsistent logic. Moreover, it can be proved by the following argument that the logic $\mathbf{L}(\bullet\!\!-\!\!\bullet)$ is the third greatest of all Kripke complete logics in $Next(\mathbf{KTB})$.

We denote the frame $\bullet\!\!-\!\!\bullet$ by $\mathcal{G}_2$, that is, $\mathcal{G}_2 = \langle V, S\rangle$, where $V := \{0,1\}$ and $S := \{(0,0),(1,1),(0,1),(1,0)\}$. Then, we can show the following.

LEMMA 23. *Let* $\mathcal{F} = \langle W, R\rangle$ *be a KTB- Kripke frame, where* $\mathcal{F}^*$ *is subdirectly irreducible and* $|W| \geq 2$. *Then* $\mathbf{L}(\mathcal{F}) \subseteq \mathbf{L}(\mathcal{G}_2)$

Proof. Since $\mathcal{F}^*$ is s.i., we have $\mathcal{F} \models \forall x, y \exists n \in \omega(xR^n y)$ by THEOREM 2. We define a partition $\{W^{(i)}\}_{i=0}^{\infty}$ of W in the following: For $i = 0$, pick up an arbitrary $x_0 \in W$ and $W^{(0)} := \{x_0\}$. For $i = n + 1$, $W^{(n+1)} := \{y \in W \mid xRy \text{ for some } x \in W^{(n)}\} - \bigcup_{i=0}^{n} W^{(i)}$. Then by this construction, $|W| \geq 2$, and our first order condition for $\mathcal{F}$, we have $W^{(0)} \neq \emptyset$, $W^{(1)} \neq \emptyset$, and $\bigcup_{i=0}^{\infty} W^{(i)} = W$. Here a map $f : W \to \{0, 1\}$ is defined as: for $a \in W$,

$$f(a) = \begin{cases} 0 & (\text{if } a \in W^{(2k)} \text{ for some } k \in \omega) \\ 1 & (\text{if } a \in W^{(2k+1)} \text{ for some } k \in \omega) \end{cases}$$

Then about this f, first, because $W^{(0)} \neq \emptyset$ and $W^{(1)} \neq \emptyset$, we can say that f is onto. Suppose aRb for $a, b \in W$. Then by our construction of $W^{(i)}$, we can see that one of the following cases occurs, that is, (1): for some m, $a \in W^{(m)}$ and $b \in W^{(m+1)}$, (2): for some m, $a \in W^{(m+1)}$ and $b \in W^{(m)}$, (3): for some m, $a, b \in W^{(m)}$. In case (1) and (2), ${}_{f(a)}S_{f(b)}$ holds due to the definition of f. In (3), ${}_{f(a)}S_{f(b)}$ also holds since S is reflexive. Last, suppose ${}_{f(a)}Sz$ for $a \in W$ and $z \in \{0, 1\}$. Then, because $W^{(0)}, W^{(1)} \neq \emptyset$ and R is reflexive, it is obvious that there exists $b \in W$ such that aRb and $f(b) = z$. Thus we have checked that this f is a p-morphism from $\mathcal{F}$ to $\mathcal{G}_2$, and so, by COROLLARY 17 , $\mathbf{L}(\mathcal{F}) \subseteq \mathbf{L}(\mathcal{G}_2)$ holds. ■

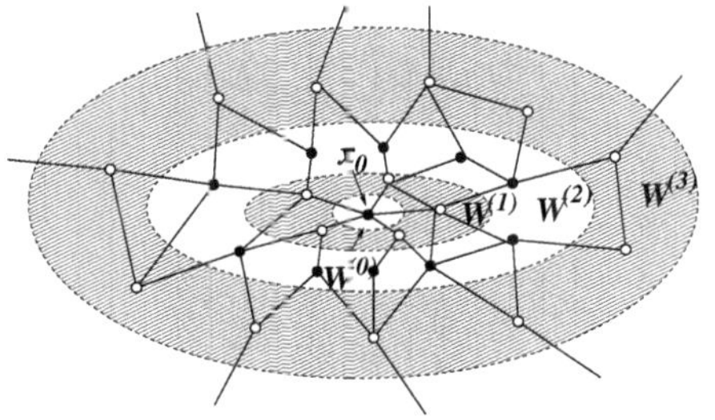

Figure 5. The partition of $\mathcal{F}$

From a graph theoretical point of view, this lemma can be reformulated into the following way. Namely, every countable, undirected, and connected graph (here *connected* means the graph theoretical sense), whose number of nodes are greater than one, can be colored in two colors, say, black and white, in such a way that any black node can see at least a white node, and vice a versa. This fact suggests that an existence of a p-morphism from a KTB- Kripke frame $\mathcal{F}$ to a KTB- Kripke frame $\mathcal{G}$ has something to do with the possibility of coloring the undirected graph $\mathcal{F}$ in some *coloring pattern* determined by the frame $\mathcal{G}$. By the above lemma, we have the following.

THEOREM 24. *The KTB-logic* $\mathbf{L}(\bullet\!-\!\bullet)$ *is the third greatest logics of all Kripke complete KTB-logics.* ■

In fact, by using Jónsson's lemma for some finite frames, we can draw a picture of the upper part of the lattice $Next(\mathbf{KTB})$. The figure below shows how the lattice structure of the upper part of $Next(\mathbf{KTB})$, which consists of connected frames of, at most, four reflexive points looks like. Each of ten circles represents the logic determined by the Kripke frame in it. In Blok's paper ([2]), the following result is established on the lattice

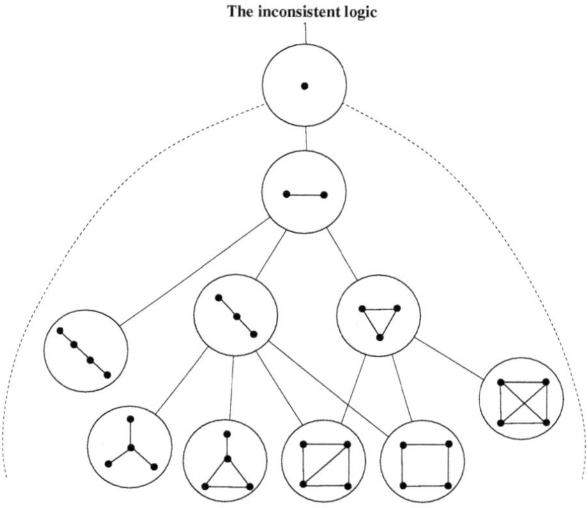

Figure 6. The upper part of $Next(\mathbf{KTB})$

$Next(\mathbf{KT})$.

THEOREM 25. *For any consistent logic* $\mathbf{L} \in Next(\mathbf{KT})$*, there exists a continuum of logics in* $Next(\mathbf{KT})$ *that are covered by* $\mathbf{L}$. ■

Of course this theorem can be applied to the logic $\mathbf{L}(\bullet)$ to show that there are uncountably many predecessors of this logic in $Next(\mathbf{KT})$. When the axiom **B** comes in their sight, it is not clear how many logics can survive of all predecessors of $\mathbf{L}(\bullet)$ in $Next(\mathbf{KT})$. But there might be a possibility that THEOREM 25 can be transformed, in some way, into the fact about the class $Next(\mathbf{KTB})$ and that the fact will show there is no third greatest logic in $Next(\mathbf{KTB})$.

6 Remarks and questions

The author cannot help saying that the study of the lattice $Next(\mathbf{KTB})$ here is very far from complete at present, because this paper treats only a very restricted class of the whole KTB-logics. But the approach in this study of assuming some conditions in order to make the object logics tractable seems to be reasonable, in the face of the situation that there is very few known facts about this class of logics.

The logics in $Next(\mathbf{KTB})$ are extremely difficult to investigate by nature, which is mentioned in several texts and papers in modal logics (for instance,[5],[8],[10]). In fact, there are only few facts known about this class of logics at present. Namely,

(1) There exists a continuum in $Next(\mathbf{KTB})$ ([14]).

(2) The cardinality of pretabular members in $Next(\mathbf{KTB})$ is at least countably infinite ([12]).

(3) The universal frame of rank 2 for $\mathbf{KTB} \oplus \Box^2 p \supset \Box^3 p$ is infinite ([4]).

To finish off this paper, we will list up some open questions about the KTB-logics and the structure of the lattice $Next(\mathbf{KTB})$. We expect that many researchers are getting interested in this area of modal logics, and they will find answers to the following questions. Moreover, we hope that the investigation of this class of logics will lead to a discovery of some new techniques and new perspectives for modal logics of broader classes rather than logics of transitive frames.

(Q1): Is the logic $\mathbf{L}(\bullet\!\!-\!\!\bullet)$ the third greatest element of $Next(\mathbf{KTB})$?

(Q2): What is the degree of incompleteness of $\mathbf{L}(\bullet)$ with respect to $\mathbf{KTB}$?

(Q3): Is there any KTB-logic that is finitely axiomatizable but not Kripke complete?

(Q4): Is there any KTB-logic that is finitely axiomatizable but not decidable?

(Q5): Is there any KTB-logic that is decidable but without the finite model property? ([8])

(Q6): Is it possible to show that for any $n \in \omega$, every normal modal logic over $\mathbf{KTBD_n}$ has the finite model property?

(Q7): How are locally tabular members in $Next(\mathbf{KTB})$ characterized? Can the diameter be the index of characterizing locally tabular members in $Next(\mathbf{KTB})$?

(Q8): Does there exist a continuum of pretabular members in $Next(\mathbf{KTB})$? ([12])

BIBLIOGRAPHY

[1] Birkhoff G., *On the structure of abstract algebras*, Proceedings of the Cambridge Philosophical Society, 31, 433–454, 1935.

[2] Blok W.J., *On the degree of incompleteness in modal logics and the covering relation in the lattice of modal logics*, Technical Report 78-07, Department of Mathematics, University of Amsterdam, 1978.

[3] Blok W.J., *The lattice of modal logics : an algebraic investigation*, Journal of Symbolic Logic, 45, 221–236 (1980).

[4] Byrd M.,*On the addition of weakened L-reduction axioms to the Brouwer system*, Zeitschrift für Mathematische Logik und Grundlagen der Mathematik, 24, 405–408 (1978).

[5] Chagrov A., Zakharyaschev M., *Modal logic*, Oxford University Press, 1997.

[6] Fine K., *Logics containing K4, part I*, Journal of Symbolic Logic, 39, 229–237 (1974).

[7] Fine K., *Logics containing K4, part II*, Journal of Symbolic Logic, 50, 619–651 (1985).

[8] Gabbay D.M.,*On decidable, finitely axiomatizable modal and tense logics without the finite model property II*, Israel Journal of Mathematics, 10, 496–503 (1971).

[9] Jónsson B., *Algebras whose congruence lattices are distributive*, Mathematica Scandinavica, 21, 110–121 (1967).

[10] Makinson D.C., *A normal modal calculus between T and S4 without the finite model property*, Journal of Symbolic Logic, 34, 35–39 (1969).

[11] Maksimova L.L., *Modallogics of finite slices*, Algebra and Logic, 14, 188–197, 1975.

[12] Meskhi V.Yu., *Critical modal logics containing the Brouwer axiom*, Mathematical Notes, 33, 65–69 (1983).

[13] Miyazaki Y., *Some properties of orthologics*, To appear in Studia Logica (2005).

[14] Miyazaki Y., *Binary logics, orthologics and their relations to normal modal logics*, in: Advances in Modal Logic, 4, eds. by P. Balbiani et al, King's College Publications (2003), 313–333.

[15] Rautenberg W., *Splitting lattices of logics*, Archiv für Mathematische logik, 20, 155–159 (1980).

[16] Segerberg S., *An essay in classical modal logic*, Philosophical Studies, Uppsala, 13, 1971.

[17] Zakharyaschev M., *Canonical formulas for K4, part I:Basic results*, Journal of Symbolic Logic, 57, 1377–1402 (1992).

[18] Zakharyaschev M., *Canonical formulas for K4, part II:Cofinal subframe logics*, Journal of Symbolic Logic, 61, 421–449 (1996)

Yutaka Miyazaki
Meme Media Laboratory, Hokkaido University
N13 W8 Kita-ku Sapporo, 060-8628 Japan
y-miya@meme.hokudai.ac.jp

On the Formal Structure of Continuous Action

THOMAS MÜLLER

ABSTRACT. Analytical investigations of agency are mostly concerned with a description *ex post acto*. However, continuous action (being doing something) needs to be considered as well. The paper shows that while the modal-logical treatment of agency in branching time-based stit theory is currently unable to handle continuous action, the stit framework can be extended such as to handle these cases as well. Our new operator, *istit*, provides for an adequate expression of the notion of being doing something, and we present a simple axiomatisation. In our extended framework, agency, ability, and refraining are linked to an agent's current strategy.

1 Introduction

Since the 1980ies, a number of agency-related concepts have been explored using the resources of modal logic. The key idea, sometimes dubbed the "Anselmian approach" since there is textual evidence for the analysis in some writings of St. Anselm's, is that acting is best described in terms of an agent's bringing about some state of affairs. Thus, the concept of agency is seen to give rise to a family of (agent-indexed) modal operators. A natural reading for these modalities is "α sees to it that ϕ", abbreviated as "α *stit* : ϕ". The stit-modalities have been given a formally rigorous semantics in the framework of branching time; the approach is laid out and well argued for by Belnap *et al.* in their recent book, *Facing the Future* [3].

The present paper is an attempt at extending that formally rigorous modal-logical treatment of agency to cover the case of continuous action, which present stit theory does not cover. We will present two related, but differently motivated challenges, one action-theoretic, the other linguistic. From these challenges, which point out a striking feature of the phenomenology of continuous action, we will derive conditions of adequacy for formal analysis. While current stit theory does not meet these conditions, we argue that they can be met by adding a new stit modality.

The paper is structured as follows: In section 2, we introduce the stit framework, giving both philosophical motivation and the key formal defini-

Advances in Modal Logic, Volume 5. © *2005, the author.*

tions. In section 3, we present the two mentioned challenges, derive formal conditions of adequacy, and argue that the existing stit framework does not meet these conditions. In section 4, we enrich the growing zoo of stit modalities by introducing our new operator, *istit* ("... is seeing to it that"), and we show that this operator meets the conditions of adequacy. The new operator involves a new index of evaluation for the formal language, s (for "strategy"), which provides a link between formal modeling on the one hand and our everyday mentalistic vocabulary on the other hand. Finally, in section 5 we present an axiomatisation for the new *istit* modality.

2 A brief outline of *stit* theory

2.1 A metaphysical presupposition

How should one set out to analyse agency, or to defend some specific analysis? Questions of methodology always loom large in philosophy, and in understanding agency, some of these questions are crucial. Before considering the phenomenon of continuous action, it is therefore important to spell out a metaphysical presupposition of this paper: We assume that agency presupposes indeterminism. This means that there can be no agency if the future is not open, if it does not contain more than one possible course of events. To some this may seem like a bold assumption, taking sides in the endless debate about compatibilism. A full defence is certainly out of place here. However, the assumption is inevitable from the methodological standpoint of descriptive metaphysics that we adopt. Descriptive metaphysics, so-called by Strawson [11], means that in deciding metaphysical questions (such as the one about determinism or indeterminism), one needs to stick to the conceptual scheme that one is actually using, as shown by a broad range of linguistic, cultural, legal and other practices. Surely a metaphysical notion of an open future is deeply entrenched in all our agency-related concepts, from deciding to attributing responsibility, praise, and blame. Thus, our formal analysis needs to honour the concept of an open future. The framework, which is based on the indeterministic theory of branching time, is therefore the most natural starting point.[1]

2.2 The formalities of *stit*

A structure is an indeterministic branching time model with agents and choices, $\mathfrak{S} = \langle M, \leq, Agents, Choice\rangle$. The substructure $\langle M, \leq\rangle$ is a branching time structure, i.e., M is a nonempty set of moments partially ordered by $\leq$, which ordering is tree-like, so it satisfies the axiom of "no backward

[1] Cf. [4] for a proposed extension to branching space-times.

branching",

$$\forall x \forall y \forall z \left((x < z \wedge y < z) \rightarrow (x \leq y \vee y \leq x) \right),$$

and any two moments have a common lower bound:

$$\forall x \forall y \exists z (z \leq x \wedge z \leq y).$$

Further postulates are discussed in [3, Chap. 7].[2] Maximal linear subsets of $\mathfrak{S}$ are called *histories.* In $\mathfrak{S}$, branching occurs where histories diverge. At each moment $m \in M$, the set $H_{(m)}$ of histories containing m is partitioned via the equivalence relation $\equiv_m$ of being undivided at m, where $h_1 \equiv_m h_2$ iff $m \in h_1 \cap h_2$ and there is $m' \in h_1 \cap h_2$ such that $m < m'$. Π_m is the respective partition of $H_{(m)}$. If Π_m has more than one element, then at m there is (indeterministic) splitting of histories.

The set *Agents* describes which agents there are in the model. In this paper, we consider a single agent for simplicity only, so the set $Agents = \{\alpha\}$. The function $Choice^\alpha$ determines the choices open for agent α at any moment. This information is needed since the metaphysical basis for agency is not just indeterminism, as encoded by Π_m, but agent-related indeterminism. Thus, $Choice^\alpha_m$ is a partition of $H_{(m)}$ that describes the set of choices open for agent α at moment m. The partition $Choice^\alpha_m$ may be more coarse-grained, but not more fine-grained, than the partition Π_m: An agent has at most as fine a "control" over what will happen as nature's indeterminism allows. If $Choice^\alpha_m$ has only the one element $H_{(m)}$, this means that at moment m, agent α has no choice. Metaphysically, the structure $\mathfrak{S}$ is taken to be a formal picture of the ontology. This means that the partitions $Choice^\alpha_m$ are given by nature and cannot be changed by the agent.[3]

As we consider a propositional formal language based on stit structures only, a stit model $\mathfrak{M} = \langle \mathfrak{S}, V \rangle$ consists of a stit structure $\mathfrak{S}$ and a valuation V that maps moment-history pairs to subsets of the set of propositional letters. In accord with standard two-dimensional semantics, sentences are evaluated at an index of evaluation, which usually consists of a context of utterance and some more indexes. In one-agent stit theory, the context is taken to be a moment of utterance m_c, and formulae are evaluated additionally at a moment-history pair m, h, where $m \in h$, to allow for standard Prior-Thomason tense operators P ("it was the case that") and F ("it will

[2]Besides in-depth coverage of the formalities of , the book [3] also presents further uses of the framework and gives historical notes as well as an extensive bibliography.

[3]If you wish, you may read an existentialist note into this: According to stit theory, our freedom is not just given, but forced upon us by the way the world is.

be the case that"). The semantics of the weak Occamist future tense operator F is the following:

$$\mathfrak{M}, m_c, m, h \models F\phi \quad \text{iff there is } m' \in h \text{ with } m < m' \text{ s.t. } \mathfrak{M}, m_c, m', h \models \phi.$$

The weak past tense operator P ("it was the case that") employs the mirror image of this, exchanging "$m' < m$" for "$m < m'$". The corresponding strong operators are denoted G ("it is always going to be the case that") and H ("it has always been the case that"). Apart from these two pairs of modal operators for shifting the m parameter, there is also a pair of so-called historical modalities, $[h]$ and $\langle h \rangle$, that shift the h parameter. The semantic clause for the strong operator $[h]$, called "historical necessity" or "settled truth", is:

$$\mathfrak{M}, m_c, m, h \models [h]\phi \quad \text{iff for all } h' \in H_{(m)}, \quad \mathfrak{M}, m_c, m, h' \models \phi.$$

The dual weak operator $\langle h \rangle$ is called "historical possibility". Note that the truth of $F\phi$ depends on the history of evaluation index h, whereas the truth of $P\phi$ depends only on the moment of evaluation m—if $P\phi$ is true at m, h, then so is $[h]P\phi$. Accordingly, operators like P are called *moment-determinate*, while F is called *moment-indeterminate*.

There is a choice of two stit operators available: the "deliberative stit", *dstit*, due to von Kutschera [14], and the "achievement stit", *astit*, due to Belnap and Perloff [2], which is based on extended stit structures.

dstit To start with the simpler of the two, the semantics for *dstit* is as follows:

$\mathfrak{M}, m_c, m, h \models \alpha\, dstit : \phi$ iff

1. for all $h' \in Choice^{\alpha}_{m}(h)$ we have $\mathfrak{M}, m_c, m, h' \models \phi$ and
2. there is a "counter" $h' \in H_{(m)}$ such that $\mathfrak{M}, m_c, m, h' \not\models \phi$.

The first clause is positive: It states that the current choice of α (singled out from $Choice^{\alpha}_{m}$ through the history h in the index of evaluation) secures the truth of ϕ. The negative second clause, on the other hand, excludes those ϕ that are true under any circumstances: nobody sees to it that $2 + 2 = 4$.

Sometimes a simplified version of *dstit* called *cstit* is considered: The semantics for *cstit* employs only the first clause of the *dstit* semantics (cf., e.g., [8]). Note that both *cstit* and *dstit* support the inference from $\alpha\, stit : \phi$ to ϕ (as $h \in Choice^{\alpha}_{m}(h)$).

astit The *dstit* operator considers a current choice that brings about ϕ. The *astit* operator, on the other hand, expresses the idea that, as compared to simultaneous possibilities in other histories, a previous choice of α (or a previous chain of such choices) secured some present truth. In order to express the notion of simultaneous possibilities across histories, *astit* requires an equivalence relation of "same clock time" across histories. Without going into details of the semantics (for which cf. [3, chap. 2]), we note that *astit* also supports the inference from $\alpha\, astit : \phi$ to ϕ. Thus, all the existing stit operators satisfy that inference:

$$\text{If } \mathfrak{M}, m_c, m, h \models \alpha\, stit : \phi, \text{ then also } \mathfrak{M}, m_c, m, h \models \phi.$$

3 Continuous action: Two challenges

3.1 Action theory: Two kinds of examples of agency

In analyses of agency, the examples used point to two differing approaches.[4] Consider the following examples from Davidson's "The logical form of action sentences" [6]:

- Jones buttered the toast in the bathroom with a knife at midnight. (107)
- The doctor removed the patient's appendix. (111)
- Amundsen flew to the north pole. (115)

In *Intention* [1], Anscombe mostly uses examples of the following kind:

- I'm pumping. (38)
- He is replenishing the water-supply. (39)
- She is making tea. (40)

The first difference that one may notice is one of tense: Davison's examples are in the past tense, whereas Anscombe's are in the present tense. Still more importantly, the examples differ with respect to verbal *aspect*: Davidson considers actions in the perfective aspect (from a point of view after the action is finished). Anscombe uses the imperfective aspect (from a point of view while the action is occurring), as marked by the present continuous. Analytical investigations of agency have mostly been concerned with Davidson-type examples. Stit theory is no exception in this respect: the examples considered are usually of the Davidsonian variety. Certainly these examples are important, but a full account of agency needs to consider Anscombe-type examples of continuous action, too. The two classes of examples differ with respect to what may be called their "phenomenology".

[4] This point is made by Thompson [12].

Davidson: When an action is finished, the following account will be adequate for many purposes: First a certain initial state of affairs obtained. Due to the agent's action, some outcome state of affairs obtained later. The agent saw to it that a transition from initial to outcome occurred. Consider apple peeling.[5] *Ex post acto*, the situation seems to be simple enough: First the apple wasn't peeled, then it was peeled, and the agent did it.[6] Stit theory is able to handle many cases of that kind smoothly, e.g., via the *dstit* operator introduced above.

Anscombe: While an agent is acting, the phenomenology is different. E.g., the ongoing action of apple peeling may be described like this: Some time in the past, the agent decided to peel, and from then until the present moment, she chose (when she had a choice) in such a way as to continue peeling. Before the peeling will be finished, the agent will have more choices. In particular, if the agent is acting freely, we need to allow for "dropping out", i.e., not finishing. The later choices need to be made when they are due; they cannot be made now ("no choice before its time"). But still, when the agent is really peeling, there will be "defaults" for the later choices. We may say that the agent is now committed to the future defaults, but still she cannot *choose* these defaults now. Thus, the following points are crucial for continuous action:

1. If an agent is doing something, she has defaults set for her future choices.
2. The future choices are real choices nonetheless: When the time comes, the default is not forced upon the agent.
3. An agent may be truly said to be doing something even though it may turn out later that she didn't finish.

If stit theory is to provide for a general analysis of agency, it must allow for such cases.

3.2 Linguistics: The "imperfective paradox"

As we have just pointed out, action theory has mostly ignored the phenomenon of continuous action, at least as far as formal analysis is concerned.

---

[5]Besides having the advantage of not involving manslaughter in the way most action-theoretic examples (including St. Anselm's!) do, apples have a venerable tradition in practical philosophy, as witnessed, e.g., by Kant [9] and, more recently, by Segerberg [10] and Xu [16].

[6]Certainly a transition from apple not peeled to peeled involves an agent's continuous action in any case. However, *ex post acto* it is often feasible to ignore this as an unimportant detail.

In linguistics, the situation is different: The progressive tense, which is employed to describe continuous action, is so pervasive in English that it could not be ignored for long. Soon after tense logic had been established as a modal-logical framework for analysing tenses, attempts were made to tackle the progressive in a formally rigorous, modal-logical way as well.[7] It was soon pointed out that the phrase "progressive tense" is misleading, as "the progressive is not simply a temporal operator, but a kind of mixed modal-temporal operator" [7, p. 146]. As mentioned above, the distinguishing feature of the progressive is the imperfective verbal aspect.

One main linguistic observation about the imperfective aspect takes up a point that was noted above. The so-called imperfective paradox consists of the observation that a sentence like

(1) Carlos is building a house.

can be true even though, in fact, Carlos does not finish, and no house comes into existence.[8] There are too many instances of this phenomenon to attempt to "explain it away". One might be inclined to argue that if no house comes into existence, that will show that the sentence (1) was uttered inappropriately, but this is not in accord with well-established usage and thus contrary to linguistic methodology. We understand perfectly well what goes on in such cases. In fact we can further differentiate, as cases of not finishing can be due to "external" or "internal" factors. Consider the following sentences:

(2) John was crossing the street when a bus hit him.

(3) God was creating a unicorn, when he changed his mind.

In sentence (2), there is outside intervention: We picture a situation in which John did everything he could be expected to do in order to be crossing the street, but still a bus hit him (unexpectedly).[9] In contrast, the situation pictured in sentence (3) is such that it is entirely up to the agent whether the result will be achieved or not. Note that this is not something exclusive to god: We all perform tasks from which we may drop out, and we all show

[7] Dowty [7, p. 145] points to early work by Montague and Scott, among others.

[8] Our examples are taken from [13], but they have been around in the linguistics literature for longer ([13] gives detailed references).

[9] The way the situation is described already shows that context will play an important role in assessing outside influence: What may count as crossing under normal, lax standards will not count as crossing (but as "running into the street" or some other action) when standards are raised. Cf. the next note.

what we call in our case "weakness of the will" every so often.[10]

Dowty's early and influential analysis of the "imperfective paradox" [7, chap. 3] stays within the modal-logical framework of tense logic, introducing the additional concept of so-called "inertia worlds" to account for the modal implications of the progressive. This addition has been quite controversial; we will discuss it in section 4.3 below. A different reaction to the phenomenon has been to abandon the modal-logical approach in favour of some event-based calculus. A sophisticated framework of that kind has recently been proposed by van Lambalgen and Hamm [13]. We wish to show that it is possible to stick to a simple and transparent modal-logical approach, but this requires doing justice to the "imperfective paradox".

3.3 Common conditions of adequacy

The action-theoretic and the linguistic challenge can be taken together under a single heading: A modal-logical analysis of agency must (a) be able to account for Anscombe-type examples and (b) offer a solution to the "imperfective paradox". In formal terms, this means that we wish to express "α is ϕ-ing" via some modal operator (or combination of operators) $O_\alpha \phi$ such that (a) O_α makes intuitive sense as a "progressive operator" and (b) it is possible to have a model $\mathfrak{M}$ and index of evaluation I s.t.

$$\mathfrak{M}, I \models (P\,O_\alpha\,\phi) \;\wedge\; (\neg\phi \wedge \neg P\phi \wedge \neg F\phi),$$

i.e., while it was true that α was ϕ-ing, α never finished, nor will finish. Figure 1 illustrates such a case. Read ϕ as "the apple will be peeled": At m, you may say that earlier (at m') you were peeling the apple, but now, it will not be peeled after all.

3.4 Why current stit theory fails the test of adequacy

As was pointed out at the end of section 2, the existing stit operators support the inference from $\alpha\, stit : \phi$ to ϕ. Thus in particular, for any stit model $\mathfrak{M}$ and all of the three stit operators, *dstit*, *cstit*, and *astit*, we have

$$\mathfrak{M}, m_c, m, h \models P\,\alpha\, stit : \phi \rightarrow P\phi.$$

This means that the existing stit operators fail the test of adequacy in every model. Current stit theory simply does not have the resources for expressing defeasible defaults.

[10] Context plays a much weaker role in assessing those cases.—While discussions of agency often assume that "weakness of the will" is a specially problematic phenomenon, in the light of our formal analysis these cases are simpler than cases of "outside influence", since the latter imply that the agent somehow misrepresented the situation (albeit in a way that is held to be excusable in the given context). Cf. note 14 below for some further remarks.

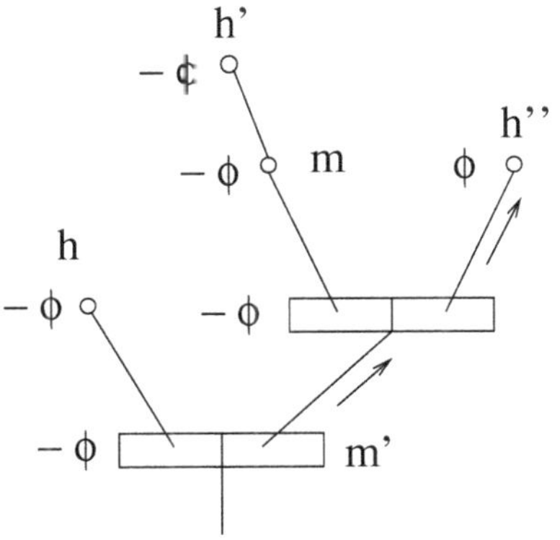

Figure 1. A picture of the conditions for adequacy: The framework needs to allow for models like the one shown, where at m, it was true (namely, at m') that α is ϕ-ing (arrows indicating defaults), but still, at m, ϕ was, is, and will be false. Histories are denoted h, h', and h''.

4 A proposed solution

How should a positive formal account of continuous action look like? In the previous section it was argued that the stit framework so far does not have the necessary resources. Thus there are two options—either to abandon the stit framework or to extend it. In our view, the second option is to be preferred, since the branching time basis of stit theory gives a formally perspicious and well-understood picture of the metaphysical basis of agency. It would be a good thing to stay within that framework. In what follows we argue that it can be done.

4.1 The *istit* operator

The target of analysis is a modal operator that we will call *istit* ("is seeing to it that"), which needs to do justice to the requirements laid down in section 3.3. Thus, we need to be able to affix "defaults" to future choices while preserving their status as genuine choices, i.e., without changing the underlying branching structure. A single element of a partition $Choice^{\alpha}_{m}$ (or of Π_m) at m is not fine-grained enough to describe future defaults—it only carries one a single step, so to speak. On the other hand, specifying a single history is too fine-grained, since an agent cannot normally guarantee a single history—there are usually other sources of indeterminism, be they nature or other agents. This means that the resources of the current stit framework do not allow one to specify sets of histories of the right level of granularity to express future defaults. More resources are needed.

Our proposed solution is to add a new parameter s to the index of evaluation. That new index is to stand for the agent's strategy with respect to which a sentence is to be evaluated.[11] The concept of a strategy is firmly grounded in the framework of branching time.[12] Roughly, a strategy s is a partial function from moments m to subsets of $H_{(m)}$ that are closed under $Choice^{\alpha}_{m}$, and $s(m)$ may be interpreted as "what the strategy advises to choose at m". For simplicity's sake, we will assume that s is total (defined on all of M) and strict (so that $s(m) \in Choice^{\alpha}_{m}$, i.e., s always gives the most specific kind of advice). It follows that s is consistent, i.e., for all $m \in M$, we have $s(m) \neq \emptyset$.

The concept of a strategy thus incorporates the required notion of "default choice" without altering the sets $Choice^{\alpha}_{m}$ (which, as we assume, are given by nature). A strategy s *admits* a history h iff by following the advice of s, history h may be reached,[13] and it admits a history h from moment m on iff h may be reached by following the advice of s starting at m. The set of histories admitted by s from m on is denoted $Admh(s, m)$:

$$h \in Admh(s,m) \quad \text{iff} \quad \text{for all } m' \text{ for which } m < m',\ h \in s(m').$$

This set of histories captures the required level of granularity.

An interesting question is whether for a total, consistent strategy, we always have $Admh(s, m) \neq \emptyset$, as one might expect. The answer in fact depends on a fine point that we have not mentioned before, namely, on the existence of "busy choosers". A busy chooser is an agent who faces infinitely many non-trivial choices in a finite time. Since we wish to study simple cases first, we explicitly exclude busy choosers from consideration. It can then be shown [3, p. 374] that in the absence of busy choosers, $Admh(s, m) \neq \emptyset$ for all $m \in M$.

Based on the new index s, we now define a new pair of modal operators, $[h_s]$ and $\langle h_s \rangle$. The strong operator $[h_s]$, the strategic necessity operator, has the semantics

$$\mathfrak{M}, m_c, m, h, s \models [h_s]\phi \text{ iff for all } h' \in Admh(s,m),\ \mathfrak{M}, m_c, m, h', s \models \phi.$$

[11] In the case of multiple agents, there will be a strategy parameter s_α for each agent α. In what follows, we will use "s" to stand for "s_α", α being the only agent under consideration.

[12] For a detailed introduction that describes various concepts of strategies, cf. [3, chap. 13].

[13] Normally a strategy will not be able to *guarantee* a single history—apart from the indeterminism over which α has control, there is usually also other indeterminism in a model, as mentioned above.

Thus $[h_s]\phi$ is true iff by following the advice of s from m on, the agent can secure the truth of ϕ, no matter how other sources of indeterminism play out. Note that this does not imply that ϕ is historically necessary (even though the inference from $[h]\phi$ to $[h_s]\phi$ is valid). Nor does the truth of $[h_s]\phi$ imply that the agent will in fact be following her strategy. Thus, the truth of $[h_s]\phi$ captures the sought-for notion of defeasible defaults.[14]

Now taking up the idea that a stit operator should contain a positive clause that describes the agent's securing an outcome and a negative clause excluding trivial cases, our semantics for the *istit* operator ("is seeing to it that") is the following:

$\mathfrak{M}, m_c, m, h, s \models \alpha\, istit : \phi$ iff

1. $\mathfrak{M}, m_c, m, h, s \models [h_s]\phi$ and
2. there is a "counter" $h' \in H_{(m)}$ such that $\mathfrak{M}, m_c, m, h', s \not\models \phi$.

Alternatively, we can introduce "$\alpha\, istit : \phi$" as an abbreviation for "$[h_s]\phi \wedge \neg[h]\phi$". Our new operator supports the inference from $\alpha\, istit : \phi$ to $\langle h\rangle\phi$ and to $\langle h\rangle\neg\phi$: If α is ϕ-ing, then the outcome is possible, but not necessary. However, in contrast to the other stit operators, the inference from $\alpha\, istit : \phi$ to ϕ is invalid—Figure 1 provides a countermodel (at index m', h', considering the formula $F\phi$). This model shows that *istit* meets the conditions of adequacy laid out in section 3.3 above. Note that *istit* is moment-determinate: If $\alpha\, istit : \phi$ is true, then so is $[h]\alpha\, istit : \phi$; the history of evaluation index plays no role.

4.2 Status and initialization of the strategy parameter s

In the usual treatment of Prior-Thomason tense logic, there is a difference between the indexes m and h: If a stand-alone sentence is to be evaluated, m

[14] We need to mention a certain complication that arises from the strictness of the semantic clause for $[h_s]$, especially in connection with "outside influence" (cf. notes 9 and 10 above): For $[h_s]$ to be true, α has to have a strategy that guarantees the truth of ϕ no matter how the other sources of indeterminism described in $\mathfrak{M}$ play out. If $\mathfrak{M}$ is taken to be a picture of the real world, we have to concede that perhaps nobody ever has a strategy that really guarantees any non-trivial ϕ—one can almost always cook up a story that would prevent ϕ, and not all of those stories have to involve alien abduction. E.g., if you are peeling an apple you normally disregard a number of real, but very remote possibilities—the knife breaking apart, earthquakes, or other emergencies. As was noted above, what counts as a salient possibility often depends on context. We believe that the best way of dealing with this complication is to follow van Lambalgen and Hamm [13] in taking models not to be mirror images of the real world, but some type of minimal models dependent on some decription of a situation. Spelling all this out in detail must be deferred to a separate paper.—It is for these reasons that "internal influences", i.e., just dropping out of something by one's own free decision, appears to be the structurally simpler case, since it is even possible in cases where $\mathfrak{M}$ is the real world, and no extra epistemic layer allowing for an agent's misrepresentation of a situation is needed.

is assigned an initial value from the context via the moment of utterance m_c, but there is no context-initialization for the history parameter, h: Assuming a "history of the context" would mean falling prey to the myth of "the real future".[15] What is the status of the s parameter?

In the *istit* picture, s functions as an interface between our everyday mentalistic vocabulary that describes an agent's plans and intentions and the formal branching time framework. We talk about an agent's current plans or intentions as something objective (they can play a role in court, for example), and this is what s should capture. Thus, s is an initialized parameter like m, not an uninitialized one like h. Paralleling the treatment of m means that we should add a "current strategy" index s_c to the context.

4.3 Comparison with "inertia worlds"

Dowty [7, pp. 151ff.] gives a reading of the progressive in a branching time framework that is similar to our treatment here. His fundamental notion is that of an "inertia world", or "inertia future". Formally, Dowty singles out a set of histories as "inertia futures", $Inr(m) \subsetneq H_{(m)}$, that are supposed to represent "what is happening *now*, what is the outcome of events as they could be expected to transpire without [...] interference" [7, p. 149]. This concept aims at providing a set of histories at exactly the same level of granularity as our set $Admh(s, m)$. There are two less important differences bewteen Dowty's treatment and *istit*: His semantics is interval-based, and he does not consider a negative clause to exclude trivial cases of "is ϕ-ing". The substantial difference between his approach and ours, which we see as the main advantage of *istit*, is that while *istit* grounds the set $Admh(s, m)$ on the formally rigorous and intuitively clear notion of a strategy, Dowty does not provide further analysis of his concept of an "inertia world" or an "inertia future". In the quote above, he may be interpreted as saying in effect that "an agent is now doing what an agent is doing now", and the inertia approach has consequently been accused of emptyness, or even of incoherence in view of the unclear notion of "intervention". By spelling out "inertia" in terms of the defeasible defaults set by strategies, our analysis goes one crucial step further. Furthermore, our analysis meshes well with the already established stit framework.

4.4 Further extensions

So far we have introduced the bare minimum for handling continuous action: The modal operator *istit* meets the adequacy conditions spelled out in section 3.3. In the extended framework that includes the strategy index, a number of other modalities and agency-related concepts can be defined,

[15] For a decisive criticism of that notion (under the name of "the thin red line"), cf. [3, chap. 6].

showing the flexibility of the framework. We briefly mention three such extensions.

Strategic modalities: In two-dimensional semantics, to each index that is not part of the context there corresponds a number of one-place modal operators for shifting that parameter. E.g., the (weak) tense operators P and F (and their strong duals H and G) shift the moment parameter m backward or forward in time. For the history parameter h there are the corresponding "historical" modalities $\langle h\rangle$ (weak) and dually $[h]$ (strong). Along the same lines one can define "strategic" modalities $\langle s\rangle$.

We call a strategy s' an *alternative* to s at m iff for all $m' \in M$ for which $m \not\leq m'$, $s'(m') = s(m')$; i.e., the two strategies have to agree everywhere outside the future of possibilities of m. The clause for the weak strategic modality $\langle s\rangle$ then reads:

$$\mathfrak{M}, m_c, m, h, s \models \langle s\rangle\phi \quad \text{iff there is } s' \text{ s.t. } \quad \mathfrak{M}, m_c, m, h, s' \models \phi,$$

where s' is a total, strict strategy that is an alternative to s at m.

Ability: The weak strategic modality $\langle s\rangle$ allows one to express the concept of ability: If $\langle s\rangle[h_s]\phi$ is true at m, h, s, then this means that the agent could change her current strategy s such as to guarantee strategically the outcome ϕ. If ϕ is not guaranteed anyway, $\langle s\rangle\alpha\, istit : \phi$ is true, meaning that the agent could be ϕ-ing.

Refraining: Refraining means not doing something while one could do it (cf., e.g., von Wright's analysis [15]). Thus, we may say that α refrains from ϕ-ing iff α is not ϕ-ing ($\neg\alpha\, istit : \phi$) and still could be ϕ-ing ($\langle s\rangle\alpha\, istit : \phi$), so (with "*iref*" the companion "refraining" operator for *istit*)

$$\alpha\, iref : \phi \;\Leftrightarrow_{df}\; \langle s\rangle[\alpha\, istit : \phi] \wedge \neg[\alpha\, istit : \phi].$$

There is a difference between our treatment of refraining and the usual stit analysis: The notion of refraining that corresponds to the other stit operators is itself agentive, i.e., can itself be phrased as a stit sentence:

$$\alpha\, ref : \phi \;\Leftrightarrow_{df}\; [\alpha\, stit : \neg[\alpha\, stit : \phi]].$$

For *dstit* this can be shown to be equivalent to our analysis in terms of ability and not doing [3, p. 438]:

$$\alpha\, ref : \phi \;\Leftrightarrow\; \langle h\rangle[\alpha\, dstit : \phi] \wedge \neg[\alpha\, dstit : \phi].$$

This equivalence does not hold for *istit*.

The crucial question is which of the two analyses is more fundamental. While we do not have a final answer to this question, it appears that the analysis championed by von Wright, which the *istit* framework is able to capture, is the more basic one. It is true that refrainings, or omissions, are generally agentive—in some countries, one can even be legally liable, e.g., for refraining from helping somebody. However, it is rare to speak of countinuous refraining,[16] so there seems to be no complelling reason to analyse "α refrains from ϕ-ing" in terms of "α is ψ-ing". Thus, the non-agentive *istit* analysis of refraining should not count against the theory.[17]

5 Axiomatising *istit*

In this final section we take up the task of axiomatising the new stit modality.

5.1 The semantics

We assume a single agent α who is not a busy chooser; α's strategy will be denoted s. (The agent α will not be mentioned in what follows.)[18] At a moment m, we have to consider the sets of all histories containing m, $H = H_{(m)}$, and the set of histories admitted by s at m, $A = Admh(s, m)$; we have

$$A \subseteq H; \quad A \neq \emptyset.$$

In these terms, the semantic clause for *istit* reads:

$\mathfrak{M}, m_c, m, h, s \models \alpha\, istit : \phi$ iff

1. for all $h' \in A$, $\mathfrak{M}, m_c, m, h', s \models \phi$ and
2. there is $h' \in H$ s.t. $\mathfrak{M}, m_c, m, h', s \not\models \phi$.

For the purpose of axiomatisation, the rest of the language that we are using (including tense operators and so forth) only plays a role by giving rise to formulae that are true at some index m, h, s and false at some other index m, h', s. (The ϕ in the above definitions is treated as a "black box".) Thus the problem of axiomatising *istit* amounts to axiomatising what we will call *set-inclusion structures* $\langle H, A\rangle$, where $A \subseteq H$ and $A \neq \emptyset$. A *set-inclusion model* is such a structure together with a valuation V. We thus consider a simple propositional language L_1 with just two modalities $\Box_H$

[16] In conversation, Michael Perloff came up with the example "I was actively refraining from hitting the guy", where the progressive is used for emphasis, stressing how hard it was to refrain.

[17] Certainly the question of continuous refraining merits further study.

[18] A multi-agent extension will have to include assumptions about the independence of simultaneous choices by different agents; the technical side of that will follow the *dstit* analysis presented in [3, chap. 17].

and $\Box_A$, and the index of evaluation is $h \in H$. The semantic clauses for the modalities are

$$\mathfrak{M}, h \models \Box_H \phi \text{ iff for all } h' \in H, \mathfrak{M}, h' \models \phi,$$

$$\mathfrak{M}, h \models \Box_A \phi \text{ iff for all } h' \in A, \mathfrak{M}, h' \models \phi.$$

Here, the modality $\Box_H$ corresponds to $[h]$ above, $\Box_A$ corresponds to $[h_s]$, and $\alpha\, istit : \phi$ is treated as an abbreviation for $\Box_A \phi \wedge \neg \Box_H \phi$. The task of axiomatising *istit* is thus reduced to the task of axiomatising the logic of set-inclusion structures.

5.2 The axiomatisation

It is obvious that by the semantic clause, $\Box_H$ is an S5 modality. The only thing that needs to be investigated is the logic of $\Box_A$ and the interrelation between these two modalities. We proceed by first finding a description of set-inclusion models in the familiar framework of relational Kripke models, employing a language L_2 with two modal operators $\Box_1$ and $\Box_2$ that have the standard Kripke semantics in terms of two relations R_1 and R_2 on a set of worlds W, $w \in W$ being the index of evaluation:

$$\mathfrak{M}, w \models \Box_1 \phi \text{ iff for all } w' \in W \text{ for which } R_1(w, w'), \mathfrak{M}, w' \models \phi$$

$$\mathfrak{M}, w \models \Box_2 \phi \text{ iff for all } w' \in W \text{ for which } R_2(w, w'), \mathfrak{M}, w' \models \phi$$

The following Lemma describes the interrelation between the two semantic approaches.

Lemma 1
For every set-inclusion model $\mathfrak{M}_1 = \langle H, A, V \rangle$ ($\emptyset \neq A \subseteq H$), there is an equivalent relational model $\mathfrak{M}_2 = \langle U, R_1, R_2, V' \rangle$, where $U = H$ and

1. *R_1 is an equivalence relation,*
2. *R_2 is serial, transitive, and Euclidean,*
3. *for all $u, u' \in U$, if $R_2(u, u')$, then also $R_1(u, u')$, and*
4. *for all $u, u', u'' \in U$, if $R_2(u, u'')$ and $R_1(u, u')$, then also $R_2(u', u'')$.*

In the other direction, given a relational model as just described, for every point-generated submodel there is an equivalent set-inclusion model.[19]

[19] A point-generated submodel of a relational model $\mathfrak{M}_2 = \langle U, R_1, R_2, V' \rangle$ is a submodel all of whose points can be reached from one point $u \in U$ by following R_1. This ensures that R_1 will be the universal relation on U.

By "equivalent" we mean that for ϕ a L_1-wff and ϕ' the L_2-wff derived from ϕ by replacing $\Box_H$ with $\Box_1$ and $\Box_A$ with $\Box_2$ everywhere, we have

$$\mathfrak{M}_1, w \models \phi \quad \textit{iff} \quad \mathfrak{M}_2, w \models \phi'.$$

Proof: "$\Rightarrow$": Let $\mathfrak{M}_1$ be given. Set $U = H$, $V' = V$, $R_1 = H \times H$ and $R_2 = H \times A$. (1) is immediate, since R_1 is the universal relation on H. For (2), seriality follows from $A \neq \emptyset$, and the other two conditions are also immediate (if $R_2(u, u'')$, then $R_2(u', u'')$ for any u'). (3) follows from $A \subseteq H$, and (4) is verified analogously to (2). The equivalence of the two models follows directly from the translation employed.

"$\Leftarrow$": Let $\mathfrak{M}_2$ be a point-generated submodel of a given relational model. Set $H = \{w' \in U \mid \exists w \in U\, R_1(w, w')\}$, $A = \{w' \in U \mid \exists w \in U\, R_2(w, w')\}$. We need to show that (a) $A \neq \emptyset$ and (b) $H \subseteq W$. Furthermore, if we can show that (c) $H = U$, (d) $R_1 = H \times H$ and (e) $R_2 = H \times A$, we can set $V = V'$, and the equivalence of the two models will follow directly, finishing the proof.

Condition (a) follows from the seriality of R_2. The fact that $\mathfrak{M}_2$ is a generated submodel ensures that R_1, being an equivalence relation, is the universal relation on U, from which we have (c) and (d). Condition (b) then follows from clause (3) above. To establish (e), let $s \in A$, so that by definition there is $w \in H$ s.t. $R_2(w, s)$. Let $w' \in H$. By (d), we have $R_1(w, w')$, and now (4) above gives us indeed $R_2(w', s)$. $\Box$

We now proceed to axiomatise the logic of set-inclusion models by axiomatising the logic of the corresponding relational models. We start with a normal bimodal logic with boxes $\Box_1$ and $\Box_2$ (i.e., propositional logic, the K-axioms for the boxes, and modus ponens, substitution and necessitation). Furthermore, we posit the following axioms for the logic L_{istit}:[20]

T_1 $\Box_1\phi \to \phi$

4_1 $\Box_1\phi \to \Box_1\Box_1\phi$

5_1 $\Diamond_1\phi \to \Box_1\Diamond_1\phi$

D_2 $\Box_2\phi \to \Diamond_2\phi$

4_2 $\Box_2\phi \to \Box_2\Box_2\phi$

5_2 $\Diamond_2\phi \to \Box_2\Diamond_1\phi$

I1 $\Diamond_2\phi \to \Diamond_1\phi$

[20]Thanks to Yuko Murakami for helpful discussions of the axiomatisation.

I2 $\Diamond_2\phi \rightarrow \Box_1\Diamond_2\phi$

α *istit* : ϕ is an abbreviation for $\Box_2\phi \wedge \neg\Box_1\phi$.

Here, the first three axioms make $\Box_1$ an S5 modality, the next three make $\Box_2$ a KD45 modality, and the axioms I1 and I2 describe the interaction between the modalities. Note that *istit* itself is not a normal modal operator, even though it is defined from two normal modal operators.

5.3 Soundness and completeness

We can now establish the sought-for result:

Theorem 1
L_{istit} is sound and complete w.r.t the class of set-inclusion structures.

Proof: The soudness of the axioms w.r.t. the set-inclusion structures can be verified immediately from the semantic clauses (I1 corresponds to $A \subseteq H$, while I2 captures the semantic clause for $\Box_A$).

In order to establish completeness, we first show that L_{istit} is complete w.r.t. the class of frames described by clauses (1)–(4) above. Since satisfiability is invariant under generated submodels, Lemma 1 then yields the desired result.

Completeness follows easily from the fact that all the axioms are Sahlqvist formulae. Since the correspondence of the first six axioms to clauses (1) and (2) above is so well known, we only treat the "interaction" axioms *I*1 and *I*2 briefly. From the Sahlqvist-van Benthem algorithm [5, chap. 3.6] we get the following:

I1 The corresponding second order formula in the variable x is
$\forall P((\exists y\, (R_2(x,y) \wedge P(y)) \rightarrow \exists z\, (R_1(x,z) \wedge P(z)))$.
Pulling out the first $\exists$ and instantiating $P(u)$ as $\lambda u.u = y$ gives
$\forall y\, ((R_2(x,y) \wedge y = y) \rightarrow \exists z\, (R_1(x,z) \wedge z = y))$,
which is equivalent to clause (3) above.

I2 The corresponding second order formula in the variable x is
$\forall P((\exists y\, (R_2(x,y) \wedge P(y)) \rightarrow \forall z\, (R_1(x,z) \rightarrow \exists z'\, (R_2(z,z') \wedge P(z')))$.
Pulling out the first $\exists$ and instantiating $P(u)$ as $\lambda u.u = y$ gives
$\forall y\, ((R_2(x,y) \wedge y = y) \rightarrow \forall z\, (R_1(x,z) \rightarrow \exists z'\, (R_2(z,z') \wedge z' = y))$,
which is equivalent to clause (4) above.

The Sahlqvist completeness theorem [5, chap. 4.3] then establishes the completeness of our axiomatisation. $\Box$

In order to bind this completeness result back to the propositional fragment of the language of *istit* described in section 4, we can use the following construction: Let ϕ be a non-theorem of L_{istit} ($L_{istit} \not\vdash \phi$), so that by Theorem 1 there is a set-inclusion model $\mathfrak{M} = \langle H, A, V \rangle$ and $h^* \in H$ s.t. $\mathfrak{M}, h^* \not\models \phi$. We wish to find a stit model $\mathfrak{M}' = \langle \mathfrak{S}, V' \rangle$ and an index m, h, s s.t. $\mathfrak{M}', m, h, s \not\models \phi$.[21] Recall that a stit structure $\mathfrak{S}$ consists of a branching time structure $\langle M, \leq \rangle$ and the two sets *Agents* and *Choices*. We set

$$M = \{m_0, m_1, m_2\} \cup \{m_h \mid h \in H\},$$

and for the partial ordering $\leq$ we take the reflexive and transitive closure of the following:

$$m_0 < m_1;\ m_0 < m_2;\ m_1 < m_h \text{ for } h \in A;\ m_2 < m_h \text{ for } h \in H - A.$$

Thus, m_0 is a minimal node with two direct successors m_1 and m_2, and the histories are in one-to one correspondence with the elements of H (as the $h \in H$ correspond to maximal elements m_h); we will denote the histories by h as well. We now set $Agents = \{\alpha\}$ and $Choice^{\alpha}_{m_0} = \Pi_{m_0}$, i.e., at m_0, the agent α has two choices corresponding to the two direct successors, m_1 and m_2. For the valuation, we set $V'(m_0, h) = V(h)$. For the strategy s, we set $s(m_0) = \Pi_{m_0}\langle m_1 \rangle$: at m_0, the strategy advises to choose the bundle of histories going through m_1, which corresponds to the set A in the set-inclusion model. It follows that

$$\mathfrak{M}', m_0, h^*, s \not\models \phi.$$

6 Conclusion

The phenomenon of continuous action poses a challenge for formal theories of agency. The branching time-based stit framework in its present form, comprising the operators *astit* and *dstit*, is unable to capture that phenomenon. However, by incorporating an agent's current strategy as a new index of evaluation it is possible to extend the stit framework such as to capture continuous action. The new operator, *istit*, forms the basis of a rich structure of agency-related modalities, and it can be axiomatised in a straightforward way.

---

[21] We ignore the moment of utterance index m_c, which plays no role for the semantics here.

Acknowledgements

I wish to thank audiences at the Center for the Philosophy of Science, University of Pittsburgh (lunchtime talk, 19 September 2003) and at Manchester (AiML 5, 10 September 2004) for helpful discussions and suggestions. Special thanks are due to Nuel Belnap, Yuko Murakami, Michael Perloff, and to two anonymous referees for AiML 5.

BIBLIOGRAPHY

[1] Anscombe, G.E.M. (1963). *Intention*, 2nd ed. Oxford.
[2] Belnap, N. and Perloff, M. (1988). Seeing to it that: A canonical form for agentives. *Theoria* 54:175–199.
[3] Belnap, N., Perloff, M., and Xu, M. (2001). *Facing the Future*. Oxford.
[4] Belnap, N. (2002). Branching histories approach to indeterminism and free will. Preprint at the Pittsburgh Philosophy of Science Archive, URL=http://philsci-archive.pitt.edu/archive/00000890/index.html.
[5] Blackburn, P., de Rijke, M., and Venema, Y. (2001). *Modal Logic*. Cambridge.
[6] Davidson, D. (1967). The Logical Form of Action Sentences. In his *Essays on Actions and Events*, Oxford 1980, 105–122.
[7] Dowty, D.R. (1979). *Word Meaning and Montague Grammar*. Dordrecht.
[8] Horty, J.F. (2001). *Agency and Deontic Logic*. Oxford.
[9] Kant, I. (1797). *Metaphysische Anfangsgründe der Rechtslehre*. Ed. B. Ludwig. Hamburg ^{2}1998.
[10] Segerberg, K. (1989). Bringing it about. *Journal of Philosophical Logic* 18:327–347.
[11] Strawson, P.F. (1959). *Individuals*. London.
[12] Thompson, M. (2004). *Life and Action*. Cambridge, MA, *to appear*.
[13] van Lambalgen, M. and Hamm, F. (2005). *The Proper Treatment of Events*. Oxford.
[14] von Kutschera, F. (1986). Bewirken. *Erkenntnis* 24:253–281.
[15] von Wright, G.H. (1963). *Norm and Action: A Logical Inquiry*. London.
[16] Xu, M. (1996). *An Investigation into the Logics of Seeing-to-it-that*. Unpublished Dissertation, University of Pittsburgh, Pittsburgh, PA.

Thomas Müller
Philosophisches Seminar, LFB III, University of Bonn,
Lennéstr. 39, 53113 Bonn, Germany
Thomas.Mueller@uni-bonn.de

Utilitarian Deontic Logic

Yuko Murakami[1]

ABSTRACT. This paper aims to examine Horty's proposal of utilitarian deontic logic [7]. It will focus on his dominance operators by way of simplified semantics. An axiomatization of the logic of the simplified semantics will be shown by the construction of a finite countermodel.

Analysis of the proof suggests possibilities of further investigations. The presented version of Horty's proposal does not make essential differences from standard deontic logic: the deontic operators are normal, and thus paradoxes of deontic logic occur. Non-monotonic agency operators should be considered.

1 Introduction

Horty [7], emphasizing game-theoretical motivations in the see-to-it-that (*stit*) theory of agency, proposes semantics of deontic operators with identifying duty as action according to dominant strategy. He introduces to stit semantics a utilitarian value assignment over branches, which represent possibilities at a moment, so as to define preference and dominance orderings among options at each moment. Defined with the dominance relations on choices, deontic operators are intended to label "actions at the dominant choice." He proposes several deontic modalities including operators according to properties of value assignment; "$\cdots$ holds in every dominant choice" and "$\cdots$ holds in every optimal choice," for example.

[1]Long-term discussion with Professor Ming Xu was essential for production of the paper. Professor Nuel Belnap allowed me to participate in his seminar in University of Pittsburgh. Professor John F. Horty saved time for me for discussion. Those visits were partially supported by a research fellowship for young scientists and grants from Japan Society for the Promotion of Science. I would like to thank those who have helped me, including anonymous referees. The author is responsible for any unclear points and mistakes, however.

It is natural to ask whether logic on the semantics is axiomatizable, and if yes, whether it is decidable. This paper answers yes to both questions.

Construction of a finite countermodel in this paper takes a simpler form than Horty's original semantics as argued in Section 3.1. It relies on the two related facts. First, the approach here follows Horty to take multi-S5 with an interactive axiom (independence of agents[1]) as agency operators, not operators argued in Belnap et al. [1]. The choice of agency operators makes the presentation considerably easier, while it does not incorporate essential properties of agency modalities required through conceptual analysis of the notion of real choices in the stit theory of agency.

Second, semantics in the paper is simplified than that in Horty's orignal presentation with branching time semantics.

Figure 1. Utilitarian frame for a single agent

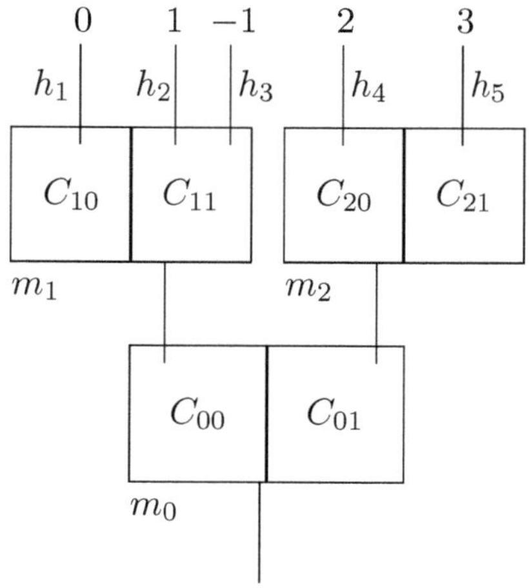

Figure 1 illustrates an utilitarian frame for a single agent in Horty's presentation. The tree structure represents a branching structure for the indeterministic time. It is open to the future, where each branch

[1]The corresponding semantic condition intuitively states that any combination of agents' choices can be realized.

(or *history*) represents a possible growth of the world. There are three moments where the agent can make a choice among two options C_{i0} and C_{i1} available at each m_i. A real number is assigned to each history in the structure to represent its utilitarian value, based on which the dominance ordering and the optimal ordering are defined. While readers should consult the following section for the exact definitions, it should also fit to a game-theoretical intuition that C_{01} is dominant over C_{00} at m_0 and C_{21} over C_{20} at m_2, while neither C_{01} nor C_{11} is dominant over each other at m_1.

Taking a language with only agency and deontic modalities whose truth conditions do not involve reference of any past or future moment, multi-S5 structures associated with a utilitarian value function preserve model and frame validity in branching time semantics as explained in Section 3.1.

2 Preliminary

2.1 Language

The language $\mathfrak{L}$ contains denumerably many propositional variables $p_0, p_1, \ldots$, denumerably many terms for agents $\alpha_0, \alpha_1, \ldots$, an identity symbol $=$, truth-functional operators $\neg$ and $\wedge$, and non-truth-functional operators $\Box$, [*cstit*:] and $\odot$. Formulas are defined as usual, except $\alpha = \beta$, $\Box A$, $[\alpha$ *cstit*: $A]$, $\odot[\alpha$ *cstit*: $A]$ are formulas whenever α and β are agent terms for agents and A is a formula. Abbreviations such as $\rightarrow$, $\leftrightarrow$, $\vee$, $\top$, $\bot$ and $\Diamond$ are introduced as usual. The following abbreviations are also defined:

$$\alpha \neq \beta =_{df} \neg(\alpha = \beta),$$

$$\triangle\alpha A =_{df} [\alpha \textit{ cstit: } A],$$

$$\ominus\alpha A =_{df} \odot[\alpha \textit{ cstit: } A].$$

An additional abbreviation $\textit{diff}(\beta_0, \ldots, \beta_n)$ (where $n > 0$) is defined as follows:

$$\textit{diff}(\beta_0, \beta_1) =_{df} \beta_0 \neq \beta_1,$$

$$\textit{diff}(\beta_0, \beta_1, \ldots, \beta_{n+1}) =_{df} \textit{diff}(\beta_0, \ldots, \beta_n) \wedge (\textstyle\bigwedge_{0 \leqslant k \leqslant n} \beta_k \neq \beta_{n+1}).$$

A *stit formula* is a formula whose major operator is $\triangle\alpha$ for some α, and an *ought formula* is a formula whose major operator is $\ominus\alpha$ for some α.

2.2 Branching-time semantics

Horty [7] proposes a semantics for *stit* and dominant ought based on the branching semantics for indeterminist time proposed by Prior and Thomason (see [9], [11] and [12]) .

A tree structure (Branching-time structure, BT) is a pair $\mathfrak{F} = \langle \mathsf{T}, < \rangle$, where T is a nonempty set, whose members m, m' etc. are called *moments*; $<$ is a tree relation on T, i.e., a partial order on T being (1) linear to the past and (2) connected. A *history* in $\mathfrak{F}$ is a maximal linear subset of T. A history h is said to *go through a moment* m when $m \in h$. A *moment-history pair* is a pair of a moment m and a history which goes through m. H_m denotes the set of histories which go through m; $\mathsf{H}_{\mathfrak{F}}$ denotes the set of all the histories in $\mathfrak{F}$; and $Moment\text{-}History$ denotes the set of all moment-history pairs in $\mathfrak{F}$.

The stit analysis adds a set of agents and a choice function for each agents on a tree structure. A *stit frame* (Branching-time and agent-choice structure, $BT + AC$) is a quadruple $\mathfrak{F} = \langle \mathsf{T}, <, \mathsf{Agent}, \mathsf{choice} \rangle$ satisfying the following. Agent is an nonempty set of *agents*; choice is a function that assigns to each $a \in \mathsf{Agent}$ and $m \in T$ a partition choice_a^m of $\mathsf{H_m}$ satisfying the following conditions.

Independence of Agents: For each function f that assigns to each pair of $a \in \mathsf{Agent}$ and $m \in \mathsf{T}$ a member of choice_a^m,

$$\bigcap_{a \in \mathsf{Agent}} f(a) \neq \varnothing$$

.

In other words, let select^m be the set of all functions f on Agent such that $f(a) \in \mathsf{choice}_a^m$ for every $a \in \mathsf{Agent}$. For each proper subset A of Agent, let $\mathsf{State}_{\mathsf{A}}^m = \{\bigcap_{a \in \mathsf{Agent}-\mathsf{A}} f(a) \mid f \in \mathsf{select}^m\}$. Then, for any $m \in T$, $\varnothing \notin \mathsf{State}_{\varnothing}^m$.

No choice between undivided histories: For any $h, h' \in$

$\mathsf{H}_{\mathfrak{F}}$, $m_0 \in h \cap h'$, and $\alpha \in \mathsf{Agent}$, if there is $m \in h \cap h'$ such that $m_0 < m$, $h \in \mathsf{choice}_{\alpha}^{m_0}$ if and only if $h' \in \mathsf{choice}_{\alpha}^{m_0}$.

A *utilitarian stit frame* is a 5-tuple $\mathfrak{F} = \langle \mathsf{T}, <, \mathsf{Agent}, \mathsf{choice}, \mathsf{value} \rangle$, where $\langle \mathsf{T}, <, \mathsf{Agent}, \mathsf{choice} \rangle$ is a stit frame, and value is a function from $\mathsf{H}_{\mathfrak{F}}$ to the set of real numbers.

Let $\mathfrak{F} = \langle \mathsf{T}, <, \mathsf{Agent}, \mathsf{choice}, \mathsf{value} \rangle$ be a utilitarian stit frame. We will use K, K' etc. to range over subsets of $\mathsf{H}_{\mathfrak{F}}$. For each $a \in \mathsf{Agent}$, $m \in \mathsf{T}$, and $h \in \mathsf{H}_m$, we use $\mathsf{choice}_a^m(h)$ for the member of choice_a^m to which h belongs. For each $m \in \mathsf{T}$, $h_0, h_1 \in \mathsf{H}_m$, let $R_a^m h_0 h_1$ iff $\mathsf{choice}_a^m(h_0) = \mathsf{choice}_a^m(h_1)$.

2.3 States, preference, and dominance

The notion of *states* plays the main role in preference and dominance. Let $\Gamma \subseteq \mathsf{Agent}$ and $m \in \mathsf{T}$. Define $\mathsf{State}_{\Gamma}^m(h) = \bigcap_{a \notin \Gamma} \mathsf{choice}_a^m(h)$. Let $\mathsf{State}_{\Gamma}^m = \{\mathsf{State}_{\Gamma}^m(h) : h \in \mathsf{H}_m\}$.

For each $a \in \mathsf{Agent}$ and $m \in \mathsf{T}$, we will write State_a^m for $\mathsf{State}_{\{a\}}^m$. Let $K, K' \subseteq \mathsf{H_m}$. K' is *weakly preferred to* K, written $K \leq K'$, iff for each $h_0 \in K$ and each $h_1 \in K'$, $\mathsf{value}(h_0) \leq \mathsf{value}(h_1)$. K' is *strongly preferred to* K, written $K < K'$, iff $K \leq K'$ and not $K' \leq K$. Let $a \in \mathsf{Agent}$ and let $K, K' \in \mathsf{choice}_a^m$. K' *weakly dominates* K, written $K \preceq K'$, iff for each $S \in \mathsf{state}_a^m$, $K \cap S \leq K' \cap S$. K' *strongly dominates* K, written $K \prec K'$, iff $K \preceq K'$ and not $K' \preceq K$.

Horty [7] extablishes the following, which should aid the reader's intuition.

PROPOSITION 1. *Let* $\langle \mathsf{T}, <, \mathsf{Agent}, \mathsf{choice}, \mathsf{value} \rangle$ *be a utilitarian stit frame, let* $m \in \mathsf{T}$, $a \in \mathsf{Agent}$, *and let* State_a^m *be as defined above. Then the following hold.*

1. $X < Y$ *iff* $\mathsf{value}(h') \leq \mathsf{value}(h)$ *for each* $h' \in X$ *and* $h \in Y$, *and* $\mathsf{value}(h_2) < \mathsf{value}(h_1)$ *for some* $h_2 \in X$ *and* $h_1 \in Y$.

2. *The weak preference ordering* $\leq$ *is transitive.*

3. *The strong preference ordering* $<$ *is a strict partial ordering.*

4. $X < Y$ *and* $Y \leq Z$ *only if* $X < Z$.

5. *$X \prec Y$ iff $X \cap S \leq Y \cap S$ for every $S \in \mathsf{State}^m_a$, and $X \cap S < Y \cap S$ for some $S \in \mathsf{State}^m_a$.*

6. *The weak dominance ordering $\preceq$ is transitive.*

7. *The strong dominance ordering $\prec$ is a strict partial ordering.*

8. *$X \prec Y$ and $Y \preceq Z$ only if $X \prec Z$.*

2.4 Utilitarian stit models, truth conditions, and semantic notions

Let $\mathfrak{F} = \langle \mathsf{T}, <, \mathsf{Agent}, \mathsf{choice}, \mathsf{value} \rangle$ be a utilitarian stit frame. A *valuation τ on $\mathfrak{F}$* is a function that assigns each agent term a member of Agent and each atomic formula a subset of *Moment-History*. A *utilitarian stit model* is a pair $\mathfrak{M} = \langle \mathfrak{F}, \tau \rangle$ where $\mathfrak{F}$ is a utilitarian stit frame and τ is a valuation on $\mathfrak{F}$.

That A is *true in $\mathfrak{M} = \langle \mathfrak{F}, \tau \rangle$ at m/h*, written $\mathfrak{M}, m/h \models A$, is defined recursively as follows, where p is any atomic formula and α is any agent term.

$\mathfrak{M}, m/h \models p$	iff	$m/h \in \tau(p)$;
$\mathfrak{M}, m/h \models \alpha = \beta$	iff	$\alpha^\tau = \beta^\tau$;
$\mathfrak{M}, m/h \models \neg A$	iff	$\mathfrak{M}, m/h \not\models A$ (i.e., not $\mathfrak{M}, m/h \models A$);
$\mathfrak{M}, m/h \models A \wedge B$	iff	$\mathfrak{M}, m/h \models A$ and $\mathfrak{M}, m/h \models B$;
$\mathfrak{M}, m/h \models \Box A$	iff	$\mathfrak{M}, m/h' \models A$ for every $h' \in H_m$;
$\mathfrak{M}, m/h \models \triangle\alpha A$	iff	$\mathfrak{M}, m/h' \models A$ for every $h' \in \mathsf{choice}^m_{\alpha^\tau}(h)$;
$\mathfrak{M}, m/h \models \ominus\alpha A$	iff	for each $K \in \mathsf{choice}^m_{\alpha^\tau}$ such that $K \not\subseteq \|A\|^\tau_m$, there is a $K' \in \mathsf{choice}^m_{\alpha^\tau}$ such that (i) $K \prec K'$, (ii) $K' \subseteq \|A\|^\tau_m$, and (iii) $K'' \subseteq \|A\|^\tau_m$ for every $K'' \in \mathsf{choice}^m_{\alpha^\tau}$ with $K' \preceq K''$.

where $\|A\|^\tau_m = \{h \in \mathsf{H}_m \mid \mathfrak{M}, m/h \models A\}$. Notions of validity and satisfiability are defined the same way as in modal logic. A is *valid in $\mathfrak{M}$* if $\mathfrak{M}, m/h \models A$ for every $m/h \in$ *Moment-History*, and A is *valid in $\mathfrak{F}$* if A is valid in every model on $\mathfrak{F}$. A is *valid* if A is valid in every utilitarian stit frame. Φ is *satisfiable in $\mathfrak{M}$* if for some m/h in *Moment-History*, $\mathfrak{M}, m/h \models A$ for every $A \in \Phi$, and Φ is *satisfiable* if Φ is satisfiable in a utilitarian stit model.

2.5 Classification of utilitarian stit frames

$\mathfrak{F} = \langle \mathsf{T}, <, \mathsf{Agent}, \mathsf{choice}, \mathsf{value}\rangle$ is *two-valued* if the range of value is $\{0, 1\}$. $\mathfrak{F}$ is *at-most-n-ary* (where $n > 0$) if $|\mathsf{choice}_a| \leqslant n$ for every $a \in \mathsf{Agent}$, and $\mathfrak{F}$ is a *finite-choice utilitarian stit frame* if choice_a^m is finite for every $a \in \mathsf{Agent}$. Counterpart notions on models are defined as usual.

Let $\mathfrak{F} = \langle \mathsf{T}, <, \mathsf{Agent}, \mathsf{choice}, \mathsf{value}\rangle$ be a utilitarian stit frame. For each $a \in \mathsf{Agent}$ and $m \in \mathsf{T}$, we define

$$\mathsf{optimal}_a^m = \{K \in \mathsf{choice}_a^m \quad \neg\exists K'(K' \in \mathsf{choice}_a \wedge K \prec K')\}.$$

$\mathfrak{F}$ is *optimal* at m if for each $a \in \mathsf{Agent}$, $K \in \mathsf{choice}_a^m - \mathsf{optimal}_a^m$ only if $K \prec K'$ for some $K' \in \mathsf{optimal}_a^m$. A utilitarian stit model $\mathfrak{M} = \langle \mathfrak{F}, \tau\rangle$ is *optimal* if $\mathfrak{F}$ is. It is easy to see that $\mathfrak{F}$ is optimal only if $\mathsf{optimal}_a^m \neq \varnothing$ for every $m \in \mathsf{T}$ and every $a \in \mathsf{Agent}$. It has been shown in Horty [7] that $\mathfrak{F}$ is a finite-choice utilitarian frame only if $\mathfrak{F}$ is optimal. It is also easy to verify the following.

PROPOSITION 2. *Let* $\mathfrak{F} = \langle \mathsf{T}, <, \mathsf{Agent}, \mathsf{choice}, \mathsf{value}\rangle$ *be any optimal utilitarian stit frame, and let* $\mathfrak{M} = \langle \mathfrak{F}, \sigma\rangle$ *be any model on* $\mathfrak{F}$. *Then for each* m, h *in* $\mathfrak{F}$ *with* $m \in h$ *and each formula* $\ominus\alpha A$,

(1) $\quad \mathfrak{M}, m/h \vDash \ominus\alpha A$ *iff* $X \subseteq \|A\|_m^\sigma$ *for every* $X \in \mathsf{optimal}_{\alpha^\sigma}^m$.

Proof. Suppose on one hand that $\mathfrak{M}, m/h \vDash \ominus\alpha A$. Then, if $X \nsubseteq \|A\|_m^\sigma$ for any $X \in \mathsf{choice}_{\alpha^\sigma}^m$, we know by the truth-definition of $\ominus$ that there is an $Y \in \mathsf{choice}_{\alpha^\sigma}^m$ such that $X \prec Y$, and hence $X \notin \mathsf{optimal}_{\alpha^\sigma}^m$. It follows that $X \subseteq \|A\|_m^\sigma$ for every $X \in \mathsf{optimal}_{\alpha^\sigma}^m$.

Suppose on the other hand that $\mathfrak{M}, m/h \nvDash \ominus\alpha A$. Then by the truth-definition of $\ominus$, there is an $X \in \mathsf{choice}_{\alpha^\sigma}^m$ such that $X \nsubseteq \|A\|_m^\sigma$. Suppose for reductio that

(2) $\quad X' \subseteq \|A\|_m^\sigma$ for every $X' \in \mathsf{optimal}_{\alpha^\sigma}^m$.

Then, since $\mathfrak{M}$ is optimal, there is an $Y \in \mathsf{optimal}_{\alpha^\sigma}^m$ such that $X \prec Y$, and then $Y \subseteq \|A\|_m^\sigma$. Applying the truth-definition of $\ominus$ again, there is a $Z \in \mathsf{choice}_{\alpha^\sigma}^m$ such that

(3) $\quad Z \nsubseteq \|A\|_m^\sigma$

and $Y \preceq Z$. If $Z \notin \mathsf{optimal}^m_{\alpha^\sigma}$, there is an $X_0 \in \mathsf{Choice}^m_{\alpha^\sigma}$ such that $Z \prec X_0$, and then, since $Y \preceq Z$, $Y \prec X_0$ by Proposition 1 (8), and hence $Y \notin \mathsf{optimal}^m_{\alpha^\sigma}$, a contradiction. It follows that $Z \in \mathsf{optimal}^m_{\alpha^\sigma}$, and hence $Z \subseteq \|A\|^\sigma_m$ by (2), contrary to (3). From this reductio we conclude that $X' \nsubseteq \|A\|^\sigma_m$ for some $X' \in \mathsf{optimal}^m_{\alpha^\sigma}$. ■

2.6 Axiomatic systems

The logic L_0 takes as axioms all substitution instances of truth-functional tautologies and all formulas of the following forms, where in A8, $A(\alpha/\beta)$ is any formula obtained from A by replacing one or more occurrences of β with occurrences of α, and AIA is a scheme of schemes which holds for all $n > 0$:

A1 $\Box(A \to B) \to (\Box A \to \Box B)$, $\Box A \to A$, $\Diamond A \to \Box\Diamond A$

A2 $\triangle\alpha(A \to B) \to (\triangle\alpha A \to \triangle\alpha B)$, $\triangle\alpha A \to A$, $\neg\triangle\alpha A \to \triangle\alpha\neg\triangle\alpha A$

A3 $\ominus\alpha(A \to B) \to (\ominus\alpha A \to \ominus\alpha B)$

A4 $\Box A \to \triangle\alpha A \wedge \ominus\alpha A$

A5 $\ominus\alpha A \to \Diamond\triangle\alpha A$

A6 $\Box\ominus\alpha A \vee \Box\neg\ominus\alpha A$

A7 $\Box(\triangle\alpha A \to \triangle\alpha B) \to (\ominus\alpha A \to \ominus\alpha B)$

A8 $\alpha = \alpha$, $\alpha = \beta \to (A \to A(\alpha/\beta))$

AIA $\mathit{diff}(\beta_0, \ldots, \beta_n) \wedge (\bigwedge_{0 \leqslant k \leqslant n} \Diamond\triangle\beta_k B_k) \to \Diamond(\bigwedge_{0 \leqslant k \leqslant n} \triangle\beta_k B_k)$

and takes as rules of inference *modus ponens* and *necessitation*, i.e.,

RN from A to infer $\Box A$.

For each $n > 0$, let L_n be the logic obtained by adding APC_n below to L_0:

APC_n $\bigwedge_{0 \leqslant k \leqslant n-1} \Diamond((\bigwedge_{0 \leqslant i \leqslant k-1} \neg A_i) \wedge \triangle\alpha A_k) \to \bigvee_{0 \leqslant k \leqslant n-1} A_k$

It has been shown in Horty [7] that for each agent term α, $\ominus\alpha$ is a normal modal operator. It is easy to show that A4–A8 are all valid in all utilitarian stit frames. Xu [15] shows that each instance of AIA is valid for all stit frames[2].

In the rest of this paper, when we use L without subscript, we presuppose that all axioms of L_0 are theorems of L, and *modus ponens* and RN hold in L. In particular, L could be any L_n with $n \geqslant 0$.

By ordinary modal logic, we know that the following are derivable in each logic mentioned above:

DR1 $A \to B \;/\; \Box A \to \Box B,\ \triangle\alpha A \to \triangle\alpha B,\ \ominus\alpha A \to \ominus\alpha B$

DR2 $A \;/\; \triangle\alpha A,\ \ominus\alpha A$

T1 $\alpha = \beta \to \Box(\alpha = \beta)$

T2 $\alpha \neq \beta \to \Box(\alpha \neq \beta)$

T3 $\ominus\alpha A \leftrightarrow \ominus\alpha\triangle\alpha A$

T4 $\ominus\alpha A \to \neg\ominus\alpha\neg A$

3 Canonical modal frames

3.1 Modal utilitarian stit frames

The semantics presented in Section 2.2 is based on $BT + AC$ structures on p. 214. The formal language in this paper contains no operators whose interpretation involves temporal reference (such as

[2]Some technical discussions related to AIA and the postulate *independence of agents* can be found in Xu [15] and Xu [16], including examples showing that the postulate cannot be fully expressed in a language like ours. It is easy to see that our AIA can be replaced by the scheme $\mathit{diff}(\beta_0, \ldots, \beta_n) \wedge (\bigwedge_{0\leqslant k\leqslant n} \Diamond\triangle\beta_k B_k) \to \Diamond(\bigwedge_{0\leqslant k\leqslant n} B_k)$. Schemes that aim at expressing *independence of agents*, using *dstit* instead of *cstit*, with or without $\Box$, can also be found in Xu [15] and Xu [16] and that for each $n > 0$, APC_n is valid in a stit frame iff the frame is at-most-n-ary. Hence by our definition of utilitarian stit frames, each instance of AIA is valid in every utilitarian stit frame, and each APC_n is valid in every at-most-n-ary utilitarian stit frame. It follows that all axioms of L_0 are valid in all utilitarian stit frames, and all axioms of $\mathsf{L}_n = \mathsf{L}_0 + \mathsf{APC}_n$ with $n > 0$ are valid in at-most-n-ary utilitarian stit frames, and hence, since *modus ponens* and *necessitation* are obviously validity preserving, we have soundness theorems for all L_n with $n \geqslant 0$.

tense operators or achievement stit operators). Therefore, the temporal relation in utilitarian stit frames can be eliminated.

Let $\mathfrak{M} = \langle \mathsf{T}, <, \mathsf{Agent}, \mathsf{choice}, \mathsf{value}, \tau \rangle$ be an arbitrary utilitarian stit model and an arbitrary moment $m \in \mathcal{T}$. Then $\mathfrak{M}(m/h) = \langle \mathsf{H}_m, \mathsf{Agent}, \mathsf{choice}^m, \mathsf{value}^m, \tau^m \rangle$, where choice^m is the restriction of choice to m, value^m the restriction of value to m, and τ^m is the function which assigns $\{h \in \mathsf{H}_m : m/h \in \tau(p)\}$ to each propositional variable p. Counterpart notions of states, preference, and dominance on $\mathfrak{M}(m/h)$ are defined by taking a history instead of the corresponding moment-history pair. Then, by induction:

OBSERVATION 3. For each formula A of $\mathcal{L}$ and for each moment-history pair m/h, $\mathfrak{M}, m/h \models A$ iff $\mathfrak{M}(m/h), h \models A$.

Let us call $F = \langle W, Agent, \{R_a\}_{a \in Agent}, v, V \rangle$ a *modal utilitarian stit frame*, where $\langle W, Agent, \{R_a\}_{a \in Agent}, v \rangle$ be an arbitrary multi-S5 frame with $Agent$ being the (non-empty) index set for modalities, and V be a function from W to real numbers (a value assignment). A *modal utilitarian stit model* is $\langle F, \tau \rangle$ where τ is a truth-value assignment in the usual sense.

Given a modal utilitarian stit model $M = \langle W, Agent, R, v, \tau \rangle$, an equivalent utilitarian stit model can be easily constructed. That is, with $r \notin W$, $M_r = \langle W \cup \{r\}, Agent, \mathsf{choice}_r, v_r, \tau_r \rangle$ is a utilitarian stit model, where choice_r is a function which assigns to each member a of $Agent$ the partition generated from R_a.

Notions such as truth and validity on modal utilitarian stit frames and models are defined in the same way as in the branching-time (BT) version.

The construction in the next subsection aims a modal utilitarian stit model to satisfy a consistent formula of L.

3.2 Construction of a modal frame

Let Φ be any maximal L-consistent set. We use W_Φ to denote the set of all maximal L-consistent sets w such that $\{\Box A \mid \Box A \in \Phi\} \subseteq w$. It is easy to verify, by ordinary modal logic (applying $\mathsf{A1}$, $\mathsf{A8}$ and $\mathsf{A4}$), that the following holds:

REMARK 4. Let Φ be any maximal L-consistent set. Then the following hold.

1. for each $w \in \mathsf{W}_\Phi$, $\{A \mid \Box A \in w\} = \{\Box A \mid \Box A \in w\} = \{\Box A \mid \Box A \in \Phi\}$;
2. for each $w \in \mathsf{W}_\Phi$, $\Box A \in w$ iff $A \in w'$ for every $w' \in \mathsf{W}_\Phi$;
3. for each α and β, $\alpha = \beta \in \Phi$ iff $\alpha = \beta \in w$ for every $w \in \mathsf{W}_\Phi$;
4. for each $w \in \mathsf{W}_\Phi$ and each α, $\{A \mid \Box A \in w\} \subseteq \{A \mid \triangle\alpha A \in w\}$.

For each set Ξ of formulas, we use $\mathsf{Sub}(\Xi)$ for the closure of Ξ under subformulas. A set Ξ of formulas is *suitable* if the following conditions hold:

1. $\Xi = \mathsf{Sub}(\Xi)$
2. α and β occur in Ξ only if $\alpha = \beta \in \Xi$,
3. $\ominus\beta B \in \Xi$ only if $\triangle\beta B \in \Xi$.

Note that each set of formulas has a smallest suitable extension. For each suitable set Ξ of formulas, we fix Ξ^+ for $\Xi \cup \{\neg A \mid A \in \Xi\}$.

Let Φ be any masimal L-consistent set, and let Ξ be any non-empty set of formulas. Ξ is *coherent with* Φ if for each A, and for each α and β with $\alpha = \beta \in \Phi$, $\triangle\alpha A \in \Xi$ iff $\triangle\beta A \in \Xi$, and $\ominus\alpha A \in \Xi$ iff $\ominus\beta A \in \Xi$.

Let Φ be any maximal L-consistent set, and let Ξ be any suitable set coherent with Φ. We define the "regular modal frame for L with respect to Φ and Ξ" as follows.

First, when Ξ contains no modal formula, consider the model $\langle\{\Phi\}, \{a\}, \{=\}, v_\Phi, V_\Phi\rangle$, where $v_\Phi(p) = \{\Phi\}$ if $p \in \Phi$ and $v_\Phi = \emptyset$ otherwise, and $V_\Phi(\Phi) = 1$. The argument in Section 4 will work for such a case.

Now, the following construction in this section assumes that Ξ contains modal formulas. Let $\mathsf{W} = \{w \cap \Xi^+ \mid w \in \mathsf{W}_\Phi\}$. We will use x, y etc. to range over members of W.

For each α and β occurring in Ξ, $\alpha \cong_{\Phi,\Xi} \beta$ iff $\alpha = \beta \in \Phi \cap \Xi^+$. Let us use $\langle\alpha\rangle_{\Phi,\Xi}$ for the $\cong_{\Phi,\Xi}$-equivalence class to which α belongs, and let Agent be the set of all $\cong_{\Phi,\Xi}$-equivalence classes of terms for agents. We will use $\mathbf{a}, \mathbf{b}$ etc. to range over members of Agent.

For each $\mathbf{a} \in \mathsf{Agent}$, let $R_{\mathbf{a}}$ be the relation on W such that for each $x, y \in \mathsf{W}$, $R_{\mathbf{a}}xy$ iff $\{\triangle\alpha A \in x \mid \alpha \in \mathbf{a}\} = \{\triangle\alpha A \in y \mid \alpha \in \mathbf{a}\}$. Obviously, each $R_{\mathbf{a}}$ is an equivalence relation. We use choice to denote the function on Agent that assigns each $\mathbf{a} \in \mathsf{Agent}$ the set $\mathsf{choice}_{\mathbf{a}}$ of all $R_{\mathbf{a}}$-equivalence classes, and use $[x]_{\mathbf{a}}$ for the $R_{\mathbf{a}}$-equivalence class to which x belongs, i.e., $[x]_{\mathbf{a}} = \mathsf{choice}_{\mathbf{a}}(x)$.

Finally, $\langle \mathsf{W}, \mathsf{Agent}, \mathsf{choice}\rangle$ is the *regular modal frame for* L *with respect to* Φ *and* Ξ if W, Agent and choice are as specified above. Let, for each $\mathbf{a} \in \mathsf{Agent}$, $\Sigma_{\mathbf{a}} = \{\triangle\alpha A \mid \ominus\alpha A \in \Phi \cap \Xi^+ \wedge \alpha \in \mathbf{a}\}$, and let $\Sigma = \bigcup_{\mathbf{a}\in\mathsf{Agent}} \Sigma_{\mathbf{a}}$. It is easy to see that $\Sigma = \{\triangle\alpha A \mid \ominus\alpha A \in \Phi \cap \Xi^+\}$. The *canonical modal frame for* L *with respect to* Φ *and* Ξ is $\langle \mathsf{W}, \mathsf{Agent}, \mathsf{choice}, \mathsf{value}\rangle$, where $\langle \mathsf{W}, \mathsf{Agent}, \mathsf{choice}\rangle$ is the regular modal frame for L with respect to Φ and Ξ, and value is the function on W such that for each $x \in \mathsf{W}$, $\mathsf{value}(x) = 1$ if $\Sigma \subseteq x$, and $\mathsf{value}(x) = 0$ otherwise. When Φ and Ξ are given explicit in the context, we will drop the references to Φ or Ξ as much as we can, and thus use $\langle\alpha\rangle$ for $\langle\alpha\rangle_{\Phi,\Xi}$, and $[x]_{\langle\alpha\rangle}$ for $[x]_{\langle\alpha\rangle_{\Phi,\Xi}}$ etc. From now on, when we speak of "the regular (canonical) modal frame for L with respect to Φ and Ξ", we presuppose that Φ is a maximal L-consistent set and Ξ is a suitable set of formulas coherent with Φ. We say that $\mathfrak{F}$ is a *regular (canonical) modal frame for* L if it is a regular (canonical) modal frame for L with respect to some Φ and Ξ.

We want to consider both the compactness and the finite model property at the same time, and thus, consider the ordinary canonical frames and their filtrations at the same time—this way we can avoid going over similar arguments twice. It is easy to see that when $\langle \mathsf{W}, \mathsf{Agent}, \mathsf{choice}, \mathsf{value}\rangle$ is the canonical modal frame for L with respect to Φ and Ξ, if Ξ is the set of all formulas, $\mathsf{W} = \mathsf{W}_\Phi$, and thus we have a subframe (generated by Φ) of the ordinary canonical frame; and if Ξ is ("logically") finite, we have a filtration.

The following remark is an immediate consequence of A8, our definition of coherence and ordinary modal logic:

REMARK 5. Let $\langle \mathsf{W}, \mathsf{Agent}, \mathsf{choice}, \mathsf{value}\rangle$ be the regular modal frame for L with respect to Φ and Ξ. Then the following hold.

1. for each $x \in \mathsf{W}$ and each $A \in \Xi$, $A \in x$ or $\neg A \in x$;
2. for each $x, y \in \mathsf{W}$, $\{\triangle\alpha A \mid \triangle\alpha A \in x\} = \{\triangle\alpha A \mid \triangle\alpha A \in y\}$ only if $R_{\langle\alpha\rangle}xy$;
3. for each $x \in \mathsf{W}$, $\{\triangle\alpha A \mid \ominus\alpha A \in \Phi \cap \Xi^{+}\} \subseteq x$ only if $\Sigma_{\langle\alpha\rangle} \subseteq x$.

The lemma below handles the desired properties of $\Box$ and $\triangle$, on the basis of which we can start dealing with *independence of agents* and the condition for the operator of dominant ought.

LEMMA 6. *Let* $\langle \mathsf{W}, \mathsf{Agent}, \mathsf{choice}\rangle$ *be the regular modal frame for* L *with respect to* Φ *and* Ξ, *and let* $x \in \mathsf{W}$. *Then the following hold.*

1. *for each* $\Box A \in \Xi$, $\Box A \in x$ *iff* $A \in y$ *for every* $y \in \mathsf{W}$;
2. *for each* α *and* β *occurring in* Ξ, $\alpha = \beta \in x$ *iff* $\langle\alpha\rangle = \langle\beta\rangle$;
3. *for each* $\triangle\alpha A \in \Xi$, $\triangle\alpha A \in x$ *iff* $A \in y$ *for every* $y \in [x]_{\langle\alpha\rangle}$.

The following lemma shows that for each L, all regular modal frames satisfy the required version of *independence of agents.*

LEMMA 7. *Let* $\langle \mathsf{W}, \mathsf{Agent}, \mathsf{choice}\rangle$ *be the regular modal frame for* L *with respect to* Φ *and* Ξ. *Then for each function* f *on* Agent *such that* $f(\mathbf{a}) \in \mathsf{choice}_{\mathbf{a}}$ *for every* $\mathbf{a} \in \mathsf{Agent}$, $\bigcap_{\mathbf{a}\in\mathsf{Agent}} f(\mathbf{a}) \neq \varnothing$.

The theorem below follows immediately from our definition and Lemma 7.

THEOREM 8. *Each regular modal frame for* L *is a stit frame, and each canonical modal frame for* L *is a two-valued utilitarian stit frame.*

The lemma below gives us a property all regular modal frames have, which will be used to deal with the semantic condition for the dominant ought operators.

LEMMA 9. *Let* $\langle \mathsf{W}, \mathsf{Agent}, \mathsf{choice}\rangle$ *be the regular modal frame for* L *with respect to* Φ *and* Ξ. *Then for each* $\ominus\alpha A \in \Xi$ *and each* $x \in \mathsf{W}$, $\ominus\alpha A \in x$ *iff* $\triangle\alpha A \in y$ *for every* $y \in \mathsf{W}$ *such that* $\Sigma_{\langle\alpha\rangle} \subseteq y$.

From now on, for each regular modal frame $\mathfrak{F} = \langle \mathsf{W}, \mathsf{Agent}, \mathsf{choice}\rangle$ for L, let $\mathsf{D}_{\mathfrak{F}}$ stand for the set $\{y \in \mathsf{W} \mid \Sigma \subseteq y\}$. For a canonical modal frame $\mathfrak{F} = \langle \mathsf{W}, \mathsf{Agent}, \mathsf{choice}, \mathsf{value} \rangle$ for L, let $\mathsf{D}_{\mathfrak{F}} = \mathsf{D}_{\mathfrak{F}'}$, where $\mathfrak{F}' = \langle \mathsf{W}, \mathsf{Agent}, \mathsf{choice} \rangle$. Note that for each canonical modal frame $\mathfrak{F}$ for L, $\mathsf{D}_{\mathfrak{F}} = \{y \in \mathsf{W} \mid \mathsf{value}(y) = 1\}$ by definition.

LEMMA 10. *Let* $\langle \mathsf{W}, \mathsf{Agent}, \mathsf{choice} \rangle$ *be the regular modal frame for* L *with respect to* Φ *and* Ξ. *Then for each* $\ominus\alpha A \in \Xi$ *and each* $x \in \mathsf{W}$, $\ominus\alpha A \in x$ *iff* $\triangle\alpha A \in y$ *for every* $y \in \mathsf{D}_{\mathfrak{F}}$.

Applying Proposition 10, Lemma 9 and $\mathsf{T4}$, one can easily verify the following.

COROLLARY 11. *Let* $\langle \mathsf{W}, \mathsf{Agent}, \mathsf{choice} \rangle$ *be the regular modal frame for* L *with respect to* Φ *and* Ξ. *Then* $\mathsf{D}_{\mathfrak{F}} \neq \varnothing$.

Because the canonical modal frame $\mathfrak{F} = \langle \mathsf{W}, \mathsf{Agent}, \mathsf{choice}, \mathsf{value} \rangle$ for L with respect to Φ and Ξ is a utilitarian stit frame, $\mathsf{optimal}_a$ is defined for every $a \in \mathsf{Agent}$ (recall that $\mathsf{optimal}_a = \{K \in \mathsf{choice}_a \mid \neg\exists K'(K' \in \mathsf{choice}_a \wedge K \prec K')\}$). Our next goal is to show that each canonical modal frame is optimal, but first we need to show the following key lemma.

LEMMA 12. *Let* $\mathfrak{F} = \langle \mathsf{W}, \mathsf{Agent}, \mathsf{choice}, \mathsf{value} \rangle$ *be the canonical modal frame for* L *with respect to* Φ *and* Ξ. *Then for each* $x \in \mathsf{W}$ *and each* $\mathbf{a} \in \mathsf{Agent}$, $\Sigma_{\mathbf{a}} \subseteq x$ *iff* $[x]_{\mathbf{a}} \in \mathsf{optimal}_{\mathbf{a}}$.

Proof. Let $x \in \mathsf{W}$ and $\mathbf{a} \in \mathsf{Agent}$. First suppose that $\Sigma_{\mathbf{a}} \subseteq x$. We show below that $[x]_{\mathbf{a}} \in \mathsf{optimal}_{\mathbf{a}}$, i.e., $[x]_{\mathbf{a}} \prec K$ for no $K \in \mathsf{choice}_{\mathbf{a}}$. To that end, we observe that by definitions of $\Sigma_{\mathbf{a}}$ and $R_{\mathbf{a}}$,

(4) $\Sigma_{\mathbf{a}} \subseteq y$ for every $y \in [x]_{\mathbf{a}}$.

Consider any $K \in \mathsf{choice}_{\mathbf{a}}$. We first show that for each $S \in \mathsf{state}_{\mathbf{a}}$,

(5) $S \cap K \leq S \cap [x]_{\mathbf{a}}$.

By definition of value, it is sufficient to show that

if $\mathsf{value}(y) = 1$ for some $y \in S \cap K$, $\mathsf{value}(z) = 1$ for all $z \in S \cap [x]_{\mathbf{a}}$.

Let $S \in \mathsf{state}_{\mathbf{a}}$, $K \in \mathsf{choice}_{\mathbf{a}}$ and $x_0 \in S \cap K$ with $\mathsf{value}(x_0) = 1$. We know by definition of $\mathsf{state}_{\mathbf{a}}$ that $S = \bigcap_{\mathbf{b} \in \mathsf{Agent}-\{\mathbf{a}\}} [x_0]_{\mathbf{b}}$. For each $\mathbf{b} \in \mathsf{Agent} - \{\mathbf{a}\}$, since $\mathsf{value}(x_0) = 1$, $\Sigma_{\mathbf{b}} \subseteq \Sigma \subseteq x_0$, and then by definitions of $\Sigma_{\mathbf{b}}$ and $R_{\mathbf{b}}$, $\Sigma_{\mathbf{b}} \subseteq y$ for every $y \in [x_0]_{\mathbf{b}}$. It follows from (4) that $\Sigma \subseteq z$ for every $z \in [x]_{\mathbf{a}} \cap (\bigcap_{\mathbf{b} \in \mathsf{Agent}-\{\mathbf{a}\}}[x_0]_{\mathbf{b}})$, and hence $\mathsf{value}(z) = 1$ for every $z \in S \cap [x]_{\mathbf{a}}$. It follows that (5) holds. Now if $[x]_{\mathbf{a}} \prec K$, $K \npreceq [x]_{\mathbf{a}}$ by definition, i.e., there is an $S' \in \mathsf{state}_{\mathbf{a}}$ such that $S' \cap K \nleq S' \cap [x]_{\mathbf{a}}$, contrary to (5). Hence $[x]_{\mathbf{a}} \nprec K$. It follows that $[x]_{\mathbf{a}} \in \mathsf{optimal}_{\mathbf{a}}$.

Next suppose that $\Sigma_{\mathbf{a}} \nsubseteq x$. Then there are $\alpha \in \mathbf{a}$ and $\triangle\alpha A \in \Sigma_{\mathbf{a}}$ such that $\triangle\alpha A \notin x$. By definition of $R_{\mathbf{a}}$, $\triangle\alpha A \notin y$ for each $y \in [x]_{\mathbf{a}}$, and thus $\Sigma_{\mathbf{a}} \nsubseteq y$ for each $y \in [x]_{\mathbf{a}}$, and hence

(6) $[x]_{\mathbf{a}} \cap \mathsf{D}_{\mathfrak{F}} = \varnothing$.

There is, by Corollary 11, a $z \in \mathsf{D}_{\mathfrak{F}}$. Since $\Sigma \subseteq z$, an argument similar to the one given above will show that for each $\mathbf{b} \in \mathsf{Agent}$, $\Sigma_{\mathbf{b}} \subseteq y$ for every $y \in [z]_{\mathbf{b}}$. Letting $S_z = \bigcap_{\mathbf{b} \in \mathsf{Agent}-\{\mathbf{a}\}}[z]_{\mathbf{b}}$, we know that $S_z \in \mathsf{state}_{\mathbf{a}}$ and $z \in S_z \cap [z]_{\mathbf{a}}$. Since $\mathsf{value}(z) = 1$ and $S_z \cap [x]_{\mathbf{a}} \neq \varnothing$ (by Theorem 8), $S_z \cap [z]_{\mathbf{a}} \nleq S_z \cap [x]_{\mathbf{a}}$ by (6), and hence $[z]_{\mathbf{a}} \npreceq [x]_{\mathbf{a}}$. It follows, also from (6), that for each $S \in \mathsf{state}_{\mathbf{a}}$, $S \cap [x]_{\mathbf{a}} \leq S \cap [z]_{\mathbf{a}}$, i.e., $[x]_{\mathbf{a}} \preceq [z]_{\mathbf{a}}$. Hence $[x]_{\mathbf{a}} \prec [z]_{\mathbf{a}}$, and therefore $[x]_{\mathbf{a}} \notin \mathsf{optimal}_{\mathbf{a}}$. ■

THEOREM 13. *Each canonical modal frame for* L *is optimal.*

Proof. Let $\mathbf{a} \in \mathsf{Agent}$, and let $K \in \mathsf{choice}_{\mathbf{a}} - \mathsf{optimal}_{\mathbf{a}}$. We show as follows that there is a $K' \in \mathsf{optimal}_{\mathbf{a}}$ such that $K \prec K'$, which is sufficient. Let $x \in K$. We know that $K = [x]_{\mathbf{a}}$ and, by Lemma 12, that $\Sigma_{\mathbf{a}} \nsubseteq x$. An argument similar to that in Lemma 12 shows that there is a $z \in \mathsf{D}_{\mathfrak{F}}$ such that $[x]_{\mathbf{a}} \prec [z]_{\mathbf{a}}$. Since $\Sigma_{\mathbf{a}} \subseteq \Sigma \subseteq z$, $[z]_{\mathbf{a}} \in \mathsf{optimal}_{\mathbf{a}}$ by Lemma 12. ■

For each $n > 0$, we know that the characteristic axiom of L_n, i.e., APC_n, corresponds to the semantic condition that the stit frames are at-most-n-ary. The following lemma shows that all regular (canonical) modal frames for L_n are at-most-n-ary.

LEMMA 14. *Let* $n > 0$, *and let* $\langle \mathsf{W}, \mathsf{Agent}, \mathsf{choice} \rangle$ *be the regular modal frame for* L_n *with respect to* Φ *and* Ξ. *Then for each* $\mathbf{a} \in \mathsf{Agent}$, $|\mathsf{choice}_{\mathbf{a}}| \leqslant n$.

Proof. Let $\mathbf{a} \in \mathsf{Agent}$. Suppose for reductio that $|\mathsf{choice}_{\mathbf{a}}| > n$. Then there are $U_0, \ldots, U_n \in \mathsf{choice}_{\mathbf{a}}$ such that $U_0, \ldots, U_n$ are all different. Select $\alpha \in \mathbf{a}$, select $x_0 \in U_0, \ldots, x_n \in U_n$, and select $w_0, \ldots, w_n \in \mathsf{W}_\Phi$ such that $x_k = w_k \cap \Xi^+$ for every k with $0 \leqslant k \leqslant n$. Consider any i and k such that $1 \leqslant k \leqslant n$, $0 \leqslant i \leqslant n$ and $i \neq k$. By definition of $R_{\mathbf{a}}$, there is an A such that either $\neg\triangle\alpha A \in x_k$ and $\triangle\alpha A \in x_i$, or $\triangle\alpha A \in x_k$ and $\neg\triangle\alpha A \in x_i$; and then by A2, either $\triangle\alpha\neg\triangle\alpha A \in w_k$ and $\neg\triangle\alpha\neg\triangle\alpha A \in w_i$, or $\triangle\alpha A \in w_k$ and $\neg\triangle\alpha A \in w_i$; and hence there is an A' such that $\triangle\alpha A' \in w_k$ and $\neg\triangle\alpha A' \in w_i$. It follows that for each k with $1 \leqslant k \leqslant n$, there are $\triangle\alpha A_{k,0}, \ldots, \triangle\alpha A_{k,n} \in w_k$ such that $\neg\triangle\alpha A_{k,i} \in w_i$ for every i with $0 \leqslant i \leqslant n$ and $i \neq k$. For each k with $1 \leqslant k \leqslant n$, letting $B_k = \triangle\alpha A_{k,0} \wedge \ldots \wedge \triangle\alpha A_{k,n}$, it is easy to see by ordinary modal logic (applying A2, RN and A4) that $\triangle\alpha B_k \in w_k$ and $\neg B_k \in w_i$ for every i with $0 \leqslant i \leqslant n$ and $i \neq k$. It follows that $\triangle\alpha B_1 \in w_1$, $\neg B_1 \wedge \triangle\alpha B_2 \in w_2, \ldots, \neg B_1 \wedge \ldots \wedge \neg B_{n-1} \wedge \triangle\alpha B_n \in w_n$, and then by Remark 4 (2), $\Diamond\triangle\alpha B_1 \in w_0$, $\Diamond(\neg B_1 \wedge \triangle\alpha B_2) \in w_0, \ldots, \Diamond(\neg B_1 \wedge \ldots \wedge \neg B_{n-1} \wedge \triangle\alpha B_n) \in w_0$, and hence by APC_n, $B_1 \vee \ldots \vee B_n \in w_0$. But, since $\neg B_i \in w_0$ for every i with $0 \leqslant i \leqslant n$, $\neg(B_1 \vee \ldots \vee B_n) \in w_0$, contrary to the assumption of L-consistency on w_0. We conclude from this reductio that $|\mathsf{choice}_{\mathbf{a}}| \leqslant n$. ■

4 Main results

Let $\mathfrak{F} = \langle \mathsf{W}, \mathsf{Agent}, \mathsf{choice}, \mathsf{value}\rangle$ be the canonical frame for L with respect to Φ and Ξ. $\mathfrak{M} = \langle \mathfrak{F}, \tau\rangle$ is the *canonical model for* L *with respect to* Φ *and* Ξ if τ is the valuation on $\mathfrak{F}$ such that for each agent term α, $\alpha^\tau = \langle\alpha\rangle$, and for each atomic formula $p \in \Xi$, $\tau(p) = \{x \in \mathsf{W} \mid p \in x\}$.

THEOREM 15. *Let* $\mathfrak{M} = \langle \mathfrak{F}, \tau\rangle$ *be the canonical model for* L *with respect to* Φ *and* Ξ, *where* $\mathfrak{F} = \langle \mathsf{W}, \mathsf{Agent}, \mathsf{choice}, \mathsf{value}\rangle$. *Then for each* $A \in \Xi$ *and for each* $x \in \mathsf{W}$, $\mathfrak{M}, x \models A$ *iff* $A \in x$.

Proof. By a routine induction we can show that for each $A \in \Xi$ and $x \in \mathsf{W}$,

(7) $\mathfrak{M}, x \models A$ iff $A \in x$.

It is straightforward that (7) holds when A is an atomic formula, or $\neg B$ or $B \wedge C$. We know by Lemma 6 (1)–(3). that (7) holds when

A is $\alpha = \beta$, $\Box B$ or $\triangle\alpha B$. Now let A be $\ominus\alpha B \in \Xi$. Consider any $x \in \mathsf{W}$. By Lemma 9, $\ominus\alpha B \in x$ iff $\triangle\alpha B \in y$ for every $y \in \mathsf{W}$ such that $\Sigma_{\langle\alpha\rangle} \subseteq y$, and then by Lemma 12,

$$(8)\quad \ominus\alpha B \in x \quad \text{iff} \quad \triangle\alpha B \in y \text{ for every } y \in \mathsf{W} \text{ such that } [y]_{\langle\alpha\rangle} \in \mathsf{optimal}_{\langle\alpha\rangle}.$$

Consider any $y \in \mathsf{W}$ such that $[y]_{\langle\alpha\rangle} \in \mathsf{optimal}_{\langle\alpha\rangle}$. We know by Lemma 6 (3) that $\triangle\alpha B \in y$ iff $B \in z$ for every $z \in [y]_{\langle\alpha\rangle}$, and then by induction hypothesis, $\triangle\alpha B \in y$ iff $[y]_{\langle\alpha\rangle} \subseteq \|B\|^{\tau}$. It follows from (8) that $\ominus\alpha B \in x$ iff $[y]_{\langle\alpha\rangle} \subseteq \|B\|^{\tau}$ for every $y \in \mathsf{W}$ such that $[y]_{\langle\alpha\rangle} \in \mathsf{optimal}_{\langle\alpha\rangle}$. Since each $K \in \mathsf{optimal}_{\langle\alpha\rangle}$ is $[y]_{\langle\alpha\rangle}$ for some $y \in \mathsf{W}$, we know that $\ominus\alpha B \in x$ iff $K \subseteq \|B\|^{\tau}$ for every $K \in \mathsf{optimal}_{\langle\alpha\rangle}$. Hence by Theorem 13 and the definition of optimal on p. 217, $\ominus\alpha B \in x$ iff $\mathfrak{M}, x \models \ominus\alpha B$. ■

THEOREM 16. L_0 *is complete and compact.*

Proof. For each L_0-consistent set Ψ of formulas, let Φ be any maximal L_0-consistent set including Ψ, let Ξ be the set of all formulas, and let $\mathfrak{M} = \langle \mathsf{W}, \mathsf{Agent}, \mathsf{choice}, \mathsf{value}, \tau\rangle$ be the canonical model for L_0 with respect to Φ and Ξ. It is easy to see that $\Phi \in \mathsf{W}$. Then by Theorem 15, $\mathfrak{M}, \Phi \models \Psi$. ■

By Theorem 13, each canonical modal frame for L_0 is an optimal utilitarian stit frame, and hence the only formulas valid in all optimal utilitarian stit frames are theorems of L_0. It follows that the theorem below holds, which shows a certain limitation of the language.

THEOREM 17. *There is no set Θ of formulas such that for each utilitarian stit frame $\mathfrak{F}$, $\mathfrak{F}$ is optimal iff all members of Θ are valid in $\mathfrak{F}$; and if L is the set of all formulas valid in all optimal utilitarian stit frames, $\mathsf{L} = \mathsf{L}_0$ (identifying L_0 with the set of all its theorems).*

Similarly, because canonical modal frames for L_0 are all two-valued, the following holds.

THEOREM 18. *There is no set Θ of formulas such that for each utilitarian stit frame $\mathfrak{F}$, $\mathfrak{F}$ is two-valued iff all members of Θ are valid in $\mathfrak{F}$; and if L is the set of all formulas valid in all two-valued* *utilitarian stit frames,* $\mathsf{L} = \mathsf{L}_0$ *(identifying* L_0 *with the set of all its theorems).*

The theorem below follows from Theorem 15 and Lemma 14, whose proof is similar to that of Theorem 16, with an application of Lemma 14.

THEOREM 19. *Let* $n > 0$. *Then the following hold:*

1. L_n *is complete; and*

2. *for each set* Θ *of formulas, if every finite subset of* Θ *has an at-most-*n*-ary utilitarian stit model, so does* Θ.

In the following, we prove the decidability of all L_n with $n \geqslant 0$ by way of the finite model property. For each L-consistent formula A, let Φ be a maximal L-consistent set containing A, and let

$\Pi_0 = \{\alpha = \beta \mid \alpha \text{ and } \beta \text{ occur in } A\}$,

$\Pi_1 = \{\triangle\beta B, \ominus\beta B \mid \ominus\alpha B \in \mathsf{Sub}(\{A\}),\ \alpha = \beta \in \Phi,\ \beta \text{ occurs in } A\}$,

$\Pi_2 = \{\triangle\beta B \mid \triangle\alpha B \in \mathsf{Sub}(\{A\}), \alpha = \beta \in \Phi,\ \beta \text{ occurs in } A\}$, and

$\Xi_A^\Phi = \mathsf{Sub}(\{A\}) \cup \Pi_0 \cup \Pi_1 \cup \Pi_2$.

It is easy to verify that for each A, Ξ_A^Φ is finite, suitable, and coherent with Φ.

THEOREM 20. *For each* $n \geqslant 0$, L_n *has the finite model property, and hence is decidable.*

Proof. Let $n \geqslant 0$ and let A be any L_n-consistent formula. Select a maximal L_n-consistent set Φ containing A, let $\Xi = \Xi_A^\Phi$. Then, let $\mathfrak{M} = \langle \mathsf{W}, \mathsf{Agent}, \mathsf{choice}, \mathsf{value}, \tau \rangle$ be the canonical model for L_n with respect to Φ and Ξ. Since Ξ is finite, so is W. Because $A \in \Phi \cap \Xi^+ \in \mathsf{W}$, we know by Theorem 15 that $\mathfrak{M}, x \vDash A$, where $x = \Phi \cap \Xi^+$. When $n > 0$, we apply Lemma 14 to conclude that $\mathfrak{M}$ is at-most-n-ary. ■

5 Conclusion

There are several classes of frames proposed by Horty and Thomason as argued in Section 2.5. Some logics of them share the same axiomatic system L_0. In particular, both the logic of the class of all optimal frames and that of the class of all two-valued frames coincide with L_0. It implies that, despite of Horty's introduction of the real-number valued value assignment, the resulting logic fails to be essentially different from standard deontic logic.

The considerations suggest that Horty's proposal needs subtler investigations than presented here. In fact, his original presentation with branching time semantics allows to take non-monotonic agency modalities instead of multi-S5. For example, the *deliberative stit* (dstit) operators [ ] induces the dstit-based dominant ought operator $\oplus\alpha$ can be defined as follows: $[\alpha]A =_{df} \triangle\alpha A \wedge \neg\Box A$, and $\oplus\alpha A =_{df} \ominus\alpha A \wedge \neg\Box A$. [ ] can be taken primitive, and let $\triangle\alpha A =_{df} [\alpha]A \vee \Box A$, $\oplus\alpha A =_{df} \odot[\alpha]A$ and $\ominus\alpha A =_{df} \oplus\alpha A \vee \Box A$. Different choices of agency operators would lead to different logics, which are to be examined.

BIBLIOGRAPHY

[1] Nuel Belnap, Michael Perloff, and Ming Xu. *Facing the Future: Agents and Choices in Our Indeterminist World.* Oxford Univerity Press, Oxford, 2001.
[2] Dagfinn Føllesdal and Risto Hilpinen. Deontic logic: an introduction. In Hilpinen (ed.) *Deontic Logic: introductory and systematic readings*, 1–35. D. Reidel, Dordrecht, 1971.
[3] Dov Gabbay and F. Guenthner, editors. *Handbook of Philosophical Logic*, volume 2. D. Reidel Publishing Company, Dordrecht, 1984.
[4] Dov M. Gabbay and Hans J. Ohlbach, editors. *Temporal Logic, First International Conference, ICTL'94, Bonn, Germany, Proceedings*, volume 827. Springer-Verlag, 1994.
[5] Peter Geach. What happened in deontic logic? *Philosophia*, 11:1–11, 1982.
[6] Risto Hilpinen, editor. *New Studies in Deontic Logic.* D. Reidel Publishing Company, Dordrecht, 1981.
[7] John F. Horty. *Agency and Deontic Logic.* Oxford University Press, Oxford, 2000.
[8] John F. Horty and Nuel Belnap The Deliberative *Stit*: a study of action, omission, ability, and obligation *Journal of Philosophical Logic*, 24:583–644 1995.
[9] Arthur Prior. *Past, Present and Future.* Oxford Univerity Press, Oxford, 1967.
[10] Krister Segerberg. Getting started: beginnings in the logic of action. *Studia Logica*, 51:347–378, 1992.
[11] Richmond H. Thomason. Indeterminist time and truth-value gaps. *Theoria*, 36:264–281, 1970.
[12] Richmond H. Thomason. Combination of tense and modality. In Gabbay and Guenthner [3], pages 135–165.
[13] Georg Henrik von Wright. *Norm and Action.* Routledge and Kegan Paul, London, 1963.

[14] Georg Henrik von Wright. Problems and prospects of deontic logic: A survey. In Evandro Agazzi (ed.) *Modern Logic: A Survey*, pages 399–423. D. Reidel, Dordrecht, 1980.

[15] Ming Xu. Decidability of deliberative *stit* theories with multiple agents. In Gabbay and Ohlbach [4], pages 332–348.

[16] Ming Xu. Axioms for Deliberative *Stit*. *Journal of Philosophical Logic*, 27:505–552, 1998.

Yuko Murakami

Department of Philosophy, Indiana University, Bloomington Indiana 47405 USA

yuko.murakami@nifty.com

Resolution for Synchrony and No Learning

CLÁUDIA NALON, CLARE DIXON, AND MICHAEL FISHER

ABSTRACT. We present a clausal resolution method for temporal logics of knowledge with synchrony and no learning. This and related logics admit axioms which include operators from both the temporal and epistemic logics, which allow the description of how knowledge evolves over time. Instead of proposing new resolution rules, further information is added to the set of clauses in order to deal with this particular interaction.

1 Introduction

Logics have been used in Computer Science for many years as a natural way for describing properties of complex systems. More recently, there has been an increasing interest in combined modal logics, as different logical languages are more suitable to specify different properties within a system. Typical examples are the specification and verification of distributed [6] and agent-based systems [12, 13]. Given such logical characterisation of a system, it is then desirable to have the appropriate tools in order to verify whether a particular property holds for this system. By verifying that a property holds, we mean to *prove* that the property is a logical consequence of the specification.

There is a wide range of logics that could be chosen in order to model and characterise such systems. Moreover, there is a variety of ways of combining the chosen logics. In the following, we concentrate on a particular combination that has been proved useful in modelling distributed and agent-based systems, namely, we are looking at *Propositional Temporal Logics of Knowledge* ($\mathsf{KL}_{(n)}$, for short). In such logics, the dynamic component is described by a propositional linear temporal logic and the informational component is described by a propositional logic of knowledge. When the combined logics are *independent*, i.e. the combination is given by the union of the axiomatic systems of both logics, proof methods can be obtained by taking the union of proof methods for the logics considered alone and making sure that enough information is passed to each component. Proof methods for combined logics cannot be obtained in a straightforward way, however, when the logics *interact*, i.e. when further axioms, including operators of both logics, are needed in order to model a specific situation. Contexts where we are particularly interested in how the knowledge of an agent evolves over time is a typical example where *interaction axioms* are required. Interactions often increase the complexity of the validity problem for the language and proof methods for such logics are, to our knowledge, rare.

In this paper, we introduce a proof method for a particular interaction between time and knowledge: *synchrony and no learning*. This property was first discussed in the context of blindfold games [10]. Recently, a similar characterisation has proved useful in the description of non-decreasing domains [9]. In such systems, once two situations are indistinguishable to an agent, the agent will never acquire any knowledge that would allow her to distinguish between such situations. Although the complexity of the interacting logic is high (non-elementary, for the multi-agent case), the axiom that expresses this property has a simple form, which allowed us to investigate in detail the requirements for its proof method.

The structure of the paper is as follows. In Section 2, we review $\mathsf{KL}_{(n)}$. In Section 3, we present a resolution method for synchronous systems with no learning. The method introduces additional information into the set of clauses, instead of introducing new (possibly complicated) inference rules. The multi-agent case is non-trivial, therefore we discuss the granularity of the information that needs to be provided in order to obtain completeness for these systems. Correctness results for the multi-agent case are given in Section 4. We discuss our results and future research in Section 5.

2 Temporal Logics of Knowledge

The syntax of $\mathsf{KL}_{(n)}$ comprises a set of modal operators and a set of temporal operators. Formulae are constructed from a denumerable set, $\mathcal{P} = \{p, q, p', q', \ldots\}$, of propositional symbols; nullary connectives, **true** and **false**; propositional connectives, $\neg$, $\wedge$, $\vee$, $\Rightarrow$, and $\Leftrightarrow$; temporal connectives, $\Diamond$, $\Box$, $\bigcirc$, $\mathcal{U}$, and $\mathcal{W}$; and a set of unary modal operators K_i, for all $i \in \mathcal{A}$, where $\mathcal{A} = \{1, \ldots, n\}$ is the set of agents. The set of well-formed formulae WFF is defined as usual: the nullary connectives and propositional symbols are in WFF; if ϕ and φ are in WFF, then so are $\neg\varphi$, $(\varphi \wedge \phi)$, $(\varphi \vee \phi)$, $(\varphi \Rightarrow \phi)$, $(\varphi \Leftrightarrow \phi)$, $\Diamond\varphi$, $\Box\varphi$, $\bigcirc\varphi$, $(\varphi \mathcal{U} \phi)$, $(\varphi \mathcal{W} \phi)$ and $\mathrm{K}_i\, \varphi$, $\forall i \in \mathcal{A}$. A *literal* is p or $\neg p$, where $p \in \mathcal{P}$; a *modal literal* is $\mathrm{K}_i\, l$ or $\neg\mathrm{K}_i\, l$, where l is a literal and $i \in \mathcal{A}$; and an *eventuality* is in the form $\Diamond l$, where l is a literal.

The semantics of $\mathsf{KL}_{(n)}$ interprets formulae over a set of temporal lines, each of which corresponds to a discrete, linear model of time with finite past and infinite future, together with the agents' accessibility relations $\mathcal{K}_i$. We define a *timeline* t as an infinitely long, linear, discrete sequence of states, indexed by the natural numbers. Let $TLines$ be the set of all timelines. A *point* q is a pair $q = (t, u)$, where $t \in TLines$ and $u \in \mathbb{N}$ is a temporal index to t. Let $Points$ be the set of all points. A *model* is a structure $M = \langle TL, \mathcal{K}_1, \dots \mathcal{K}_n, \pi \rangle$ where $TL \subseteq TLines$ is a set of timelines with a distinguished timeline t_0; $\mathcal{K}_i$, for all $i \in \mathcal{A}$, is an equivalence relation over points, i.e., $\mathcal{K}_i \subseteq Points \times Points$; and π is a function $\pi : Points \times \mathcal{P} \rightarrow \{true, false\}$. Truth of a formula is given as follows:

- $\langle M, (t, u) \rangle \models \mathbf{true}$
- $\langle M, (t, u) \rangle \models p$ iff $\pi(t, u)(p) = true$, where $p \in \mathcal{P}$
- $\langle M, (t, u) \rangle \models \neg\varphi$ iff $\langle M, (t, u) \rangle \not\models \varphi$
- $\langle M, (t, u) \rangle \models (\varphi \wedge \phi)$ iff $\langle M, (t, u) \rangle \models \varphi$ and $\langle M, (t, u) \rangle \models \phi$
- $\langle M, (t, u) \rangle \models \bigcirc\varphi$ iff $\langle M, (t, u + 1) \rangle \models \varphi$
- $\langle M, (t, u) \rangle \models \varphi\mathcal{U}\phi$ iff $\exists k \in \mathbb{N}$, $k \geq u$, $\langle M, (t, k) \rangle \models \phi$ and $\forall j \in \mathbb{N}$, $u \leq j < k$, $\langle M, (t, j) \rangle \models \varphi$
- $\langle M, (t, u) \rangle \models \mathrm{K}_i\, \varphi$ iff $\forall t', u'$, such that $((t, u), (t', u')) \in \mathcal{K}_i$, $\langle M, (t', u') \rangle \models \varphi$.

The semantics of the other connectives are given by $\mathbf{false} \equiv \neg\mathbf{true}$, $(\varphi \vee \psi) \equiv \neg(\neg\varphi \wedge \neg\psi)$, $(\varphi \Rightarrow \psi) \equiv (\neg\varphi \vee \psi)$, $(\varphi \Leftrightarrow \psi) \equiv ((\varphi \Rightarrow \psi) \wedge (\psi \Rightarrow \varphi))$, $\Diamond\varphi \equiv (\mathbf{true}\mathcal{U}\varphi)$, $\Box\varphi \equiv \neg\Diamond\neg\varphi$, and $(\varphi\mathcal{W}\psi) \equiv (\Box\neg\varphi \vee \varphi\mathcal{U}\psi)$. We write $(t, u) \sim_i (t', u')$, if $((t, u), (t', u')) \in \mathcal{K}_i$. A formula φ is said to be *satisfiable* if there is a model M such that $\langle M, (t_0, 0) \rangle \models \varphi$; φ is *valid* if $\langle M, (t_0, 0) \rangle \models \varphi$, for every model M.

The resolution-based proof method for $\mathsf{KL}_{(n)}$ in [3] combines the inference rules for temporal and multi-modal knowledge logics when considered alone. A formula is first translated into a normal form, called *Separated Normal Form for Logics of Knowledge* (SNF_K). A nullary connective, **start**, which intuitively represents the beginning of time, is introduced. Formally, $\langle M, (t, u) \rangle \models \mathbf{start}$ if, and only if, $t = t_0$ and $u = 0$, where M is a model and (t, u) is a point. Formulae are represented by a conjunction of clauses, which are true in all states, i.e. they have the general form $\Box^* \bigwedge_i A_i$, where the universal operator is defined as $\Box^*\varphi \Leftrightarrow \Box^{\pm}(\varphi \wedge C\,\Box^*\varphi)$ (with $\Box^{\pm}\varphi \Leftrightarrow \Box\varphi \wedge \Box^{-}\varphi$ and $\langle M, (t, u) \rangle \models \Box^{-}\varphi$ if, and only if, $\forall k, k \in \mathbb{N}$, if $0 \leq k \leq u$, then $\langle M, (t, k) \rangle \models \varphi$), C is the common knowledge operator (i.e. $C\varphi \Leftrightarrow E(\varphi \wedge C\varphi)$, where $E\varphi \Leftrightarrow \bigwedge_{i \in \mathcal{A}} \mathrm{K}_i\, \varphi)$, and A_i is a clause which is in one of the following forms, where l, l_i, k_i are literals, m_{i_j} are literals or modal literals in the form $\mathrm{K}_i\, l$ or $\neg\mathrm{K}_i\, l$:

$$
\begin{array}{llcl}
\textbf{Initial clause:} & \mathbf{start} & \Rightarrow & \bigvee_{b=1}^{r} l_b \\
\textbf{Sometime clause:} & \bigwedge_{a=1}^{g} k_a & \Rightarrow & \Diamond l \\
\textbf{Step clause:} & \bigwedge_{a=1}^{g} k_a & \Rightarrow & \bigcirc \bigvee_{b=1}^{r} l_b \\
\mathbf{K}_i\textbf{-clause:} & \mathbf{true} & \Rightarrow & \bigvee_{b=1}^{r} m_{i_b} \\
\textbf{Literal clause:} & \mathbf{true} & \Rightarrow & \bigvee_{b=1}^{r} l_b
\end{array}
$$

Transformation into the SNF_K, whose satisfiability preserving transformation rules are given in [4] and [3], depends on three main operations: the renaming of complex subformulae; the removal of temporal operators; and classical style rewrite operations. Once a formula has been transformed into SNF_K, the resolution method can be applied. The method consists of two main procedures: the first performs initial, modal and step resolution; the second performs temporal resolution. Each procedure is performed until a contradiction (either **true** $\Rightarrow$ **false** or **start** $\Rightarrow$ **false**) is generated or no new clauses can be generated. In the following l, l_i are literals; m_i are literals or modal literals; D, D' are disjunctions of literals; M, M' are disjunction of literals or modal literals; and C, C' are conjunctions of literals.

Initial Resolution is applied to clauses that hold at the beginning of time:

$$
[\text{IRES1}]\ \begin{array}{rcl} \mathbf{true} & \Rightarrow & (D \vee l) \\ \mathbf{start} & \Rightarrow & (D' \vee \neg l) \\ \hline \mathbf{start} & \Rightarrow & (D \vee D') \end{array}
\qquad
[\text{IRES2}]\ \begin{array}{rcl} \mathbf{start} & \Rightarrow & (D \vee l) \\ \mathbf{start} & \Rightarrow & (D' \vee \neg l) \\ \hline \mathbf{start} & \Rightarrow & (D \vee D') \end{array}
$$

Modal Resolution is applied between clauses of same index (i.e. two K_i-clauses; a literal and a K_i-clause; or two literal clauses):

$$
[\text{MRES1}]\ \begin{array}{rcl} \mathbf{true} & \Rightarrow & (M \vee m_i) \\ \mathbf{true} & \Rightarrow & (M' \vee \neg m_i) \\ \hline \mathbf{true} & \Rightarrow & (M \vee M') \end{array}
\qquad
[\text{MRES2}]\ \begin{array}{rcl} \mathbf{true} & \Rightarrow & (M \vee \mathrm{K}_i\, l) \\ \mathbf{true} & \Rightarrow & (M' \vee \mathrm{K}_i \neg l) \\ \hline \mathbf{true} & \Rightarrow & (M \vee M') \end{array}
$$

$$
[\text{MRES3}]\ \begin{array}{rcl} \mathbf{true} & \Rightarrow & (M \vee \mathrm{K}_i\, l) \\ \mathbf{true} & \Rightarrow & (M' \vee \neg l) \\ \hline \mathbf{true} & \Rightarrow & (M \vee M') \end{array}
\qquad
[\text{MRES5}]\ \begin{array}{rcl} \mathbf{true} & \Rightarrow & (D \vee \mathrm{K}_i\, l_1 \vee \mathrm{K}_i\, l_2 \vee \ldots) \\ \hline \mathbf{true} & \Rightarrow & (D \vee l_1 \vee l_2 \vee \ldots) \end{array}
$$

$$
[\text{MRES4}]\ \begin{array}{rcl} \mathbf{true} & \Rightarrow & (M \vee \neg\mathrm{K}_i\, l) \\ \mathbf{true} & \Rightarrow & (M' \vee l) \\ \hline \mathbf{true} & \Rightarrow & (M \vee mod_i(M')) \end{array}
\quad \text{where} \quad
\begin{array}{rcl} mod_i(A \vee B) & = & mod_i(A) \vee mod_i(B) \\ mod_i(\mathrm{K}_i\, l) & = & \mathrm{K}_i\, l \\ mod_i(\neg\mathrm{K}_i\, l) & = & \neg\mathrm{K}_i\, l \\ mod_i(l) & = & \neg\mathrm{K}_i \neg\, l \end{array}
$$

MRES1 corresponds to classical resolution. MRES2 is justified by the axiom D, i.e. $\vdash \mathrm{K}_i\, \varphi \Rightarrow \neg\mathrm{K}_i \neg\, \varphi$, for any formula φ. The rules MRES3 and

MRES5 are justified by the axiom T, i.e. $\vdash \mathrm{K}_i\,\varphi \Rightarrow \varphi$, for any formula φ. The rule MRES4 is justified by the external universal operator surrounding each clause. By modal and propositional reasoning, the clause $\Box^*(\mathbf{true} \Rightarrow (D' \vee l))$ implies $\Box^*(\mathbf{true} \Rightarrow \neg\mathrm{K}_i \neg D' \vee \mathrm{K}_i\, l)$. The resolution inference rule is, then, applied between the clauses containing the complementary modal literals $\neg\mathrm{K}_i\, l$ (from the first premise) and $\mathrm{K}_i\, l$ (from the transformation of the second premise). The function mod_i makes use of the axioms K (for distributing the knowledge operator over D'), 4, and 5 (for modal simplification) to generate the clausal form of $\neg\mathrm{K}_i \neg D'$.

Step Resolution is applied to clauses that hold at the same moment in time:

$$[\text{SRES1}]\quad \frac{\begin{array}{rcl} C & \Rightarrow & \bigcirc(D \vee l) \\ C' & \Rightarrow & \bigcirc(D' \vee \neg l) \end{array}}{(C \wedge C') \Rightarrow \bigcirc(D \vee D')} \qquad [\text{SRES2}]\quad \frac{\begin{array}{rcl} \mathbf{true} & \Rightarrow & (D \vee l) \\ C & \Rightarrow & \bigcirc(D' \vee \neg l) \end{array}}{C \Rightarrow \bigcirc(D \vee D')}$$

together with the following simplification rule:

$$[\text{SIMP1}]\quad \frac{C \Rightarrow \bigcirc\,\mathbf{false}}{\mathbf{true} \Rightarrow \neg C}$$

Temporal Resolution is applied between an eventuality $\Diamond l$ and a *set of clauses* which forces l always to be false. In detail, the temporal resolution rule is (where A_j is a conjunction of literals, B_j is a disjunction of literals, and C and l are as above):

$$[\text{TRES}]\quad \frac{\begin{array}{c} A_0 \Rightarrow \bigcirc B_0 \\ \vdots \\ A_n \Rightarrow \bigcirc B_n \\ C \Rightarrow \Diamond l \end{array}}{C \Rightarrow \left(\bigwedge_{i=0}^{n} (\neg A_i)\right) \mathcal{W}\, l} \quad \text{where} \quad \begin{array}{rcl} \forall i, 0 \leq i \leq n, \vdash B_i & \Rightarrow & \neg l \\ \forall i, 0 \leq i \leq n, \vdash B_i & \Rightarrow & \bigvee_{j=0}^{n} A_j \end{array}$$

The set of clauses that satisfy the side conditions are together known as *a loop in* $\neg l$. Algorithms for finding such a loop can be found in [1]. We note that each $A_j \Rightarrow \bigcirc B_j$ are step clauses in *merged* SNF_K, that is, they correspond to a conjunction of step clauses in SNF_K. A translation of the resolvent into the normal form is given by the following clauses (where t is a new proposition): $\mathbf{true} \Rightarrow (\neg C \vee \neg A_i \vee l)$, $t \Rightarrow \bigcirc(\neg A_i \vee l)$, $\mathbf{true} \Rightarrow (\neg C \vee t \vee l)$, and $t \Rightarrow \bigcirc(t \vee l)$. Clauses are kept in their simplest form by performing classical style simplification. Classical subsumption is also applied and valid formulae can be removed during simplification as they cannot contribute to the generation of a contradiction.

3 Synchronous Systems with No Learning

We now describe a clausal resolution system for temporal logics of knowledge in synchronous systems with no learning ($\mathsf{KL}^{snl}_{(n)}$). A system is synchronous if the agent has access to a common external clock. Intuitively, if the system is synchronous, the agent *knows* the time, which is common to all agents. The agent has the property of *no learning*, if her knowledge does not increase over time. Formally, in the class of models for synchronous systems with no learning, if two points, (s, m) and (t, n), are in the accessibility relation of agent i, i.e. $(s, m) \sim_i (t, n)$, then, because of synchrony, they share the same time index ($m = n$) and, because of no learning, their successors are also indistinguishable to agent i, i.e. $(s, m + 1) \sim_i (t, n + 1)$.

The syntax and semantics for $\mathsf{KL}^{snl}_{(n)}$ are the same as for $\mathsf{KL}_{(n)}$. A complete axiomatisation for $\mathsf{KL}^{snl}_{(n)}$ comprises the set of axioms of both PTL and $\mathsf{S5}_{(n)}$, together with the axioms $\vdash \bigcirc \mathrm{K}_i\, \varphi \Rightarrow \mathrm{K}_i \bigcirc \varphi$ (SNL), for all agents $i \in \mathcal{A}$ [8]. The validity problem for such systems is EXPSPACE for $n = 1$ and non-elementary space for $n \geq 2$ [8].

3.1 Proof Method

The general approach for dealing with synchrony and no learning is as follows. Given a set of clauses in SNF_K, we first add some new clauses which ensure that the constraints expressed by the SNL axiom are made explicit before applying the rules given in Section 2. As making such constraints explicit is essential part of the method, we explain better its motivation.

In resolution-based proof methods, generally speaking, one has to identify complementary formulae (or sets of formulae) in order to apply the inference rules. This procedure can be relatively easy for basic logics. For instance, for propositional logics, there is only one resolution inference rule, which is applied to clauses containing complementary literals, l and $\neg l$. However, for more complex logics, trying to identify complementary formulae can be non-trivial, costly, and often achieved by the introduction of new inference rules. This is the case, for instance, in the modal epistemic case, where several modal (resolution) inference rules are introduced in order to resolve a literal l with its possible complements, namely $\neg l$ and $\mathrm{K}_i \neg\, l$. The apparent simplicity of the method for the combined logics of knowledge and time comes from the separation of the different dimensions (via the normal form) and from making sure that all relevant information is made available to these different dimensions (through the propositional language, which is shared by all logics, via simplification rules). Thus, there is no need for new inference rules: separation provides an elegant way to deal with the combined logic.

Although elegance and simplicity are desirable features for any proof method, this cannot be achieved in a straightforward way when dealing with interactions. In this case, by definition, different dimensions are not separated. We have chosen to adopt the same set of inference rules of $\mathsf{KL}_{(n)}$, as the proof method for the interacting logic must still comprise all the inference rules for the underlying languages, so that we are still able to provide refutations for formulae in those languages. Having chosen that, some extra mechanism should be added to the proof method in order to deal with the interactions. For synchrony and no learning, instead of adding rather complex inference rules or trying to identify two complementary sets of clauses, we have chosen to add further information to the set of clauses.

We remark that we use the contrapositive form of the SNL axiom: $\vdash \neg\mathrm{K}_i\neg \bigcirc\varphi \Rightarrow \bigcirc\neg\mathrm{K}_i\neg\varphi$. A set of clauses satisfying its antecedent is written into the normal form as (at least) two clauses: a clause (or a set of clauses which imply) **true** $\Rightarrow \psi \vee \neg\mathrm{K}_i\neg l$ and a step clause (or a set of step clauses which imply) $l \Rightarrow \bigcirc\varphi$, where ψ is a disjunction of literals or modal literals, and φ and l are literals. That is, those clauses together imply $\neg\psi \Rightarrow \neg\mathrm{K}_i\neg \bigcirc\varphi$. Because of the SNL axiom, those clauses also imply $\neg\psi \Rightarrow \bigcirc\neg\mathrm{K}_i\neg\varphi$. This is the extra information that we make available by introducing the new clauses. Instead of looking for such a set of clauses, we introduce new clauses for every step clause.

Recall that a step clause is in the general form $X \Rightarrow \bigcirc Y$, where X is a conjunction and Y is a disjunction of literals. The information we wish to make explicit is $\neg\mathrm{K}_i\neg X \Rightarrow \bigcirc\neg\mathrm{K}_i\neg Y$. The general approach to generating the new clauses consists of taking the contrapositive form of a step clause and distributing the knowledge operator K_i through this clause. Then, we take the contrapositive form of the resulting clause, exchange the knowledge and temporal operators, and rename the modal literals to keep the normal form. That is, if X is a conjunction, then we replace X by a new propositional symbol new_X, called $\wedge$-proposition; then, the modal literals in the temporal clause are renamed by new propositional symbols, $nkn_i(X)$ and $nkn_i(Y)$, called SNL_i propositions. The resulting clause, with the new propositional symbols representing the modal literals, is called a SNL_i clause. The SNL_i clauses and the clauses defining the new propositional symbols are those added to the set of clauses.

3.2 Generating New Clauses

Here we give formal definitions for the new literals and clauses informally discussed in the previous section. First, we make the distinction between the already existing literals and the new ones to be added. *Basic literals* are any literals in the original formula and any new literals introduced during translation into SNF_K. The other literals are defined later. We use the term *literal* alone, if there is no need to distinguish which type of literal we are referring.

We firstly rename conjunctions of basic literals, adding the corresponding definitions to the set of clauses. To aid this process, we define a function, *NEW*, which takes a conjunction of literals, φ, as its argument and returns the new name for this conjunction, new_φ. Those new propositions are called $\wedge$-propositions.

DEFINITION 1. Let $c_1 \wedge \ldots \wedge c_n$, $d_1 \wedge \ldots \wedge d_m$, $d'_1 \wedge \ldots \wedge d'_{m'}$, $n, m, m' \geq 2$, be conjunctions literals in the language of $\mathsf{KL}^{snl}_{(n)}$. Assume there is an order over the set of literals, such that $l_i < \neg l_i < l_j < \neg l_j$, if $i < j$, for all positive literals l_i and l_j. Let $SIMP(\varphi)$ be the result of applying simplification rules to φ and of ordering the conjuncts. We define the function *NEW* as follows:

- $NEW(\mathbf{false}) = \mathbf{false}$
- $NEW(\mathbf{true}) = \mathbf{true}$
- $NEW(l) = l$, for any literal l
- $NEW(c_1 \wedge \ldots \wedge c_n) = new_{SIMP(c_1 \wedge \ldots \wedge c_n)}$
- $NEW(c_1 \wedge \ldots \wedge c_n \wedge new_{d_1 \wedge \ldots \wedge d_m} \wedge \ldots \wedge new_{d'_1 \wedge \ldots \wedge d'_{m'}}) = NEW(SIMP(c_1 \wedge \ldots \wedge c_n \wedge d_1 \wedge \ldots \wedge d_m \wedge d'_1 \wedge \ldots \wedge d'_{m'}))$

The new proposition is labelled by the simplified, ordered form of the conjunction it is renaming. Simplification is given by usual rules, i.e. by deleting repeated literals and/or **true** from conjunctions, and by reducing contradictions to **false** and tautologies to **true**. Note that we do not need to rename either a constant, a literal or conjuncts which are $\wedge$-propositions (e.g. $NEW(a \wedge new_{b \wedge c})$ is $new_{a \wedge b \wedge c}$).

DEFINITION 2. For each $\wedge$-proposition, $new_{c_1 \wedge \ldots \wedge c_n}$, $n \geq 2$, we add $\mathbf{true} \Rightarrow new_{c_1 \wedge \ldots \wedge c_n} \vee \neg c_1 \vee \ldots \vee \neg c_n$ to the set of clauses.

These clauses, called $\wedge$-clauses, correspond to the normal form of one direction of the double implication $new_{c_1 \wedge \ldots \wedge c_n} \Leftrightarrow c_1 \wedge \ldots \wedge c_n$, which defines the $\wedge$-propositions. As we rename conjunctions on the left-hand side of step clauses (i.e. formulae of negative polarity), we only need the equivalent (in SNF_K) to $c_1 \wedge \ldots \wedge c_n \Rightarrow new_{c_1 \wedge \ldots \wedge c_n}$.

Once the $\wedge$-propositions have been generated, we define the new names for modal literals, which are the result of distributing the knowledge operator through step clauses. Renaming is used here in order to retain the normal form. We define a set of renaming functions, REN_i, one for each agent $i \in \mathcal{A}$, each of which takes as its argument a conjunction or a disjunction of literals, say φ, returning the new name for $\neg \mathrm{K}_i \neg \varphi$, that is, $nkn_i(\varphi)$. Because conjunctions are firstly renamed and the knowledge operator can be distributed over disjunctions, these functions will only be applied to literals.

The new names, $nkn_i(l)$, where l is a literal, are called SNL_i-propositions. A SNL_i-literal is a SNL_i-proposition or its negation.

DEFINITION 3. Let $\bigvee_j l_j$ be a disjunction of literals and $\bigwedge l_b$, $\bigwedge l_{s_i}$, $\bigwedge l_{s_j}$, and $\bigwedge l_{\wedge_k}$ be conjunctions of basic, SNL_i, SNL_j $(j \neq i)$, and $\wedge$-literals respectively.

- $REN_i(l) = nkn_i(l)$, if l is either a basic, SNL_j $(j \neq i)$ or $\wedge$-literal;
- $REN_i(l) = l$, if l is a SNL_i-literal;
- $REN_i(\bigvee_j l_j) = \bigvee_j REN_i(l_j)$, for any literal l_j;
- $REN_i(\bigwedge l_{s_i} \wedge \bigwedge l_{s_j} \wedge \bigwedge l_b \wedge \bigwedge l_{\wedge_k}) = \bigwedge l_{s_i} \wedge REN_i(NEW(\bigwedge l_{s_j} \wedge \bigwedge l_b \wedge \bigwedge l_{\wedge_k}))$, where $j \neq i$.

The last case says that we rename the conjunctions which involve SNL_j literals, for $j \neq i$ by the corresponding $\wedge$-proposition before renaming the modal literal, but we do not need to rename the SNL_i literals in the conjunction (because $\vdash \neg K_i \neg (\neg K_i \neg \varphi \wedge \neg K_i \neg \psi) \Leftrightarrow (\neg K_i \neg \varphi \wedge \neg K_i \neg \psi)$, for any formulae φ and ψ). For instance, $REN_i(a \wedge new_{b \wedge c} \wedge nkn_i(d) \wedge nkn_j(e)) = nkn_i(d) \wedge nkn_i(new_{a \wedge b \wedge c \wedge nkn_j(e)})$. We also remark that clauses are kept in their simplest form (e.g. $REN_i(a \wedge nkn_i(a)) = nkn_i(a)$).

DEFINITION 4. Let l be a basic, a SNL_j $(j \neq i)$ or a $\wedge$-literal l. We add $SNL_i^{\Rightarrow}(l)$: $\mathbf{true} \Rightarrow \neg nkn_i(l) \vee \neg K_i \neg l$ and $SNL_i^{\Leftarrow}(l)$: $\mathbf{true} \Rightarrow nkn_i(l) \vee K_i \neg l$ to the set of clauses.

These clauses, called SNL_i definition clauses, correspond to the definitions of the SNL_i literals, i.e. the equivalence $nkn_i(l) \Leftrightarrow \neg K_i \neg l$ for each literal l. We need both sides of the double implication, because SNL_i literals can occur with both negative and positive polarities in the set of clauses.

DEFINITION 5. Given a step clause $X \Rightarrow \bigcirc Y$, the corresponding SNL_i-clause is $REN_i(X) \Rightarrow \bigcirc(REN_i(Y))$, where X is a conjunction and Y is a disjunction of literals, and REN_i is the function defined above.

Note that the SNL_i-clauses are defined for both the initial set of step clauses and those step clauses generated while performing resolution.

Note also that these definitions alone could lead to the generation of an infinite number of new literals. If we consider only one agent, it is clear that this process terminates, because once the $\wedge$-propositions have been generated, due to simplification, we can determine all the SNL_1 literals that need to be generated. However, when we consider multiple agents, it is not clear where we could stop generating new literals. Suppose, for instance, that $\mathcal{A} = \{1, 2\}$ and the set of basic literals is $\{a, b\}$. In this case, we generate (among others) the $\wedge$-proposition $new_{a \wedge b}$, and the SNL_1 and SNL_2 literals, $nkn_1(new_{a \wedge b})$ and $nkn_2(new_{a \wedge b})$. We might need, now, to consider these new propositions as part of possible conjunctions (e.g. $new_{a \wedge nkn_1(new_{a \wedge b})}$)

and generate the respective SNL_i literals (e.g. $nkn_2(new_{a \wedge nkn_1(new_{a\wedge b})})$), as simplification might not apply in this case.

We can prove that the number of literals that need to be generated depends on the structure of the original formula that we are trying to refute. We define the nesting depth of a SNL_i literal, $|nkn_i(l)|_{snl}$, as being the number of times that different REN_i renaming functions have been applied to any literal: $|l|_{snl} = 0$, if l is a basic literal; $|new_{l_1 \wedge \ldots \wedge l_n}|_{snl} = max(|l_1|_{snl}, \ldots, |l_n|_{snl})$, if $new_{l_1 \wedge \ldots \wedge l_n}$ is a $\wedge$-literal; and $|nkn_i(l)|_{snl} = 1 + |l|_{snl}$, otherwise. The maximum nesting depth of SNL_i-literals needed in the resolution method is at most the same as the number of alternations of distinct knowledge operators in the original formula, that is, the *alternating modal depth* of the formula. Thus, by allowing only a finite number of literals, termination of the method is guaranteed.

We call SNF_{snl} the set of clauses resulting from the transformation of a formula into SNF_K, the SNL_i, the $\wedge$, and the SNL_i definition clauses. The resolution method applied to a set of SNF_{snl} clauses is essentially the same as that described in Section 2, except that we extend the function mod_i so that $mod_i(l) = l$, if l is a SNL_i-literal.

Below, we illustrate the use of the method. The example is the proof that $\Box \mathrm{K}_1 \mathrm{K}_2 \varphi \Rightarrow \mathrm{K}_1 \mathrm{K}_2 \Box \varphi$ is valid in $\mathsf{KL}_{(n)}^{snl}$. We start by transforming the negation of this formula into its normal form.

1.	$\mathbf{start} \Rightarrow x$			6.	$\mathbf{true} \Rightarrow \neg y \vee \mathrm{K}_1 w$		
2.	$\mathbf{true} \Rightarrow \neg x \vee y$			7.	$\mathbf{true} \Rightarrow \neg w \vee \mathrm{K}_2 \varphi$		
3.	$\mathbf{true} \Rightarrow \neg x \vee z$			8.	$\mathbf{true} \Rightarrow \neg x \vee \neg \mathrm{K}_1 \neg r$		
4.	$z \Rightarrow \bigcirc y$			9.	$\mathbf{true} \Rightarrow \neg r \vee \neg \mathrm{K}_2 \neg s$		
5.	$z \Rightarrow \bigcirc z$			10.	$s \Rightarrow \Diamond \neg \varphi$		

Then, we add the new SNL_i clauses:

11.	$nkn_1(z)$	$\Rightarrow$	$\bigcirc nkn_1(y)$	$[4, SNL_1]$
12.	$nkn_1(z)$	$\Rightarrow$	$\bigcirc nkn_1(z)$	$[5, SNL_1]$
13.	$nkn_2(nkn_1(z))$	$\Rightarrow$	$\bigcirc nkn_2(nkn_1(y))$	$[11, SNL_2]$
14.	$nkn_2(nkn_1(z))$	$\Rightarrow$	$\bigcirc nkn_2(nkn_1(z))$	$[12, SNL_2]$

and also the SNL_i definition clauses that will be needed in the refutation:

$SNL_1^{\Rightarrow}(y)$:	$\mathbf{true}$	$\Rightarrow$	$\neg nkn_1(y) \vee \neg \mathrm{K}_1 \neg y$
$SNL_1^{\Leftarrow}(z)$:	$\mathbf{true}$	$\Rightarrow$	$nkn_1(z) \vee \mathrm{K}_1 \neg z$
$SNL_2^{\Rightarrow}(nkn_1(y))$:	$\mathbf{true}$	$\Rightarrow$	$\neg nkn_2(nkn_1(y)) \vee \neg \mathrm{K}_2 \neg nkn_1(y)$
$SNL_2^{\Leftarrow}(nkn_1(z))$:	$\mathbf{true}$	$\Rightarrow$	$nkn_2(nkn_1(z)) \vee \mathrm{K}_2 \neg nkn_1(z)$

The refutation now proceeds as follows

15.	**true**	$\Rightarrow$	$\neg nkn_1(y) \vee \mathrm{K}_1\, w$	$[6, SNL_1^{\Rightarrow}(y), \mathrm{MRES4}]$
16.	**true**	$\Rightarrow$	$\neg nkn_1(y) \vee w$	$[15, \mathrm{MRES5}]$
17.	**true**	$\Rightarrow$	$\neg nkn_2(nkn_1(y)) \vee \neg \mathrm{K}_2 \neg\, w$	$[16, SNL_2^{\Rightarrow}(nkn_1(y)), \mathrm{MRES4}]$
18.	**true**	$\Rightarrow$	$\neg nkn_2(nkn_1(y)) \vee \mathrm{K}_2\, \varphi$	$[17, 7, \mathrm{MRES4}]$
19.	**true**	$\Rightarrow$	$\neg nkn_2(nkn_1(y)) \vee \varphi$	$[18, \mathrm{MRES5}]$
20.	$nkn_2(nkn_1(z))$	$\Rightarrow$	$\bigcirc \varphi$	$[13, 19, \mathrm{SRES2}]$
21.	s	$\Rightarrow$	$\neg nkn_2(nkn_1(z))\, \mathcal{W} \neg \varphi$	$[20, 14, 10, \mathrm{TRES}]$
22.	**true**	$\Rightarrow$	$\neg s \vee \neg nkn_2(nkn_1(z)) \vee \neg \varphi$	$[21, \mathsf{SNF}]$
23.	**true**	$\Rightarrow$	$\neg r \vee \neg nkn_2(nkn_1(z)) \vee \neg \mathrm{K}_2\, \varphi$	$[22, 9, \mathrm{MRES4}]$
24.	**true**	$\Rightarrow$	$\neg r \vee \neg nkn_2(nkn_1(z)) \vee \neg w$	$[23, 7, \mathrm{MRES1}]$
25.	**true**	$\Rightarrow$	$\neg r \vee \mathrm{K}_2 \neg nkn_1(z) \vee \neg w$	$[24, SNL_2^{\Leftarrow}(nkn_1(z)), \mathrm{MRES1}]$
26.	**true**	$\Rightarrow$	$\neg r \vee \neg nkn_1(z) \vee \neg w$	$[25, \mathrm{MRES5}]$
27.	**true**	$\Rightarrow$	$\neg x \vee \neg nkn_1(z) \vee \neg \mathrm{K}_1\, w$	$[26, 8, \mathrm{MRES4}]$
28.	**true**	$\Rightarrow$	$\neg x \vee \neg nkn_1(z) \vee \neg y$	$[27, 6, \mathrm{MRES1}]$
29.	**true**	$\Rightarrow$	$\neg x \vee \neg nkn_1(z)$	$[28, 2, \mathrm{MRES1}]$
30.	**true**	$\Rightarrow$	$\neg x \vee \mathrm{K}_1 \neg\, z$	$[29, SNL_1^{\Leftarrow}(z), \mathrm{MRES1}]$
31.	**true**	$\Rightarrow$	$\neg x$	$[30, 3, \mathrm{MRES3}]$
32.	**start**	$\Rightarrow$	**false**	$[31, 1, \mathrm{IRES1}]$

4 Correctness

Now, we give the results for soundness, termination, and completeness of the resolution method for $\mathsf{KL}_{(n)}^{snl}$. Full proofs can be found in [11].

Soundness. The soundness proof consists of showing that, given a formula φ in $\mathsf{KL}_{(n)}^{snl}$, its transformation into SNF_{snl} is satisfiability preserving and that the application of the inference rules to the set of clauses in the normal form is also satisfiability preserving.

The proof that the transformation into SNF_{snl} is satisfiability preserving consists of showing that (a) the transformation into SNF_K; (b) the addition of $\wedge$-clauses; (c) the addition of SNL_i definition clauses; and (d) the addition of SNL_i clauses are all satisfiability preserving. A proof for (a) is given in [3]. The proofs for (b) is by construction: given a model M for a set of clauses T, we build a model M' for T', the set of clauses augmented with $\wedge$-clauses. For every state (t, u) in M, let the corresponding state (t', u') in M' be exactly as (t, u), except that $\pi(t', u')(new_{l_1 \wedge \ldots \wedge l_n}) = true$, if $\langle M, (t, u)\rangle \models l_1 \wedge \ldots \wedge l_n$ for all possible conjunctions of literals. Temporal and equivalence relations are kept as in the original model Obviously, M' satisfies all clauses in T; also, it follows from its construction that M' satisfies all definition clauses for the $\wedge$-literals. A model M for T is obtained from M' by ignoring the values of the $\wedge$-literals. The proof for (c) is similar, except that we take $\pi(t', u')(nkn_i(l)) = true$ if, and only if, $\langle M, (t, u)\rangle \models \neg \mathrm{K}_i \neg\, l$ for all agents $i \in \mathcal{A}$ and literals l. The proof for (d) is by the semantics of the universal operator, $\Box^*$, which surrounds all clauses, propositional reasoning and applications of the axioms K and SNL.

The inference rules are the same as those in [3], except that we add to the definition of the function mod_i that $mod_i(l) = l$, if l is a SNL_i-literal. Redefining the function is not essential, but it saves steps in the refutation. Thus, soundness for $\mathsf{KL}_{(n)}^{snl}$ follows from the results in [3], that is, that all inference rules are sound, and from the observation that the same resolvent from the modified MRES4 can also be obtained from successive applications of the original MRES4 and MRES1 to the original resolvent and the SNL_i definition clauses.

Termination. The method presented here is based on that for $\mathsf{KL}_{(n)}$. The difference is that SNL_i and $\wedge$-literals, together with their corresponding definition clauses, are introduced before starting the application of the resolution method. Also, SNL_i-clauses, corresponding to existing or newly generated step clauses, are introduced when applying the method. It has been shown in [3] that the method for $\mathsf{KL}_{(n)}$ terminates, i.e. given a finite number of clauses only a finite number of clauses (modulo order and simplification) can be generated, so at some point either **false** is generated or no new clauses are generated. In order to transfer the termination results from $\mathsf{KL}_{(n)}$ to $\mathsf{KL}_{(n)}^{sync,nl}$, we show that all propositional symbols needed in the refutation can be defined before the resolution rules are applied.

Firstly, it has been shown in [7] that the new propositional symbols required for translating the resolvent obtained by an application of the temporal resolution rule can be added at the beginning of the proof. No other inference rule requires the introduction of new symbols. As there is a finite number of symbols, due to simplification, only a finite number of clauses is generated. Secondly, simplification is applied when generating the $\wedge$-literals and SNL_i-literals, so the number of definition clauses for these literals is also finite. Thirdly, the number of step clauses is (at any point) finite, and so it is the number of SNL_i clauses. Finally, given the alternating modal depth of the original formula, the maximum nesting depth of SNL_i-literals is determined, and so the number of new literals that might be needed in the refutation is finite. Given that only a finite number of symbols and clauses is introduced, a finite number of clauses that is defined. Thus, as the resolution method applied to the set of clauses in the SNF_{snl} is the same as the method for $\mathsf{KL}_{(n)}$, termination follows from the results in [3].

Completeness. This proof is based on that given in [7], where a graph is built from a set of clauses. The construction of the graph is given in more detail in [11]. The proof consists in showing that an empty graph corresponds to an unsatisfiable set of clauses and that, in this case, there is a refutation by the resolution method presented here.

Let T be a set of clauses into SNF_{snl}. We construct a finite directed *graph* $G = \langle N, E \rangle$ *for* T, where N is a set of nodes and E is a set of labelled edges, as follows. A node $\eta = (V, Y)$ is a pair, where V is a maximal consistent set of literals and modal literals; and Y is a subset of basic literals occurring on the right hand side of a sometime clause. Intuitively, V corresponds to states and Y corresponds to eventualities that have not been satisfied by the predecessors of V. There are $n+1$ types of edges: one for the temporal dimension plus one for each agent in $\mathcal{A} = \{1, \ldots, n\}$. For every set V, we construct nodes $\eta = (V, Y)$, where Y is any of the possible subsets of literals occurring on the right-hand side of sometime clauses in the set of clauses. We delete any nodes that do not immediately satisfy the literal and modal clauses, including the definition clauses, in T.

Given a non-empty set of nodes, we construct the set of labelled edges for each agent $i \in \mathcal{A}$. There is an i-edge between two nodes $\eta = (V, Y)$ and $\eta' = (V', Y')$, if, and only if, V and V' contain the same set of modal literals for that agent. We say that a node η' is i-reachable from η, if there is an i-edge between η and η'. We say that a node η' is $\{i_0, \ldots, i_m\}$-reachable from η, if there is a sequence of nodes $\eta_0, \ldots, \eta_{m+1}$, such that $\eta_0 = \eta$, $\eta_{m+1} = \eta'$, and there is a i_j-edge between every two nodes η_j and η_{j+1}, for $0 \leq j \leq m$. We define $[\eta]_i$ as the set of nodes that are i-reachable from η. Clearly, $[\eta]_i$ defines an equivalence relation over the set of nodes.

Then, we construct the temporal edges. We start with a full (temporal) graph, i.e. there is a t-edge linking every two nodes in the graph (because **true** $\Rightarrow \bigcirc$**true**). We say that a node η' is t-reachable from η, if there is a sequence of nodes $\eta_0, \ldots, \eta_m$, such that $\eta_0 = \eta$, $\eta_m = \eta'$, and there is a t-edge between every two nodes η_j and η_{j+1}, for $0 \leq j < m$. We say that η is a predecessor of η', if η' is t-reachable from η.

For every step clause $(\bigwedge l \Rightarrow \bigcirc \bigvee l') \in T$, we delete a t-edge between $\eta = (V, Y)$ and $\eta' = (V', Y')$, if $\eta \models \bigwedge l$ and $\eta' \not\models \bigvee l'$. For every sometime clause $\varphi \Rightarrow \Diamond l \in T$, we also delete a node $\eta = (V, Y)$, if $V \models \varphi$ and $l \notin Y$; and a t-edge from $\eta = (V, Y)$ to $\eta' = (V', Y')$, if (a) $l \in Y$, $l \notin Y'$, and $V \not\models l$; or (b) $l \in Y'$, $V \models l$, and $V' \not\models \varphi$. This ensures that (a) eventualities not satisfied by a predecessor are not "forgotten" and (b) there will be no edge from a node that satisfies an eventuality to another node that is "waiting" for the same eventuality to be satisfied, unless this successor satisfies the left-hand side of a sometime clause that also says that this eventuality should hold at some moment in the future. In other words, we are only waiting for eventualities to occur that originally came from satisfying the left-hand side of a sometime clause.

A node $\eta = (V, Y)$ is an *initial node* if, and only if, V satisfies all initial clauses in T and, for each sometime clause $\varphi \Rightarrow \Diamond l$, such that $V \models \varphi$, if, and only if, $l \in Y$. We say that a node η' is *reachable* from a node η, if, and only if, there is a sequence of nodes $\eta_0, \ldots, \eta_m$, such that, $\eta_0 = \eta$, $\eta_m = \eta'$, and η_{j+1} is t- or i-reachable from η_j, for $0 \leq j < m$ and $i \in \mathcal{A}$.

We further delete any nodes that are not predecessors of any node which is reachable from an initial node. This reduces the graph to (possibly disjoint) connected components which include at least one initial node. The resulting graph is called a *behaviour graph for T*. Given a behaviour graph for T, we recursively delete any nodes (V, Y) (and edges to it) that have no temporal successors (no infinite temporal line can be built from the node, so it is not part of any model); satisfy the left-hand side of a sometime clause, but there is no node satisfying the eventuality that is t-reachable from this node; and/or satisfy a formula as $\neg K_i \neg p$ but there is no i-edge to a node which satisfies p. The resulting graph is called *reduced behaviour graph*. Deletions of either nodes or edges are repeated non-deterministically until the graph is empty or no other deletion can be done. This procedure corresponds, respectively, to application of temporal simplification (SIMP1), temporal resolution (TRES), and modal resolution to the set of clauses. If the graph is empty, then the set of clauses is not satisfiable in $\mathsf{KL}^{snl}_{(n)}$.

If the reduced behaviour graph is not empty, we show that given two nodes η and μ in the same equivalence class, i.e. $[\eta]_i = [\mu]_i$, for agent $i \in \mathcal{A}$, such that η' is a successor of η, then μ has also a successor in $[\eta']_i$.

THEOREM 6. *Let G be a non-empty reduced behaviour graph for a set of clauses T in SNF_{snl}. Let η, μ, η', μ', be nodes in G, such that $[\eta]_i = [\mu]_i$ and there is a t-edge from η to η'. Then there is a node μ', $[\mu']_i = [\eta']_i$, such that there is a t-edge from μ to μ'.*

PROOF . We show, by contradiction, that if the graph is not empty and we have a temporal edge from a node η to a node η', then all nodes in $[\eta]_i$ have successors in $[\eta']_i$. Suppose that there is a temporal edge from η to η' and $\mu \in [\eta]_i$ has no successors in $[\eta']_i$. Thus, we can identify a set of step clauses that together imply $\psi \Rightarrow \bigcirc\chi$, such that μ satisfies ψ and, for all $\mu' \in [\eta']_i$, μ' satisfies $\neg\chi$. If we could not identify this set of clauses, the temporal edges would not have been removed. If all $\mu' \in [\eta']_i$ satisfies $\neg\chi$, then, by the semantics of the knowledge operator, $\mu' \models K_i \neg \chi$. The addition of the SNL_i-clause corresponding to $\psi \Rightarrow \bigcirc\chi$, i.e. $REN_i(\psi) \Rightarrow \bigcirc REN_i(\chi)$, ensures that every node must satisfy $\neg K_i \neg \psi \Rightarrow \bigcirc \neg K_i \neg \chi$. We have that μ satisfies ψ and, by the semantics of the knowledge operator, it also satisfies $\neg K_i \neg \psi$ (or the corresponding $\wedge$-proposition, new_ψ, and the formula $\neg K_i \neg new_\psi$, in the case where ψ is a conjunction). In fact, every node in $[\eta]_i$ satisfies $\neg K_i \neg \psi$. If all nodes in $[\eta']_i$ satisfy $K_i \neg \chi$, because of the SNL_i clause, then there must be no temporal edge from η to η'. This contradicts with our initial assumptions, so there must be a temporal edge between μ and μ' or none of the nodes in $[\eta]_i$ have successors in $[\eta']_i$.□

Theorem 6 shows that every two nodes in the same equivalence class, say $[\eta]_i$, have successors in the same equivalence class, say $[\eta']_i$. However, there might be nodes in $[\eta']_i$ that have no predecessors in $[\eta]_i$. In order to be able to construct a model, we also show that every node has a predecessor.

THEOREM 7. *Let G be a non-empty reduced behaviour graph for a set of clauses T in SNF_{snl}. Let η be a node in G. Then, there is a node η' such that η' is a predecessor of η.*

PROOF . The node $\eta' = (V', E')$, where $V' \models \neg\mathbf{start}$ and $V' \models \neg p$ for all propositional symbols occurring in the set of SNF_{snl} clauses, precedes all nodes, if the graph is not empty. Note that η' satisfies trivially all clauses in the set of SNF_{snl} clauses, either by falsifying the left-hand side of a (initial, step, or sometime) clause or by satisfying the right-hand side of a modal or literal clause (from translation, there is at least one negative literal ocurring on the right-hand side of such a clause). □

Hence, if the reduced behaviour graph is not empty, we can inductively construct a model in which, if two nodes, (s, m) and (t, m), are in the same equivalence class, i.e. $(s, m) \sim_i (t, m)$, then their successors are also in the same equivalence class, i.e. $(s, m+1) \sim_i (t, m+1)$, which suffices to prove that the model is in the class of $\mathsf{KL}_{(n)}^{snl}$ [8]. In order to build the first timeline for the agent i, we choose an initial node η_0 from the graph, which corresponds to the point $(t_0, 0)$. Let the nodes of $[\eta_0]_i$ be the initial points of other timelines for agent i. Note that only the first point at the first timeline needs to be an initial node in the graph. Then, we choose η_1 from the immediate successors (i.e. t-reachable by one t-edge) of η_0. We build the next point in each timeline, by choosing the immediate successors of each node in $[\eta_0]_i$ from the nodes in $[\eta_1]_i$. From Theorem 6, this is always possible. From Theorem 7, for points at time greater than zero, we can construct the predecessors of a node back to initial point of every timeline.

We note that the introduction of the SNL_i-clauses not only delete edges which are not part of a model for $\mathsf{KL}_{(n)}^{snl}$, but also contribute to the temporal resolution procedure. Temporal resolution between a sometime clause, say $l \Rightarrow \Diamond p$, and a set of clauses that together imply $\bigcirc \Box \neg p$, corresponds to removing from the graph subcomponents which satisfy l, but never satisfy the eventuality p. As we are considering the combined logics, a sometime clauses may be preceded by a chain of knowledge operators. For instance, the clauses (8) $\mathbf{true} \Rightarrow \neg x \vee \neg \mathrm{K}_1 \neg r$, (9) $\mathbf{true} \Rightarrow \neg r \vee \neg \mathrm{K}_2 \neg s$, and (10) $s \Rightarrow \Diamond \neg \varphi$ (from the example on Page 240) show that the left-hand side of the sometime clause is preceded by $\neg \mathrm{K}_1 \neg$ and $\neg \mathrm{K}_2 \neg$. In $\mathsf{KL}_{(n)}^{snl}$, $\neg \mathrm{K}_1 \mathrm{K}_2 \neg \Diamond \neg \varphi$ implies $\Diamond \neg \mathrm{K}_1 \mathrm{K}_2 \neg \varphi$, which should be resolved with a set of clauses that together imply $\bigcirc \Box \mathrm{K}_1 \mathrm{K}_2 \varphi$. Note that because of the axiom T, the same loop implies $\bigcirc \Box \varphi$. Although this loop can be resolved with the sometime clause, the resolvents generated by temporal resolution do not contribute to removing the chain of knowledge operators preceding the eventuality. The

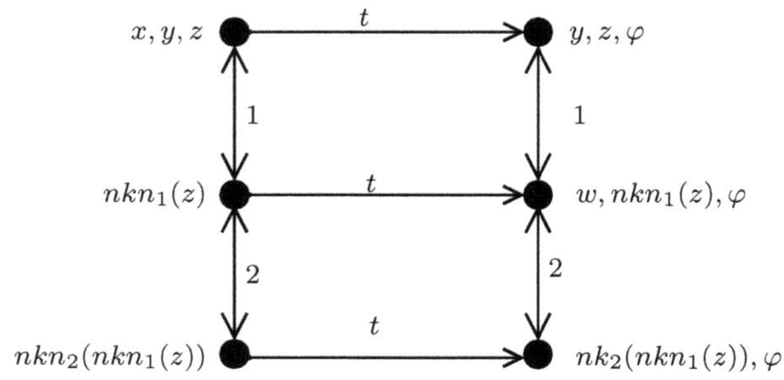

Figure 1. Some temporal loops related to the second example.

sometime clause occurs in a timeline that is $\{1, 2\}$-reachable from the initial timeline. Therefore, we are interested in a loop that is also $\{1, 2\}$-reachable from that timeline. In Figure 1, we show this loop, where the states are represented by circles; each state is labelled by the propositional symbols which hold at that state; and the edges are labelled either by t (indicating the temporal successor) or by the agent index (indicating that those states are in the same equivalence class). We note that we could have introduced SNL_i-literals in sometime clauses in the same way that those literals where introduced in step clauses, but this is not necessary for completeness. The next theorem ensures that, in this situation, the appropriate loop is found.

THEOREM 8. *Let C be the set of SNF_{snl} clauses that together imply $\neg K_{i_1} K_{i_2} \ldots K_{i_m} \neg \Diamond \neg \varphi$ where φ is a basic literal and every two consecutive knowledge operators differ in their index. If there is a set of clauses that imply $\bigcirc \Box K_{i_1} K_{i_2} \ldots K_{i_m} \varphi$, then there is a set of nested SNL_i clauses C' that imply $\bigcirc \Box \varphi$.*

Specifically, if χ is the conjunction of literals on the left-hand side of the step clauses which satisfy the loop conditions, then we can prove that there is a loop given by $nkn_{i_{m\ldots1}}(\chi) \Rightarrow \bigcirc\varphi$ (where $nkn_{i_{m\ldots1}}(l)$ is an abbreviation for successively applying the renaming functions $REN_1, \ldots, REN_m$ to a literal l). The temporal resolvents, the set of clauses that represent the chain of knowledge operators preceding the sometime clause, and the (nested) SNL_i definitions of χ can be successively resolved by applying the modal inference rules. We can assume without loss of generality that subsequent knowledge operators in the chain preceding a temporal formula had different indices. The maximum size of this chain corresponds to the alternating modal depth of the knowledge operators in the original formula and gives an upper bound for the maximum nesting of SNL_i literals.

5 Conclusions

We have shown how to extend the method presented in [3] to deal with $\mathsf{KL}_{(n)}^{snl}$. The addition of new information to the set of clauses, together with the use of renaming, provides an intuitive way of dealing with this particular interaction. Although the introduction of new clauses is costly, we avoid the introduction of new inference rules and, potentially, expensive searches over the set of clauses. As there is no change in the set of inference rules, implementation is relatively easily obtained by adapting and re-using existing theorem-provers. Correctness for the multi-agent case has been established. Results for the multi-agent case for synchrony and perfect recall [2] can be established in a similar way and it is ongoing work.

In [5], tableaux for synchronous systems with perfect recall or no learning for the single-agent case are presented. As in the resolution methods for those systems, step clauses representing the constraints imposed by the particular interacting axiom are added to the set of clauses before the construction of the tableaux and complex formulae in the scope of the knowledge operator are renamed. Future work involves the extension to the multi-agent case, the investigation of strategies for efficient construction of the tableau, and comparison with the resolution methods for those systems.

Synchronous systems with either perfect recall or no learning have complete axiomatisations given by axioms with simple structures. We have been investigating how the ideas behind our proof method could be applied to logics that include axioms of the (simple) form $\mathrm{K}_i \bigcirc^p \varphi \Rightarrow \bigcirc^q \mathrm{K}_i \varphi$ or $\bigcirc^p \mathrm{K}_i \varphi \Rightarrow \mathrm{K}_i \bigcirc^q \varphi$ where p and q are integers $(p, q > 0)$, and the iterated next operator is defined as $\bigcirc^0 \varphi = \varphi$, and $\bigcirc^p \varphi = \bigcirc^{p-1} \bigcirc \varphi$, for $p > 0$, where φ is a formula. Although the generation of new clauses would not be straightforward, we have also been investigating other interactions expressed by finite axiomatisations which include more complex axioms, as, for instance, the no learning axiom, $\vdash (\mathrm{K}_i \varphi_1 \mathcal{U} \mathrm{K}_i \varphi_2) \Rightarrow \mathrm{K}_i (\mathrm{K}_i \varphi_1 \mathcal{U} \mathrm{K}_i \varphi_2)$, when synchrony is not required.

BIBLIOGRAPHY

[1] C. Dixon. Temporal Resolution using a Breadth-First Search Algorithm. *Annals of Mathematics and Artificial Intelligence*, 22:87–115, 1998.

[2] C. Dixon and M. Fisher. Clausal Resolution for Logics of Time and Knowledge with Synchrony and Perfect Recall. In *Proceedings of ICTL 2000*, Leipzig, Germany, 2000.

[3] C. Dixon and M. Fisher. Resolution-Based Proof for Multi-Modal Temporal Logics of Knowledge. In S. Goodwin and A. Trudel, editors, *Proceedings of TIME'00*, pages 69–78, Cape Breton, Nova Scotia, Canada, July 2000. IEEE Computer Society Press.

[4] C. Dixon, M. Fisher, and M. Wooldridge. Resolution for Temporal Logics of Knowledge. *Journal of Logic and Computation*, 8(3):345–372, 1998.

[5] C. Dixon, C. Nalon, and M. Fisher. Tableaux for temporal logics of knowledge: Synchronous systems of perfect recall or no learning. In *Proceedings of TIME-ICTL 2003*, IEEE, Cairns, Queensland, Australia, July 2003.

[6] R. Fagin, J. Y. Halpern, Y. Moses, and M. Y. Vardi. *Reasoning About Knowledge.* MIT Press, 1995.
[7] M. Fisher, C. Dixon, and M. Peim. Clausal Temporal Resolution. *ACM Transactions on Computational Logic*, 2(1), January 2001.
[8] J. Y. Halpern, R. van der Meyden, and M. Y. Vardi. Complete Axiomatizations for Reasoning About Knowledge and Time. *SIAM Journal on Computing*, 33(3):674–703, 2004.
[9] B. Heinemann. Linear Tense Logics of Increasing Sets. *Journal of Logic and Computation*, 12(4):583–606, 2002.
[10] R. E. Ladner and J. H. Reif. The logic of distributed protocols (preliminary report). In Joseph Y. Halpern, editor, *Theoretical Aspects of Reasoning about Knowledge: Proceedings of the First Conference*, pages 207–222, Los Altos, California, 1986. Morgan Kaufmann Publishers, Inc.
[11] C. Nalon. *Resolution for Synchrony and No Learning.* PhD thesis, Univ. of Liverpool, March 2004.
[12] A. S. Rao and M. P. Georgeff. Modeling Rational Agents within a BDI-Architecture. In R. Fikes and E. Sandewall, editors, *Proceedings of Knowledge Representation and Reasoning (KR&R-91)*, pages 473–484. Morgan-Kaufmann, April 1991.
[13] B. van Linder, W. van der Hoek, and J. J. Ch. Meyer. Formalising Motivational Attitudes of Agents: On Preferences, Goals and Commitmentes. In M. Wooldridge, J. P.Püller, and M. Tambe, editors, *Intelligent Agents II*, volume 1037 of *LNAI*, pages 17–32. Springer-Verlag, 1996.

Acknowledgments

This work was partially supported by CAPES (BEX 1158-99-6), FINATEC (Proc. 8820), and the EPSRC research grants GR/R45376 and GR/S63182.

Cláudia Nalon
Departmento de Ciência da Computação
Universidade de Brasília
Caixa Postal 4466
Brasília - DF - Brazil
CEP: 70.910-090
nalon@unb.br

Clare Dixon & Michael Fisher
Department of Computer Science
University of Liverpool
Liverpool, L69 7ZF, UK.
{C.Dixon, M.Fisher}@csc.liv.ac.uk

On the Complexity of Fragments of Modal Logics

LINH ANH NGUYEN

ABSTRACT. We study and give a summary of the complexity of 15 basic normal monomodal logics under the restriction to the Horn fragment and/or bounded modal depth. As new results, we show that: a) the satisfiability problem of sets of Horn modal clauses with modal depth bounded by $k \geq 2$ in the modal logics $K4$ and $KD4$ is PSPACE-complete, in K is NP-complete; b) the satisfiability problem of modal formulas with modal depth bounded by 1 in $K4$, $KD4$, and $S4$ is NP-complete; c) the satisfiability problem of sets of Horn modal clauses with modal depth bounded by 1 in K, $K4$, $KD4$, and $S4$ is PTIME-complete.

We also study the complexity of the multimodal logics L_n under the mentioned restrictions, where L is one of the 15 basic monomodal logics. We show that, for $n \geq 2$: a) the satisfiability problem of sets of Horn modal clauses in $K5_n$, $KD5_n$, $K45_n$, and $KD45_n$ is PSPACE-complete; b) the satisfiability problem of sets of Horn modal clauses with modal depth bounded by $k \geq 2$ in K_n, KB_n, $K5_n$, $K45_n$, $KB5_n$ is NP-complete, and in KD_n, T_n, KDB_n, B_n, $KD5_n$, $KD45_n$, $S5_n$ is PTIME-complete.

1 Introduction

In the field of modal logics, a lot of works are devoted to monomodal logics that extend the modal logic K by some of the axioms D, T, B, 4, and 5. The reason is not that those logics are useful in practice, but because they are *basic* modal logics. Many useful multimodal logics, e.g. ones for reasoning about knowledge and belief, are also formed using the mentioned axioms and are extensions of some basic monomodal logics.

Advances in Modal Logic, Volume 5. © *2005, Linh Anh Nguyen.*

Decidability and complexity are important aspects of logics. In [15], Ladner proved that the complexity of the satisfiability problem in the modal logics K, T, B, and $S4$ is PSPACE-complete, and in $S5$ is NP-complete. This means that the satisfiability problem is NP-hard in all of those logics. In order to reduce the complexity to PTIME, one must focus on fragments of the considered logic. Such fragments are often specified by restrictions on the language. There are of course many kinds of restrictions, but the obtained fragments may be useful or not. The Horn fragment is very useful in logic programming, and in many logics it significantly reduces the complexity of the problem. For modal logics, the restriction of bounded modal depth is also acceptable, because in practice modal formulas often have small modal depth. We can also combine these two restrictions. Given an "acceptable" restriction and a modal logic, one may want to study the complexity of the satisfiability problem in the obtained fragment of the logic. The result may be positive (PTIME) or negative (NP-hard, PSPACE-hard, etc). Both of the cases are useful: the positive case is good for the fragment itself, while the negative case implies that every multimodal logic containing the fragment is hard at least as the fragment.

In this work, we study and give a summary of the complexity of the satisfiability problem in the basic normal monomodal logics (which are obtained from the logic K by adding an arbitrary combination of the axioms D, T, B, 4, and 5) under the restriction to the Horn fragment and/or bounded modal depth.

In [11], Halpern studied the effect of bounding modal depth on the complexity of modal logics and showed that the complexity of the satisfiability problem of formulas with modal depth bounded by $k \geq 2$ in K and T is NP-complete, and in $S4$ is PSPACE-complete. His arguments for K and T can also be applied for the logics KB, KDB, and B, to obtain the NP-completeness.

In [6], Fariñas del Cerro and Penttonen showed that the satisfiability problem of sets of Horn modal clauses in $S5$ is decidable in PTIME. In [4], Chen and Lin showed that the similar problem for a normal modal logic L being an extension of $K5$ (write $K5 \leq L$) is also decidable in PTIME. Chen and Lin also proved that for a normal

(in this column, $k \geq 2$)	K	KD, T KB, KDB, B	$K4$, $KD4$, $S4$	$K5$, $KD5$, $KB5$ $K45$, $KD45$, $S5$
no restrictions	PS-cp [15]	PS-cp [15]	PS-cp [15]	NP-cp [15]
mdepth $\leq k$	NP-cp [11]	NP-cp [11]	PS-cp [11] [$*$]	NP-cp [15]
mdepth $= 1$	NP-cp [11]	NP-cp [11]	NP-cp [$\star$]	NP-cp [15]
Horn	PS-cp [4]	PS-cp [4]	PS-cp [4]	PT-cp [6, 4]
Horn, *mdepth* $\leq k$	NP-cp [$\star$]	PT-cp [17]	PS-cp [4] [$*$]	PT-cp [6, 4]
Horn, *mdepth* $= 1$	PT-cp [$\star$]	PT-cp [17]	PT-cp [$\star$]	PT-cp [6, 4]

Table 1. The complexity of the satisfiability problem for modal logics

modal logic L such that $K \leq L \leq S4$ or $K \leq L \leq B$, the problem is PSPACE-hard. They also made a comment that the problem is still PSPACE-hard for $S4$ even when the modal depth is restricted to 2.

In [17], we showed that the complexity of the satisfiability problem of sets of Horn modal clauses with finitely bounded modal depth in KD, T, KB, KDB, and B is decidable in PTIME. These PTIME results can further be categorized as PTIME-complete, because the satisfiability problem of sets of Horn clauses in the classical propositional logic is PTIME-complete, as proved by Jones and Laaser [13].

In this work, we show that the satisfiability problem of sets of Horn modal clauses with modal depth bounded by $k \geq 2$ in the modal logics $K4$ and $KD4$ is PSPACE-complete, and in K is NP-complete. We also show that the satisfiability problem of modal formulas with modal depth bounded by 1 in $K4$, $KD4$, and $S4$ is NP-complete; the satisfiability problem of sets of Horn modal clauses with modal depth bounded by 1 in K, $K4$, $KD4$, and $S4$ is PTIME-complete.

In Table 1, we summarize the complexity of the basic monomodal logics under the mentioned restrictions. There, *mdepth* stands for "modal depth"; PS-cp, NP-cp, and PT-cp respectively stand for PSPACE-complete, NP-complete, and PTIME-complete. The marks [$\star$] and [$*$] indicate the results of this work, where [$*$] involves with $K4$ and $KD4$.

As an extension to the preliminary version, we also study the complexity of the multimodal logics L_n under the mentioned restrictions, where L is one of the basic monomodal logics. Some results were established by Halpern and Moses [12] and Halpern [11]. Some of our results are:

- The satisfiability problem of sets of Horn modal clauses in $K5_n$, $KD5_n$, $K45_n$, and $KD45_n$ is PSPACE-complete.

- The satisfiability problem of sets of Horn modal clauses with modal depth bounded by $k \geq 2$ in K_n, KB_n, $K5_n$, $K45_n$, and $KB5_n$ is NP-complete, and in KD_n, T_n, KDB_n, B_n, $KD5_n$, $KD45_n$, and $S5_n$ is PTIME-complete.

This paper is structured as follows: In Section 2, we give preliminaries for monomodal logics. In Section 3, we present our results for monomodal logics. In Section 4, we discuss the complexity and give some results for multimodal logics. We conclude in Section 5.

2 Preliminaries

In this section we give preliminaries for monomodal logics. For abbreviation, we will ignore the prefix "mono" in this section and the next one.

2.1 Syntax and Semantics of Propositional Modal Logics

A modal formula, hereafter simply called a *formula*, is any finite sequence obtained by applying the following rules: any primitive proposition p_i is a formula, and if φ and ψ are formulas then so are $\neg\varphi$, $\varphi \wedge \psi$, $\varphi \vee \psi$, $\varphi \rightarrow \psi$, $\Box\varphi$, and $\Diamond\varphi$. We use letters p and q to denote primitive propositions, and Greek letters φ, ψ, ζ to denote formulas.

A *Kripke frame* is a triple $\langle W, \tau, R\rangle$, where W is a nonempty set of possible worlds, $\tau \in W$ is the actual world, and R is a binary relation on W, called the accessibility relation. If $R(w, u)$ holds then we say that the world u is accessible from the world w.

A *Kripke model* is a tuple $\langle W, \tau, R, h\rangle$, where $\langle W, \tau, R\rangle$ is a Kripke frame and h is a function mapping worlds to sets of primitive propositions. For $w \in W$, $h(w)$ is the set of primitive propositions which are "true" at w.

We call $\langle W, \tau, R, h\rangle$ a *flat model* if $W = \{\tau\}$ and $R = \emptyset$.

A *model graph* is a tuple $\langle W, \tau, R, H\rangle$, where $\langle W, \tau, R\rangle$ is a Kripke frame and H is a function mapping worlds to formula sets. We sometimes treat model graphs as models with H being restricted to the set of primitive propositions.

Given a Kripke model $M = \langle W, \tau, R, h\rangle$ and a world $w \in W$, the *satisfaction relation* $\models$ is defined as follows:

$M, w \models p$	iff $p \in h(w)$;
$M, w \models \neg\varphi$	iff $M, w \not\models \varphi$;
$M, w \models \varphi \wedge \psi$	iff $M, w \models \varphi$ and $M, w \models \psi$;
$M, w \models \varphi \vee \psi$	iff $M, w \models \varphi$ or $M, w \models \psi$;
$M, w \models \varphi \rightarrow \psi$	iff $M, w \not\models \varphi$ or $M, w \models \psi$;
$M, w \models \Box\varphi$	iff for all $v \in W$ s.t. $R(w, v)$, $M, v \models \varphi$;
$M, w \models \Diamond\varphi$	iff there exists $v \in W$ s.t. $R(w, v)$ and $M, v \models \varphi$.

We say that φ *is satisfied at* w *in* M if $M, w \models \varphi$, and that φ *is satisfied in* M, write $M \models \varphi$ and call M a *model of* φ, if $M, \tau \models \varphi$.

The *size* of a finite Kripke model $\langle W, \tau, R, h\rangle$ is $|W| + |R| + \Sigma_{w \in W} |h(w)|$. The *length* of a formula φ is the number of occurrences of connectives and primitive propositions in φ. The *modal depth* of a formula φ is the maximal nesting depth of modalities occurring in φ, e.g. $mdepth(p \wedge \Box(\Diamond q \vee \Diamond r)) = 2$.

The following lemma is well known and can be proved easily.

LEMMA 1. *Given a finite model* M *and a formula* φ*, the problem of checking whether* $M \models \varphi$ *is decidable in polynomial time (in the size of* M *and the length of* φ*).*

If as the class of admissible interpretations we take the class of all Kripke models (with no restrictions on the accessibility relations) then we obtain a normal modal logic which has a standard Hilbert-style axiomatization denoted by K. Other normal modal logics are obtained by adding to K certain axioms. The most popular axioms used for extending K are D, T, B, 4, and 5, whose schemata are listed in Table 2. These axioms respectively correspond to seriality, reflexiveness, symmetry, transitiveness, and euclideaness of the accessibility relation. A modal logic L is *serial* if it contains the axiom D.

In this work, we consider all of the 15 basic modal logics that are obtained from K by adding an arbitrary combination of the above axioms, namely K, KD, T, KB, KDB, B, $K4$, $KD4$, $S4$, $K5$, $KD5$, $K45$, $KD45$, $KB5$, $S5$. The names of these logics often consist of K and the names of the added axioms, e.g. KDB is the logic which extends K with the axioms D and B. The special cases are T, B,

Axiom	Schema	Corresponding Condition on R
D	$\Box\varphi \to \Diamond\varphi$	$\forall w\ \exists u\ R(w,u)$
T	$\Box\varphi \to \varphi$	$\forall w\ R(w,w)$
B	$\varphi \to \Box\Diamond\varphi$	$\forall w,u\ R(w,u) \to R(u,w)$
4	$\Box\varphi \to \Box\Box\varphi$	$\forall w,u,v\ R(w,u) \land R(u,v) \to R(w,v)$
5	$\Diamond\varphi \to \Box\Diamond\varphi$	$\forall w,u,v\ R(w,u) \land R(w,v) \to R(u,v)$

Table 2. Modal logics and frame restriction

$S4$, and $S5$, which stand for KT, KTB, $KT4$, and $KT5$, respectively. For a further reading about modal logics, see, e.g., [2, 3].

We refer to the properties of the accessibility relation of a modal logic L as the *L-frame restrictions.* We call a model M an *L-model* if the accessibility relation of M satisfies all L-frame restrictions. We say that φ is *L-satisfiable* if there exists an L-model of φ. A formula is *L-valid* if it is satisfied in every L-model. We write $\varphi \models_L \psi$ to denote that ψ is satisfied in every L-model of φ.

2.2 Modal Horn Formulas and Positive Modal Logic Programs

We call formulas of the form p or $\neg p$, where p is a primitive proposition, *classical literals* and use letters a, b, c to denote them. We call formulas of the form p, $\Box p$, or $\Diamond p$ *atoms* and use letters A, B, C to denote them.

A *clause* is a formula of the form $\Box^s(A_1 \lor \ldots \lor A_n \lor \neg B_1 \lor \ldots \lor \neg B_m)$, where $s, m, n \geq 0$. The sequence $\Box^s$ is called the *modal context* of the clause[1]. If $s = 0$ then the clause is called a *simple clause.* Note that the modal depth of a clause is not greater than the length of its modal context plus 1.

A formula set is sometimes considered as the conjunction of its formulas, in particular when we are talking about length, modal depth, or satisfiability.

A formula is in *negative normal form* if it does not contain the connective $\to$, and the connective $\neg$ can occur only immediately before a primitive proposition. Every formula can be transformed to

[1]Assume that the modal context of $\Box^s\Box p$ is $\Box^{s+1}$.

the equivalent negative normal form in the usual way. A formula is called *negative* if in its negative normal form every primitive proposition is prefixed by negation. A formula is called *non-negative* if it is not negative, and *positive* if its negation is a negative formula.

A formula φ is a *Horn formula* if it is of one of the following forms:

- a primitive proposition or a negative formula,
- $\Box\psi$, $\Diamond\psi$, or $\psi \wedge \zeta$, where ψ and ζ are Horn formulas,
- $\psi \to \zeta$, where ψ is a positive formula and ζ is a Horn formula,
- a disjunction of a negative formula and a Horn formula.

A clause is called a *Horn clause* if it is a Horn formula.

Our definitions of Horn clauses/formulas are different than the one of Chen and Lin [4]. A Horn clause by our definition is also a Horn clause by the definition of Chen and Lin, and the latter is a Horn formula by our definition, but not vice versa. These definitions, however, are equivalent. As stated by Lemma 2 given below, every Horn formula φ can be translated to a set X of Horn clauses such that for any normal modal logic L, φ is L-satisfiable iff X is L-satisfiable.

A *positive propositional modal logic program* is a finite set of rules of the following form: $\Box^s(B_1 \wedge \ldots \wedge B_k \to A)$, where $s \geq 0$, $k \geq 0$, and $A, B_1, \ldots, B_k$ are atoms of the form p, $\Box p$, or $\Diamond p$, where p is a primitive proposition.

Formula sets X and Y are said to be *equisatisfiable* in a logic L (or L-equisatisfiable) iff (X is L-satisfiable iff Y is L-satisfiable).

LEMMA 2. *For any formula set X, there exists a clause set Y s.t.:*

- *X and Y are equisatisfiable in any normal modal logic.*
- *If X is a set of Horn formulas, then Y is a set of Horn clauses.*
- *The modal depth of Y is equal to the modal depth of X, and the length of Y is of quadratic order in the length of X.*

Moreover, if X is a set of Horn formulas and Y is divided into P and Q such that P contains only non-negative clauses and Q contains only negative clauses, then P can be treated as a positive program, and X

is L-satisfiable iff $P \not\models_L \neg Q$, *where L is any normal modal logic. The translation from X to Y is computable in polynomial time.*

The proof for the case when X is a set of Horn formulas can be found in [17]. The proof for the other case is similar. The translation technique is based on replacing a complicated formula by a fresh primitive proposition and "defining" that primitive proposition by the formula. For example, $\Box^s(\Diamond\varphi \vee \psi)$, where $s \geq 0$ and φ is not a primitive proposition, is replaced by $\Box^s(\Diamond p \vee \psi)$ and $\Box^{s+1}(\neg p \vee \varphi)$, where p is a fresh primitive proposition.

2.3 Ordering Kripke Models

Let $M = \langle W, \tau, R, h\rangle$ and $N = \langle W', \tau', R', h'\rangle$ be Kripke models. We say that M is *less than or equal to* N *w.r.t.* $r \subseteq W \times W'$, and write $M \leq N$ w.r.t. r, if the following conditions hold:

1. $r(\tau, \tau')$
2. $\forall x, x', y \;\; R(x, y) \wedge r(x, x') \rightarrow \exists y' \; R'(x', y') \wedge r(y, y')$
3. $\forall x, x', y' \;\; R'(x', y') \wedge r(x, x') \rightarrow \exists y \; R(x, y) \wedge r(y, y')$
4. $\forall x, x' \;\; r(x, x') \rightarrow (h(x) \subseteq h'(x'))$.

The first three conditions state that r is a bisimulation of the frames of M and N. Intuitively, $r(x, x')$ states that the world x is less than or equal to x'.

We say that a model M is *less than or equal*[2] *to* N, and write $M \leq N$, if $M \leq N$ w.r.t. some r. This relation is a pre-order [17]. Also see [17] for the proof of the following lemma.

LEMMA 3. *Suppose that* $M \leq N$. *Then* $M \models \varphi$ *implies* $N \models \varphi$ *for every positive formula* φ.

Let P be a positive program in a normal modal logic L. We say that M is a *least L-model* of P if M is an L-model of P and M is less than or equal to every L-model of P. Observe that if P is a positive program in a normal modal logic L, and M is a least L-model of P, then for any positive formula φ, $M \models \varphi$ iff $P \models_L \varphi$.

[2]This kind of "equality" is induced by the pre-order $\leq$. By Lemma 3, if $M \leq N$ and $N \leq M$ then for every positive formula φ, $M \models \varphi$ iff $N \models \varphi$.

A model M is called the *least flat model* of a positive program P if it is a flat model of P and is less than or equal to any flat model of P. In [17], we showed that any positive modal logic program that has some flat model has the least flat model, which can be constructed in polynomial time and has polynomial size.

3 New Results for Monomodal Logics

We first consider the complexity of the satisfiability problem of sets of Horn formulas with modal depth bounded by $k \geq 2$ in the logics $K4$, $KD4$, and $S4$.

If X and Y are formula sets then we write $X; Y$ to denote the union of them. We write $X; \varphi$ for $X; \{\varphi\}$. We need the two following auxiliary lemmas. The first one is used to reduce lengths of modal contexts of clauses.

LEMMA 4. *In the following, let p and q be new primitive propositions (i.e. p and q occur only at the indicated positions) and φ a simple clause. Then the following pairs of formula sets are equisatisfiable in any normal modal logic that is an extension of $K4$.*

(1)	$X; \Box^2\varphi$	*and*	$X; \Box^2 p; \Box(\neg p \vee \varphi)$	
(2)	$X; \Box^{2k}\varphi$	*and*	$X; \Box^k q; \Box(\neg q \vee \Box^k\varphi)$	*where* $k \geq 2$
(3)	$X; \Box^{2k+1}\varphi$	*and*	$X; \Box^{k+1} q; \Box(\neg q \vee \Box^k\varphi)$	$k \geq 1$
(4)	$X; \Box(a \vee \Box^{2k}\varphi)$	*and*	$X; \Box(a \vee \Box^k q); \Box(\neg q \vee \Box^k\varphi)$	$k \geq 1$
(5)	$X; \Box(a \vee \Box^{2k+1}\varphi)$	*and*	$X; \Box(a \vee \Box^{k+1} q); \Box(\neg q \vee \Box^k\varphi)$	$k \geq 0$

Proof. $\rightarrow$) Choose one of the pairs. Suppose that the LHS set is satisfied in a model $M = \langle W, \tau, R, h\rangle$. Let $M' = \langle W, \tau, R, h'\rangle$ with $x \in h'(u)$ iff $x \in h(u)$ for $x \neq p$ and $x \neq q$, $p \in h'(u)$ iff $M, u \models \varphi$, and $q \in h'(u)$ iff $M, u \models \Box^k\varphi$, where p and q are the new primitive propositions. It is easily seen that the RHS set is satisfied in M'.

$\leftarrow$) Choose one of the pairs. We show that the RHS formula set implies the LHS set in any modal logic that is an extension of $K4$.

The assertion holds for the pair (1) because that the formulas $\Box(\neg p \vee \varphi) \rightarrow \Box^2(\neg p \vee \varphi)$ and $\Box^2 p \wedge \Box^2(\neg p \vee \varphi) \rightarrow \Box^2\varphi$ are $K4$-valid.

The assertion holds for the pair (2) because that the formulas $\Box(\neg q \lor \Box^k\varphi) \to \Box^k(\neg q \lor \Box^k\varphi)$ and $\Box^k q \land \Box^k(\neg q \lor \Box^k\varphi) \to \Box^{2k}\varphi$ are $K4$-valid.

The assertion holds for the pair (4) because the following formulas are $K4$-valid: $\Box(\neg q \lor \Box^k\varphi) \to \Box^{k+1}(\neg q \lor \Box^k\varphi)$ and

$$\Box(a \lor \Box^k q) \land \Box^{k+1}(\neg q \lor \Box^k\varphi) \to \Box(a \lor \Box^{2k}\varphi)$$

Analogously, the assertion holds for the pairs (3) and (5). ■

LEMMA 5. *Let L be a normal modal logic that is an extension of $K4$. Every formula set X can be translated to an L-equisatisfiable set Y of clauses with modal depth bounded by 2. Furthermore, if X is a set of Horn formulas then Y is a set of Horn clauses. The translation can be done in polynomial time and the length of Y is bounded by a polynomial in the length of X.*

Proof. By Lemma 2, we can translate X in polynomial time to a clause set Z such that: X and Z are L-equisatisfiable; if X is a set of Horn formulas then Z is a set of Horn clauses; the modal depth of Z is equal to the modal depth of X, and the length of Z is of quadratic order in the length of X.

We refer to the pairs of equisatisfiable formula sets given in Lemma 4 as translation rules (with left to right direction of application). We then apply[3] these translation rules to Z. We apply the rule (1) only when the modal depth of φ is 1, and the rule (5) only when $k \geq 1$, or $k = 0$ and φ is not a classical literal. We apply the rules until no more changes can be made to the set. Let Y be the resulting set. Observe that the modal depth of Y is bounded by 2.

Observe also that each of the applications decreases the modal depth of some formula of the set by a half (with an inaccuracy up to 2) and increases the length of the set by a constant number (of symbols). Hence there exists a constant h such that we can decrease the modal depth of the set by a half (with an inaccuracy up to 2) while the length of the set increases not more than h times. Hence the process terminates in polynomial time. It is easily seen that the length of Y is bounded by a polynomial in the size of Z, and Y is a set of Horn clauses if so is Z.

[3]Each application of a rule is done for the whole formula set but not a fragment.

Hence, the translation from X to Y (via Z) is done in polynomial time, the length of Y is bounded by a polynomial in the length of X, and Y is a set of Horn clauses if X is a set of Horn formulas. ■

As a consequence we have the following result:

THEOREM 6. *The complexity of the satisfiability problem of sets of Horn formulas with modal depth bounded by $k \geq 2$ in the logics $K4$, $KD4$, and $S4$ is PSPACE-complete.*

This theorem follows from the above lemma and the reason that the similar problem without bounding modal depth is PSPACE-complete [4]. The assertion for $S4$ has been previously proved by Chen and Lin [4].

By this theorem, the complexity of the satisfiability problem of formula sets (without the Horn restriction) with modal depth bounded by $k \geq 2$ in $K4$, $KD4$, and $S4$ is PSPACE-complete (the upper bound follows from [15]).

THEOREM 7. *The complexity of the satisfiability problem of sets of Horn formulas with modal depth bounded by $k \geq 2$ in the logic K is NP-complete.*

Proof. The upper bound follows from Halpern [11]. For the lower bound, we use a reduction from the 3SAT problem, which is known to be NP-hard. The 3SAT problem is to check satisfiability of a clause set $X = \{C_1, \ldots, C_n\}$, where $C_i = c_{i1} \vee c_{i2} \vee c_{i3}$ and c_{i1}, c_{i2}, c_{i3} are classical literals. Given such a set X, we construct in polynomial time a set Y of Horn formulas with modal depth bounded by 2 such that X is satisfiable iff Y is K-satisfiable.

Let t and f be new propositions, which informally stand for "true" and "false". The presence of the formula $\Box f$ (resp. $\Diamond t$) at a world w informally says that there are no worlds (resp. there is some world) accessible from w. Let Y be the set consisting of the formulas

$$\Diamond p_i, \Diamond q_i, \neg\Diamond(p_i \wedge q_i), \neg\Diamond^2 f, \Box^2 t,$$
$$\Diamond(p_i \wedge \Box f) \wedge \Diamond(q_i \wedge \Box f) \to c_{i1},$$
$$\Diamond(p_i \wedge \Diamond t) \to c_{i2},$$
$$\Diamond(q_i \wedge \Diamond t) \to c_{i3},$$

for $1 \leq i \leq n$, and p_i and q_i are new propositions. Denote the set Y also by $\pi_{3SAT}(X)$. Note that Y contains only Horn formulas with modal depth bounded by 2.

Suppose that X is satisfied by a variable assignment V. We show that Y is K-satisfiable. Let $M = \langle W, \tau, R, h \rangle$ be a model defined as: $W = \{\tau, w_{1p}, w_{1q}, \ldots, w_{np}, w_{nq}, u\}$, $h(\tau) = \{p \mid V(p)\}$, $h(u) = \{t\}$, and for $1 \leq i \leq n$, $h(w_{ip}) = \{p_i\}$ and $h(w_{iq}) = \{q_i\}$, and
$R = \{(\tau, w_{ip}), (\tau, w_{iq}) \mid 1 \leq i \leq n\} \cup \{(w_{ip}, u) \mid 1 \leq i \leq n \text{ and } V(c_{i2})\}$
$\cup \{(w_{iq}, u) \mid 1 \leq i \leq n \text{ and } V(c_{i3})\}$.

It is easy to verify that $M \vDash Y$. Therefore Y is K-satisfiable.

Now suppose that Y is K-satisfiable. We show that X is satisfiable. Let M be a model of Y. Let w_{ip}, w_{iq} be worlds accessible from τ such that $M, w_{ip} \vDash p_i$ and $M, w_{iq} \vDash q_i$, for $1 \leq i \leq n$. If there exists a world accessible from w_{ip}, then $M, \tau \vDash \Diamond(p_i \wedge \Diamond t)$, and hence $M, \tau \vDash c_{i2}$. Similarly, if there exists a world accessible from w_{iq}, then $M, \tau \vDash c_{i3}$. If there are no worlds accessible from w_{ip} or w_{iq}, then $M, \tau \vDash \Diamond(p_i \wedge \Box f) \wedge \Diamond(q_i \wedge \Box f)$, and hence $M, \tau \vDash c_{i1}$. Consequently, $M, \tau \vDash C_i$, for $1 \leq i \leq n$. Hence $M, \tau \vDash X$, and X is satisfiable. ■

In the remainder of this section, we study the satisfiability problem of modal formulas with modal depth bounded by 1. The problem is NP-complete, and for the Horn fragment it is PTIME-complete, for all of the monomodal logics considered in this work. Some parts of these results immediately follow from known ones. We complete the picture by the two following theorems.

THEOREM 8. *The complexity of the satisfiability problem of formulas with modal depth bounded by 1 in the logics K4, KD4, and S4 is NP-complete.*

Proof. The lower bound NP-hard follows from the fact that the satisfiability problem in the classical propositional logic is NP-complete. For the upper bound, let L be one of the logics $K4$, $KD4$, $S4$, and let X be any L-satisfiable formula set with modal depth bounded by 1.

It can be proved that X has an L-model $M = \langle W, \tau, R, h \rangle$ such that for any u and v different to τ, if $R(\tau, u)$ and $R(u, v)$ hold, then $u = v$. In fact, if $M' = \langle W, \tau, R', h \rangle$ is an L-model of X, then by deleting edges (u, v) with $u \neq \tau$ from R' and adding edges (u, u) for $u \neq \tau$ to the frame, we obtain such a mentioned L-model M of X.

An L-model M of X with the mentioned frame restriction can be nondeterministically constructed in polynomial time by building an L-model graph for X (see, e.g., [21, 10, 18] for the technique). Therefore the satisfiability problem of formulas with modal depth bounded by 1 in $K4$, $KD4$, and $S4$ belongs to the NP class. ■

THEOREM 9. *The complexity of the satisfiability problem of sets of Horn formulas with modal depth bounded by 1 in K, K4, KD4, and S4 is PTIME-complete.*

Proof. The lower bound PTIME-hard follows from that the complexity of the satisfiability problem of sets of Horn formulas in the classical propositional logic is PTIME-complete (Jones and Laaser [13]).

By the result of [17], every positive modal logic program with modal depth bounded by 1 has the least $KD4$-model and the least $S4$-model, which can be constructed in polynomial time and have polynomial size. Consequently, by Lemmas 2 and 1, the problem of checking satisfiability of sets of Horn formulas with modal depth bounded by 1 in $KD4$ and $S4$ is decidable in PTIME.

It remains to show that the similar problem for the logics K and $K4$ is decidable in PTIME. Let L denote K or $K4$, and P be any positive modal logic program with modal depth bounded by 1. Let $M = \langle W, \tau, R, H \rangle$ be the model graph constructed as follows.

1. Let $W = \{\tau, \rho\}$, $R = \{(\tau, \rho)\}$, $H(\tau) = P$, $H(\rho) = \emptyset$.

2. For every $w \in W$, and every $\varphi \in H(w)$,

   (a) Case $\varphi = (B_1 \wedge \ldots \wedge B_k \rightarrow A)$: if $M, w \models B_i$ for all $1 \leq i \leq k$, then add A to $H(w)$;

   (b) Case $\varphi = \Box\psi$: add ψ to every world u accessible from w;

   (c) Case $\varphi = \Diamond p$: if $M, w \not\models p$ then add a new world u with content $\{p\}$ to W and connect w to u (i.e. let $W = W \cup \{u\}$, $H(u) = \{p\}$, $R = R \cup \{(w, u)\}$).

3. While some change occurred, repeat step 2.

Observe that, for any w and u, $R(w, u)$ holds only when $w = \tau$ (since the modal depth is bounded by 1). Hence, the above algorithm terminates in polynomial time. It can be shown by induction on the structure of φ that for any $w \in W$ and any $\varphi \in H(w)$, $M, w \models \varphi$. Hence M is a K-model of P. By the mentioned property of R, M is also a $K4$-model of P.

If $N = \langle W', \tau', R', h' \rangle$ is a model of P such that $R' \neq \emptyset$ and for any x, y, $R'(x, y)$ holds only when $x = \tau$, then $M \leq N$. This claim can be proved by showing that it is an invariant of the loop of the above algorithm that there exists a relation $r \subseteq W \times W'$ such that the following assertions hold:

$r(\tau, \tau')$
$\forall x \;\; R(\tau, x) \rightarrow \exists x' \; R'(\tau', x') \wedge r(x, x')$
$\forall x' \; R'(\tau', x') \rightarrow \exists x \; R(\tau, x) \wedge r(x, x')$
$\forall x, x' \; \forall \varphi \in H(x) \;\; r(x, x') \rightarrow N, x' \models \varphi$

Such relations r can be built as follows: After the execution of step 1, let $r = \{(\tau, \tau')\} \cup \{(\rho, w') \mid R'(\tau', w')\}$, and after each execution of step 2c, let $r = r \cup \{(u, u') \mid R'(\tau', u') \text{ and } p \in h'(u')\}$.

If P has a flat model, then let M' be the least flat model of P, else let $M' = M$. Both M and M' can be constructed in polynomial time and have size bounded by a polynomial in the size of P.

We claim that for any positive formula φ with modal depth bounded by 1, $P \not\models_L \varphi$ iff $M \not\models \varphi$ or $M' \not\models \varphi$. The "if" part clearly holds. For the "only if" part, suppose that $P \not\models_L \varphi$, where φ is a positive formula with modal depth bounded by 1. It follows that there exists an L-model N of P such that $N \not\models \varphi$. Let $N_{|1}$ be the model obtained from N by deleting all edges not starting from τ. We have $N_{|1} \models P$ and $N_{|1} \not\models \varphi$, because the modal depths of P and φ are bounded by 1. If $N_{|1}$ is a flat model, then M' is the least flat model of P, and hence $M' \not\models \varphi$. Otherwise, $M \leq N_{|1}$, and hence $M \not\models \varphi$.

By Lemmas 2 and 1, we conclude that checking satisfiability of sets of Horn formulas with modal depth bounded by 1 in K and $K4$ is decidable in PTIME. ■

4 On the Complexity of Multimodal Logics

A language for multimodal logics uses n pairs of modal operators $\Box_i$ and $\Diamond_i$, for $1 \leq i \leq n$, where n is a fixed number *greater than* 1.

Formulas in multimodal logics are formed in the usual way. Interpretations used for multimodal logics are usually Kripke models with n accessibility relations (one for each of the pairs $\Box_i$ and $\Diamond_i$, $1 \leq i \leq n$). The satisfaction relation is also defined in the usual way.

Multimodal logics can be formed by combining modal logics. The combination of modal logics has been intensively studied in the last decade (see, e.g., [1, 7, 14, 9, 16, 5, 22, 19, 8]). A simple way to combine modal logics is to make their fusion and we can consider fusions of variants (by renaming modal operators) of the same logic. Given a monomodal logic L, the fusion L_n is the multimodal logic axiomatized by the axioms of the classical propositional logic, the modus ponens rule, the modal axioms and modal rules of L with $\Box$ and $\Diamond$ replaced respectively by $\Box_i$ and $\Diamond_i$, for each $1 \leq i \leq n$. Note that there are no interaction axioms between different kinds of modal operators in L_n.

In this section, we discuss the complexity of the multimodal logics L_n under the restriction to the Horn fragment and/or bounded modal depth, where L is one of the 15 basic monomodal logics. We show how Table 1 changes when each logic L is replaced by L_n.

In [12], Halpern and Moses showed that the satisfiability problem in the multimodal logics K_n, T_n, $S4_n$, $KD45_n$, and $S5_n$ is PSPACE-complete. Halpern in [11] claimed that the PSPACE-complete complexity also holds for $K45_n$, as its proof does not differ much from the proof for $KD45_n$.

The complexity PSPACE-complete also holds for KD_n, $K4_n$, $KD4_n$, $K5_n$, and $KD5_n$. The reasons are as follows:

- Nondeterministic PSPACE algorithms for checking satisfiability in KD_n, $K4_n$, $KD4_n$, $K5_n$, and $KD5_n$ can be developed, e.g., in a similar way as for K_n, $S4_n$, and $KD45_n$ in [12]. Hence we have the upper bound PSPACE.

- If $L_n \in \{KD_n, K4_n, KD4_n\}$, then the lower bound for L_n follows from the lower bound for L (PSPACE-hard). The lower bound PSPACE-hard for $K5_n$ and $KD5_n$ will be shown later in this section.

As a corollary (of the upper bound), every PSPACE-completeness result in Table 1 for $L \in \{K, KD, T, K4, KD4, S4\}$ also holds for L_n. Note that Theorem 6 is useful here for $K4_n$ and $KD4_n$.

We guess that the satisfiability problem in the multimodal logics KB_n, KDB_n, B_n, and $KB5_n$ is also PSPACE-complete. For KB_n, KDB_n, and B_n, it suffices to show the upper bound. If our prediction is true, then the satisfiability problem of sets of Horn clauses in KB_n, KDB_n, and B_n is also PSPACE-complete.

In [11], Halpern showed that the satisfiability problem of formulas with modal depth bounded by $k \geq 2$ in K_n, T_n, $K45_n$, $KD45_n$, and $S5_n$ is NP-complete. (The lower bound NP-hard follows from the lower bound of the monomodal case, the upper bound NP can be seen not difficultly.) Using similar argumentations, one can claim that the assertion also holds for KD_n, KB_n, KDB_n, B_n, $K5_n$, $KD5_n$, $KB5_n$.

Next, we claim that the satisfiability problem of formulas with modal depth bounded by 1 in L_n is NP-complete for L being any one the 15 basic monomodal logics. The only point that needs justification is the case of $K4_n$, $KD4_n$, and $S4_n$. For this case, use similar argumentations as the proof of Theorem 8.

We now consider the restriction to the Horn fragment.

Let Horn formulas be defined similarly as in the case of monomodal logics. A *clause* is a formula of the form $\Delta(A_1 \vee \ldots \vee A_h \vee \neg B_1 \vee \ldots \vee \neg B_k)$, where Δ is a sequence of universal modal operators, A_i and B_j are atoms of the form p, $\Box_t p$, or $\Diamond_t p$. Let Horn clauses and positive logic programs be defined similarly as in the case of monomodal logics. It can be seen that Lemma 2 still holds for normal multimodal logics.

In the following, we show that the satisfiability problem of sets of Horn clauses in $K5_n$, $KD5_n$, $K45_n$, and $KD45_n$ is PSPACE-complete.

Let X be a set of clauses in the language of monomodal logics. Let $\pi_{bi}(X)$ be the set of clauses obtained from X as follows: modal operators at odd modal nesting depths are subscripted by 1, and modal operators at even modal nesting depths are subscripted by 2. For example, the clause $\Box\Box\Box(p \vee \Box q \vee \Diamond r)$ is replaced by $\Box_1\Box_2\Box_1(p \vee \Box_2 q \vee \Diamond_2 r)$. It is clear that if X is a set of Horn clauses then $\pi_{bi}(X)$ is also a set of Horn clauses.

LEMMA 10. *Let X be a set of Horn clauses in the language of monomodal logics, L_n be either $K5_n$ or $K45_n$, and LD_n be either $KD5_n$ or $KD45_n$. Then X is K-satisfiable iff $\pi_{bi}(X)$ is L_n-satisfiable; and X is KD-satisfiable iff $\pi_{bi}(X)$ is LD_n-satisfiable.*

The proof of this lemma is not included due to the lack of space.

THEOREM 11. *The satisfiability problem of sets of Horn clauses in $K5_n$, $KD5_n$, $K45_n$, and $KD45_n$ is PSPACE-complete.*

Proof. The upper bound PSPACE has been justified earlier (for the case without restrictions). The lower bound PSPACE-hard follows from the above lemma and the facts that the satisfiability problem of sets of Horn clauses in K and KD is PSPACE-complete [4] and $\pi_{bi}(X)$ can be obtained from X in linear time and has a linear size (in the size of X). ■

COROLLARY 12. *The satisfiability problem (without restrictions) in $K5_n$ and $KD5_n$ is PSPACE-complete.*

We now consider the combination of the restriction to the Horn fragment and the restriction to bounded modal depth.

THEOREM 13. *The satisfiability problem of sets of Horn clauses with modal depth bounded by $k \geq 2$ in the multimodal logics K_n, KB_n, $K5_n$, $K45_n$, $KB5_n$ is NP-complete, and in KD_n, T_n, KDB_n, B_n, $KD5_n$, $KD45_n$, $S5_n$ is PTIME-complete.*

The two groups of modal logics mentioned in this theorem differ at the aspect that logics in the first group are non-serial, while logics in the second group are serial. For $L \in \{KB, K5, K45, KB5\}$, the complexity jumps from PTIME-complete for L to NP-complete for L_n because that L is *almost serial*[4] while L_n does not have such a similar property.

Sketch of the proof Consider the case of K_n, KB_n, $K5_n$, $K45_n$, $KB5_n$. The essential point here is the lower bound NP-hard. Let X

[4]A frame $\langle W, \tau, R\rangle$ is *connected* if W contains only worlds reachable directly or indirectly from τ via R. A monomodal logic L is *almost serial* if every connected L-frame $\langle W, \tau, R\rangle$ with $W \neq \{\tau\}$ is serial (i.e. $\forall x \exists y\, R(x, y)$).

be a set of clauses of the form $c_1 \vee c_2 \vee c_3$, where c_1, c_2, c_3 are classical literals. We use the translation π_{3SAT} as in the proof of Theorem 7. Consider the set $\pi_{bi}(\pi_{3SAT}(X))$. We claim that for $L_n \in \{K_n, KB_n, K5_n, K45_n, KB5_n\}$, X is satisfiable iff $\pi_{bi}(\pi_{3SAT}(X))$ is L_n-satisfiable. The proof of this is more or less the same as the proof of Theorem 7. Hence the 3SAT problem is reducible to the satisfiability problem of sets of Horn clauses with modal depth bounded by $k \geq 2$ in the multimodal logics K_n, KB_n, $K5_n$, $K45_n$, $KB5_n$. Therefore the latter problem is NP-hard.

For the case of $KD_n, T_n, KDB_n, B_n, KD5_n, KD45_n, S5_n$, the essential point is the upper bound PTIME. The proof of this is similar to the proof given in [17] of that the problem of checking satisfiability of sets of Horn clauses with modal depth bounded by $k \geq 2$ in *KD*, *T*, *KDB*, and *B* is in PTIME. The key of the proof is that if P is a positive logic program (in the language of multimodal logics) consisting of clauses whose modal depths are bounded by some constant k, then a least[5] L_n-model of P, for $L_n \in \{KD_n, T_n, KDB_n, B_n, KD5_n, KD45_n, S5_n\}$, can be constructed in polynomial time, and has a polynomial size if $L_n \in \{KD_n, T_n, KD5_n, KD45_n, S5_n\}$, or can be encoded in polynomial space if $L_n \in \{KDB_n, B_n\}$ (similarly as in the case of KDB and B [17]). ■

Analogously as for the proof of Theorem 9, one can show that the complexity of the satisfiability problem of sets of Horn clauses with modal depth bounded by 1 in K_n, KB_n, $K4_n$, $KD4_n$, $S4_n$, $K5_n$, $K45_n$, and $KB5_n$ is PTIME-complete. The main change is that, for K_n, KB_n, $K4_n$, $K5_n$, $K45_n$, $KB5_n$ we need to use 2^n models instead of 2 models as in the proof of Theorem 9. More precisely, for each set $I \subseteq \{1, \ldots, n\}$, consider the case when $\exists x\, R_i(\tau, x)$ holds iff $i \in I$ and construct a "minimal" model of the considered program for that case. We conclude that the satisfiability problem of sets of Horn clauses with modal depth bounded by 1 in L_n is PTIME-complete for every L being one the 15 basic monomodal logics.

In summary, there are open problems on the complexity of the satisfiability problem in KB_n, KDB_n, B_n, $KB5_n$ and the satisfia-

[5] An ordering of Kripke models in multimodal logics is defined similarly as for the monomodal case. See [20] for details.

bility problem of sets of Horn modal clauses in $KB5_n$ and $S5_n$. It is probable that these problems are PSPACE-complete. Under this assumption, Table 1 changes as follows when every logic L in that table is replaced by L_n:

- The satisfiability problem in $K5_n$, $KD5_n$, $K45_n$, $KD45_n$, $KB5_n$, $S5_n$ is PSPACE-complete for both of the cases: without restrictions or with the restriction to the Horn fragment.
- The satisfiability problem of sets of Horn clauses with modal depth bounded by $k \geq 2$ in the multimodal logics KB_n, $K5_n$, $K45_n$, and $KB5_n$ is NP-complete.

5 Conclusions

We have summarized the complexity of the satisfiability problem in all of the 15 basic normal monomodal logics under the restriction to the Horn fragment and/or bounded modal depth. To fulfill the complexity table, we have given some new results. Our Theorems 6 and 7 show that the modal logics K, $K4$, and $KD4$ are hard even under the mentioned restrictions. The restriction of modal depth to 1 is quite tight and the corresponding fragments are rather useless. However, our results for that case are still interesting from the theoretical point of view.

We have also discussed and given some results on the complexity of the multimodal logics L_n under the mentioned restrictions, where L is one of the 15 basic monomodal logics. There remain some open problems.

Acknowledgements

The author would like to thank professor Andrzej Szałas and the anonymous reviewers for helpful comments.

BIBLIOGRAPHY

[1] F. Baader and H.-J. Ohlbach. A multi-dimensional terminological knowledge representation language. *Journal of Applied Non-Classical Logics*, 5:153–197, 1995.

[2] P. Blackburn, M. de Rijke, and Y. Venema. *Modal Logic*. Cambridge University Press, 2002.

[3] A. Chagrov and M. Zakharyaschev. *Modal Logic*. Clarendon Press, Oxford, Oxford Logic Guides 35, 1997.

[4] C.C. Chen and I.P. Lin. The computational complexity of the satisfiability of modal Horn clauses for modal propositional logics. *Theoretical Computer Science*, 129:95–121, 1994.
[5] S. Demri. Complexity of simple dependent bimodal logics. In *Roy Dyckhoff (Ed.): TABLEAUX 2000, LNAI 1847*, pages 190–204. Springer, 2000.
[6] L. Fariñas del Cerro and M. Penttonen. A note on the complexity of the satisfiability of modal Horn clauses. *Logic Programming*, 4:1–10, 1987.
[7] D. Gabbay. Fibred semantics and the weaving of logics part 1: Modal and intuitionistic logics. *The Journal of Symbolic Logic*, 61(4):1057–1120, 1996.
[8] D. Gabbay, A. Kurucz, F. Wolter, and M. Zakharyaschev. *Many-Dimensional Modal Logics: Theory and Applications*, volume 148 of *Studies in Logic*. Elsevier, 2003.
[9] D. Gabbay and V. Shehtman. Products of modal logics, part 1. *Logic Journal of the IGPL*, 6(1):73–146, 1998.
[10] R. Goré. Tableau methods for modal and temporal logics. In D'Agostino, Gabbay, Hähnle, and Posegga, editors, *Handbook of Tableau Methods*, pages 297–396. Kluwer Academic Publishers, 1999.
[11] J.Y. Halpern. The effect of bounding the number of primitive propositions and the depth of nesting on the complexity of modal logic. *Artificial Intelligence*, 75(2):361–372, 1995.
[12] J.Y. Halpern and Y. Moses. A guide to completeness and complexity for modal logics of knowledge and belief. *Artificial Intelligence*, 54:319–379, 1992.
[13] N.D. Jones and T.W. Laaser. Complete problems for deterministic polynomial time. *Theoretical Computer Science*, 3:105–112, 1976.
[14] M. Kracht and F. Wolter. Simulation and transfer results in modal logic - a survey. *Studia Logica*, 59:149–177, 1997.
[15] R. Ladner. The computational complexity of provability in systems of modal propositional logic. *SIAM Journal of Computing*, 6:467–480, 1977.
[16] M. Marx. Complexity of products of modal logics. *Journal of Logic and Computation*, 9(2):221–238, 1999.
[17] L.A. Nguyen. Constructing the least models for positive modal logic programs. *Fundamenta Informaticae*, 42(1):29–60, 2000.
[18] L.A. Nguyen. Sequent-like tableau systems with the analytic superformula property for the modal logics KB, KDB, K5, KD5. In Roy Dyckhoff, editor, *Proceedings of TABLEAUX 2000, LNAI 1847*, pages 341–351. Springer, 2000.
[19] L.A. Nguyen. Analytic tableau systems for propositional bimodal logics of knowledge and belief. In U. Egly and C.G. Femüller, editors, *Proceedings of TABLEAUX 2002, LNAI 2381*, pages 206–220. Springer-Verlag Berlin Heidelberg, 2002.
[20] L.A. Nguyen. Multimodal logic programming and its applications to modal deductive databases. Manuscript, available at `http://www.mimuw.edu.pl/~nguyen/papers.html`, 2003.
[21] W. Rautenberg. Modal tableau calculi and interpolation. *Journal of Philosophical Logic*, 12:403–423, 1983.
[22] M. Reynolds and M. Zakharyaschev. On the products of linear modal logics. *Journal of Logic and Computation*, 11:909–931, 2001.

Linh Anh Nguyen
Institute of Informatics, University of Warsaw
ul. Banacha 2, 02-097 Warsaw, Poland
nguyen@mimuw.edu.pl

On PSPACE-decidability in Transitive Modal Logics

I. SHAPIROVSKY

ABSTRACT. In this paper we describe a new method, allowing us to prove PSPACE-decidability for transitive modal logics. We apply it to $\mathbf{L}_1$ and $\mathbf{L}_2$, modal logics of Minkowski spacetime. We also show how to extend this method to some other transitive logics.

1 Introduction

Computational complexity of modal logics was first studied by Ladner in [11]. To obtain upper complexity bounds, he modified the *tableau method* from [10][1]. Later various tableau-based methods were used in PSPACE-decidability proofs for a number of monomodal logics (like $\mathbf{K}, \mathbf{K4}, \mathbf{S4}$, etc. [11],[14]), and also for multimodal and tense logics, cf. [8],[15].

In this paper we propose an alternative proof for PSPACE upper bounds in transitive modal logics. The satisfiability problem is reduced to satisfiability in some "standard" finite frames. To obtain these frames, we apply *selective filtration* (see e.g. [4]) and extract a finite submodel from the canonical model. The height of this submodel is polynomially bounded, due to the *maximality property* [6] of the canonical model.

This construction allows us to give a rather simple description of the decision procedure. The method happens to be "robust" – after adding extra axioms (such as density or McKinsey axiom), only a slight modification is sufficient.

To illustrate our method, we consider two particular logics, $\mathbf{L}_1$ and $\mathbf{L}_2$. These logics were introduced in the study of chronological future modalities in Minkowski spacetime [7],[12]. In [12] the finite model property (FMP) of these logics was proved. "Standard" finite frames for $\mathbf{L}_1$ and $\mathbf{L}_2$ can be obtained following the lines of [12]. Basing on this construction, we describe the deciding deterministic algorithm working within a polynomial space. It follows that $\mathbf{L}_1$ and $\mathbf{L}_2$ are PSPACE-complete (the lower bounds can be obtained by Ladner's reduction of the QBF-validity problem to the modal satisfiability problem [11]).

[1]The tableau method for the propositional calculi was first developed in [1].

We show how to apply our method to some other transitive logics. In particular, for the logic **K4** and its extensions by density, reflexivity, confluence, McKinsey axiom, we propose a new proof of PSPACE-decidability.

2 Preliminaries

In this paper we consider propositional normal monomodal logics containing **K4**.

We assume that $\Diamond$, $\rightarrow$, $\bot$ are the basic connectives, and $\Box$, $\neg$, $\vee$, $\wedge$, $\top$ are derived. PV denotes the countable set of propositional variables. For a modal logic Λ and a modal formula φ, the notation $\Lambda + \varphi$ denotes the smallest modal logic containing $\Lambda \cup \{\varphi\}$; $\Lambda \vdash \varphi$ means $\varphi \in \Lambda$. $Sub(\varphi)$ denotes the set of all subformulas of φ, $PV(\varphi) := PV \cap Sub(\varphi)$. Here are the names for some particular axioms:

$A4 := \Diamond\Diamond p \rightarrow \Diamond p$	*transitivity,*
$AT := p \rightarrow \Diamond p$	*reflexivity,*
$AD := \Diamond\top$	*seriality,*
$A1 := \Box\Diamond p \rightarrow \Diamond\Box p$	*McKinsey axiom,*
$A2 := \Diamond\Box p \rightarrow \Box\Diamond p$	*confluence,*
$Ad = Ad_1 := \Diamond p \rightarrow \Diamond\Diamond p$	*density,*
$Ad_2 := \Diamond p_1 \wedge \Diamond p_2 \rightarrow \Diamond(\Diamond p_1 \wedge \Diamond p_2)$	*2-density;*

and the names for some logics:

$$\mathbf{K4} := \mathbf{K} + A4, \qquad \mathbf{K4d} := \mathbf{K4} + Ad, \quad \mathbf{S4} := \mathbf{K4} + AT,$$
$$\mathbf{L}_1 := \mathbf{K4} + AD + Ad_2, \quad \mathbf{L}_2 := \mathbf{L}_1 + A2.$$

For a logic Λ let $\Lambda.1 := \Lambda + A1$, $\Lambda.2 := \Lambda + A2$.

As usual, a *(Kripke) frame* is a pair (W, R), where $W \neq \varnothing$, $R \subseteq W \times W$. We consider only transitive frames. A *(Kripke) model* is a Kripke frame with a valuation: $M = (W, R, \theta)$, where $\theta : PV \longrightarrow 2^W$, 2^W denotes the power set of W. For a model $M = (W, R, \theta)$ or a frame $F = (W, R)$, the notation $x \in M$ or $x \in F$ means $x \in W$. As usual, for $x \in W$, $V \subseteq W$ let $R(x) := \{y \mid xRy\}$, $R(V) := \bigcup_{x \in V} R(x)$, $R|V := R \cap (V \times V)$. We also put $W^x := \{x\} \cup R(x)$, $F^x := (W^x, R|W^x)$.

A model $M_1 = (W_1, R_1, \theta_1)$ is a *(weak) submodel* of $M = (W, R, \theta)$ (notation: $M_1 \subseteq M$) if

$$W_1 \subseteq W,\ R_1 \subseteq R,\ \theta_1(p) = \theta(p) \cap 2^{W_1}$$

for every $p \in PV$. If $R_1 = R|W_1$, then M_1 is called the *restriction* of M to W_1 and denoted by $M|W_1$. The submodel $M^x := M|W^x$ is called a *cone* in M.

The sign $\models$ denotes the truth at a point of a model and also the validity in a frame. For a class of frames $\mathcal{F}$, $\mathbf{L}(\mathcal{F})$ denotes the set of all formulas that are valid in all frames from $\mathcal{F}$. For a single frame F, $\mathbf{L}(F)$ abbreviates $\mathbf{L}(\{F\})$. For a logic Λ, if $\mathbf{L}(F) \supseteq \Lambda$, then we say that F is Λ*-frame*. Recall that Λ is *Kripke-complete* if $\Lambda = \mathbf{L}(\mathcal{F})$ for some class of frames $\mathcal{F}$.

A formula φ is *satisfiable in a model* M if for some $x \in M$ we have $M, x \models \varphi$; φ is *satisfiable in a frame* F if φ is satisfiable in some model over F. For a class of frames $\mathcal{F}$, φ is $\mathcal{F}$*-satisfiable* if φ is satisfiable in some $F \in \mathcal{F}$. φ is Λ*-satisfiable* if φ is satisfiable in some Λ-frame. Note that if Λ is Kripke-complete, then we have: φ is Λ-satisfiable $\Leftrightarrow \Lambda \nvdash \neg\varphi$.

The *disjoint union* $F_1 \sqcup F_2$ and the *ordinal sum* $F_1 + F_2$ of frames F_1, F_2 are defined in a standard way. The notation $f : F_1 \twoheadrightarrow F_2$ means that f is a p-morphism from F_1 onto F_2, and $F_1 \twoheadrightarrow F_2$ means that $f : F_1 \twoheadrightarrow F_2$ for some f.

Recall that a *cluster* in (W, R) is an equivalence class under the relation $\sim_R := (R \cap R^{-1}) \cup Id_W$, where Id_W is the equality relation on W. For a point x, $\overline{x}$ denotes its cluster. C_0 denotes a *degenerate cluster*, i.e. an irreflexive singleton; C_1 denotes a reflexive singleton; C_n denotes an n-element cluster for $n \geq 2$. Let $W/{\sim_R} := \{\overline{x} \mid x \in W\}$. For clusters $C, D \in W/{\sim_R}$ we put

$$C \leqq_R D := D \subseteq R(C), \; C <_R D := C \leqq_R D \text{ and } C \neq D.$$

Note that the relations $\leqq_R, <_R$ are transitive and antisymmetric, $<_R$ is irreflexive, and $C \leqq C$ iff C is non-degenerate. A cluster D is a *successor* of C, if $C <_R D$ and there is no cluster C' such that $C <_R C' <_R D$. For $x, y \in F$ we say that y is a *successor* of x, if $\overline{y}$ is a successor of $\overline{x}$. The frame $F/{\sim_R} := (W/{\sim_R}, \leqq_R)$ is called the *skeleton* of F (and of every model over F). A point $x \in F$ is called *maximal* (*minimal*) if its cluster is maximal (minimal) in $F/{\sim_R}$.

In this paper a frame (W, R) is called *rooted* if for some x $W = W^x$, and the cluster $\overline{x}$ is one-element. A *tree* is a rooted frame (W, R) such that R is transitive and antisymmetric, and $R^{-1}(x)$ is a chain for every $x \in W$. A frame F is called a *quasitree* if its skeleton $F/{\sim_R}$ is a tree.

Let us recall the notion of *selective filtration* [12], cf. [2],[4].

DEFINITION 1. Let M be a Kripke model, Ψ a set of formulas closed under subformulas. A submodel $M_1 \subseteq M$ (with the relation R_1) is called a *selective filtration of* M *through* Ψ (notation: $M_1 \in \mathcal{SF}(M, \Psi)$), if for any

$x \in M_1$, for any formula φ

$$\Diamond\varphi \in \Psi \ \& \ M, x \vDash \Diamond\varphi \Rightarrow \exists y \in R_1(x) \ M, y \vDash \varphi.$$

The following lemma is proved easily by induction on the length of a formula φ.

LEMMA 2. *If $M_1 \in \mathcal{SF}(M, \Psi)$, then for any $x \in M_1$, for any $\varphi \in \Psi$*

$$M, x \vDash \varphi \Leftrightarrow M_1, x \vDash \varphi.$$

The following lemma states the "maximality property" of a canonical model, cf. [6],[12].

LEMMA 3. *Let $\mathfrak{M}$ be the canonical model of a logic Λ, and assume that $\mathfrak{M}, x \vDash \varphi$. Consider the set of all those clusters in $\mathfrak{M}^x$, in which φ is satisfied:*[2]

$$\Gamma := \{C \subseteq \mathfrak{M}^x \mid \exists y \in C \ \varphi \in y\}.$$

Then the model $\mathfrak{M}|\bigcup\Gamma$ contains a maximal cluster.

3 Decrease of thickness and branching

Let $|V|$ denote the cardinality of a set V. Consider a finite frame $F = (W, R)$. For a cluster C let $next(C)$ denote the set of all successors of C, $\mathbf{b}(C) := |next(C)|$. We put:

$$\begin{array}{ll} \mathbf{h}(F) := \max\{\ |\Sigma| \ \mid \Sigma \text{ is a } <_R\text{-chain in } W/{\sim_R}\} & \textit{height,} \\ \mathbf{b}(F) := \max\{\mathbf{b}(C) \mid C \in W/{\sim_R}\} & \textit{branching,} \\ \mathbf{t}(F) := \max\{\ |C| \ \mid C \in W/{\sim_R}\} & \textit{thickness.} \end{array}$$

Note that $\mathbf{h}(F) = \mathbf{h}(F/{\sim_R})$, $\mathbf{b}(F) = \mathbf{b}(F/{\sim_R})$.
For a model M over F we put

$$\mathbf{h}(M) := \mathbf{h}(F),\ \mathbf{b}(M) := \mathbf{b}(F),\ \mathbf{t}(M) := \mathbf{t}(F).$$

Two following simple lemmas allow us to decrease the thickness and the branching of a given model.

LEMMA 4. *Assume that $M, y \vDash \varphi$, $n = |Sub(\varphi)|$. Then there exists a restriction M' of M such that $M', y \vDash \varphi$, $\mathbf{t}(M') \leq n$, and the skeletons of M, M' are isomorphic.*

---

[2]Recall that in the canonical model $\mathfrak{M}, y \vDash \varphi$ iff $\varphi \in y$.

Proof. For a cluster C let

$$\Psi(C) := \{\psi \in Sub(\varphi) \mid \exists x \in C\ M, x \vDash \psi\}.$$

Now define the set V_C as follows.

If $|C| \leq n$, then $V_C := C$; if $|C| > n$, then for every $\psi \in \Psi(C)$ we choose a point x_ψ such that $M, x \vDash \psi$ (in the particular case, when $C = \overline{y}$, we put $x_\varphi := y$), and let $V_C := \{x_\psi \mid \psi \in \Psi(C)\}$.

We put $W' := \bigcup_{C \in F/\sim_R} V_C$, $M' := M|W'$. Obviously, $\mathbf{t}(M') \leq n$ and the skeletons of M and M' are isomorphic.

It is easy to see that $M' \in \mathcal{SF}(M, Sub(\varphi))$, so $M', y \vDash \varphi$. ■

LEMMA 5. *Let M be a finite*[3] *model over a quasitree, $M, y \vDash \varphi$, $n = |Sub(\varphi)|$. Then there exists a restriction M' of M such that $M', y \vDash \varphi$, $\mathbf{b}(M') \leq n$.*

Proof. It is sufficient to consider the case $M = M^y$.

Induction on the number of clusters such that $\mathbf{b}(C) > n$.

For the induction step, suppose $\mathbf{b}(C) > n$ for some cluster C. Let

$$\Phi(C) := \{\Diamond\psi \in Sub(\varphi) | \text{ for some } \overline{z} \in next(C)\ M, z \vDash \psi \vee \Diamond\psi\}.$$

Suppose that $\Phi(C) := \{\Diamond\psi_1, \ldots, \Diamond\psi_k\}$. For every formula $\Diamond\psi_i$ we choose $\overline{x}_i \in next(C)$ such that ψ_i is satisfiable in M^{x_i}. If $\overline{z} \in next(C) - \{\overline{x}_1, \ldots, \overline{x}_k\}$, then we say that the cone M^z is *redundant* for C.

Let M_1 be the restriction of M, obtained after elimination of all cones redundant for C. One can see that $M_1 \in \mathcal{SF}(M, Sub(\varphi))$, thus $M_1, y \vDash \varphi$. By the induction hypothesis, there exists a restriction M' of M_1 such that $M', y \vDash \varphi$, $\mathbf{b}(M') \leq n$. ■

4 Completeness results for $\mathbf{L}_1$ and $\mathbf{L}_2$

The following lemmas are proved rather easily, cf. [7],[12].

LEMMA 6. *For all n*

$$\mathbf{K4} + Ad_2 \vdash \Diamond p_1 \wedge \ldots \wedge \Diamond p_n \to \Diamond(\Diamond p_1 \wedge \ldots \wedge \Diamond p_n).$$

LEMMA 7.

(i) $F \vDash Ad_2$ iff F is 2-dense, *i.e.,*

$$\forall x \forall y_1 \forall y_2 (xRy_1\ \&\ xRy_2 \to \exists z(xRz\ \&\ zRy_1\ \&\ zRy_2)).$$

[3] We assume that M is finite only for the sake of simplicity.

(ii) A finite frame F is 2-dense iff every degenerate non-maximal $C \in F/{\sim_R}$ has a unique successor D, and D is non-degenerate.

By Sahlqvist's Theorem we obtain

LEMMA 8. *The logics $\mathbf{L}_1, \mathbf{L}_2$ are canonical.*

Recall that the chronological future relation $\prec$ in Minkowski spacetime $\mathbb{R}^n$, $n \geq 2$ is defined as follows:

$$(x_1, \ldots, x_n) \prec (y_1, \ldots, y_n) \Leftrightarrow \sum_{i=1}^{n-1} (x_i - y_i)^2 < (x_n - y_n)^2 \;\&\; x_n < y_n.$$

Let us quote the main completeness results for the logics $\mathbf{L}_1$ and $\mathbf{L}_2$ [12].

THEOREM 9. $\mathbf{L}(\mathbb{R}^n, \prec) = \mathbf{L}_2$, $n \geq 2$.

THEOREM 10. *Let X be an open connected domain in $\mathbb{R}^2$ bounded by a closed smooth curve. Then $\mathbf{L}(X, \prec) = \mathbf{L}_1$.*

These logics can also be interpreted as fragments of the interval logic of the real line. Let I be the set of all open intervals on $\mathbb{R}$:

$$\mathrm{I} := \{\ ]a, b[\subseteq \mathbb{R} \mid a < b\}.$$

Consider the following relation between intervals:

$$]a_1, b_1[\ \sqsubset\ ]a_2, b_2[\ := a_2 < a_1 \& b_1 < b_2,$$

and its converse $\sqsupset$.

THEOREM 11. $\mathbf{L}(\mathrm{I}, \sqsubset) = \mathbf{L}_2$, $\mathbf{L}(\mathrm{I}, \sqsupset) = \mathbf{L}_1$.

5 Strong finite model property of $\mathbf{L}_1$, $\mathbf{L}_2$

In this section we show how to reduce the $\mathbf{L}_1$- and $\mathbf{L}_2$-satisfiability of a given formula to satisfiability in appropriate finite frames.

Let $\mathcal{F}_1$ be the class of all finite $\mathbf{L}_1$-frames,

$$\mathcal{F}_2 := \{F + C \mid F \in \mathcal{F}_1,\ C \text{ is a finite non-degenerate cluster}\},$$

and let μ_i be the class of Kripke models over frames from $\mathcal{F}_i$, $i = 1, 2$.

For 2-dense frames it is convenient to modify the function $\mathbf{h}$. Namely, for a $<_R$-chain Σ in $W/{\sim_R}$ let $\mathbf{h}_r(\Sigma)$ be the number of all non-degenerate clusters in Σ. The *r-height* of F (and of a model over F) is defined as follows:

$$\mathbf{h}_r(F) := \max\{\mathbf{h}_r(\Sigma) \mid \Sigma \text{ is a } <_R\text{-chain in } W/{\sim_R}\}$$

In [12] it was proved that the logics $\mathbf{L}_1$ and $\mathbf{L}_2$ have the FMP. This proof actually yields the following

LEMMA 12. *Consider a formula* φ, $n = |Sub(\varphi)|$.

(i) *If* φ *is* $\mathbf{L}_1$*-satisfiable, then* φ *is satisfiable in a frame* $F \in \mathcal{F}_1$ *such that* $\mathbf{h}_r(F) \leq n$.

(ii) *If* φ *is* $\mathbf{L}_2$*-satisfiable then* φ *is satisfiable in a frame* $F \in \mathcal{F}_2$ *such that* $\mathbf{h}_r(F) \leq n+1$.

Proof.

(i) Let $\mathfrak{M}$ be the canonical model of $\mathbf{L}_1$ with the accessibility relation R. For some x_0 we have $\mathfrak{M}, x_0 \models \varphi$. We will construct a model $M \subseteq \mathfrak{M}$ such that $M \in \mu_1$, $\mathbf{h}_r(M) \leq n$, $M, x_0 \models \varphi$.

Let $\Phi := Sub(\varphi) \cup \{\Diamond\top\}$. For every $x \in \mathfrak{M}$ we put:

$$\Phi_x := \{\Diamond\psi \mid \Diamond\psi \in \Phi \cap x\}, \quad \phi_x := \bigwedge_{\Diamond\psi \in \Phi_x} \Diamond\psi,$$

$$\Phi_x^{\sim} := \{\Diamond\psi \in \Phi_x \mid \exists t \sim_R x \ \ \psi \in t\}, \ \Phi_x^{\uparrow} := \Phi_x - \Phi_x^{\sim},$$

$$Y_x := \{y \mid xRy, \ \phi_x \in y\}.$$

Due to the seriality, $\phi_x \in x$, and by Lemma 6, $\Diamond\phi_x \in x$. Thus Y_x is non-empty, and by Lemma 3 Y_x contains a maximal point.

For every $x \in \mathfrak{M}$ we choose a point x', which is maximal in Y_x (we put $x' := x$ if x already is maximal in Y_x). It is easy to see that x' is reflexive. Indeed, $\Diamond\phi_x \in x'$, thus for some $y \in R(x')$ we have $y \in Y_x$. Since x' is maximal in Y_x, we have $y \in \overline{x'}$, and so $\overline{x'}$ is non-degenerate, i.e. x' is reflexive. Note that $\Phi_x = \Phi_{x'}$ and for every $z \in \mathfrak{M}$ if $\overline{x'} <_R \overline{z}$ then $|\Phi_z| < |\Phi_x|$.

Now by induction we construct a filtration $M \in \mathcal{SF}(\mathfrak{M}, \Psi)$. We also define an auxiliary set X_k at every stage k.

Stage 0. We put

$$W_0 := \{x_0, x_0'\}, \ R_0 := \{(x_0, x_0'), (x_0', x_0')\}, X_0 := \{x_0'\}.$$

Let M_0 be the submodel of $\mathfrak{M}$ over the frame (W_0, R_0).

Stage k+1. Assume that on stage k we have a model M_k over a frame (W_k, R_k) such that $M_k \subseteq \mathfrak{M}$, $M_k \in \mu_1$, $X_k \neq \varnothing$ and the following holds:

(1) if $x' \in W_k - X_k$, $\Diamond\psi \in \Phi_x$, then $\psi \in y$ for some $y \in R_k(x)$;

(2) if $x \in X_k$, then $|\Phi_x| \leq n - k$;

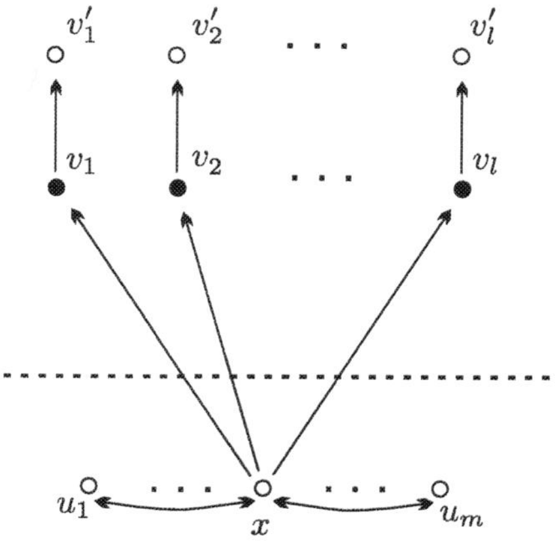

Figure 1.

(3) if $\overline{x}$ is non-degenerate in W_k and $\overline{x} <_R \overline{y}$, then $|\Psi_y| < |\Psi_x|$.

Now let us construct M_{k+1}, X_{k+1}.

For every $x \in X_k$ we define the sets of points $U_x^\sim$, $U_x^\uparrow$ and U_x' as follows. If $\Phi_x^\sim = \{\Diamond\psi_1, \ldots, \Diamond\psi_m\}$, then there exist points $u_1, \ldots, u_m$ such that $u_i \ni \psi_i$, $u_i \sim_R x$. We put $U_x^\sim := \{u_1, \ldots, u_m\}$ (note that $\Phi_x^\sim \neq \varnothing$, since $\Diamond\top \in \Phi_x$). For $\Phi_x^\uparrow = \{\Diamond\chi_1, \ldots, \Diamond\chi_l\}$ we put $U_x^\uparrow := \{v_1, \ldots, v_l\}$ and $U_x' := \{v_1', \ldots, v_l'\}$, where $v_i \ni \chi_i$, $\overline{x} <_R \overline{v_i}$ (if $\Phi_x^\uparrow = \varnothing$, we put $U_x^\uparrow := U_x' := \varnothing$), Figure 1. Let

$$R_x := \bigcup_{1 \le i \le l} \{(x, v_i), (v_i, v_i'), (v_i', v_i')\} \ \cup \ (U_x^\sim \cup \{x\})^2;$$

$$W_{k+1} := \bigcup_{x \in X_k} (U_x^\sim \cup U_x^\uparrow \cup U_x') \cup W_k, \quad X_{k+1} := \bigcup_{x \in X_k} U_x'.$$

Let R_{k+1} be the transitive closure of $R_k \cup \bigcup_{x \in X_k} R_x$.

One can see that $M_{k+1} \subseteq \mathfrak{M}$, $M_{k+1} \in \mu_1$. The property (1) holds due to the construction. The property (2) holds, since $|\Phi_y| < |\Phi_x|$ for any $x \in X_k$, $y \in U_x'$. If $\overline{x}$ is non-degenerate in M_{k+1}, then $\overline{x}$ contains a point of some X_i, $1 \le i \le k+1$, so the property (3) holds.

Due to the property (2), it follows that $X_{k+1} = \varnothing$ at some stage k. The construction terminates at this stage, and we put $M := M_{k+1}$. Due to the property (1), $M \in \mathcal{SF}(\mathfrak{M}, \Phi)$, so $M, x_0 \models \varphi$.

For every $x \in M$ $|\Psi_x| \leq n, |\Psi_x| \geq 1$, so by the property (3), we obtain $\mathbf{h}_r(M) \leq n$.

(ii) Let $\mathfrak{M}$ be the canonical model of $\mathbf{L}_2$, $\mathfrak{M}, x_0 \vDash \varphi$. As well as in (i), we construct a finite submodel M of $\mathfrak{M}$ such that $M \in \mu_1$, $M, x_0 \vDash \varphi$, and $\mathbf{h}_r(M) \leq n$.

Since $\mathfrak{M}$ is serial, $\mathfrak{M}^{x_0} = \{y | y \in \mathfrak{M}^{x_0} \& \Diamond\top \in y\}$. By Lemma 3, $\mathfrak{M}^{x_0}$ contains a maximal cluster C. By confluence and seriality, C is the non-degenerate final cluster in $\mathfrak{M}^{x_0}$.

If M contains points from C, then the frame F of M is confluent. Let C' be the copy of the final cluster of F, $F' := F + C'$. Obviously, $F' \in \mathcal{F}_2$, $\mathbf{h}_r(F) \leq n+1$. Since $F' \twoheadrightarrow F$, φ is satisfiable in F.

Assume that M does not contain points from C. Consider the set of formulas $\Psi = \{\psi \in Sub(\varphi) \mid \exists x \in C\ \psi \in x\} \cup \{\top\}$. For $\Psi = \{\psi_1, \ldots, \psi_k\}$ we put $C' := \{x_1, \ldots, x_k\}$, where $\mathfrak{M}, x_i \vDash \psi_i$, $x_i \in C$. Then the submodel $M' \subseteq \mathfrak{M}$ obtained by putting the cluster C' on the top of M is in μ_2. One can see that $M', x_0 \vDash \varphi$ and $\mathbf{h}_r(M') \leq n+1$. ∎

Let $\mathcal{G}_1$ be the class of all quasitrees from $\mathcal{F}_1$,

$$\begin{aligned}
\mathcal{G}_2 &:= \{G + C \mid G \in \mathcal{G}_1,\ C \text{ is a finite non-degenerate cluster}\},\\
\mathcal{G}_1(n) &:= \{G \in \mathcal{G}_1 \mid \mathbf{h}_r(G) \leq n,\ \mathbf{b}(G) \leq n,\ \mathbf{t}(G) \leq n\},\\
\mathcal{G}_2(n) &:= \{G \in \mathcal{G}_2 \mid \mathbf{h}_r(G) \leq n+1,\ \mathbf{b}(G) \leq n,\ \mathbf{t}(G) \leq n\}.
\end{aligned}$$

Let us formulate the following *strong finite model property* (SFMP) of $\mathbf{L}_1$ and $\mathbf{L}_2$.

LEMMA 13. *Consider a formula φ, $n = |Sub(\varphi)|$.*

(i) φ is $\mathbf{L}_1$-satisfiable $\Leftrightarrow$ φ is $\mathcal{G}_1(n)$-satisfiable.

(ii) φ is $\mathbf{L}_2$-satisfiable $\Leftrightarrow$ φ is $\mathcal{G}_2(n)$-satisfiable.

Proof.

(i) Suppose φ is $\mathbf{L}_1$-satisfiable. By Lemma 12, φ is satisfiable in some frame $F \in \mathcal{F}_1$ such that $\mathbf{h}_r(F) \leq n$. Obviously, we can assume that F has the initial cluster (in which φ is satisfied). By standard unravelling argument, F is a p-morphic image of some quasitree $F' \in \mathcal{G}_1$, and $\mathbf{h}_r(F) = \mathbf{h}_r(F')$ (see [12] for more details). By the p-morphism lemma φ is satisfiable in F', and by Lemma 4, φ is satisfiable in some $F'' \in \mathcal{G}_1$ such that $\mathbf{t}(F'') \leq n$, $\mathbf{h}_r(F'') \leq n$.

To decrease the branching, we proceed in the same way as in Lemma 5. Note that the transformation described in Lemma 5 preserves 2-density:

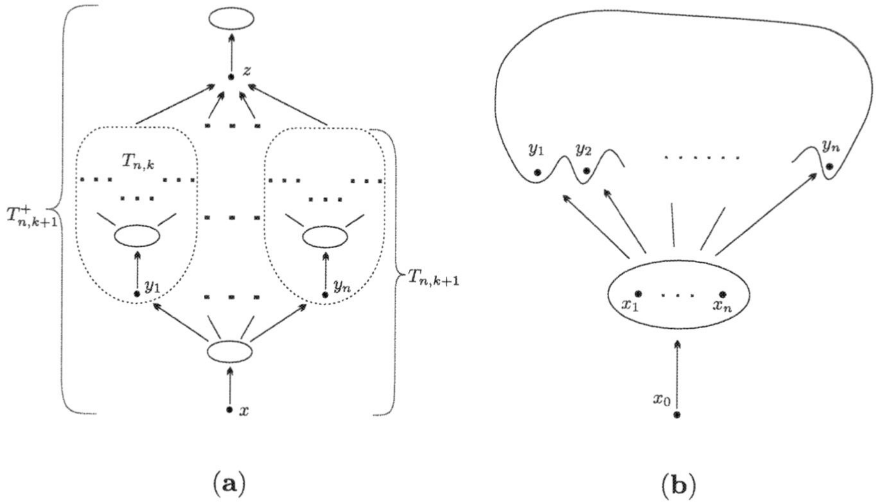

Figure 2.

every degenerate cluster still has at most one non-degenerate successor. The following slight modification allows us to preserve seriality: if for some cluster C we have $\mathbf{b}(C) > n$ and $\Phi(C) = \varnothing$ (in notation of Lemma 5), then we put $\Phi(C) := \{\Diamond\top\}$.

By applying this transformation to F'' we obtain $\mathbf{L}_1$-quasitree G such that $\mathbf{b}(G) \leq n$ and φ is satisfiable in G. Obviously, $\mathbf{h}_r(G) \leq \mathbf{h}_r(F'')$, $\mathbf{t}(G) \leq \mathbf{t}(F'')$, thus $G \in \mathcal{G}_1(n)$.

(ii) Suppose φ is $\mathbf{L}_2$-satisfiable. By Lemma 12, φ is satisfiable in some frame $F^+ \in \mathcal{F}_2$ such that $\mathbf{h}_r(F^+) \leq n+1$, i.e. $F^+ = F + C$, where $F \in \mathcal{G}_1(n)$, C is a non-degenerate cluster. By Lemma 4, we can assume that $|C| \leq n$.

As well as in (i), we modify F into $G \in \mathcal{G}_1(n)$. Then $G + C \in \mathcal{G}_2(n)$. It is not difficult to check that φ is satisfiable in $G + C$. ∎

Actually, this lemma is sufficient to show that the logics $\mathbf{L}_1$ and $\mathbf{L}_2$ are PSPACE-decidable. Indeed, it is possible to describe an algorithm checking satisfiability in all frames from $\mathcal{G}_1(n)$ (or from $\mathcal{G}_2(n)$) within the space polynomial of n. However, to simplify the algorithm, we can only check a single frame, as explained below.

Let $\mathcal{T}_{n,1}$ be the class of all frames isomorphic to $C_0 + C_n$. We put

$$\mathcal{T}_{n,k+1} := \{F + (G_1 \sqcup \ldots \sqcup G_n) \mid F \in \mathcal{T}_{n,1},\ G_1, \ldots, G_n \in \mathcal{T}_{n,k}\},$$

$$\mathcal{T}_{n,k}^{+} := \{F + F' \mid F \in \mathcal{T}_{n,k},\ F' \in \mathcal{T}_{n,1}\}.$$

Let $T_{n,k} \in \mathcal{T}_{n,k}$, $T_{n,k}^{+} \in \mathcal{T}_{n,k}^{+}$ (Figure 2a).

LEMMA 14. *Consider a formula φ, $n = |Sub(\varphi)|$.*

(i) φ is $\mathbf{L}_1$-satisfiable $\Leftrightarrow$ φ is satisfiable at the root of $T_{n,n}$.

(ii) φ is $\mathbf{L}_2$-satisfiable $\Leftrightarrow$ φ is satisfiable at the root of $T_{n,n}^{+}$.

Proof.

(i) ($\Rightarrow$). By induction on r-height it is easy to check that $T_{n,n} \twoheadrightarrow G$ for every $G \in \mathcal{G}_1(n)$. The statement follows from Lemma 13 and the p-morphism lemma.
($\Leftarrow$). Note that $T_{n,n}$ is $\mathbf{L}_1$-frame.

(ii) Similar to (i). ■

6 PSPACE-completeness for $\mathbf{L}_1$ and $\mathbf{L}_2$

In this section we prove the PSPACE-completeness for the logics $\mathbf{L}_1$ and $\mathbf{L}_2$.

First we show that $\mathbf{L}_1, \mathbf{L}_2 \in$ PSPACE. By Lemma 14, it is sufficient to describe the algorithms deciding whether a given formula is satisfiable at the roots of $T_{n,n}$ and $T_{n,n}^{+}$, using space polynomial of n.

Consider a formula φ and assume that $Sub(\varphi) = \{\psi_1, \ldots, \psi_n\}$, $PV(\varphi)=\{p_1, \ldots, p_m\}$. Let us order $Sub(\varphi)$ as follows: for $i \leq m$ we put $\psi_i := p_i$, and if ψ_i is a subformula of ψ_j then $i \leq j$. We can achieve this within $O(n \log n)$ units of space by making an array of pointers. Note that $\psi_n = \varphi$.

Consider a boolean vector $\mathbf{v} = (\mathrm{v}_1, \ldots, \mathrm{v}_n) \in \{0,1\}^n$. We put $\varphi^{\mathbf{v}} := \bigwedge\limits_i \psi_i^{\mathrm{v}_i}$, where $\psi^0 := \neg\psi$, $\psi^1 := \psi$.

DEFINITION 15. A boolean vector $\mathbf{v} \in \{0,1\}^n$ is called *φ-consistent* in a frame F at a point x (notation: $F, x \Vdash_{\varphi} \mathbf{v}$) if for some valuation θ we have $F, \theta, x \models \varphi^{\mathbf{v}}$. Boolean vectors $\mathbf{v}^1, \ldots, \mathbf{v}^l \in \{0,1\}^n$ are called *simultaneously φ-consistent* in F on a tuple $\mathbf{y} = (y_1, \ldots, y_l) \in W^l$ (notation: $F, \mathbf{y} \Vdash_{\varphi} (\mathbf{v}^1, \ldots, \mathbf{v}^l)$) if for some valuation θ for all $i = 1 \ldots l$ we have $F, \theta, y_i \models \varphi^{\mathbf{v}^i}$.

If x is the root of F and y is the root of G, then $F \Vdash_{\varphi} \mathbf{v}$ abbreviates $F, x \Vdash_{\varphi} \mathbf{v}$ and $F + G \Vdash_{\varphi} (\mathbf{v}, \mathbf{u})$ abbreviates $F + G, (x, y) \Vdash_{\varphi} (\mathbf{v}, \mathbf{u})$.

Let us reformulate Lemma 14.

LEMMA 16.

(i) *φ is $\mathbf{L}_1$-satisfiable $\Leftrightarrow$ there exists $\mathbf{v} \in \{0,1\}^n$ such that $T_{n,n} \Vdash_\varphi \mathbf{v}$ and $\mathrm{v}_n = 1$.*

(ii) *φ is $\mathbf{L}_2$-satisfiable $\Leftrightarrow$ there exist $\mathbf{v}, \mathbf{u} \in \{0,1\}^n$ such that $T^+_{n,n} \Vdash_\varphi (\mathbf{v}, \mathbf{u})$ and $\mathrm{v}_n = 1$.*

Consider the frame $F = T_{n,1} + G$, where G has exactly n minimal clusters $\overline{y}_1, \dots, \overline{y}_n$, and let x_0 be the root of $T_{n,1}$ (Figure 2b). The truth value of a formula φ at x_0 in a model over F is fully determined by the truth values of its variables in $T_{n,1}$ and the truth values of its subformulas at $y_1, \dots, y_n$. In Appendix we describe the algorithm *SatLoc* working within a space polynomial of n, deciding whether $F, (x, y_1, \dots, y_n) \Vdash_\varphi (\mathbf{v}, \mathbf{v}^1, \dots, \mathbf{v}^n)$, provided $F, (y_1, \dots, y_n) \Vdash_\varphi (\mathbf{v}^1, \dots, \mathbf{v}^n)$.

LEMMA 17. *Let $\mathbf{v}, \mathbf{u} \in \{0,1\}^n$.*

(i) *$T_{n,k+1} \Vdash_\varphi \mathbf{v} \Leftrightarrow$ there exist $\mathbf{v}^1, \dots, \mathbf{v}^n \in \{0,1\}^n$ such that $T_{n,k} \Vdash_\varphi \mathbf{v}^i$, $i = 1 \dots n$, and $SatLoc(\varphi, \mathbf{v}, \mathbf{v}^1, \dots, \mathbf{v}^n)$ =true.*

(ii) *$T^+_{n,k+1} \Vdash_\varphi (\mathbf{v}, \mathbf{u}) \Leftrightarrow$ there exist $\mathbf{v}^1, \dots, \mathbf{v}^n \in \{0,1\}^n$ such that $T^+_{n,k} \Vdash_\varphi (\mathbf{v}^i, \mathbf{u})$, $i = 1 \dots n$, and $SatLoc(\varphi, \mathbf{v}, \mathbf{v}^1, \dots, \mathbf{v}^n)$ =true.*

Proof. By definition,

$$T_{n,k+1} = F + (G_1 \sqcup \ldots \sqcup G_n), \; T^+_{n,k+1} = T_{n,k+1} + F',$$

where $F, F' \in \mathcal{T}_{n,1}$, $G_1, \dots, G_n \in \mathcal{T}_{n,k}$. Let $x, y_1, \dots, y_n, z$ be the roots of $F, G_1, \dots, G_n, F'$ respectively, $\mathbf{x} := (x, y_1, \dots, y_n)$, $\mathbf{y} := (y_1, \dots, y_n)$ (Figure 2a).

(i) ($\Rightarrow$). For some θ we have: $T_{n,k+1}, \theta, x \vDash \varphi^{\mathbf{v}}$. Thus $T_{n,k+1}, \theta, y_i \vDash \varphi^{\mathbf{v}^i}$ for some vectors $\mathbf{v}^1, \dots, \mathbf{v}^n \in \{0,1\}^n$. Obviously, $T_{n,k} \Vdash_\varphi \mathbf{v}^i$ for all $i = 1 \dots n$. Since $T_{n,k+1}, \mathbf{y} \Vdash_\varphi (\mathbf{v}^1, \dots, \mathbf{v}^n)$ and $T_{n,k+1}, \mathbf{x} \Vdash_\varphi (\mathbf{v}, \mathbf{v}^1, \dots, \mathbf{v}^n)$, we have $SatLoc(\varphi, \mathbf{v}, \mathbf{v}^1, \dots, \mathbf{v}^n)$=true.
($\Leftarrow$). It is not difficult to see that $T_{n,k+1}, \mathbf{y} \Vdash_\varphi (\mathbf{v}^1, \dots, \mathbf{v}^n)$.
Since $SatLoc(\varphi, \mathbf{v}, \mathbf{v}^1, \dots, \mathbf{v}^n)$=true, we have $T_{n,k+1}, \mathbf{x} \Vdash_\varphi (\mathbf{v}, \mathbf{v}^1, \dots, \mathbf{v}^n)$, so $T_{n,k+1} \Vdash_\varphi \mathbf{v}$.

(ii) ($\Rightarrow$). Similar to (i).
($\Leftarrow$). For some valuations $\theta_1, \dots, \theta_n$ we have:

$$T^+_{n,k+1}, \theta_i, y_i \vDash \varphi^{\mathbf{v}^i}, \; T^+_{n,k+1}, \theta_i, z \vDash \varphi^{\mathbf{u}}.$$

We define a valuation θ as follows:

$$\theta(p) := \bigcup_i \{y \in G_i \mid y \in \theta_i(p)\} \cup \{y \in F' \mid y \in \theta_-(p)\}.$$

A straightforward argument shows that $T_{n,k+1}, \theta, y_i \vDash \varphi^{\mathbf{v}^i}$ for all $i = 1 \dots n$, so $T_{n,k+1}, \mathbf{y} \Vdash_\varphi (\mathbf{v}^1, \dots, \mathbf{v}^n)$. $SatLoc(\varphi, \mathbf{v}, \mathbf{v}^1, \dots, \mathbf{v}^n)=\mathsf{true}$ implies $T^+_{n,k+1}, \mathbf{x} \Vdash_\varphi (\mathbf{v}, \mathbf{v}^1, \dots, \mathbf{v}^n)$, i.e. for some valuation η we have

$$T^+_{n,k+1}, \eta, x \vDash \varphi^{\mathbf{v}},\ T^+_{n,k+1}, \eta, y_i \vDash \varphi^{\mathbf{v}^i} \text{ for all } i = 1 \dots n.$$

We put

$$\eta'(p) := \{y \in F \mid x \in \eta(p)\} \cup \{x \notin F \mid x \in \theta(p)\}.$$

One can check that $T^+_{n,k+1}, \eta', x \vDash \varphi^{\mathbf{v}}$ and $T^+_{n,k+1}, \eta', z \vDash \varphi^{\mathbf{u}}$, that is $T_{n,k+1} \Vdash_\varphi (\mathbf{v}, \mathbf{u})$. ■

Now let us give a recursive description of the algorithms $SatTree$ and $SatTree^+$ determining whether $T_{n,k} \Vdash_\varphi \mathbf{v}$ and $T^+_{n,k} \Vdash_\varphi (\mathbf{v}, \mathbf{u})$ (for the basic case $k = 1$ these algorithms - SAT_1 and SAT_1^+ are constructed in Appendix).

```
Function SatTree(φ, v, k) returns boolean
Begin
  if k = 1 then return(SAT_1(φ, v));
  for all v^1, ..., v^n ∈ {0,1}^n:
    if ⋀_{1≤i≤n} SatTree(φ, v^i, k − 1) ⋀ SatLoc(φ, v, v^1, ..., v^n)
      then return(true);
  return(false);
End.
```

```
Function SatTree^+(φ, v, u, k) returns boolean
Begin
  if k = 1 then return(SAT_1^+(φ, v, u));
  for all v^1, ..., v^n ∈ {0,1}^n:
    if ⋀_{1≤i≤n} SatTree^+(φ, v^i, u, k − 1) ⋀ SatLoc(φ, v, v^1, ..., v^n)
      then return(true);
  return(false);
End.
```

By Lemma 17, we obtain

LEMMA 18.

(i) $T_{n,k} \Vdash_\varphi \mathbf{v} \Leftrightarrow SatTree(\varphi, \mathbf{v}, k) =$ true.

(ii) $T^+_{n,k} \Vdash_\varphi (\mathbf{v}, \mathbf{u}) \Leftrightarrow SatTree^+(\varphi, \mathbf{v}, \mathbf{u}, k) =$ true.

```
Function SATL1(φ) returns boolean
Begin
  for all v ∈ {0,1}^n, v_n = 1:
    if SatTree(φ, v, n) then return(true);
  return(false);
End.
```

```
Function SATL2(φ) returns boolean
Begin
  for all v, u ∈ {0,1}^n, v_n = 1:
    if SatTree+(φ, v, u, n) then return(true);
  return(false);
End.
```

By Lemma 16 and Lemma 18, we obtain

THEOREM 19. *φ is $\mathbf{L}_i$-satisfiable* $\Leftrightarrow$ $SAT\mathbf{L}_i(\varphi)$=true, $i = 1, 2$.

One can see that the space used on each level of recursion is $O(n^2)$. The depth of recursion is n, and the total amount of space required is $O(n^3)$.

Now let us show that the satisfiability problems for $\mathbf{L}_1$ and $\mathbf{L}_2$ are PSPACE-hard.

Since satisfiability problem for all logics between **K** and **S4** is PSPACE-hard (Ladner's Theorem [11]), we obtain that $\mathbf{L}_1$ is PSPACE-hard. Since **S4** $\nvdash A2$, the logic $\mathbf{L}_2 \not\subseteq$ **S4**. However, the following slight modification of Ladner's construction [11] proves the PSPACE-hardness for all logics between **K4** and **S4.1.2**.

Let A be a propositional logic formula, $PV(A) = \{p_1, \ldots, p_n\}$, and $B = Q_1 p_1 \ldots Q_n p_n A$, where $Q_1, \ldots, Q_n \in \{\forall, \exists\}$. We put

$$\phi_1(B) := \bigwedge_{0 \leq i \leq n} (q_i \to \Diamond(\neg q_i \wedge q_{i+1})),$$

$$\phi_2(B) := \bigwedge_{\{i \mid Q_i = \forall\}} (q_{i-1} \to \Diamond(q_i \wedge p_i) \wedge \Diamond(q_i \wedge \neg p_i)),$$

$$\phi_3(B) := \bigwedge_{1 \leq i \leq n} ((q_i \wedge p_i \to \Box(q_n \to p_i)) \wedge (q_i \wedge \neg p_i \to \Box(q_n \to \neg p_i))),$$

$$\varphi(B) := q_0 \wedge \Box(q_n \to A) \wedge \Box(\phi_1(B) \wedge \phi_2(B) \wedge \phi_3(B)).$$

A straightforward argument shows that

B is valid $\Rightarrow \varphi(B)$ is **S4.1.2**-satisfiable;

$\varphi(B)$ is **K4**-satisfiable $\Rightarrow B$ is valid.

Since the validity problem for prenex quantified boolean formulas is PSPACE-complete [16], we obtain

THEOREM 20. *If* $\mathbf{K4} \subseteq \mathbf{L} \subseteq \mathbf{S4.1.2}$, *then the satisfiability problem for* $\mathbf{L}$ *is PSPACE-hard.*[4]

Note that $\mathbf{K4} \subset \mathbf{L}_2 \subset \mathbf{S4.1.2}$. So by Theorems 19, 20, we obtain

THEOREM 21. $\mathbf{L}_1$, $\mathbf{L}_2$ *are PSPACE-complete.*

7 Examples

In this section we illustrate our method with some examples. We consider the logics $\mathbf{K4}, \mathbf{K4d}, \mathbf{S4}$ and their extensions by confluence and McKinsey axiom. These logics are known to be in PSPACE[5] (cf. [11],[3],[9]). Our method yields an alternative proof of this fact.

Consider a logic $\Lambda = \mathbf{L}(\mathcal{F}^\Lambda)$. To prove the PSPACE-decidability of Λ it is sufficient to show that for any formula φ there exists a class $\mathcal{F}^\Lambda_\varphi \subseteq \mathcal{F}^\Lambda$ such that:

- φ is Λ-satisfiable $\Rightarrow \varphi$ is $\mathcal{F}^\Lambda_\varphi$-satisfiable;
- It is possible to decide whether φ is $\mathcal{F}^\Lambda_\varphi$-satisfiable within the space polynomial of $|Sub(\varphi)|$.

In many cases, it is possible to present $\mathcal{F}^\Lambda_\varphi$ as a finite class of quasitrees (or quasitrees with some additional clusters on the top) with appropriate restriction of height, branching and thickness. Actually, the main problem is how to restrict the height of frames in $\mathcal{F}^\Lambda_\varphi$.

Let us reformulate Lemma 3:

LEMMA 22. *Let* $\mathfrak{M}$ *be the canonical model of a logic* $\Lambda \supseteq \mathbf{K4}$, *and assume that a formula* $\Diamond\psi$ *is satisfied at some* $x \in \mathfrak{M}$. *Let*

$$Y := \{y \mid xRy \ \&\ \mathfrak{M}, y \vDash \Diamond\psi\} \cup \{x\}.$$

Then the model $\mathfrak{M}|Y$ *contains a maximal cluster.*

[4] This statement actually holds for all logics between **K** and **S4.1.2**. The proof is by an easy modification of φ.

[5] Moreover, they are PSPACE-complete.

Using this lemma, it is not difficult to check that for every logic $\Lambda \supseteq \mathbf{K4}$ and formula φ it is possible to extract a selective filtration M from the canonical model of Λ through $Sub(\varphi)$ such that $\mathbf{h}(M) = O(|Sub(\varphi)|)$.

However, it is necessary to show that we obtain a Λ-frame. For example, consider $\Lambda \supseteq \mathbf{K4d}$. It is not difficult to see that in this case every maximal cluster in Y is non-degenerate. It allows us to obtain a dense selective filtration, which implies the result for **K4d**. For the logics **K4d.1**, **K4d.2**, **K4d.1.2** we modify the construction as we did in Lemma 12 (ii) for the logic $\mathbf{L}_2$, i.e. we extract an additional final cluster (or clusters) from the canonical model.

Let $\mathcal{C}(n) := \{C \mid C$ is a non-degenerate cluster, $|C| \leq n\}$. For a frame G let G^{MK} denote the frame obtained by putting a reflexive singleton above each maximal cluster in G.

For $\Lambda = \mathbf{K4}, \mathbf{K4d}$ we put:

$$\mathcal{G}^{\Lambda}(n) := \{G \mid G \text{ is } \Lambda\text{-quasitree}, \ \mathbf{h}(G) \leq 2n, \ \mathbf{b}(G) \leq n, \ \mathbf{t}(G) \leq n\};$$
$$\mathcal{G}^{\Lambda.1}(n) := \{G^{MK} \mid G \in \mathcal{G}^{\Lambda}(n)\};$$
$$\mathcal{G}^{\Lambda.2}(n) := \{G + C \mid G \in \mathcal{G}^{\Lambda}(n), \ C \in \mathcal{C}(n)\} \cup \{C_0\};$$
$$\mathcal{G}^{\Lambda.1.2}(n) := \{G + C \mid G \in \mathcal{G}^{\Lambda}(n), \ C \text{ is a reflexive singleton}\};$$

Similar to Lemma 13, one can check:

LEMMA 23. *Consider the logic*
$\Lambda \in \{\mathbf{K4}, \ \mathbf{K4.1}, \ \mathbf{K4.2}, \ \mathbf{K4.1.2}, \ \mathbf{K4d}, \ \mathbf{K4d.1}, \ \mathbf{K4d.2}, \ \mathbf{K4d.1.2}\}$.
For a formula φ *such that* $|Sub(\varphi)| = n$ *we have:*
φ *is* Λ*-satisfiable* $\Leftrightarrow$ φ *is* $\mathcal{G}^{\Lambda}(n)$*-satisfiable.*

Note that this lemma is sufficient for establishing the PSPACE-decidability of these logics, so by Theorem 20 they are PSPACE-complete.

Sometimes, to check the satisfiability, one can use a single frame (as it was in the case of $\mathbf{L}_1$, $\mathbf{L}_2$). For example, consider the logic **S4**. It is easy to modify the construction in Lemma 12 (i) to obtain an appropriate model for **S4**: in the case of 2-dense logics degenerate clusters arise, and in the reflexive case these clusters are reflexive singletons. To obtain the frame with McKinsey property (or confluence, or both), we proceed as in Lemma 12 (ii). Similar to Lemmas 13,14, we obtain the following construction.

Let $\mathcal{T}^{S4}_{n,1}$ be the class of all frames isomorphic to $C_1 + C_n$. We put

$$\mathcal{T}^{S4}_{n,k+1} := \{F + (G_1 \sqcup \ldots \sqcup G_n) \mid F \in \mathcal{T}^{S4}_{n,1}, \ G_1, \ldots, G_n \in \mathcal{T}^{S4}_{n,k}\},$$

$$\mathcal{T}^{S4.1}_{n,k} := \{F^{Mc} \mid F \in \mathcal{T}^{S4}_{n,k}\},$$

$$\mathcal{T}^{S4.2}_{n,k} := \{F + F' \mid F \in \mathcal{T}^{S4}_{n,k},\ F' \in \mathcal{T}^{S4}_{n,1}\},$$

$$\mathcal{T}^{S4.1.2}_{n,k} := \{F + C \mid F \in \mathcal{T}^{S4}_{n,k},\ C \text{ is a reflexive singleton}\}.$$

Let $T^{\Lambda}_{n,k} \in \mathbf{T}^{\Lambda}_{n,k}$.

LEMMA 24. *Consider the logic* $\Lambda \in \{\mathbf{S4},\ \mathbf{S4.1},\ \mathbf{S4.2},\ \mathbf{S4.1.2}\}$.
For a formula φ *such that* $|Sub(\varphi)| = n$ *we have:*
φ *is* Λ*-satisfiable* $\Leftrightarrow$ φ *is satisfiable at the root of* $T^{\Lambda}_{n,n}$.

A slight modification of the algorithms $SatLoc$, $SatTree$, $SatTree^{+}$ allows us to check the Λ-satisfiability in $O(|Sub(\varphi)|^3)$ amount of space, so we obtain that $\mathbf{S4}, \mathbf{S4.1}, \mathbf{S4.2}, \mathbf{S4.1.2}$ are PSPACE-complete.

In this paper we consider only transitive logics. However, sometimes our method can be used in the non-transitive case. For example, consider the logic of weak transitivity

$$\mathbf{K4^0} := \mathbf{K} + \Diamond\Diamond p \to \Diamond p \vee p,$$

axiomatizing derivation in arbitrary topological spaces [5]. The canonical model of $\mathbf{K4^0}$ has the maximality property, similarly to the transitive canonical model. This allows us to obtain a finite weakly transitive selective filtration of $\mathbf{K4^0}$ satisfying a given formula. Moreover, basing on the ideas of this paper, we showed that $\mathbf{K4^0}$ is in PSPACE, the proof will be published in the sequel.

8 Appendix

Function $SatLoc(\varphi, \mathbf{v}, \mathbf{v}^1, \ldots, \mathbf{v}^n)$ returns boolean
Begin
REM{ *For* $\theta \in \{0,1\}^{(n+1)\times m}$ *we construct* $\eta \in \{0,1\}^{(n+1)\times n}$, where
θ^i_j *is the trues value of* p_j *at* x_i,
η^i_j *is the trues value of* ψ_j *at* x_i, *Figure 2b.*}

```
  for all θ ∈ {0,1}^{(n+1)×m}:
  begin
    for j := 1...n:                                   REM{ ψ_j ∈ Sub(φ)}
    begin
      for i := 0...n:                                 REM{ x_i ∈ C_0 + C_n}
      begin
        η^i_j := 0;
        if j ≤ m then η^i_j := θ^i_j;                 REM{ ψ_j is a variable}
        if ψ_j = ψ_s → ψ_l then                       REM{ note that s,l < j}
          if η^i_s = 0 or η^i_l = 1 then η^i_j := 1;
        if ψ_j = ◇ψ_s then                            REM{ note that s < j}
```

begin
for $l := 1 \ldots n$: if $\eta_s^l = 1$ then $\eta_j^i := 1$;
for $l := 1 \ldots n$: if $\mathrm{v}_s^l = 1$ or $\mathrm{v}_j^l = 1$ then $\eta_j^i := 1$;
end;
end;
end;
if $(\eta_1^0, \ldots, \eta_n^0) = \mathbf{v}$ then return(true);
end;
return(false);
End.

Function $SAT_1(\varphi, \mathbf{v})$ returns boolean
Begin return($SatLoc(\varphi, \mathbf{v}, \bar{0}, \ldots, \bar{0})$); End.

Function $SAT_1^+(\varphi, \mathbf{v}, \mathbf{u})$ returns boolean
Begin return($SatLoc(\varphi, \mathbf{v}, \mathbf{u}, \bar{0}, \ldots, \bar{0}) \bigwedge SAT_1(\varphi, \mathbf{u})$); End.

9 Acknowledgements

The author is grateful to prof. Valentin Shehtman for his help and also to the anonymous referees for their useful comments.

The work on this paper was supported by Poncelet Laboratory (UMI 2615 of CNRS and Independent University of Moscow), RFBR (project No.02-01-22003), and by CNRS (ECO-NET 2004, project No. 08111TL).

BIBLIOGRAPHY

[1] E. W. Beth. The foundation of mathemathics. North-Holland, Amsterdam, 1959.
[2] P. Blackburn, M. de Rijke and Y. Venema. Modal logic. Cambridge University Press, 2001
[3] L. Farinas del Cerro and O. Gasquet. A general framework for pattern-driven modal tableaux. Logic Journal of the IGPL, 10(1):51-83, 2002.
[4] A. Chagrov, M. Zakharyaschev. Modal logic. Oxford University Press, 1997.
[5] L. Esakia. Weak transitivity - a restitution (in Russian). In: Logical investigations, v.8, p.244-245. Moscow, Nauka, 2001.
[6] K. Fine. Logics containing K4, part II. Journal of Symbolic Logic, 50: 619-651, 1985.
[7] R. Goldblatt. Diodorean modality in Minkowski spacetime. Studia Logica, v. 39 (1980), 219-236.
[8] J. Halpern and Y.Moses. A guide to completeness and complexity for modal logics of knowledge and belief. Artificial Intelligence, 54:319-379, 1992.
[9] Marcus Kracht. Notes on the space requirements for checking satisfiability in modal Logics. Advances in Modal Logic, Volume 4, 243-264. King's College Publications, 2003.
[10] S. Kripke. A semantical analysis of modal logic I: Normal modal propositional calculi. Z. Math. Logik Grundl. Math., 9:67-96, 1963.
[11] R. Ladner. The computational complexity of provability in systems of modal propositional logic. SIAM Journal of Computing, 6:467-480, 1977.

[12] I. Shapirovsky, V. Shehtman. Chronological future modality in Minkowski spacetime. Advances in Modal Logic, Volume 4, 437-459. King's College Publications, 2003.
[13] V.B. Shehtman. Modal logics of domains on the real plane. Studia Logica, v. 42 (1983), 63-80.
[14] E. Spaan. Complexity of modal logics. PhD thesis, University of Amsterdam, 1993.
[15] E. Spaan. The complexity of propositional tense logics. In de Rijke, M. (Ed.), Diamonds and Defaults, pp. 287-307. Kluwer Academic Publishers, Dordrecht, 1993.
[16] L. J. Stockmeyer and A. R. Meyer. Word problems requiring exponential time: preliminary report. In Proc. 5th ACM Symp. on Theory of Computing, pp.1-9, 1973.

Ilya Shapirovsky
Institute for Information Transmission Problems
Russian Academy of Sciences,
B.Karetny 19, Moscow, Russia, 101447
E-mail: ilshapir@netscape.net

Filtration via Bisimulation

VALENTIN SHEHTMAN

ABSTRACT. We develop a new version of the well-known filtration method in modal logic, allowing us to construct large countermodels and to solve some open problems on the finite model property for products of modal logics. This filtration is based on the bisimilarity relation between parts of the original model; it generalizes earlier versions of the filtration method introduced by E. Lemmon, K. Segerberg, D. Gabbay, and the author.

1 Introduction

The filtration method is the oldest and the most well-known method of finite model property proofs in modal logic. First developed in the 1960s by S. Kripke, E. Lemmon, K. Segerberg, and D. Gabbay, it was afterwards modified and successfully applied to different types of nonclassical logics.

However, in the field of many-dimensional modal logic, the traditional filtration method is not very popular in decidability proofs. Indeed, decidable many-dimensional logics may be of high complexity, but standard filtrations yield rather moderate upper complexity bounds. Other methods seem to be more successful here, like the method of quasimodels (or mosaics), which is widely used in the recent monograph [4].

Still, traditional filtrations seem so simple, that it is worth making them work for complex logics as well. In this paper we propose such a modification allowing us to simplify some earlier proofs and also to prove new fmp results — for example, for the logic $\mathbf{K} \times \mathbf{S4}$ and related ones. So we hope for further applications of this method.

Let us briefly describe the main idea. A filtration constructs a finite Kripke model, which is in some sense equivalent to a given infinite model. There are two main types of filtrations: selective filtrations (Kripke – Gabbay) and "epi-filtrations" (Lemmon – Segerberg). Selective filtrations extracting finite submodels, are not discussed here, details can be found in [3], for a more recent application cf. [10].

An epi-filtration identifies possible worlds by some "faithful" equivalence relation. In the simplest case [8], given a set of formulas Ψ, the equivalence $\equiv_\Psi$ is defined as the truth of the same formulas from Ψ. If Ψ is finite, we readily obtain a finite model, but the validity of the original logic may be lost. To avoid this, we can try to refine the equivalence relation.

The first modification of this kind was proposed in [2]: the new relation is $\equiv_\Phi$ for some $\Phi \supseteq \Psi$. Our modification is based on a more general construction from [9], [5]. We start from a finite set Ψ and a Kripke model $M = (W, R_1, ..., R_n, \theta)$. The worlds of M are identified by equivalence relation $\approx \subseteq \equiv_\Psi$ with finitely many equivalence classes. An appropriate choice of $\approx$ allows us to preserve the original logic in the filtrated frame.

In [5] this method is applied to the logic $\mathbf{S5} \times \mathbf{K}$. In this case every $R_1(x)$ is a cluster; we put $x \approx y$ iff $x \equiv_\Psi y$ and the same $\equiv_\Psi$-classes are presented in $R_1(x)$ and $R_1(y)$. But such a definition is no good if R_1 is not an equivalence (e.g. for the logic $\mathbf{K} \times \mathbf{K}$), because $x \approx y$ should somehow take the original R_1 into account. So we require that the corresponding generated R_1-submodels are *bisimilar*. To obtain finitely many $\approx$-classes, one should first restrict the depth of all these submodels. This step is crucial for the whole proof and it is not always possible. In fact, the method works quite well for intransitive R_1, but it may fail when all relations in M are transitive. For example, this happens for the logic $\mathbf{K4} \times \mathbf{K4}$, which lacks the fmp at all [7].

2 Basic definitions and facts

First let us recall some well-known material. Our terminology and notation mainly follow [5]. $\mathcal{L}_n$ denotes the set of propositional formulas built from a countable set $PL = \{p_1, p_2, \dots\}$ of proposition letters, classical connectives $\rightarrow, \bot$, and modal connectives $\Box_1, \dots, \Box_n$. Let $\mathcal{L}_n\lceil k$ be the set of all formulas in $\mathcal{L}_n$ using only proposition letters from the set $PL\lceil k = \{p_1, p_2, \ldots p_k\}$. *Closed formulas* do not contain proposition letters.

As usual, an *n-modal logic* is a set of $\mathcal{L}_n$-formulas containing the classical tautologies, the axiom $\Box(p_1 \to p_2) \to (\Box p_1 \to \Box p_2)$, closed under Substitution, Modus Ponens, and Necessitation $(A/\Box_i A)$, $1 \leq i \leq n$. For a set of $\mathcal{L}_n$-formulas Γ and an n-modal logic $\mathbf{\Lambda}$, the smallest n-modal logic containing

$(\mathbf{\Lambda} \cup \Gamma)$ is denoted by $\mathbf{\Lambda} + \Gamma$. $\mathbf{K}_n$ denotes the minimal n-modal logic, and

$$\mathbf{K}_{\pm n} := \mathbf{K}_{2n} + \{\Diamond_i \Box_{n+i} p \to p,\ \Diamond_{n+i} \Box_i p \to p \mid 1 \leq i \leq n\}$$

is the minimal n-temporal logic[1]; in the latter case $\Box_{n+i}$ is denoted by $\Box_i^{-1}$; similarly for $\Diamond$.

Recall that the *fusion* of two logics, n-modal $\mathbf{L}_1$ and m-modal $\mathbf{L}_2$ is $\mathbf{L}_1 * \mathbf{L}_2 := \mathbf{K}_{n+m} + \mathbf{L}_1 + \mathbf{L}_2^{+n}$, where $\mathbf{L}_2^{+n}$ is obtained from $\mathbf{L}_2$ by replacing every occurrence of any $\Box_j$ with $\Box_{j+n}$.

Kripke semantics is defined in a standard way. An *n-modal (Kripke) frame* is a tuple $F = (W, R_1, \ldots, R_n)$, where $W \neq \varnothing$ is a set of possible worlds, $R_i \subseteq W \times W$ are accessibility relations. A *Kripke model* over F is a pair $M = (F, \theta)$, where $\theta : PL \longrightarrow 2^W$ is a *valuation*.

Valuations are extended to all formulas as usual:

$\theta(\bot) = \varnothing$, $\theta(A \to B) = (W - \theta(A)) \cup \theta(B)$,

$\theta(\Box_i A) = \{x \mid R_i(x) \subseteq \theta(A)\}$.

Similarly, a *k-restricted Kripke model* is $M = (F, \theta)$, where $\theta : PL\lceil k \longrightarrow 2^W$; in this case θ is extended to $\mathcal{L}_n\lceil k$. A formula A is called *true at a world* w of M if $w \in \theta(A)$ (in another notation: $M, w \vDash A$). Since the truth value of a formula depends only on proposition letters occurring in it, we may fix k and assume that all Kripke models are k-restricted.

A formula A is *valid in a frame* F (notation: $F \vDash A$) if it is true at every world of every Kripke model over F. A set of formulas Γ is *valid* in F (notation: $F \vDash \Gamma$) if every $A \in \Gamma$ is valid. In the latter case we also say that F is a Γ-*frame*. A logic $\mathbf{\Lambda}$ is *determined* by a class of frames $\mathcal{C}$ if $\mathbf{\Lambda}$ is the set of all formulas valid in all frames from $\mathcal{C}$.

---

[1] $\mathbf{K}_{\pm 1}$ is the well known logic $\mathbf{K.t}$.

DEFINITION 1. Let $F = (W, R_1, \ldots, R_n)$ be a frame, $u, v \in W$, $m \geq 1$. A *path of length m* from u to v is a sequence $(u_0, j_0, u_1, \ldots, j_{m-1}, u_m)$ such that $u = u_0$, $v = u_m$ and for all $i < m$, $u_i R_{j_i} u_{i+1}$. A singleton sequence (u) is the *path of length 0* (from u to u). Recall that the *subframe of F generated by u* (notation: F^u) is the restriction of F to the set of all v such that there exists a path from u to v; similarly a *generated Kripke submodel* M^u is defined.

DEFINITION 2. A *tree* with root u is a frame F such that $F = F^u$ and for every $v \in F$ there exists a unique path from u to v. The length of this path is called the *height* of v and denoted by $h(v)$. The height of F $(h(F))$ is the maximal $h(v)$ (if it exists), or ∞ otherwise.

DEFINITION 3. For a $2n$-modal tree $G = (W, S_1, \ldots, S_{2n})$, the frame $F = (W, R_1, \ldots, R_n, R_1^{-1}, \ldots, R_n^{-1})$, where $R_i = S_i \cup S_{n+i}^{-1}$, is called the *n-temporal tree (with the pattern G)*. The height function in F is then defined as the height function in G.

Speaking informally, a temporal tree is a modal tree, in which some of the arrows are inverted.

DEFINITION 4. A 1-tree $(W, \sqsubset)$ is called *standard* if its worlds are (some) finite sequences of natural numbers, its root is the void sequence λ and $\alpha \sqsubset \beta$ iff β is obtained by adding a single element at the end of α. An n-tree $(W, R_1, \ldots, R_n)$ is called *standard* if the 1-tree $(W, R_1 \cup \cdots \cup R_n)$ is standard. An n-temporal tree is called *standard* if its pattern is standard.

DEFINITION 5. Let $M = (W, R_1, \ldots, R_n, \theta)$ be an n-modal Kripke model, Ψ a set of n-modal formulas closed under subformulas. For $x \in W$ let $\Psi(x) := \{A \in \Psi \mid M, x \vDash A\}$. Two worlds $x, y \in W$ are called *Ψ-equivalent in M* (notation: $(M, x) \equiv_\Psi (M, y)$, or just $x \equiv_\Psi y$) if $\Psi(x) = \Psi(y)$.

DEFINITION 6. (cf. [5]) Under the assumptions of Definition 5, let $\approx$ be an equivalence relation on W. Let $x^\sim$ denote the $\approx$-class of x. A Kripke model $M' = (W', R'_1, ..., R'_n, \theta')$ is called a *filtration of M through* $\Psi, \approx$ if for any $x, y \in W$, for any formula A, $1 \leq i \leq n$:

(f1) $\approx \subseteq \equiv_\Psi$;

(f2) $W' = W/\approx$;

(f3) $xR_iy \Longrightarrow x^\sim R'_i y^\sim$;

(f4) $x^\sim R'_i y^\sim$ & $M, x \models \Box_i A$ & $\Box_i A \in \Psi \Longrightarrow M, y \models A$;

(f5) if $q \in \Psi \cap PL$, then $M, x \models q \Longleftrightarrow M', x^\sim \models q$.[2]

LEMMA 7. *(Filtration Lemma). Let M' be a filtration of M through $\Psi, \approx$. Then for any $x \in W$, for any $A \in \Psi$*
$M, x \models A$ iff $M', x^\sim \models A$.

Proof. Standard, by induction on the length of A, cf. [5], [9]. In the case $A = \Box_i B$ use (f3) for 'if' and (f4) for 'only if'. ■

LEMMA 8. *Let $M = (W, R_1, \dots, R_n, \theta)$, Ψ be the same as in Definition 5, $\approx$ an equivalence relation on W such that $\approx \subseteq \equiv_\Psi$; and let $W' = W/\approx$. Then the model $\underline{M} = (W', \underline{R}_1, \dots, \underline{R}_n, \theta')$ such that*

- *for any $x, y \in W$, $x^\sim \underline{R}_i y^\sim$ iff $\exists x_1 \in x^\sim \exists y_1 \in y^\sim\ x_1 R_i y_1$;*
- *for any $q \in \Psi \cap PL$, $\theta'(q) = \{x^\sim \mid M, x \models q\}$*

is a filtration of M through $\Psi, \approx$ (the least filtration).

Proof. Also standard; cf. [5], [9]. (f5) follows from the definition of θ'. To check (f4), assume $x \models \Box_i A$, $\Box_i A \in \Psi$, $x^\sim R'_i y^\sim$. Then $x_1 R_i y_1$ for some $x_1 \approx x$, $y_1 \approx y$, and thus $x_1 \models \Box_i A$, $y_1 \models A$, and hence $y \models A$. ■

DEFINITION 9. Let J be a finite set of positive integers. A binary relation R is called *J-quasitransitive* if $R^{j+1} \subseteq R$ for any $j \in J$.

LEMMA 10. *For any binary relation R, the smallest J-quasitransitive relation containing R (the J-*quasitransitive closure*) is $R^+ := \bigcup_{h \in H} R^{h+1}$, where H is the additive closure of $J \cup \{0\}$ in ω.*

Proof. Assume that $R \subseteq S$ and S is J-quasitransitive. Let $H_0 = \{h \mid R^{h+1} \subseteq S\}$, then obviously, $0 \in H_0$. Next, assume $h \in H_0$, $j \in J$.

[2] [5] contains misprints in this item.

Then $(h+j) \in H_0$; in fact, $R^{h+j+1} = R^{h+1} \circ R^j \subseteq S \circ R^j \subseteq S^{j+1} \subseteq S$. Therefore $H \subseteq H_0$, which readily implies $R^+ \subseteq S$.

It remains to show that R^+ is J-quasitransitive. So let us check that $(R^+)^{j+1} \subseteq R^+$ for $j \in J$. Assume $x(R^+)^{j+1}y$; then $x(R^{h_1+1} \circ \cdots \circ R^{h_{j+1}+1})y$ for some $h_1, \ldots, h_{j+1} \in H$. But this means $xR^{h_1+\cdots+h_{j+1}+j+1}y$, while $(h_1 + \cdots + h_{j+1} + j) \in H$ (remember that $J \subseteq H$ and H is additive closed). Hence xR^+y. ■

Remark. R^+ is a particular case of Horn closure, cf. [5].

LEMMA 11. *(cf. [5])* [3] *Let M, Ψ be the same as in Definition 5, and suppose that R_i is J_i-quasitransitive for $1 \leq i \leq n$. Also let*

$$\Phi \supseteq \Psi \cup \{\Box_i^j C \mid 1 \leq i \leq n,\ J_i \neq \varnothing,\ 1 \leq j \leq \max(J_i),\ \Box_i C \in \Psi\},$$

be a set of formulas closed under subformulas, and let $\underline{M} := (W', \underline{R}_1, \ldots, \underline{R}_n, \theta')$ be the least filtration of M through $\Phi, \approx$. If $R_i' = (\underline{R}_i)^+$ is the J_i-quasitransitive closure of $\underline{R}_i$, then $M' := (W', R_1', \ldots, R_n', \theta')$, is a filtration of M through $\Psi, \approx$. Moreover, if R_i is reflexive (respectively, symmetric), then R_i' also is.

Proof. Obviously, $\approx \subseteq \equiv_\Phi \subseteq \equiv_\Psi$. Since (f3), (f5) hold for $\underline{M}$, they also hold for M' (note that $\underline{R}_i \subseteq R_i'$). So let us check (f4) for R_i', provided $J_i \neq \varnothing$. To simplify notation, we drop the subscript i.

Assume $x^\sim R' y^\sim$, $x \vDash \Box A$, $\Box A \in \Psi$. Let H be the additive closure of $J \cup \{0\}$, and let us show that for any $h \in H$

(*) $x^\sim \underline{R}^h y^\sim$ implies $y \vDash \Box A$.

One can argue by induction on the number r of summands in the representation $h = j_1 + \cdots + j_r$, with $j_1, \ldots, j_r \in J$, and it suffices to prove that

(**) $y \vDash \Box A$ & $j \in J$ & $y^\sim \underline{R}^j z^\sim$ implies $z \vDash \Box A$.

So assume $y \vDash \Box A$, $y^\sim \underline{R}^j z^\sim$. Then we have $y \approx y_0 R z_1 \approx y_1 R \ldots y_{j-1} R z_j \approx z$, and thus $y_0 \vDash \Box A$ (since $\approx \subseteq \equiv_\Psi$ and $\Box A \in \Psi$),

$y_0 \vDash \Box^{j+1} A$ (since R is J-quasitransitive), $z_0 \vDash \Box^j A$, $y_1 \vDash \Box^j A$ (since $\approx \subseteq \equiv_\Phi$, $\Box^j A \in \Phi$), $z_1 \vDash \Box^{j-1} A$, $y_2 \vDash \Box^{j-1} A$ (since $\Box^{j-1} A \in \Phi$), ..., $y_{j-1} \vDash \Box^2 A$, $z_j \vDash \Box A$, and finally $z \vDash \Box A$. This proves (**).

[3]But again there are misprints in the definition of Φ in [5], p. 123.

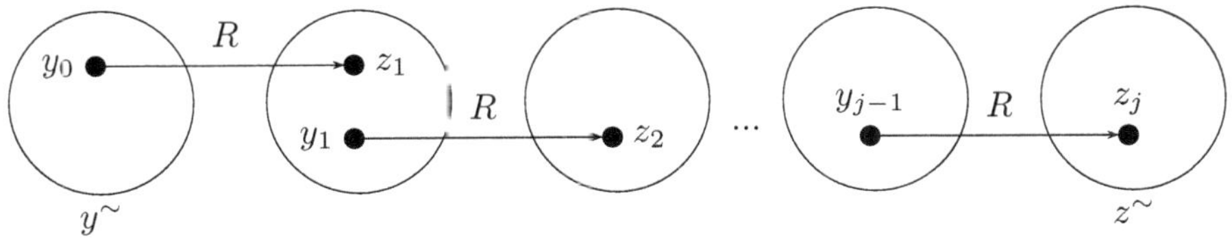

Figure 1.

Now, given $x \vDash \Box A$, $\Box A \in \Psi$, $x^\sim R' y^\sim$, we have $x^\sim \underline{R}^h z^\sim \underline{R} y^\sim$ for some z and some $h \in H$. By (*), we obtain $z \vDash \Box A$, and since (f4) holds for $\underline{R}$, this implies $y \vDash A$.

If R_i is reflexive (respectively, symmetric), then obviously, $\underline{R}_i$ is reflexive (respectively, symmetric), and thus $R'_i = \bigcup_{h \in H} \underline{R}_i^{h+1}$ is reflexive (symmetric) as well. ■

3 Filtration via bisimulation

DEFINITION 12. A *bisimulation* between Kripke models $M = (W, R_1, ..., R_n, \theta)$, $N = (V, S_1, ..., S_n, \eta)$ is a relation $E \subseteq W \times V$ with the following properties:

- $pr_1(E) = W$;
- $pr_2(E) = V$;
- $E \circ S_i \subseteq R_i \circ E$ for $1 \leq i \leq n$;
- $R_i^{-1} \circ E \subseteq E \circ S_i^{-1}$ for $1 \leq i \leq n$;
- if xEy, then for any $q \in PL\lceil k$, $M, x \vDash q$ iff $N, y \vDash q$.

$E : M, x \asymp N, y$ denotes that E is a bisimulation between M and N such that xEy. Two worlds $x \in M$, $y \in N$ are called *bisimilar* (notation: $M, x \asymp N, y$) if there exists a bisimulation $E : M, x \asymp N, y$.

If a bisimulation E is a function, it is a *p-morphism* from M onto N. In this case the condition $R_i^{-1} \circ E \subseteq E \circ S_i^{-1}$ is the *monotonicity*: $xR_iy \implies E(x)S_iE(y)$, and $E \circ S_i \subseteq R_i \circ E$ is the *lift property*: $E(x)S_iz \implies \exists y\ (xR_iy\ \&\ E(y) = z)$.

It follows easily from the definition that bisimilarity is an equivalence relation. The following Bisimulation Lemma is well-known:

LEMMA 13. *Bisimulations preserve truth values of formulas, i.e.,*
$M, x \asymp N, y$ *implies* $M, x \models A$ *iff* $N, y \models A$ *for any formula* $A \in \mathcal{L}_n\lceil k$.

Proof. Cf. [1], Theorem 2.20. ■

DEFINITION 14. Let $M = (W, R_1, \ldots, R_n, S_1, \ldots, S_m, \theta)$ be a Kripke model, Ψ a set of formulas closed under subformulas. For $x \in W$ the model $x\!\uparrow := (W, R_1, \ldots, R_n, R_1^{-1}, \ldots, R_n^{-1}, \theta)^x$ (cf. Definition 1) is called the *n-trace* of x. Put $x \approx y$ iff there exists $E : x\!\uparrow, x \asymp y\!\uparrow, y$ such that $E \subseteq \equiv_\Psi$ (*bisimilarity modulo* Ψ *with respect to* $R_1, \ldots, R_n$).

Note that $x\!\uparrow$ is $2n$-modal; the use of R_i^{-1} is essential for Lemma 15 below. It follows that $\approx$ is an equivalence relation. In fact, reflexivity and symmetry are obvious. For transitivity, note that $E_1 : x\!\uparrow, x \asymp y\!\uparrow, y$ and $E_2 : y\!\uparrow, y \asymp z\!\uparrow, z$ imply $(E_1 \circ E_2) : x\!\uparrow, x \asymp z\!\uparrow, z$; this is checked in a straightforward way: $E_1 \circ E_2 \circ R_i = E_1 \circ R_i \circ E_2 = R_i \circ E_1 \circ E_2$, and the same for R_i^{-1}; we also have $E_1 \circ E_2 \subseteq (\equiv_\Psi) \circ (\equiv_\Psi) = (\equiv_\Psi)$. By definition, $\approx \subseteq \equiv_\Psi$, and thus we can consider filtrations based on $\approx$.

LEMMA 15. *Under the conditions of Definition 14, for any* $i \leq n$, $\approx \circ R_i = R_i \circ \approx$.

Proof. First let us show that $\approx \circ R_i \subseteq R_i \circ \approx$. Assume $x \approx y R_i z$. Then x, y are equivalent modulo Ψ and there exists $E : x\!\uparrow, x \asymp y\!\uparrow, y$ such that $E \subseteq \equiv_\Psi$. So $x(E \circ R_i)z$, and thus $x(R_i \circ E)z$, i.e., $x R_i u E z$ for some u. Since $u\!\uparrow = x\!\uparrow$ and $y\!\uparrow = z\!\uparrow$, the same E yields $u \approx z$. In a similar way it follows that $\approx \circ R_i^{-1} \subseteq R_i^{-1} \circ \approx$, or $R_i \circ \approx \subseteq \approx \circ R_i$. ■

Hence we obtain filtrations preserving some commutation properties.

LEMMA 16. *Let* $\approx$ *be the bisimilarity modulo* Ψ *as in Definition 14. Then for the least filtration of* M *through* $\Psi, \approx$ *we have*

(1) $S_j \circ R_i \subseteq R_i \circ S_j \Longrightarrow \underline{S}_j \circ \underline{R}_i \subseteq \underline{R}_i \circ \underline{S}_j$;

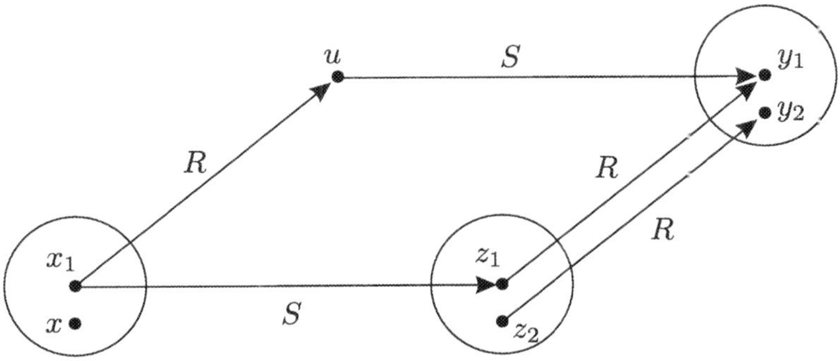

Figure 2.

(2) $R_i \circ S_j \subseteq S_j \circ R_i \Longrightarrow \underline{R}_i \circ \underline{S}_j \subseteq \underline{S}_j \circ \underline{R}_i$;

(3) $R_i^{-1} \circ S_j \subseteq S_j \circ R_i^{-1} \Longrightarrow \underline{R}_i^{-1} \circ \underline{S}_j \subseteq \underline{S}_j \circ \underline{R}_i^{-1}$.

Proof. We prove only (1); the proofs of (2),(3) are similar. To simplify the notation, we drop the subscripts i, j. Assume $S \circ R \subseteq R \circ S$, $x^\sim(\underline{S} \circ \underline{R})y^\sim$. Then for some z, $x^\sim \underline{S} z^\sim \underline{R} y^\sim$, and hence by definition in Lemma 8, we obtain $x_1 \approx x$, $z_1 \approx z_2 \approx z$, $y_2 \approx y$ such that $x_1 S z_1$, $z_2 R y_2$. So $z_1(\approx \circ R)y_2$, and thus by Lemma 15, there exists y_1 such that $z_1 R y_1$, $y_1 \approx y_2 \approx y$.

Next, $S \circ R \subseteq R \circ S$ implies $x_1(R \circ S)y_1$. So there exists u such that $x_1 R u S y_1$, and thus $x^\sim \underline{S} u^\sim \underline{R} y^\sim$. ■

LEMMA 17. *Let M, Ψ be the same as in Definition 14. Assume that R_i is J_i-quasitransitive, $1 \leq i \leq n$. Construct the set Φ as in Lemma 11. Let $\approx$ be the bisimilarity modulo Φ, and according to Lemma 11, consider the filtration $M' = (W', R_1', \ldots, R_n', S_1' \ldots, S_m')$ of M through $\Psi, \approx$ such that*

$R_i' = \underline{R}_i^+$, the J_i-quasitransitive closure of $\underline{R}_i$, and $S_i' = \underline{S}_i$.

Then the following holds:

(1) $S_k \circ R_i \subseteq R_i \circ S_k \Longrightarrow S_k' \circ R_i' \subseteq R_i' \circ S_k'$;

(2) $R_i \circ S_k \subseteq S_k \circ R_i \Longrightarrow R_i' \circ S_k' \subseteq S_k' \circ R_i'$;

(3) $R_i^{-1} \circ S_k \subseteq S_k \circ R_i^{-1} \Longrightarrow {R'}_i^{-1} \circ {S'}_k \subseteq S'_k \circ {R'}_i^{-1}$.

Proof. Follows easily from the previous Lemma. By Lemma 10, R'_i can be presented as $\bigcup_{h \in H_i} \underline{R}_i^{h+1}$, where H_i is the additive closure of $\{0\} \cup J_i$. By Lemma 16 (1), $S_k \circ R_i \subseteq R_i \circ S_k$ implies $\underline{S}_k \circ \underline{R}_i \subseteq \underline{R}_i \circ \underline{S}_k$, and hence by induction, $\underline{S}_k \circ \underline{R}_i^{j+1} \subseteq \underline{R}_i^{j+1} \circ \underline{S}_k$, which implies $\underline{S}_k \circ R'_i \subseteq R'_i \circ \underline{S}_k$. Similarly, for the claims (2), (3). ∎

4 Main results on finite model property

Let us now recall the definitions of products and relativised products.

DEFINITION 18. The *product* of Kripke frames $F = (W, R_1, \ldots, R_n)$, $G = (V, S_1, \ldots, S_m)$ is the frame

$$F \times G = (W \times V, R_{11}, \ldots, R_{n1}, S_{12}, \ldots, S_{m2})$$

such that

$$(x, y) R_{i1} (x', y') \Leftrightarrow x R_i x' \;\&\; y = y';$$

$$(x, y) S_{j2} (x', y') \Leftrightarrow x = x' \;\&\; y S_j y'.$$

A *relativised product* of F and G is an arbitrary subframe of $F \times G$.

DEFINITION 19. (cf. [5]) A *quasitranstitive (QT) formula* is one of the following kinds:

- $\Box_i p \to \Box_i^j p$ $(j \geq 0)$;
- $\Diamond_i \Box_i p \to p$.

A *QTC-logic* is a modal logic axiomatized by a finite set containing QT-formulas and maybe also closed formulas. A *$QTC_\pm$-logic* has the form $\mathbf{K}_{\pm m} + \mathbf{L}_0$, where $\mathbf{L}_0$ is a QTC-logic.

Remark. The above two kinds of QT-formulas correspond to the following conditions on frames: $R_i^j \subseteq R_i$, $R_i = R_i^{-1}$. A more general type, *pseudotransitive (PT)* formulas, was also considered in [5]: $\nabla_1 \Box_j p \to \Delta_2 p$, where ∇_1 is a sequence of diamonds, Δ_2 is a sequence of boxes; such a formula corresponds to the frame condition

$(R_{\Delta_1})^{-1}R_{\Delta_2} \subseteq R_j$, where R_{Δ_i} are the corresponding compositions of basic relations in a frame. A *PTC-logic* is axiomatized by PT-formulas plus maybe, closed formulas. From [5] it is known that any two PTC-logics are product-matching, but the question if every PTC-logic has the fmp, is open[4]. This explains why we deal with a smaller class of QTC-logics in this paper.

DEFINITION 20. A *weak product* of an n-modal logic $\mathbf{L}_1$ and an m-modal logic $\mathbf{L}_2$ is obtained from their fusion $\mathbf{L}_1 * \mathbf{L}_2$ by adding some commutation axioms of the forms
$\Diamond_i\Box_{n+j}p \to \Box_{n+j}\Diamond_i p$, $\Box_i\Box_{n+j}p \to \Box_{n+j}\Box_i p$, $\Box_{n+j}\Box_i p \to \Box_i\Box_{n+j}p$, for $1 \leq i \leq n$, $1 \leq j \leq m$.

Recall that the corresponding frame conditions are:

$$R_i^{-1} \circ R_{n+j} \subseteq R_{n+j} \circ R_i^{-1},\ R_{n+j} \circ R_i \subseteq R_i \circ R_{n+j},\ R_i \circ R_{n+j} \subseteq R_{n+j} \circ R_i.$$

THEOREM 21. *Let $\mathbf{\Lambda}$ be a weak product of $\mathbf{K}_n$ and a $QTC_\pm$-logic $\mathbf{L}_2$. Then $\mathbf{\Lambda}$ is determined by some class of relativized products $G \subseteq F_1 \times F_2$, where F_1 is an n-tree, $F_2 \vDash \mathbf{L}_2$.*

Proof. By an appropriate p-morphism construction similar to [4, Section 9.1]. First note that $\mathbf{\Lambda}$ is elementary and complete by Sahlqvist's theorem. Assume that $A \notin \mathbf{\Lambda}$, then there exists a rooted countable frame $F = (W, R_1, R_2) \vDash \mathbf{\Lambda}$ refuting A. It follows that F is a p-morphic image of some relativized product $G \subseteq F_1 \times F_2$, where F_1 is a standard (infinite) tree, F_2 is the Horn closure (corresponding to the QT-axioms of $\mathbf{L}_2$) of a standard tree. G is selected from their product by a game-theoretic argument (see below), so that it validates the commutation axioms from Definition 20. Since QT-formulas correspond to universal first order conditions, their validity is preserved for subframes. The validity of closed formulas is reflected by p-morphisms, so we obtain that $G \vDash \mathbf{\Lambda}$.

Let us describe the corresponding game for the case $n = 1$, $\mathbf{L}_2 = \mathbf{K}$. We may assume that the relations R_1, R_2 in F are non-empty — otherwise the claim is trivial. Let T_ω be the standard (intransitive

[4]The simplest unclear cases are $\mathbf{K} + \Diamond_1\Box_1 p \to \Box_2 p$ and $\mathbf{K} + \Box_1 p \to \Box_2\Box_1\Box_2 p$.

irreflexive) countable tree consisting of all finite sequences in ω, and let S_1, S_2 be the basic relations in the square $T_\omega \times T_\omega$.

An *arrow* in F is a triple (x, j, y), such that $x, y \in W$, $j \in \{1, 2\}$, and xR_jy.

A *network over* F is a function $h : N \to W$ such that

- $(\lambda, \lambda) \in N$, where λ is the empty sequence.
- $N \subseteq T_\omega \times T_\omega$ is a finite connected subset (i.e., in N the root (λ, λ) is connected by a path with every point).
- h is monotonic: $\forall a, b \in N\ (aS_ib \Rightarrow h(a)R_ih(b))$.

If h, g are two networks, we write: $h \subseteq g$ to denote that g prolongs h.

The *rectification game over* F *of length* ω (notation: $RG_\omega(F)$) is a game between two players, $\forall$ and $\exists$, who build a countable increasing sequence of networks: $h_0 \subseteq h_1 \subseteq \ldots \subseteq h_i \subseteq \ldots$, where $h_i : N_i \longrightarrow W$, according to the rules:

- $N_0 = \{(\lambda, \lambda)\}$; $h_0(\lambda, \lambda) = u_0$ (the root of F).
- $\forall$ starts the game, and $\forall$ and $\exists$ make their moves in turn.
- h_i is built from h_{i-1} at the i-th move of $\exists$.
- $\forall$ is allowed to make the i-th move of one of the two types:
  - (i) choose a 'lift enquiry':
    a quadruple (a, x, j, y), where $a \in N_{i-1}$, $x = h_{i-1}(a)$, and (x, j, y) is an arrow in F;
  - (ii) choose a 'commutation enquiry', which is a pair of arrows in N_{i-1}:
    if the axiom $\Diamond_1\Box_2 p \to \Box_2\Diamond_1 p$ is present in $\mathbf{\Lambda}$, this is $((a, 1, b), (a, 2, c))$;
    if the axiom $\Box_1\Box_2 p \to \Box_2\Box_1 p$ is in $\mathbf{\Lambda}$, this is $((a, 2, b), (b, 1, c))$, and similarly for the axiom $\Box_2\Box_1 p \to \Box_1\Box_2 p$.
- $\exists$ is allowed to respond to the moves of $\forall$ as follows:

(i) in a response to a lift enquiry (a, x, j, y) — to build a network $h_i : N_i \longrightarrow W$ such that $h_{i-1} \subseteq h_i$, and for some $b \in N_i$, $aS_jb, h_i(b) = y$, (i.e. h_i lifts the arrow (x, j, y) to (a, j, b));

(ii) in a response to a commutation enqury — to build a network $h_i : N_i \longrightarrow F$ extending h_{i-1} such that N_i contains the missing element. That is, in the case of the enquiry $((a, 1, b), (a, 2, c))$, N_i should contain d such that bS_2d and cS_1d; in the case of the enquiry $((a, 2, b), (b, 1, c))$, N_i should contain d such that aS_1d and dS_2c etc.

We assume that the player $\exists$ wins in every infinite play of the game; she loses if at some stage she cannot respond to a move of $\forall$.

A *winning strategy* for $\exists$ in the game is defined as usual; this is a function of the already constructed network and the last move of $\forall$, giving the response of $\exists$.

LEMMA 22. *If the player $\exists$ has a winning strategy in $RG_\omega(F)$, then F is a p-morhic image of a relativized product of two trees validating the appropriate commutation axioms.*

Proof.

Assume that $\exists$ has a winning strategy. Let us find a strategy for $\forall$, allowing $\exists$ to construct a required p-morphism.

Let $Ar(F)$ be the set of all arrows in F; then the elements of the countable set $\Pi = T_\omega \times Ar(F)$ and all commutation enquiries in $T_\omega \times T_\omega$ can be put into a sequence $\pi_1, \ldots, \pi_n, \ldots$ Now we choose the following strategy for $\forall$:

Every move of $\forall$ is the first occurence of a lift enquiry or a commutation enquiry in the sequence π, which is an allowed move and which has not been used in the previous moves; if it does not exist, this is a repetition of his previous move.

Obviously, $\forall$ can make the first move, since by our assumption, $R_1(u_0) \cup R_2(u_0) \neq \varnothing$.

If $\exists$ uses her winning strategy in response to these moves of $\forall$, they can build a sequence of networks $h_0 \subseteq h_1 \subseteq \ldots$.

Let $h = \bigcup\limits_{i=0}^{\infty} h_i$, $N_i = \mathrm{dom}(h_i)$, $N = \mathrm{dom}(h)$; then $N = \bigcup\limits_{i=0}^{\infty} N_i$.

We claim that h is a required p-morphism. Obviously, h is monotonic, since every h_i is a network. By the same reason, N is a connected subset of $\Phi_1 \times \Phi_2$, where the sets $\Phi_1 := pr_1(N)$, $\Phi_2 := pr_2(N)$ are standard trees. Let us show that the commutation axioms hold in N. In fact, suppose the contrary. Then there exists a commutation enquiry in N, say, $\pi_n = (a, 1, b, 2, c)$, but there does not exist $d \in N$ such that aS_2dS_1c. Take i such that $a, b, c \in N_i$. Then the chosen strategy for $\forall$ suggests that π_n should be his even move with a number $k \leq i + n$. In fact, after the i-th move is made, the only reason to postpone the move π_n is the existence of an allowed commutation enquiry with a number less than n. But all these enquiries should be exhausted by the $(i + n)$-th move.

In a similar way, let us check the lift property for h. Assume that $a \in N_{k-1}$, $h(a) = xR_jy$.

Then the lift enquiry (a, x, j, y) is an allowed move of $\forall$ with a number $\geq k$. This enquiry occurs in the sequence π as some π_n, and according to the strategy of $\forall$, this must be his move with a number $l \leq k + n$, because this enquiry, after the k-th move is made, can be postponed only in favour of π_i with $i < n$. The response of $\exists$ is a network $h_l : N_l \longrightarrow F$ lifting (x, j, y) to (a, j, b). So $h(b) = y$, aR_jb. ■

LEMMA 23. *$\exists$ has a winning strategy in $RG_\omega(F)$.*

Proof. Consider the i-th move of $\forall$, which is a lift enquiry (a, x, k, y), and assume that $k = 2$. (If $k = 1$, the argument is the same.)

If $a = (\alpha, \beta)$ then we take n such that $b = (\alpha, \beta n) \notin N_{i-1}$ (it exists because N_{i-1} is finite), and put $N_i = N_{i-1} \cup \{b\}, h_i(b) = y$. It is clear that h_i is monotonic and N_i is connected; so h_i is a correct response.

Suppose the i-th move of $\forall$ is a commutation enquiry $(a, 1, b, 2, d)$. Then $\Box_2\Box_1 p \to \Box_1\Box_2 p \in \mathbf{\Lambda}$, and thus $R_1 \circ R_2 \subseteq R_2 \circ R_1$ holds in F. It is clear that there exists a unique $c \in T_\omega \times T_\omega$ such that aS_2cS_1d. If $c \in N_{i-1}$, the response of $\exists$ will be $h_i = h_{i-1}$. Otherwise, let $h_{i-1}(a) = x$, $h_{i-1}(b) = y$, $h_{i-1}(d) = z$. Since h_{i-1} is monotonic, we have xR_1yR_2z, and due to the commutation property of F, $x(R_2 \circ R_1)z$, i.e., there exists t such that xR_2tR_1y. Then the response of $\exists$ will be $h_i = h_{i-1} \cup \{(c, t)\}$.

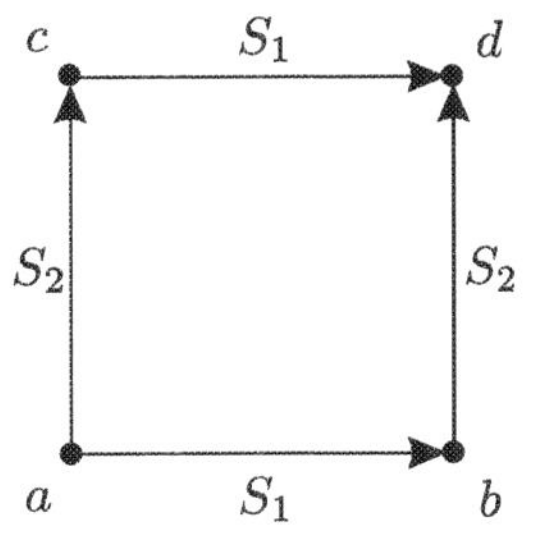

Figure 3.

For other types of commutation enquiries the argument is quite similar. ■

An analogous construction can be applied to $\mathbf{L}_2 = \mathbf{K}_n$. If $\mathbf{L}_2$ also has axioms $\Box_i p \to \Box_i^j p$ or $\Diamond_i \Box_i p \to p$, they are valid in F, and we should take the corresponding Horn closures G^+, F_2 of G, Φ_2. Then h is a p-morphism from G^+ onto F, cf. [5], Proposition 7.9.

■

LEMMA 24. *There exist finitely many equivalence classes with respect to the relation $M, x \asymp N, y$ for n-tree models* *M, N of fixed finite height with roots x, y.*

Proof. Since in trees bisimilar paths are of equal length, $M, x \asymp N, y$ implies $h(M) = h(N)$. Now the argument is by induction on $h(M)$.

If $h(M) = 0$, M, N are singletons, and $x \asymp y$ iff the same proposition letters are true at x and y. So in this case there exist 2^k $\asymp$-classes (recall that k is the fixed number of proposition letters).

Assume that $h(M) = l$ and there exist r $\asymp$-classes for n-tree models of height $< l$. We claim that $M, x \asymp N, y$ iff the same proposition letters are true at x and y, and also $\forall i \leq n$ $(\{z_\asymp \mid xR_i z\} = \{t_\asymp \mid yS_i t\})$, where R_i, S_i are the relations in M, N respectively, and $z_\asymp$ denotes the $\asymp$-class of M^z, z (or N^z, z).

In fact, the direction 'only if' is easy. To check 'if', assume that the truth values of proposition letters coincide in x and y and $\{z_\asymp \mid z \in R_i(x)\} = \{t_\asymp \mid t \in S_i(y)\}$ for any $i \leq n$. Then there

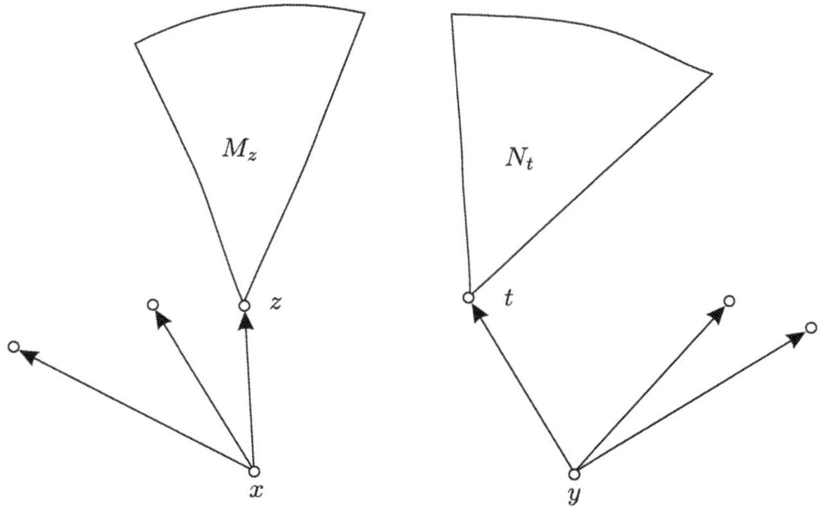

Figure 4.

exist bisimulations $E_{izt} : M^z, z \asymp N^t, t$, with $(z,t) \in V_i$, for some set $V_i \subseteq R_i(x) \times S_i(y)$ such that $pr_1(V_i) = R_i(x)$, $pr_2(V_i) = S_i(y)$.

Then

$$E := \{(x,y)\} \cup \bigcup \{E_{izt} \mid 1 \leq i \leq n,\ (z,t) \in V_i\}$$

is a bisimulation between M, x and N, y. In fact, E_{izt} works properly between M^z and N^t. Since $pr_2(V_i) = S_i(y)$, we also have $\forall z \in R_i(x)\ \exists t \in S_i(y)\ tEz$, and

similarly, every $t \in S_i(y)$ corresponds to some $z \in R_i(x)$, and finally, E sends x, the unique predecessor of z, to y, the unique predecessor of t.

Thus the $\asymp$-class of M, x is uniquely determined by the set $\{p_i \mid M, x \vDash p_i\}$ and $\{z_\asymp \mid z \in R_i(x)\}$ (for $1 \leq i \leq n$), and so there exist at most $(2^k \cdot (2^r)^n)$ $\asymp$-classes of M, x with $h(M) = l$. ■

LEMMA 25. *Let M, $\approx$ be the same as in Definition 14 and assume that every n-trace $x\uparrow$ is an n-tree of height $\leq l$ for some fixed l. Then the set $W/\approx$ is finite.*

Proof. In fact, $x \approx y$ only if $x\uparrow, x \asymp y\uparrow, y$, only if $x\uparrow, u \asymp y\uparrow, v$, where u, v are the roots of $x\uparrow$, $y\uparrow$. So the number of $\approx$-classes is finite, by Lemma 24. ■

THEOREM 26. *Every logic* $\mathbf{\Lambda}$ *from Theorem 21 has the fmp.*

Proof. By Theorem 21, every $A \notin \mathbf{\Lambda}$ is refuted in a model M over a $\mathbf{\Lambda}$-frame $F \subseteq F_1 \times F_2$, where F_1 is an n-tree; moreover, we may assume that $M, (x_0, y_0) \not\models A$, where x_0 is the root of F_1, y_0 is the root of F_2.

Let $d_1(B)$ denote the modal depth of a formula B with respect to $\Box_1, \dots, \Box_n$ (the maximal number of nested modalities of this type). Assume that $d_1(A) = r$. For $x \in F$ let $h_1(x)$ be the height of its first coordinate $pr_1(x)$ in F_1.

Let M^- (respectively, F^-) be the restriction of M (respectively, F) to the set $\{x \mid h_1(x) \leq r\}$. Then for any $v \in F^-$, for any formula B:

$$\text{(1)} \quad \text{if } h_1(v) + d_1(B) \leq r, \text{ then } M, v \models B \Longleftrightarrow M^-, v \models B.$$

This is proved by induction on $d_1(B)$ (cf. Lemma 9.11 in [5]). In fact, if $d_1(B) = 0$, the claim is obvious. If $d_1(B) > 0$, the only non-trivial case is when $B = \Box_i C$. So assume that (1) is proved for C, $M, v \models B$. Then $M, w \models C$ for any $w \in R_{i1}(v)$. Since $vR_{i1}w$ implies $h_1(w) = h_1(v) - 1$ and $d_1(C) = d_1(B) - 1$, we obtain $d_1(C) + h_1(w) = d_1(B) + h_1(v) \leq r$, and thus by induction hypothesis, $M^-, w \models C$. Thus $M^-, w \models B$.

If $B = \Box_{n+j} C$, then $M, v \models B$ means $\forall w \in S_{j2}(v)\ M, w \models C$. But $vS_{j2}w$ implies $h_1(v) = h_1(w)$, and thus $d_1(C) + h_1(w) < d_1(B) + h_1(v) \leq r$. Hence $M^-, w \models C$, by induction hypothesis, and thus $M^-, w \models B$.

The converse ($M^-, v \models B \Longrightarrow M, v \models B$) is proved in the same way.

Since $h_1(x_0, y_0) = 0$, from (1) we obtain:

$$\text{(2)} \quad M^-, (x_0, y_0) \not\models A.$$

We also have:

$$\text{(3)} \quad F^- \models \mathbf{\Lambda}.$$

In fact, for every $x \in F^-$ the subframes generated by x along the second coordinate in F and in F^- are the same.

Thus $F^- \models \mathbf{K}_n * \mathbf{L}_2$.

It is also clear that F^- inherits the commutation properties of F. For example, assume that $R_{i1} \circ S_{j2} \subseteq S_{j2} \circ R_{i1}$ holds in F. Now if $(x_1, y_1)R_{i1}(x_2, y_1)S_{j2}(x_2, y_2)$ in F^-, then $h_1(x_1, y_2) = h_1(x_1, y_1) \leq r$, and so $(x_1, y_2) \in F^-$; thus $(x_1, y_1)(S_{j2} \circ R_{i1})(x_2, y_2)$ holds in F^-.

Therefore we can take the filtration of M^- as in Lemma 17. The resulting model M' is finite, due to Lemma 25. ■

The previous Theorem can be generalized to the temporal case.

THEOREM 27. *Let $\mathbf{\Lambda}$ be a weak product of $\mathbf{K}_{\pm n}$ and a $QTC_{\pm}$-logic $\mathbf{L}_2$. Then $\mathbf{\Lambda}$ has the fmp.*

Proof. (Sketch.) The idea of the proof is the same as above, but n-trees are replaced with temporal n-trees. Note that now we do not need Church – Rosser axioms $\Diamond_i \Box_{n+j} p \to \Box_{n+j} \Diamond_i p$ — they can be replaced with $\Box_{n+j} \Box_i^{-1} p \to \Box_i^{-1} \Box_{n+j} p$. The proof of Theorem 21 is easily modified for this case. For example, if $\mathbf{\Lambda} = \mathbf{K}_{\pm 1} \times \mathbf{K}_{\pm 1}$, we take the standard temporal tree $T_\omega^\pm$ and construct $G \subseteq T_\omega^\pm \times T_\omega^\pm$ by a rectification game, in which commutation enquiries may be of the form $(a, i, b), (b, j, c))$, where $i = \pm 1,\ j = \pm 2$ or $j = \pm 1,\ i = \pm 2$.

The proof of the analogue of Lemma 24 is slightly more delicate, because now a path can use the same arrow in both directions many times. So we first make a reduction of a given temporal tree model M (with a chosen root x) as follows. We can identify two arrows (x, i, y) and (x, i, z) (or (y, i, x) and (z, i, x)) if there exists a bisimulation of M associating x with x and y with z. So we successively identify arrows in this way as long as possible. The reduced tree is bisimilar to the original one. Every point (except x) has a unique *father* (the predecessor in the unique path from x to that point); non-terminal points may also have R_i- or R_i^{-1}-*sons*. Now similar to the proof of Lemma 24, we have the criterion of bisimilarity for two reduced n-temporal trees of the same height: $M, x \asymp N, y$ iff the same proposition letters are true in M, x and N, y and the set of $\asymp$-classes of M_z, z and N_z, z are the same for R_i-sons of x and y, and also for R_i^{-1}-sons of x and y. Here M_z denotes the subtree of M with root

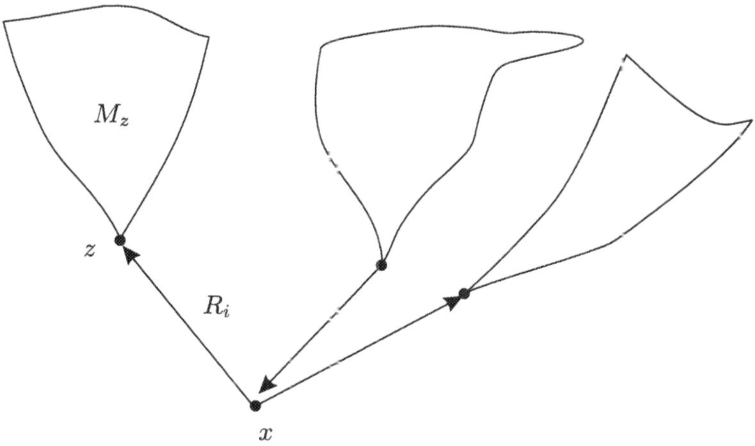

Figure 5.

z obtained from M by eliminating all points connected with z via x (i.e., those meeting x on on the shortest path to z).

■

Hence we obtain the fmp for some product logics:

COROLLARY 28. *The logics* $\mathbf{K}_{\pm n} \times \mathbf{K}_{\pm m}$, $\mathbf{K}_{\pm n} \times \mathbf{S5}_m$, $\mathbf{K}_{\pm n} \times \mathbf{K4}_{\pm m}$, $\mathbf{K}_{\pm n} \times \mathbf{S4}_{\pm m}$ *have the fmp.*

Remarks. 1. This implies the fmp of the logic $\mathbf{K.t}^2$ (which has been an open question) and therefore, of $\mathbf{K} \times \mathbf{K.t}$. The proof of the latter result given in [6], is much more complicated.

2. We also obtain the fmp for $\mathbf{K} \times \mathbf{S5}_m$, $m > 1$. Note that the earlier proof given in [5], contained a gap noticed by A. Kurucz [4], Section 5.3. Apparently, the methods of this paper can be extended to the products $\mathbf{S5}_m \times \mathbf{L}_2$, where $\mathbf{L}_2$ is a QTC-logic.

3. This also implies a positive answer to the question about the fmp of $\mathbf{K} \times \mathbf{K4}$ put in [4], p.339. However the question, whether the more powerful (but decidable) logic $\mathbf{K} \times \mathbf{PDL}$ has the fmp, remains open; our methods are not directly transferable to this case.

Acknowledgements. I would like to thank the anonymous referees for very useful comments on the paper. The work on this paper was partly supported by Russian Foundation for Basic Research, projects No. 02-01-22003, 01-01-00493, 02-01-01041, and by Jumelage (twinship) program CNRS-IUM.

BIBLIOGRAPHY

[1] P. Blackburn, M. de Rijke, Y. Venema. Modal logic. Cambridge University Press, 2001.

[2] D. Gabbay. General filtration method for modal logic. Journal of Philosophical Logic, v. 1 (1972), pp. 29-34,.

[3] D. Gabbay, I. Hodkinson, M. Reynolds. Temporal logic, v. 1. Oxford University Press, 1994.

[4] D. Gabbay, A. Kurucz, F. Wolter, M. Zakharyaschev. Many-dimensional modal logics. Theory and applications. Elsevier, 2003.

[5] D. Gabbay, V. Shehtman. Products of modal logics, I. Logic Journal of the IGPL, v.6 (1998), pp. 73-146.

[6] D. Gabbay, V. Shehtman. Products of modal logics. III. Products of modal and temporal logics. Studia Logica, v. 72 (2002), No. 2, p. 157-183.

[7] D. Gabelaia, A. Kurucz, M. Zakharyaschev. Products of 'transitive' modal logics without the (abstract) finite model property. In: AIML-2004, Advances in Modal Logic. Department of Computer Science University of Manchester, Technical Report Series, UMCS-04-91, pp. 104-115.

[8] K.Segerberg. Decidability of S4.1. Theoria, v.34 (1968), p. 7-20.

[9] V.Shehtman. A logic with progressive tenses. Diamonds and Defaults: Studies in Pure and Applied Intensional Logic. (Ed. by M. De Rijke). Kluwer Academic Publishers, 1993, p. 255-285.

[10] V. Shehtman, I. Shapirovsky. Chronological future modality in Minkowski spacetime. Advances in Modal Logic, v. 4, pp. 437-459. King's College Publications, 2003.

Institute for Information Transmission Problems, B. Karetny 19, 101447, Moscow, Russia

Department of Mathematics, Moscow State University

CNRS Laboratory Poncelet, B. Vlasyevsky 11, 119002, Moscow, Russia

shehtman@lpcs.math.msu.su

A Systematic Proof Theory for Several Modal Logics

CHARLES STEWART AND PHINIKI STOUPPA

ABSTRACT. The family of normal propositional modal logic systems is given a very systematic organisation by their model theory. This model theory is generally given using frame semantics, and it is systematic in the sense that for the most important systems we have a clean, exact correspondence between their constitutive axioms as they are usually given in a Hilbert-Lewis style and conditions on the accessibility relation on frames.

By contrast, the usual structural proof theory of modal logic, as given in Gentzen systems, is ad-hoc. While we can formulate several modal logics in the sequent calculus that enjoy cut-elimination, their formalisation arises through system-by-system fine tuning to ensure that the cut-elimination holds, and the correspondence to the axioms of the Hilbert-Lewis systems becomes opaque.

This paper introduces a systematic presentation for the systems **K**, **D**, **M**, **S**4, and **S**5 in the calculus of structures, a structural proof theory that employs *deep inference*. Because of this, we are able to axiomatise the modal logics in a manner directly analogous to the Hilbert-Lewis axiomatisation. We show that the calculus possesses a cut-elimination property directly analogous to cut-elimination for the sequent calculus for these systems, and we discuss the extension to several other modal logics.

1 Introduction

Modal logic has been one of the most fundamental advances in logic since Frege's invention of the quantifier, and indeed arguably the most important innovation of modern logic. We might attribute this success to:

- The ability of modal notations to naturally accomodate key concepts needed in formal description, most fundamentally concepts of mode and tense. While many propositional modal logics can be encoded in first-order predicate logic in a straightforward way, for many applications the modal systems are simpler and more useful;
- Most of the widely used modal logics being decidable;
- The normal modal logics possession of an elementary model theory in the form of frame semantics (due to Kripke, Hintikka and Kanger, see [8]) that provides a systematic correspondence between their constitutive axioms as they are usually given in Hilbert style and conditions on the accessibility relation on frames[1].

Let's look at the toolkit available to the classical logician for propositional modal logic compared to that for propositional logic. In propositional logic we characterise consequence relations using Hilbert deductive systems, give our semantics using Boolean algebra valued models, provide decision procedures using tableau methods, and provide techniques for proof analysis using sequent calculi in the mould of Gentzen.

In modal logic, close substitutes for the first three of these tools can be found: Hilbert systems are extended in a standard way with rules to provide what we might call Hilbert-Lewis deductive systems; Kripke or Beth frame semantics provides a sufficiently rich universe to provide models for almost all modal logics we care about; and tableau systems can be extended to accomodate modal logics. However, when it comes to proof analysis the situation is more fraught. A relatively natural extension of Gentzen-style sequent calculi provides cut-free characterisations of several important modal logics; however it fails

[1] See Garson [12], and Blackburn, de Rijke and Venema [3] for readable accounts of this relationship.

to provide characterisations of others, and even in the cases where sequent systems can be provided, these are arrived at in an ad-hoc manner[2], and the relationship to the Hilbert-Lewis calculus becomes opaque. It's not an exaggeration to say that, by and large, modal logicians have not found proof analysis to be worth the effort; this, no doubt, is responsible for the omission of structural proof theory in texts such as Blackburn, de Rijke and Venema [3].

In this paper we will provide a new proof theoretic basis for modal logic using the calculus of structures introduced by Guglielmi [14]. This proof calculus provides a structural proof theory in the sense of Gentzen for logical systems because it possesses a notion of cut-free proof directly analogous to that for the sequent calculus. It possesses several properties that give proof theory carried out in the calculus of structures important advantages over the traditional proof theory carried out in structural proof theory, and which extend to the modal logic systems we cover. It is our hope that these new possibilities for proof analysis can reinvigorate this neglected corner of the modal logician's toolkit.

The structure of the rest of this paper is as follows. In the next section we recall the characterisation of a much studied 'cube' of 15 normal modal logics in Hilbert-Lewis deductive systems, and describe the sequent calculus for just one of these, the system M (also known as system T). In section 3 we present the calculus of structures for classical logic, introduce its extension to system M, and talk about some of the most important analytical properties the calculus has in detail, including cut-elimination. In section 4 we discuss the generalisation of the structural proof theory to the other modal logics of the cube in both sequent calculus and the calculus of structures. In section 5 we provide a critical and slightly philosophical discussion comparing the achievements of this calculus to the rival calculus that is best placed to challenge us in providing a highly general approach to proof analysis, namely Belnap's display logic as applied to modal

---

[2]I do not mean by this that the designer of sequent systems has no methodology available other than trial and error, rather I mean that this methodology consists of a complex set of heuristics of low generality, and so falls short of being systematic.

logic by Wansing, as well as the more conservative but nevertheless highly interesting approach of Pottinger's hypersequent calculus as investigated by Avron. Finally, section six provides a brief, critical summary of what has been achieved in this paper.

2 Proof theory of system M

Let us begin by establishing some terminology. A *logical system* is defined over a formal language by a deductive system that determines its *theory*, a subset of the sentences of the language. Each formula in the theory of a logical system is called a *theorem*. Logical systems are *schematic*, that is they allow schematic letters that may stand in for arbitrary propositions, so the theory of a logical system for a more restrictive language can naturally be projected to richer languages by application of substitution. Two logical systems are *equivalent* if their theories are the same, they are *distinct* otherwise. If the theory of one logical system strictly contains the theory of another, then we call the former system the stronger system.

We begin by showing how propositional modal logic is defined using Hilbert-Lewis deductive systems, as these proof calculi give the oldest and most common axiomatisations, and so we may consider these axiomatisations as constituting a standard. The sentences of classical propositional logic contain an inexhaustible supply of schematic letters, letters for truth and falsity, which we indicate **tt** and **ff**, and are closed under the binary operators $\supset$, $\leftrightarrow$, $\wedge$ and $\vee$ and the unary operator $\neg$. The Hilbert style axiomatisation of this logical system proceeds in the standard way, where any substitution of propositions for schematic letters in the axioms of the system may be used to start a derivation or add a new theorem to the derivation, and non-axiomatic propositions may be obtained by modus ponens.

Hilbert-Lewis systems differ from Hilbert systems in that they possess (at least) a pair of dual modalities, and characteristically are schematic not through the schematicity of the axioms, but via the rule *uniform substitution*, which allows the schematic letters to be replaced by any proposition; the consequences of this difference are that deductions are normally a little longer, but it is slightly easier to define some metalogical apparatus; in any case the consequence relation should be unaffected by this difference, and in this paper we will be casual about the difference, defining axioms as sentences in the spirit of Hilbert-Lewis systems, but applying them as if they were schematic. In addition, all of the Hilbert-Lewis systems in which we are here interested will have the single further rule of necessitation: if $\vdash p$ then $\vdash \Box p$, and will have the duality captured by the axiom **DM**: $\neg \Box p \leftrightarrow \Diamond \neg p$.

Then the weakest normal modal logic **K** is axiomatised as a Hilbert-Lewis system by extending classical propositional logic with the axiom **K**: $\Box(p \supset q) \supset (\Box p \supset \Box q)$. Naturally, this corresponds to the class of frame models[3] where there are no conditions imposed on the frame accessibility relation.

We then obtain the stronger modal logics we are interested in by adding further axioms. All of the most studied systems of modal logic arise by adding some subset of the following axioms:

DEFINITION 1

1. *Axiom **D**:* $\Box p \supset \Diamond p$
2. *Axiom **T**:* $\Box p \supset p$
3. *Axiom **4**:* $\Box p \supset \Box \Box p$
4. *Axiom **B**:* $p \supset \Box \Diamond p$
5. *Axiom **5**:* $\Diamond p \supset \Box \Diamond p$

which correspond to constraining the accessibility relation to be serial (ie. there are always nodes accessible from any node), reflexive, transitive, symmetric and Euclidean respectively. There are a total of fifteen distinct logical systems obtainable, which we may organise

[3]We avoid the terminology of possible worlds, since we consider model systems lacking the T axiom, that cannot reasonably be considered to be about possibility or contingency, and further we agree with Thomas Forster that the language of possible world semantics is intoxicating to the careless and is best avoided[11]. Hence we call Hintikka/Kripke/Kanger style semantics, frame semantics; we avoid formalising these but will need to talk about the frame accessibility relation, which is given over nodes.

in a lattice according to the 'stronger than' relation, with the most important of these being:

1. **K** itself;
2. **D**, obtained by extending system **K** with rule **D**;
3. **M**, obtained by extending system **K** with rule **T**: note this system is often named **T**, a nomenclature we avoid to prevent confusion with Gödel's system T.
4. **S4**, obtained by extending system **M** with rule **4**;
5. **B**, obtained by extending system **M** with rule **B**;
6. **S5**, obtained by extending system **M** with rule **5**, or equivalently by extending system **M** with rules **B** and **4**.

The strongest system from these modal logics that is perfectly straightforward to formulate in a sequent system and to prove cut-free is system **G-M** (for Gentzen system **M**): we formulate[4] this using Schütte's approach of one-sided sequents. We make use of a *dualising function* in formulating the syntax of the calculus: $\overline{\phi}$ is defined recursively to be the De Morgan dual of the formula ϕ:

$$\begin{array}{cccc} \overline{\neg A} = A & \overline{A \wedge B} = \overline{A} \vee \overline{B} & \overline{\Box A} = \Diamond \overline{A} & \overline{\mathbf{tt}} = \mathbf{ff} \\ \overline{p} = \neg p & \overline{A \vee B} = \overline{A} \wedge \overline{B} & \overline{\Diamond A} = \Box \overline{A} & \overline{\mathbf{ff}} = \mathbf{tt} \end{array}$$

where p is a schematic letter, and $A \supset B$ and $A \leftrightarrow B$ are treated as derived connectives using the standard encodings. The inferences of the system are given by trees generated by the following inference rules, where the nodes are multisets of formulae (indicated by Γ, Δ, with the notations $\Box\Gamma, \Diamond\Delta$ indicating the result of prefixing each formula of the multiset appropriately):

[4]It is possible to show that, given liberal constraints on the form of the sequent calculus, 2-sided (Gentzen style) systems and 1-sided (Schütte style) systems characterise exactly the same consequence relations in the presence of De Morgan dualities [21], and this demonstration generalises to hypersequents.

Axiom and cut:

$$\frac{}{\vdash \Gamma, A, \overline{A}}\ \text{ax} \qquad \frac{\vdash \Gamma, A \qquad \vdash \Delta, \overline{A}}{\vdash \Gamma, \Delta}\ \text{cut}$$

Contraction and weakening:

$$\frac{\vdash \Gamma, A, A}{\vdash \Gamma, A}\ \text{contr} \qquad \frac{\vdash \Gamma}{\vdash \Gamma, A}\ \text{wk}$$

Truth:

$$\frac{}{\vdash \mathbf{tt}}\ \mathbf{tt}$$

Logical connectives and modal operators:

$$\frac{\vdash \Gamma, A, B}{\vdash \Gamma, A \vee B}\ \vee \qquad \frac{\vdash \Gamma, A \qquad \vdash \Gamma, B}{\vdash \Gamma, A \wedge B}\ \wedge$$

$$\frac{\vdash \Gamma, \overline{A}}{\vdash \Gamma, \neg A}\ \neg$$

$$\frac{\vdash \Gamma, A}{\vdash \diamond\Gamma, \Box A}\ \Box_1 \qquad \frac{\vdash \Gamma, A}{\vdash \Gamma, \diamond A}\ \diamond_1$$

The theorems of this system then are given by the concluding formulae of inferences where the concluding sequent contains exactly one formula. It is easily shown that the sequent formulation contains all the theorems of the Hilbert-Lewis formulation, and with the lightest touch of ingenuity, modelling the multisets by disjunctions, the reverse containment can also be demonstrated. We can also show the following theorem, where an inference rule or set of inference rules are *admissible* if all theorems of the system may be proven without their use:

THEOREM 2 *The cut rule is admissible in the sequent calculus formalisation of system **M**.*

This cut-elimination theorem is not really more complex to prove than that for its subsystem of classical logic, though there is a novelty: since the inference rules for the modal operators introduce 'side-effects'; the rule $\Box_1$ adds $\diamond$ modalities to side assumptions, which can then be used in cuts. This rule doesn't interfere with permutability of cuts, however. These side effects are are characteristic of an important sense in which these 'modal Gentzen calculi' go beyond traditional Gentzen calculi: with the traditional calculi only a formula in the consequence, and some subset of its subformulae in the premisses, are 'touched' by the logical rule, hence they are what we call *strongly focussed*; modal Gentzen calculi with rules such as $\Box_1$ we call *strongly unfocussed*, where we say a system is *weakly unfocussed* if it is not strongly focussed, and is *weakly focussed* if it is not strongly unfocussed[5]. Later we will encounter weakly unfocussed calculi.

3 The calculus of structures

Now we introduce a characterisation of the system **M** in the calculus of structures. As for the single-sided sequent calculus, we make use of De Morgan dualities between connectives in the formulation of the system. By convention, the calculus has a rather different notation for formulae than is used in Hilbert-Lewis and Gentzen proof calculi: the calculus of structures uses a bracket notation for the propositional logical connectives to suggest the associativity and commutativity properties: the calculus treats *structures* that are equivalence classes of the syntactic representations quotiented over these relations, and also quotiented over rules for units. Conjunction is represented by "(...)", so (R, S) stands for the conjunction of R and S; likewise disjunction is represented by "[...]". We also introduce a dualising operator in the same way as for the single-sided sequent calculus.

So the *formulae* of the calculus of structures are built up from the units **tt**, **ff**, the schematic letters, where for each schematic letter a we admit the complement $\overline{a}$ as a formula, and whenever $R_1, \ldots, R_n$ are formulae, so are $[R_1, \ldots, R_n]$ and $(R_1, \ldots, R_n)$. A *formula context* $S\{-\}$ is obtained from a formula by replacing any leaf (eg. a schematic letter) by '$-$'; then $S\{R\}$ is the formula obtained by re-

[5]The point of the terminology of focussing is to say that a calculus is focussed if each rule that deals with a connective is only about that connective. As such this is related to the properties such as separation as defined by Wansing [27], but neither strong nor weak focussedness is expressible in terms of Wansing's properties, and our interest here is technical, concerned with permutability of cut, rather that the meaning theoretic issues that motivated Wansing.

placing this – by the formula R. Curly braces are ommited when the formula R is precisely of the form $(R_1, \ldots, R_n)$ or $[R_1, \ldots, R_n]$.

The *duals* of formulae are defined recursively, where the dual of each schematic letter is its complement and vica versa, and

$$\overline{(R_1, \ldots, R_n)} = [\overline{R_1}, \ldots, \overline{R_n}]$$
$$\overline{[R_1, \ldots, R_n]} = (\overline{R_1}, \ldots, \overline{R_n})$$
$$\overline{\mathsf{tt}} = \mathsf{ff} \qquad \overline{\mathsf{ff}} = \mathsf{tt}$$

so for any formula R, $R = \overline{\overline{R}}$.

The *structures* are the equivalence classes of formulae obtained by quotienting over

1. (Associativity)
   $[R_1, \ldots, R_i, [T_1, \ldots, T_j], U_1, \ldots, U_k] = [R_1, \ldots, R_i, T_1, \ldots, T_j, U_1, \ldots, U_k]$
   and
   $(R_1, \ldots, R_i, (T_1, \ldots, T_j), U_1, \ldots, U_k) = (R_1, \ldots, R_i, T_1, \ldots, T_j, U_1, \ldots, U_k)$

2. (Congruence) If $R_1 = R_2$ then $S\{R_1\} = S\{R_2\}$, for any formula context $S\{-\}$;

3. (Commutativity) $[R_1, R_2] = [R_2, R_1]$ and $(R_1, R_2) = (R_2, R_1)$;

4. (Identities) $R = [R, \mathsf{ff}]$ and $R = (R, \mathsf{tt})$.

The equations can be considered as being in the spirit of the equations Schütte introduced over propositional formulae that ensure each formula is equal to its negation normal form, but they go further, representing more complex, though still tractable, equivalences over formulae. Their presence makes the calculus rather more difficult to master than the sequent calculus (especially the identities present difficulties) but they make working with the system much easier once this initial hurdle is passed, and normally ensure that important theorems avoid the trivial issues of syntax that they would otherwise be burdened with.

Inferences are chains of applications of inference rules, where each inference rule relating one premise, given above the rule, to one conclusion, below, whiere premise and conclusion are structures rather than formulae. All of the inference rules are *deep*, which means that each rule is given by a pair of formulae specifying the premise and conclusion that are both given with the same formula context. The inference rules for classical logic, the system **SKSg** are given:

Interaction and cut rules:

$$i\downarrow\frac{S\{\mathsf{tt}\}}{S[R,\overline{R}]}\qquad i\uparrow\frac{S(R,\overline{R})}{S\{\mathsf{ff}\}}$$

The switch rule:

$$s\,\frac{S([R,T],U)}{S[(R,U),T]}$$

The weakening rules:

$$w\downarrow\frac{S\{\mathsf{ff}\}}{S\{R\}}\qquad w\uparrow\frac{S\{R\}}{S\{\mathsf{tt}\}}$$

And the contraction rules:

$$c\downarrow\frac{S[R,R]}{S\{R\}}\qquad c\uparrow\frac{S\{R\}}{S(R,R)}$$

The theorems of this system are the formulae belonging to the structures that occur as conclusions of inferences whose premise is the structure tt.

This system has many desirable properties:

1. The system is self dual: each rule labelled with $\uparrow$ takes form:

$$*\uparrow\frac{S\{R\}}{S\{T\}}$$

   matching by De Morgan duality the form of the corresponding rule labelled with $\downarrow$:

$$*\downarrow\frac{S\{\overline{T}\}}{S\{\overline{R}\}}$$

   where rules not labelled with $\uparrow$ or $\downarrow$ are self-dual (so far we have seen just the switch rule).

2. The entire *up-fragment*, ie. the rules labelled with $\uparrow$, is admissible; that is, the full system is equivalent to the system obtained by removing the whole up-fragment. This is shown by Bruennler [6] and discussed in more detail below, by means of a translation from cut-free proofs of the sequent calculus into proofs of system **KSg**, that is, system **SKSg** without the rules of the up-fragment. Because the rule $i\uparrow$ closely models the cut rule, it is natural to describe this admissibility result as cut-elimination for the calculus of structures.

3. We can restrict the interaction, cut and weakening rules to atoms, by which we mean that the applications of the rules using formulae R can be restricted to the case where R is an atom (ie. a or $\bar{a}$) both for system **SKSg** and the cut-free system **KSg**. Furthermore, when adding to system **SKSg** the *medial rule*
$$m\,\frac{S[(R,U),(T,V)]}{S([R,T],[U,V])}$$
we can also restrict the contraction rule to atoms. These restrictions are achieved by simple local transformations on inferences, as opposed to the complex global transformation that we associate with cut-elimination. The system with these restrictions we call *atomic* **SKSg** (analogously atomic **KSg**), or just **SKS** (analogously **KS**). An analogous restriction to atoms can be achieved for the sequent calculus by similarly local transformations in the case of the axiom and weakening rules, but for fundamental reasons of the shallowness of inference cannot be achieved at all for contraction, and only by means of global transformations in the case of cut.

4. System **SKS** enjoys important computational properties: it is *local*, and so too is its subsystem **KS**, in the sense that looking at the inferences going either up or down, structure is rearranged, or atoms introduced, abandoned or duplicated, but arbitrarily large substructures are never introduced, abandoned or duplicated. Brünnler also discusses [op cit] an important advantage corollary to locality and atomicity of cut: using deep

inference he gets a *finitary* variant of **SKS** by simple means (i.e. without cut elimination); this means that there are only finitely different ways rules of the system can be applied to obtain a given conclusion, which in principle means the calculus can be used for proof search. However, it is fair to point out that these benefits do carry a cost: by contrast to the sequent calculus there are a great many possible rules that one may apply during proof search, so one cannot read off a tableaux algorithm from **KS** in the way one can for cut-free **LK**. Devising efficient proof search algorithms that can take advantage of these properties is an important goal of the program of research in the calculus of structures.

A thorough mathematical and conceptual examination of these properties and their importance is given in Bruennler's [6]; a shorter discussion appears in [5].

We can extend this calculus to obtain the system **M** by allowing formulae of the form $\Box R$, and $\Diamond R$, extending the equivalences defining structures with $\mathbf{tt} = \Box\mathbf{tt}$ and $\mathbf{ff} = \Diamond\mathbf{ff}$ and extending the set of inferences as follows:

$$k\downarrow\frac{S\{\Box[R,T]\}}{S[\Box R,\Diamond T]} \qquad t\downarrow\frac{S\{\Box R\}}{S\{R\}}$$

We call the system that extends system **KSg** with the above down rules **KSg-M**, and the symmetric system extending system **SKSg** with both the down and up rules **SKSg-M**.

We show the equivalence of system **SKSg-M** to **M** in two steps. Firstly we show we can map inferences of **SKSg-M** onto inferences of **M**:

DEFINITION 3 *We define the map $-^s$ from formulae of **M** onto structures of **SKSg-M** recursively:*

$$p^s = p \text{ for } p \text{ one of } \mathbf{tt},\ \mathbf{ff},\ \text{or a schematic letter;}$$

$$(\neg A)^s = \overline{A^s} \qquad (A \wedge B)^s = (A^s, B^s) \qquad (\Box A)^s = \Box A^s$$

$$(A \vee B)^s = [A^s, B^s] \qquad (\Diamond A)^s = \Diamond A^s$$

PROPOSITION 4

1. *There is a map* $-^h$ *from structures of* ***SKSg-M*** *to formulae of* ***M*** *with the properties that* $-^{hs}$ *is the identity map on structures, and for all formulae* A, B, *if* A *is a subformula of* B *then either* A^{sh} *is a subformula of* B^{sh} *or* $(\neg A)^{sh}$ *is a subformula of* $(\neg B)^{sh}$.

2. *If* $A^s = B^s$ *then* $\vdash A \leftrightarrow B$ *is a theorem of* ***M****;*

3. *If* $R = A^s$ *and* A *is a theorem of* ***M****, then* R *is a theorem of* ***SKSg-M****;*

PROOF Part 1 follows from defining the obvious recursive map from the formulae of **SKSg-M**, and observing that for any structure we may choose a lexically simplest representative formula (given some total order on atoms and their complements), to obtain the reverse map which is easily verified to have the desired properties.

Part 2 follows easily from the observation that $-^s$ is bijective when restricted to disjunctive normal forms.

To prove part 3, we must first prove that each axiom of **M** maps onto a theorem of **SKSg-M**, which is a straightforward exercise in the construction of inferences in the calculus of structures: for didactic reasons we recommend the reader works through at least two cases, for example $(A \supset B \supset C) \supset (A \supset C) \supset B \supset C$, and axiom **K**, and then show that inferences of the Hilbert-Lewis calculus map onto inferences of **SKSg-M** by considering the three inference rules. To see that inferences making use of uniform substitutivity are modelled in **SKSg-M** we need simply observe that the inference rules of **SKSg-M** are schematic. In the other two rules proceed by an induction on the length of proofs. □

THEOREM 5 ***M*** *and* ***SKSg-M*** *are equivalent.*

PROOF For each inference rule of **KSg-M**, which takes the general form:

$$\times \frac{S\{R\}}{S\{T\}}$$

we show that $R^h \supset T^h$ is a theorem of **M**. The dual rules map onto these from the theorem of classical logic: $(A \supset B) \leftrightarrow (\neg B \supset \neg A)$.

The following lemma is a straightforward exercise in theoremhood over **K**:

LEMMA 6 *If $A \supset B$ is a theorem of **M**, then so are:*

1. $A \wedge C \supset B \wedge C$;
2. $A \vee C \supset B \vee C$;
3. $\Box A \supset \Box B$;
4. $\Diamond A \supset \Diamond B$.

from which, by an induction on the makeup of formula contexts $S\{-\}$, we obtain for each inference rule $(S\{R\})^h \supset (S\{T\})^h$. Then inferences of **KSg-M** can be mapped onto chains of applications of modus ponens, starting from the theorem tt. □

THEOREM 7 *The up-fragment of **SKSg-M** is admissible.*

PROOF We show this by mapping cut-free proofs of **G-M** onto proofs of **KSg-M**, following just the same technique as [5], that is, for each rule of **G-M**

$$\frac{\Gamma_1 \qquad \ldots \qquad \Gamma_n}{\Delta}$$

we must find an inference of **KSg-M** with premise $\Box^i(\Gamma_1^s, \ldots, \Gamma_n^s)$ for some i and conclusion Δ^s, where we extend the mapping $-^s$ from formulae to sequents by considering sequents to be disjunctions of their constituent formulae, and where the notation $\Box^i A$ indicates the formulae obtained from A by prefixing $\Box$ to it i times. These proofs are straightforward; the only case where we need i to be nonzero is mapping $\Box_1$; the rule then maps onto a number of applications of the $k\downarrow$ rule equal to the number of formulae in Γ. We can then use these inferences to construct our map on cut-free proofs recursively. □

Lastly if we add the following rules:

$$j\downarrow\frac{S[\Diamond R, \Diamond T]}{S\{\Diamond[R,T]\}} \qquad uw\downarrow\frac{S\{\mathsf{ff}\}}{S\{\Box\mathsf{ff}\}} \qquad l\downarrow\frac{S[\Box R, \Box T]}{S\{\Box[R,T]\}}$$

we can restrict cut, interaction, contraction and weakening to atoms, and so obtain a local system, which we call **KS-M**.

4 Systems K, D, S4, and S5

We can extend the account of the structural proof theory that we have given of **M** to the systems **K**, **D**, and **S4** quite analogously. First we consider the axiomatisation of these systems in the sequent calculus: define two new inference rules:

$$\frac{\vdash \diamond\Gamma, A}{\vdash \diamond\Gamma, \Box A}\,\Box_2 \qquad \frac{\vdash \Gamma, A}{\vdash \diamond\Gamma, \diamond A}\,\diamond_2$$

Then we obtain our axiomatisations as follows:

1. **G-K** is obtained by dropping rule $\diamond_1$ from **G-M**;

2. **G-D** is obtained from **G-M** by substituting rule $\diamond_2$ for $\diamond_1$;

3. **G-S4** is obtained from **G-M** by substituting rule $\Box_2$ for $\Box_1$;

THEOREM 8 *Each of these systems is equivalent to the matching Hilbert system, and for each, the cut rule is admissible.*

PROOF For **G-K** and **G-D** the proofs are quite analogous to that for **G-M**. **G-S4** is similarly shown to be equivalent to the matching system **S4**; the cut-elimination proof is more subtle because many of the usual permutations on cuts fail, a fact that is unfortunately rather glossed over in the literature. Cf. Ohnishi and Matsumoto, and Valenti [17, 24, 25]. □

Our axiomatisation in the calculus of structures proceeds by defining the inference rules:

$$d\,\frac{S\{\Box R\}}{S\{\diamond R\}} \qquad 4\downarrow\,\frac{S\{\diamond\diamond R\}}{S\{\diamond R\}}$$

Then we obtain the equivalent systems:

1. **KSg-K** is **KSg-M** without the $t\downarrow$ rule;

2. **KSg-D** is **KSg-K** together with the $d\downarrow$ rule;

3. **KSg-S4** is **KSg-K** and the $t\downarrow$ and $4\downarrow$ rules;

4. **SKSg-K**, **SKSg-D**, and **SKSg-S4** are the symmetric versions of the above.

Note that inclusion of an axiom in the Hilbert system corresponds to inclusion of a down rule in the corresponding system in the calculus of structures.

THEOREM 9 *The analogs of Proposition 4 and Theorem 5 hold for each of* ***SKSg-K****,* ***SKSg-D****, and* ***SKSg-S4****.*

The axiomatisation of systems **S5** and **B** in the one-sided sequent calculus present serious difficulties, however. For example, it is possible to axiomatise **S5** by replacing the rule $\Box_1$ of **G-M** by any one of the following three rules[6]:

$$\frac{\vdash \diamond\Gamma, \Box\Delta, A}{\vdash \diamond\Gamma, \Box\Delta, \Box A}\,\Box_3$$
$$\frac{\vdash \Gamma, \Box\Delta, A}{\vdash \diamond\Gamma, \Box\Delta, \Box A}\,\Box_4$$
$$\frac{\vdash \diamond\Gamma, \Delta, A}{\vdash \diamond\Gamma, \Box\Delta, \Box A}\,\Box_5$$

This system can be mapped onto a reasonable axiomatisation of **KSg-S5** which we axiomatise as **KSg-S4** plus the following rule:

$$b\downarrow\,\frac{S\{\diamond\Box R\}}{S\{R\}}$$

We can show that the theories of the systems **S5**, the three possible formulations of **G-S5** and **SKSg-S5** are equivalent, and that cut-free proofs of **G-S5** map onto **KSg-S5**, but unfortunately cut-elimination for all three variants of **G-S5** fails:

PROPOSITION 10

1. *The axiomatisations of* ***G-S5*** *formulated using rules* $\Box_3$ *and* $\Box_5$ *have no up-fragment free proof of* $\neg A \vee \Box\diamond A$*, a theorem of* ***S5****.*

[6]The first of these three alternatives, using $\Box_3$ is the one-sided analogue of the two-sided sequent formulation of **S5** due to Ohnishi and Matsumoto, [17].

2. *The axiomatisation of **G-S5** formulated using rule $\Box_4$ has no cut-free proof of $\neg \Box \Box A \vee \Diamond A$, a theorem of **S5**.*

PROOF The are cut-bearing proofs for part 1 with cut formula $\Box A$, and for part 2 with cut formula $\Box \Diamond A$, for which in each case all the side formulae have $\Box$ as their main operator, which in each case is the principal conclusion of the appropriate $\Box_i$ rule. □

While cut-free sequent systems for **S5** have appeared in the literature; for example [4], these formulations do not take the simple form of the sequent systems treated in this paper, possessing either sophisticated rules that expose proof structure (either explicitly as in Braüner's connections between formula, or tacitly by some form of extra-logical labelling), or introduce some kind of deep inference that involves judgements with more sophisticated structure than Gentzen's sequents. We have not heard of any truly cut-free sequent formulation of **B** in a Gentzen-style sequent calculus.

We can axiomatise system **B** in the calculus of structure by adding $b\downarrow$ to **KSg-M** to obtain **KSg-B**, and likewise **SKSg-B**. Here our inference rules dealing with the modal operators find themselves in a one-one relationship to four of the five axioms given in the lattice of Hilbert systems described in Definition 1; we can extend this to rule **5** with the inference rule:

$$5\downarrow \frac{S\{\Diamond \Box R\}}{S\{\Box R\}}$$

CONJECTURE 11 *For all of the systems of the cube characterised by adding any subset of the modal down rules we have described above, cut is admissible.*

Note that though there are fifteen systems of the cube, there are thirty-two possible ways of characterising them; all of these characterisations are conjectured to allow cut elimination.

In particular we make an observation.

THEOREM 12 ***KSg-S5** is cut-free.*

We will discuss this result in section five, which provides the strongest of several pices of evidence for the conjecture, when we treat hypersequents.

The approach to proofs we have so far outlined are proofs by translation which leverage known cut elimination proofs for the sequent calculus. It is also possible to prove cut elimination directly, although non-constructively, by semantic means. The advantage of this method is that it is as systematic as the formalisation in the calculus of structures; the disadvantages are, besides non-constructivity, that such methods give only limited proof-theoretic insight, and so far are restricted to fewer axioms than are needed to describe the systems of the cube.

Let us introduce two new Hilbert-Lewis axioms:

DEFINITION 13

1. *Axiom* ***W5***: $\Diamond \Box p \supset \Box \Box p$

2. *Axiom* ***C4***: $\Box \Box p \supset \Box p$

Then we have $\mathbf{W5}, \mathbf{C4} \vdash \mathbf{5}$, $\mathbf{5} \vdash \mathbf{W5}$, and $\mathbf{T} \vdash \mathbf{C4}$; these new axioms then are weak versions of **5** and **T** that combine to provide the inferential strength of axiom **5**. Each of these can be incorporated into the calculus of structures as follows:

$$w5\downarrow \frac{S\{\Diamond \Box R\}}{S\{\Box \Box R\}} \qquad c4\downarrow \frac{S\{\Box \Box R\}}{S\{\Box R\}}$$

CONJECTURE 14 *For all 128 possibilities of adding some subset of the seven rules* $d\downarrow$, $t\downarrow$, $4\downarrow$, $b\downarrow$, $c4\downarrow$, $5\downarrow$ *and* $w5\downarrow$ *to* ***KSg*** *the up-fragment is admissible.*

Obviously this conjecture is stronger than the first, and so far the evidence for it is weaker. At least a subsystem generated by four of the above seven rules is known to be cut-free by a semantic argument. It appears to be the case that the semantic argument works for systems described by axioms that correspond to *rooted conditions* on the frame accessibility relation; that is, conditions that can be described by a Π^0_2 formula where the universal quantifiers describe a tree of arcs across nodes, and the existential quantifiers assert the existence of arcs each of which either originates from the root of the tree, or which originates from a node that is existentially asserted and distinct from any in the original tree. The rules that correspond to rooted conditions we call rooted, these that don't we call unrooted.

THEOREM 15 *For each of the 10 systems described by some subset of the four unrooted rules d $\downarrow$, t $\downarrow$, 4 $\downarrow$, and c4 $\downarrow$, the up-fragment is admissible.*

To repeat, this theorem is proven by semantic means that we believe extend to all systems described by rooted rules. Additionally the six systems that can be formalised without the c4 $\downarrow$ rule can all be proven by means of translation from known systems in the sequent calculus (four of which are the systems discussed in most depth at the beginning of this section).

5 Display logic and hypersequents

As remarked before, the most obvious rival to our calculus is the application of display logic to modal logic as investigated by Wansing [27]. Modal display logic is modular in the sense that we seek, and it has a syntactic cut elimination that ensures any properly presented system (that is, a 'properly displayable' system) is cut-free, in the true sense of guaranteeing analytic normal forms. It embraces classes of calculi that we do not know how to express satisfactorily in the calculus of structures, namely calculi without De Morgan duality, such as intuitionistic logic[7]. And indeed these strengths come from a property display logic possesses which is similar to a property of the calculus of structures, namely, the manipulations on structures allow the logical rules to be applied at effectively unbounded depths, or in other words, display logic is also a calculus of deep inference.

Given all this, one might reasonably ask – why pursue another calculus of deep inference? Our answer is that we believe that for the purposes of proof analysis, the design of the calculus of structures will ultimately lead to technically better results, essentially for the following reason: display logic wishes to combine the sophisticated notion of proof structure one obtains with deep inference with the traditional

[7]Though there are logics expressible in the calculus of structures which it appears that display logic cannot express at all, such as system NEL [13], due to the branching nature of proofs in display logic

approach to proof analysis based on the subformula property. The result is that display logic leads a double life, with the subformula property holding on formulae, the leaves of the structures display logic manipulates, but not able to say anything about the fully fledged structures which is really where all the action happens. A second consequence is that display logic seeks to embrace constraints on its presentations, such as Došen's principle, that appear to be necessary to get reasonable results from subformla-property-based proof analysis. Even so, the proof analysis does not seem to be very fruitful, when judged against the paradigmatic stadard of Gentzen's formalisations: in the sequent calculus one can simply read of a tableu mechanism from a cut-free sequent calculus; in display logic, the procedure, or perhaps better heuristic, is more fraught. To be blunt, some of the technical advantages that flow from the subformula property in the context of the sequent calculus do not flow from the property that display logicians call 'subformula property'; perhaps, by analogy, the meaning-theoretic parallels depend upon a certain amount of charitable optimististm.

By contrast the calculus of structures makes no distinction between logical and structural rules, which brings simplicity; constraints such as Došen's principle, or some of the properties of systems described by Wansing like separation, symmetry and explicitness simply make no sense in this context; and seeks to substitute a wholly novel methodology of proof analysis, with some striking properties that improve on Gentzen-family calculi, such as atomicity.

Additionally there is a technical disadvantage of display logic for our investigation into systems of modal logic: display logic appears to be most naturally a tense logic, in that in introducing a modality, one obtains not only introduction rules for that modality and its De Morgan dual, but also the reverse modalities in the tense logical sense (so four modal operators, rather than the expected two). Hence, whereas when one has a proof in **KS-S5** each structure occuring in the inference corresponds to a theorem of **S5**[8], in display logic the axiom dealing for 5 is expressed in terms of tense logic (due to the conversion of modal axioms to rules involving primitive tense formula,

[8]By means of the translation $-^h$ described in section three.

following the results of Marcus Kracht [15]), and so the judgements appearing in the tree of a proof may be assertions not of **S5** but of **S5t**, its tensed extension. By conservativity, we know that we have the right theorems, but conservativity of theoremhood does not map into conservativity of cut-free provability; there is still a problem unsolved. In particular, if we consider these kind of results adequate, then the problem of cut elimination for modal logic becomes mostly a solved problem: we simply use calculi that embed modal logics in sequent calculi for geometric theories, such as pioneered by Alex Simpson[20].

Certainly there is room for dispute over the relative merits of the two approaches to providing a structural proof theory for modal logic; however we think it is important to recognise firstly that the ultimate test of health of a proof calculus will lie in its usefulness to logicians not interested in proof theory for proof theory's sake; secondly that in providing a toolkit for proof analysis display logic and the calculus of structures are not necessarily rivals, but may instead be complementary formalisms.

We also wish to discuss hypersequents as they have been treated by Arnon Avron. These extend the sequent calculus, allowing several strands of inference to occur in parallel, and to communicate by means of certain 'external' structural rules (the familiar structural rules are called 'internal', we will call rules that act locally upon a sequent internal also). We note that for the modal logics in which we are interested, the 2-sided hypersequents Avron treats are equivalent to 1-sided hypersequents; thus we may describe the general hypersequent $\vdash \Gamma_1 | \dots | \vdash \Gamma_n$ by the structure $[\Box[\Gamma_1^s], \dots, \Box[\Gamma_1^s]]$. Avron captures the system **S5** in the hypersequent system **HS5** by giving internal rules corresponding to the rules for **S4**, together with, in addition to external contraction and weakening, a modalised splitting rule (we give our 1-sided variant):

$$\mathbf{MS}\ \frac{G| \vdash \Box\Gamma_1, \Diamond\Gamma_2, \Gamma_3}{G| \vdash \Box\Gamma_1, \Diamond\Gamma_2 | \vdash \Gamma_3}$$

All internal rules map easily onto the rules for **S4**, just in a deeper context of the form $[\Box\{-\}, R]$, and the external weakening and contraction rules are straightforwardly modelled by $w\downarrow$ and $c\downarrow$; all that remains is to capture the modalised splitting rule, which can be modelled by the $4\downarrow$, $5\downarrow$ and s rules. By these means, cut free proofs of system **HS5** are mapped to cut-free proofs of **KSg-S5**, thus establishing the theorem described in section four.

Let us note, though, that the modalised splitting rule does not fix all of the problems arising due to lack of expressiveness of the cut-free sequent calculus. In particular, the modalised splitting rule gives the inferential strength of the 5 axiom only by also giving the inferential strength of the 4 axiom as well. To be more precise, hypersequent systems with the modalised splitting rule cannot be used to axiomatise systems **B**, **KB**, **DB**, **K5** or **D5**, and it has not been shown how the inferential strength of unrooted axioms can be achieved in a hypersequent calculus other than by means of modalised splitting. Thus, while hypersequents achieve an important result, one that we depend upon for the results of this paper, the means by which it achieves the goals appears to be fundamentally limited.

6 Review

The most important achievement of this paper bas been to provide a novel, modular approach to the proof theory of modal logic that begins to allow the kind of powerful proof analysis described by Brünnler for several important modal logics (those for which we can prove up-fragment admissibility), and for which we have reason to hope can be extended to all the modal logic systems we have described. Furthermore we hope that our readers will find the system to be elegant and provocative.

The most serious defect of the account given here from the point of view of formal aesthetics is that our proofs are external: our methods for showing cut-elimination are so far a rather underpowered semantical technique or by means of translations from a quite separate theory, so though we could pedantically claim to have provided a syntactic proof of cut-elimination, the proof is quite as external as with semantical proofs of cut-elimination. Furthermore, we are mostly left without resources to deal with the cases where we do not have appropriate cut-free axiomatisations, such as for system **B**. Nonetheless, it must be emphasised that these are defects of our knowledge of how to prove results; to the best of this limited knowledge, the family of proof systems are in excellent shape: no rival calculus appears to describe as many systems so simply.

There are three principal avenues to explore in search of the missing proof techniques. The first is to try to extend the semantic proofs; Avron's paper on hypersequents succeeds in providing a semantical proof of cut elimination for **HS5** and it is possible the technique can be adpated and generalised. Second, Guglielmi [14] introduces a technique, called splitting, that may be regarded as the preferred way to give internal, syntactic proofs of cut-elimination in the calculus of structures, and in joint work with one of the authors (in preparation) has shown how this technique may be extended to deal with the case of SKS. Lastly, a very syntactically involved technique of permutability of rules can be applied to prove cut elimination; two proofs of this nature are due to Strassburger [23], as well as a proof by Brünnler [6]. Of these approaches, a splitting proof would be the most valuable.

A further issue relates to the role of proof analysis in the toolkit of the modal logician. The single most important application of proof analysis in propositional logic has been its crucial role in influencing the design of tableau methods. With modal logics, the ad hoc nature of sequent characterisations has been reflected by an equally ad hoc methodology for the design of modal tableau. Furthermore the designers of modal tableau have found it necessary to resort to ugly techniques equivalent to 'analytic' cut (a proof with an analytic cut is by virtue of this cut not analytic; this is perhaps the most misleading piece of nomenclature in proof theory), so indeed the degree of disorder among modal tableau is greater than among modal sequent calculi. The most obvious and useful test of a claim to have provided a worthy modal proof theory is leverage the theory to providing a principled approach to modal tableau. The application of the calculus of structures to the design of such systems is under intensive investigation, making use of insights flowing from Dale Miller's 'proof search as computation' slogan [7, 16].

BIBLIOGRAPHY

[1] A. Avron The method of hypersequents in the proof theory of propositional non-classical logics. In W. Hodges et al. (eds.), *Logic: From Foundations to Applications*,

pages 1–32. Oxford University Press, 1996.
[2] N. Belnap. Display Logic. *Journal of Philosophical Logic*, 11:375–417, 1982.
[3] P. Blackburn, M. de Rijke, Y. Venema. *Modal Logic*. Cambridge University Press, 2001.
[4] T. Braüner. A cut-free Gentzen formulation of the modal logic S5. In the *Logic Journal of the Interest Group in Pure and Applied Logics*, volume 8(5), pages 629–643, 2000.
[5] K. Brünnler. Locality for classical logic. Submitted to *Archive for Mathematical Logic*, 2003. Preprint available `http://www.wv.inf.tu-dresden.de/ kai/LocalityClassical.pdf`.
[6] K. Brünnler. Deep Inference and Symmetry in Classical Logic. Logos Verlag, Berlin, 2004.
[7] P. Bruscoli. A Purely Logical Account of Sequentiality in Proof Search. In *Proc. ICLP 2002*, LNCS 2401:302–316. Springer-Verlag, 2002.
[8] R. A. Bull and K. Segerberg. Basic Modal Logic. In *The Handbook of Philosophical Logic*, volume 2, pages 1–88. Kluwer, 1984.
[9] L. Cardelli and A. D. Gordon. Anytime, anywhere: Modal logics for mobile ambients. In *Proc. Principles of Programming Languages 2000*, pages 365–377. ACM Press, 2000.
[10] K. Došen. Sequent-Systems for Modal Logic. In *Journal of Symbolic Logic*, volume 50(1), pages 149–168, 1985.
[11] T. Forster. The modal aether. Manuscript.
[12] J. Garson. Modal Logic. In E. N. Zalta, editor, *The Stanford Encyclopaedia of Philosophy*, Winter 2001 edition. Available `http://plato.stanford.edu/archives/win2001/entries/logic-modal`.
[13] A. Guglielmi and L. Straßburger. Non-commutativity and MELL in the Calculus of Structures. In *Proc. Computer Science Logic 2001*, LNCS 2142:54–68. Springer-Verlag, 2001.
[14] A. Guglielmi. A System of Interaction and Structure. Submitted to *ACM Transactions on Computational Logic*, 2002. Preprint available `http://www.ki.inf.tu-dresden.de/~ guglielmi/Research/Gug/Gug.pdf`.
[15] M. Kracht Power and weakness of the modal display calculus. In, H. Wansing (ed.), *Proof Theory of Modal Logic*, pages 93–121. Kluwer, Dordrecht, 1996.
[16] R. Milner. *Communicating and mobile systems: the pi-calculus.* Cambridge University Press, 1999.
[17] M. Ohnishi and K. Matsumoto. Gentzen method in modal calculi, parts I and II. In *Osaka Mathematical Journal*, volume 9, pages 113–130, 1957, and volume 11, pages 115–120, 1959.
[18] K. Schütte. *Proof Theory.* North-Holland, 1962.
[19] H. Schwichtenberg and A. Troelstra. *Basic Proof Theory.* Cambridge University Press, 1998.
[20] A. Simpson *Proof theory of intuitionistic modal logic.* PhD thesis, University of Edinburgh, 1994.
[21] C. A. Stewart. Which 2-sided sequent systems are equivalent to 1-sided systems? Unpublished manuscript, 2004.
[22] P. Stouppa. The design of modal proof theories. MSc dissertation, Technische Universität Dresden. In preparation.
[23] L. Straßburger. A local system for linear logic. In *LPAR 02*, 2002.
[24] S. Valenti. Cut-elimination in a modal sequent calculus for K. In *Bolletino dell'Unione Mathematica Italiana*, volume 1B, pages 119–130, 1982.
[25] S. Valenti. The sequent calculus for the modal logic D. In *Bolletino dell'Unione Mathematica Italiana*, volume 7A, pages 455–460, 1993.
[26] H. Wansing. Sequent Calculi for Normal Modal Propositional Logics. In *Journal of Logic and Computation*, volume 4(2), pages 125–142, 1994.
[27] H. Wansing. *Displaying Modal Logic.* Kluwer, Dordrecht, 1998.

Charles Stewart
International Centre for Computational Logic, Technische Universität Dresden.
cas@linearity.org

Phiniki Stouppa
Institut für Informatik und angewandte Mathematik, University of Bern.
stouppa@iam.unibe.ch

Public Announcements and Belief Expansion

Hans van Ditmarsch, Wiebe van der Hoek and Barteld Kooi

ABSTRACT. In this paper we study the relation between two approaches to information change: Dynamic Epistemic Logic and Belief Revision. One of the main differences between these approaches is that higher-order information plays an important role in the field of Dynamic Epistemic Logic, whereas it does not feature in Belief Revision. In this paper we study to which extent public announcements (a particular kind of information change studied in Dynamic Epistemic Logic) can be viewed as a belief expansion (a particular kind of information change studied in Belief Revision).

1 Introduction

Since Hintikka's [9] epistemic logic, the logic of knowledge, has flourished as a research area in philosophy [10], computer science [6], artificial intelligence [12] and game theory [2]. The three mentioned application areas made it apparent that in multi-agent systems *higher-order information*, knowledge about (other) agents' knowledge, is crucial.

A natural question to ask, once the formal framework to reason about the information of agents — we will use the terms 'belief' and 'knowledge' interchangeably in this paper — is in place, is how the agent's belief changes over time, when he is confronted with new information; be it in a static, or an evolving world. The famous paper [1] by Alchourrón *et al.* put this change of information, coined as *Belief Revision (BR)*, as a topic on the philosophical and logical agenda: it was followed by a large stream of publications, on fine-tuning the notion of epistemic entrenchment [13], on revising (finite) belief bases [4], on differences between belief revision and belief updates [11], and the problem of iterated belief change [5].

This stream of research typically represents the beliefs of the (only) agent as a theory in propositional logic, and then describes, in terms of rationality postulates, how new information should be incorporated in it. Hence, in this approach the dynamics is studied on a level above the informational

Advances in Modal Logic, Volume 5.

level, not allowing reasoning about change of agents' knowledge and ignorance within the framework, let alone about the change of other agents' information from the perspective of a given agent.

Alternatively, the area of *Dynamic Epistemic Logic (DEL)* takes as its starting point modal epistemic logic in the tradition of [9], and adds a dynamic component to it. The aim here is not only to dynamize the epistemics, but also to 'epistemize the dynamics': the actions that (groups of) agents perform are epistemic actions. Different agents may have different information about which action is taking place, including higher-order information. This rather recent approach treats all of knowledge, higher-order knowledge, and its dynamics, on the same foot. Following an original contribution by Plaza in 1989 [14], a stream of publications appeared around the year 2000 [8, 3, 16, 17].

The relation between belief revision and dynamic modal logic is investigated by Segerberg in [15]. However, this study is limited to cases where higher-order information plays no role. In this paper we study a case with higher-order information, namely how a public announcement, a specific epistemic action in DEL, can be viewed as belief expansion, one of the three distinguished operations in BR. As such, our paper can be conceived as a next step in making the relation between the approaches explicit.

In Section 2 and 3 we briefly introduce public announcement logic and belief expansion, respectively. Section 4 demonstrates that belief expansion can be seen as a special case of public updates. In section 5 conclusions are drawn, and directions for further research are indicated.

2 Public announcement logic

Public announcement logic (PAL) was first developed by Plaza [14]. It extends epistemic logic to allow reasoning about information change due to public announcements: an epistemic action $[\varphi]$ where the whole group of agents are aware that they learn φ.

Definition 1 (Language of PAL). Let a finite set of propositional variables $\mathcal{P}$ and a finite set of agents $\mathcal{A}$ be given. The language $\mathscr{L}_{\mathsf{PAL}}$ is given by the following BNF:

$$\varphi ::= p \mid \neg\varphi \mid \varphi_1 \wedge \varphi_2 \mid \Box_a\varphi \mid [\varphi_1]\varphi_2$$

where $p \in \mathcal{P}$, $a \in \mathcal{A}$. Besides the usual abbreviations we use $E\varphi$ as an abbreviation for $\bigwedge_{a\in\mathcal{A}} \Box_a\varphi$.

A formula of the form $[\varphi]\psi$ has to be read as "ψ holds after the announcement of φ". This language is interpreted in models for epistemic logic. These are Kripke models with an accessibility relation for each agent.

All the logical languages presented in this paper are interpreted in these models.

Definition 2 (Epistemic models). Let a finite set of propositional variables $\mathcal{P}$ and a finite set of agents $\mathcal{A}$ be given. An epistemic model is a triple $M = (W, R, V)$ such that:

- $W \neq \emptyset$; a set of possible worlds;
- $R : \mathcal{A} \to 2^{W \times W}$; assigns an accessibility relation to each agent;
- $V : \mathcal{P} \to 2^W$; assigns a set of worlds to each propositional variable.

In epistemic logic, R is usually restricted to equivalence relations. In this paper we treat the weakest modal case where there are no restrictions on R, consequently most results also hold for stronger logics. The semantics are defined with respect to models with a distinguished world (the actual world): (M, w). We also call this a model.

The methodology used in dynamic epistemic logics is to regard epistemic actions as *state transformers*, i.e. they convey us from one model to another. The semantics are justified by showing how these state transformers work in examples, and by showing that the results agree with one's intuitions about these kinds of information change. The idea of public announcements is that all the agents simultaneously learn the announced formula, moreover it is common knowledge among them that they learn it. We assume that only *true* announcements can be made. This leads to the following semantics.

Definition 3 (Semantics of PAL). Let a model (M, w) with $M = (W, R, V)$ be given. Let $p \in \mathcal{P}$, $a \in \mathcal{A}$, and $\varphi, \psi \in \mathscr{L}_{\mathsf{PAL}}$.

$$
\begin{array}{lll}
(M,w) \models p & \text{iff} & w \in V(p) \\
(M,w) \models \neg\varphi & \text{iff} & (M,w) \not\models \varphi \\
(M,w) \models \varphi \wedge \psi & \text{iff} & (M,w) \models \varphi \text{ and } (M,w) \models \psi \\
(M,w) \models \Box_a \varphi & \text{iff} & (M,v) \models \varphi \text{ for all } v \text{ such that } (w,v) \in R(a) \\
(M,w) \models [\varphi]\psi & \text{iff} & (M,w) \models \varphi \text{ implies } (M|\varphi, w) \models \psi
\end{array}
$$

The model $M|\varphi = (W', R', V')$ is defined by restricting M to those worlds where φ holds. Let $[\![\varphi]\!] = \{v \in W | (M,v) \models \varphi\}$. Then $W' = [\![\varphi]\!]$, and the accessibility relations and valuation are restricted to $[\![\varphi]\!]$ as well: $R'(a) = R(a) \cap [\![\varphi]\!]^2$ and $V'(p) = V(p) \cap [\![\varphi]\!]$. If Γ is a set of formulas, then $Cn(\Gamma) = \{\varphi |$ for all (M,w), if $(M,w) \models \Gamma$, then $(M,w) \models \varphi\}$.

Definition 4 (Proof system for PAL). The proof system for PAL consists of all the axioms and rules of K (for modal operators $\Box_a$ and $[\varphi]$) plus the following axioms:

Atoms	$[\varphi]p \leftrightarrow (\varphi \rightarrow p)$	(atoms)
PF	$[\varphi]\neg\psi \leftrightarrow (\varphi \rightarrow \neg[\varphi]\psi)$	(partial functionality)
Distr	$[\varphi](\psi \wedge \chi) \leftrightarrow ([\varphi]\psi \wedge [\varphi]\chi)$	(distribution)
Ramsey	$[\varphi]\Box_a\psi \leftrightarrow (\varphi \rightarrow \Box_a[\varphi]\psi)$	(Ramsey axiom)

Completeness for this proof system is easy: One shows that every formula of the language of public announcements can be translated to a provably equivalent formula without announcement operators.

Definition 5 (Translation). The translation function t takes a formula from the language of PAL and yields a formula in the language of epistemic logic, in such a way that $t(p) = p$, and t distributes over $\neg$, $\wedge$ and $\Box_a$. The announcement operator is treated as follows:

$$\begin{array}{lcl} t([\varphi]p) & = & t(\varphi) \rightarrow p \\ t([\varphi]\neg\psi) & = & t(\varphi) \rightarrow \neg t([\varphi]\psi) \\ t([\varphi](\psi \wedge \chi)) & = & t([\varphi]\psi) \wedge t([\varphi]\chi) \\ t([\varphi]\Box_a\psi) & = & t(\varphi) \rightarrow \Box_a t([\varphi]\psi) \\ t([\varphi][\psi]\chi) & = & t([\varphi]t([\psi]\chi)) \end{array}$$

Lemma 1 (Plaza). $\vdash \varphi \leftrightarrow t(\varphi)$ for every formula φ.

Some have commented on the fact that every formula containing announcements is equivalent to a formula without any announcements, by remarking that announcements are merely syntactic sugar. However, applying the same line of argument to the study of propositional logic, one would be forced to say that one can make do with a language containing Sheffer's stroke as the only logical operator. Although this is technically feasible, it makes no sense philosophically, and practically it is quite bothersome. And, in the case of public announcement logic the above translation also demonstrates its succintness with respect to epistemic logic, since the translation may give a formula with a size exponentional in its input.

The next definition and lemma will be used in section 4.

Definition 6 (Epistemic depth). The epistemic depth of a formula is a function $d : \mathscr{L}_{\mathsf{PAL}} \rightarrow \mathbb{N}$ such that:

$$\begin{array}{lcl} d(p) & = & 0 \\ d(\neg\varphi) & = & d(\varphi) \\ d(\varphi \wedge \psi) & = & \max(d(\varphi), d(\psi)) \\ d(\Box_a\varphi) & = & d(\varphi) + 1 \\ d([\varphi]\psi)) & = & d(\varphi) + d(\psi) \end{array}$$

Lemma 2. For all formula $\varphi \in \mathscr{L}_{\mathsf{PAL}}$, φ and $t(\varphi)$ have the same epistemic depth.

Proof. By strong induction on n. The base case is easy. For the induction step one can simply check every clause of the translation. ■

Example 1. For an example, consider the formula $[\Box_a p]\Box_b q$. Its translation into epistemic logic is $t([\Box_a p]\Box_b q) = t(\Box_a p) \to \Box_b t([\Box_a p]q) = \Box_a p \to \Box_b(t(\Box_a p) \to q) = \Box_a p \to \Box_b(\Box_a p \to q)$. According to the definition of modal depth, $d([\Box_a p]\Box_b q) = 2$ and this indeed reflects that its translation $\Box_a p \to \Box_b(\Box_a p \to q)$ has a stack of two epistemic operators.

The following lemma will be used in the the proof of Theorem 3 in Section 4.

Lemma 3. Let $d(\psi) = n$. Then: $\vdash E^n\varphi \to ([\varphi]\psi \leftrightarrow \psi)$

Proof. If $d(\psi) = 0$, it follows straightforwardly from atoms, partial functionality and distribution.

If $d(\psi) = n + 1$, the interesting case is when ψ is of the form $\Box_a\chi$. Then by the Ramsey axiom $\vdash [\varphi]\Box_a\chi \leftrightarrow (\varphi \to \Box_a[\varphi]\chi)$. By propositional reasoning we get $\vdash (E^{n+1}\varphi \to [\varphi]\Box_a\chi) \leftrightarrow (E^{n+1}\varphi \to (\varphi \to \Box_a[\varphi]\psi))$. Observe that $\vdash E^{n+1}\varphi \to \varphi$ and $\vdash E^{n+1}\varphi \to \Box_a E^n\varphi$. With some modal reasoning we get $\vdash (E^{n+1}\varphi \to [\varphi]\Box_a\chi) \leftrightarrow (E^{n+1}\varphi \to \Box_a(E^n\varphi \wedge [\varphi]\psi)$. The induction hypothesis tells us that $\vdash E^n\varphi \to ([\varphi]\psi \leftrightarrow \psi)$. Therefore $\vdash (E^{n+1}\varphi \to [\varphi]\Box_a\chi) \leftrightarrow (E^{n+1}\varphi \to \Box_a(E^n\varphi \wedge \psi))$. With some more modal and propositional reasoning we finally get $E^n\varphi \to ([\varphi]\Box_a\chi \leftrightarrow \Box_a\chi)$. ■

3 Belief expansion

In belief revision the beliefs of an agent are not represented by means of a Kripke model, but with a set of propositional formulas, called a belief set. Therefore all the logical notions refer to propositional logic.

Definition 7 (Belief sets). A belief set K is a set of formulas closed under logical consequence, i.e. $K = Cn(K)$.

One of the types of information change studied in belief revision is *belief expansion*. When an agent accepts the information that φ, the agent is said to *expand* his or her beliefs with φ. The methodology used in this branch of logic, is to state properties that such a change of information satisfies, assuming that the agent performing the information change is rational. These are called *rationality postulates*, which can be justified by arguments that appeal to one's intuition about the information change. Then one can set out to prove that these postulates uniquely characterize a certain operation on belief sets. In [1] the expansion operation $+$ is characterized by the following rationality postulates.

Definition 8 (Rationality postulates for belief expansion).

K1 $K + \varphi$ is a belief set.

K2 $\varphi \in K + \varphi$.

K3 $K \subseteq K + \varphi$.

K4 If $\varphi \in K$, then $K + \varphi = K$.

K5 If $K \subseteq K'$, then $K + \varphi \subseteq K' + \varphi$.

K6 $K + \varphi$ is the smallest belief set that satisfies **K1-K5**.

This operation turns out to be fully characterized by these six postulates. This leads to the following theorem.

Theorem 1 (Gärdenfors [7]). The expansion function $+$ satisfies **K1-K6** iff $K + \varphi = Cn(K \cup \{\varphi\})$

So the operation of expansion is simply adding the new information to the agent's belief set and closing under logical consequence. In the next section we show that under certain conditions, public announcements can also be seen as expansions.

4 Positive knowledge

So far, we have seen two logical approaches that provide an analysis of the same kind of information change: incorporating new information. Therefore one would expect that, although technically they are worked out differently, there are great similarities. However, the similarities turn out to be limited to a very special case. The postulates of expansion, discussed in Section 3, do not straightforwardly translate to properties in the context of dynamic epistemic logic. First of all, many logics for knowledge and belief assume the properties of *positive* ($\Box\varphi \rightarrow \Box\Box\varphi$) and *negative introspection* ($\neg\Box\varphi \rightarrow \Box\neg\Box\varphi$) and this makes it very hard to compare belief sets, because these cannot even be expressed in terms of belief sets of propositional formulas. But even if we extend belief sets with modal formulas there is a problem. Suppose we take a belief set to be some set $K_{(M,w)} = \{\varphi \mid (M, w) \models \Box\varphi\}$, for a given model (M, w). Take two such sets K and K'. Now suppose that $K \subseteq K'$. This implies that $K = K'$. This can be seen as follows. Suppose it is not the case that $K' \subseteq K$. Then there is a formula $\psi \in K'$

and $\psi \notin K$. If $\psi \in K'$, then by positive introspection, it follows that $\Box\psi \in K'$. On the other hand, $\psi \notin K$ implies that in the underlying model $(M, w) \not\models \Box\psi$. Therefore by negative introspection $(M, w) \models \Box\neg\Box\psi$, and so $\neg\Box\psi \in K$. This together with the assumption that $K \subseteq K'$ also gives $\neg\Box\psi \in K'$. And so we have arrived at a contradiction. In other words, fully introspective agents can never add or remove beliefs, without making their belief set inconsistent!

Secondly, postulate **K2** of expansions ($\varphi \in K + \varphi$, also called *success*), is hard to enforce in a setting where we allow for epistemic operators in the object language. To see this, for any epistemic logic satisfying the weak variant $\Box\neg\Box\varphi \rightarrow \neg\Box\varphi$ of veridicality, it not easy to see how an expansion of K with, say, $p \wedge \neg\Box p$ should look like. By the postulate of success, we have $(p \wedge \neg\Box p) \in K$, which is impossible to fulfill in combination with our weak notion of veridicality and the requirements $K \nvdash \bot$ and $K = Cn(K)$.

In this section we show that public announcements *do* behave like belief expansions, when we restrict ourselves to *positive knowledge*, and when the formulas that are learnt are *successful*. The problem (that beliefs sets are incomparable) and the analysis below is not unlike the one presented in [18], where the driving question is what it means that an agent claims to 'only know' some fact φ.

For this purpose we look explicitly at the levels of knowledge that an agent possesses. This leads to the following inductive definition of the positive fragment of the language of public announcement logic.

Definition 9 ($\mathscr{L}_{pos}$). The languages $\mathscr{L}^n_{pos}$ (for each $n \in \mathbb{N}$) are defined inductively as follows, where k is any value smaller than or equal to $n+1$:

$$\begin{array}{lll} \varphi_0 & ::= & p \mid \neg p \mid \varphi_0 \vee \varphi_0 \quad \varphi_0 \wedge \varphi_0 \\ \varphi_{n+1} & ::= & \varphi_n \mid \Box_a \varphi_n \mid [\neg\varphi_k]\varphi_{(n+1)-k} \mid \varphi_{n+1} \vee \varphi_{n+1} \mid \varphi_{n+1} \wedge \varphi_{n+1} \end{array}$$

$\mathscr{L}_{pos} = \bigcup_{n \in \mathbb{N}} \mathscr{L}^n_{pos}$.

Note that if $\varphi \in \mathscr{L}^n_{pos}$, then $d(\varphi) \leq n$. Also note that if $\varphi \in \mathscr{L}^n_{pos}$, then there is a formula ψ without announcement operators, which is equivalent to φ, such that $\psi \in \mathscr{L}^n_{pos}$, by lemma's 1 and 2. Therefore, in the following proofs by induction on $\mathscr{L}^n_{pos}$ we omit the case for $[\varphi]\psi$.

Definition 10 (Finite simulation). Let two models $M = (W, R, V)$ and $M' = (W', R', V')$, and two worlds $w \in W$ and $w' \in W'$ be given. $S \subseteq W \times W'$ is a simulation up to 0 for w and w' iff wSw and for all $p \in \mathcal{P}$ it is the case that $w \in V(p)$ iff $w' \in V'(p)$. $S \subseteq W \times W'$ is a simulation up to $n+1$ for w and w' iff

atoms S is a simulation up to 0 for w and w', and

forth for every $a \in \mathcal{A}$ and $v \in W$, if $wR(a)v$, then there is a $v' \in W'$ such that $w'R'(a)v'$ and S is a simulation up to n for v and v'.

If there is a simulation up to n for w and w' we write $(M, w) \rightleftarrows_n (M', w')$. If there is a simulation up to every $n \in \mathbb{N}$ for we write $(M, w) \rightleftarrows_\omega (M', w')$

The following lemma holds due to the fact that we only have a finite number of propositional variables and only a finite number of agents, and because we only look at formulas up to a certain depth.

Lemma 4 (Propositional finiteness). The number of different propositions that can be expressed by formulas of $\mathscr{L}^n_{pos}$ is finite for every n. More formally: logical equivalence yields a finite partition on $\mathscr{L}^n_{pos}$.

So although there are infinitely many different formulas in the language up to n, only finitely many things can be expressed with it.

Definition 11. Let $Th^n_{pos}(M, w) = \{\varphi \in \mathscr{L}^n_{pos} | (M, w) \models \varphi\}$

Theorem 2. $(M, w) \rightleftarrows_n (M', w')$ iff $Th^n_{pos}(M', w') \subseteq Th^n_{pos}(M, w)$

Proof. By induction on n, the base case being trivial. So suppose the theorem holds up to n.

($\Rightarrow$) Suppose $(M, w) \rightleftarrows_{n+1} (M', w')$. Let $\varphi \in \mathscr{L}^{n+1}_{pos}$. We proceed by induction on φ. If $\varphi \in \mathscr{L}^n_{pos}$ we can simply apply the induction hypothesis. Suppose φ is of the form $\Box_a \psi$. By contraposition. Suppose $(M, w) \not\models \Box_a \psi$. Then there is a world v such that $wR(a)v$ and $(M, v) \not\models \psi$. From forth it follows that there is a v' such that $w'R(a)v'$ and $(M, v) \rightleftarrows_n (M', v')$. Since $\psi \in \mathscr{L}^n_{pos}$, we can apply the contrapositive of the induction hypothesis. Therefore $(M', v') \not\models \psi$, and therefore $(M', w') \not\models \Box_a \psi$.

($\Leftarrow$) Suppose that $Th^{n+1}_{pos}(M', w') \subseteq Th^{n+1}_{pos}(M, w)$. Now let S be, for all worlds: $S = \{(v, v') | Th^{n+1}_{pos}(M', v') \subseteq Th^{n+1}_{pos}(M, v)\}$. We prove that S is a simulation up to $n+1$. The **atoms** clause follows directly. Suppose toward a contradiction that there is a $wR(a)v$ and for all $v' \in W'$ such that $w'R'(a)v'$ the relation S is not a simulation up to n for v and v'. By the induction hypothesis there is a formula $\psi_{vv'} \in \mathscr{L}^n_{pos}$ such that $(M', v') \models \psi_{vv'}$ and $(M, v) \not\models \psi_{vv'}$. Let $\Gamma = \{\psi_{vv'} | w'R'(a)v'\}$. We can assume this set is finite, without loss of generality, by Lemma 4. Therefore $(M', w') \models \Box_a \bigvee \Gamma$ and $(M, w) \not\models \Box_a \bigvee \Gamma$. But this formula is in $\mathscr{L}^{n+1}_{pos}$, therefore $(M, w) \models \Box_a \bigvee \Gamma$. And so we arrive at a contradiction. ■

Example 2. The supermodel relation yields a simulation up to an arbitrary depth. This also shows that the submodel relationship is *not* a simulation beyond a certain level.

For example, consider the following four-state epistemic state (M, w): the set of possible worlds of the model M is $\{w, w', v, v'\}$; there is one fact p,

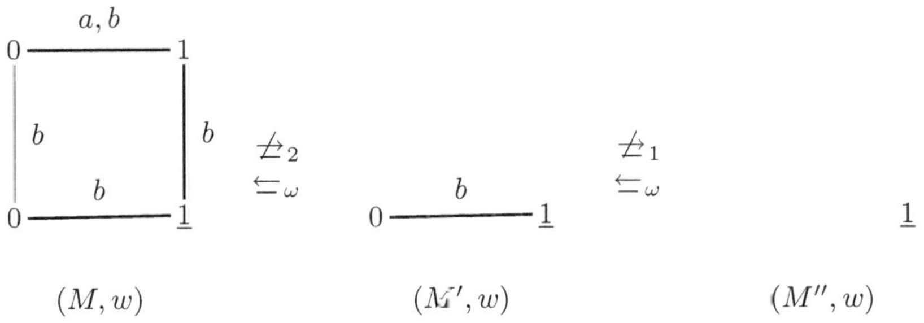

Figure 1. Simulation relations between three example models. All are defined for one atom p only and two agents a and b. The point of the model is underlined. In states named 0 fact p is false, and in states named 1 true. All access is symmetric and transitive, and reflexive access is assumed as well. For example, in the leftmost model access for b is the universal relation; and in the rightmost model access for both a and b is reflexive. The last depicts the situation where they both know that p (and where they know from each other that they know that, etc.).

which is true in w, w' only, access for agent b is universal, and access for a induces the partition $\{w, w'\}, \{v\}, \{v'\}$. The model represents an epistemic state where agent b suspects agent a of knowing the truth about p. We further consider the submodels (M', w) and (M'', w): M' is the submodel of M for restricted domain $\{v, w\}$, and M'' is the restriction of M to $\{w\}$ only. These models are visualized in Figure 1, including their relevant simulation relations. We now observe the following, which is a direct application of Theorem 2.

The formula $\Box_a p$ is true in (M, w). Therefore, by Theorem 2, and because of $(M'', w) \rightleftarrows_\omega (M', w) \rightleftarrows_\omega (M, w)$, it is also true in (M', w) and (M'', w). The formula $\Box_b \neg p$ is false in (M'', w). Therefore, because of $(M'', w) \rightleftarrows_\omega (M', w) \rightleftarrows_\omega (M, w)$, it is also false in (M', w) and (M, w). To show that there is no simulation for (M, w) and (M', w') up to 2, it suffices to observe that $\Box_b(\Box_a p \vee \Box_a \neg p) \in Th^2_{pos}(M', w)$ but $\Box_b(\Box_a p \vee \Box_a \neg p) \notin Th^2_{pos}(M, w)$. There is no simulation for (M', w) and (M'', w) up to 1, because $\Box_b p \in Th^1_{pos}(M'', w)$ but $\Box_b p \notin Th^1_{pos}(M', w)$.

Let $Cn^n_{pos}(\Gamma) = \{\varphi \in \mathcal{L}^n_{pos} |$ for all (M, w), if $(M, w) \models \Gamma$, then $(M, w) \models \varphi\}$.

Theorem 3. Let $\varphi \in \mathcal{L}^n_{pos}$, and let $\neg\varphi$ be successful, i.e. $\models [\neg\varphi]\neg\varphi$, then

$$Th^n_{pos}(M|\neg\varphi, w) = Cn^n_{pos}(Th^n_{pos}(M, w) \cup \{E^n \neg\varphi\})$$

Proof. ($\Rightarrow$) Let $\psi \in \mathcal{L}^n_{pos}$. Suppose $\psi \in Th^n_{pos}(M|\neg\varphi, w)$, i.e. $(M|\neg\varphi, w) \models \psi$. Let (M', w') be an arbitrary model such that $(M', w') \models Th^n_{pos}(M, w) \cup \{E^n\neg\varphi\}$. Therefore $Th^n_{pos}(M, w) \subseteq Th^n_{pos}(M', w')$. By Theorem 2, this is equivalent to $(M', w') \sqsupseteq_n (M, w)$. Therefore $(M'|\neg\varphi, w') \sqsupseteq_n (M|\neg\varphi, w)$. Since $\psi \in \mathcal{L}^n_{pos}$, by Theorem 2, $(M'|\neg\varphi, w') \models \psi$. Therefore $(M', w') \models [\neg\varphi]\psi$. Since $(M', w') \models E^n\neg\varphi$, by Lemma 3, $(M', w') \models \psi$.

($\Leftarrow$) It is sufficient to show that $(M|\neg\varphi, w) \models Th^n_{pos}(M, w) \cup \{E^n\neg\varphi\}$. It is easy to see that $(M|\neg\varphi, w) \sqsupseteq_n (M, w)$. Therefore by Lemma 3 $(M|\neg\varphi, w) \models Th^n_{pos}(M, w)$. Suppose that $(M|\neg\varphi, w) \not\models E^n\neg\varphi$. Then there is a world v accessible from w with a path of length at most n such that $(M|\neg\varphi, v) \models \varphi$. This contradicts $\models [\neg\varphi]\neg\varphi$. Therefore $(M|\neg\varphi, w) \models E^n\neg\varphi$. ■

This theorem shows that those positive formulas which are true after a successful announcement of a negation of a positive formula φ, are exactly those positive formulas which follow from the positive knowledge before the announcement together with knowledge of $\neg\varphi$ up to a certain depth. In order to make the link with belief expansion more clear, we state the following two corollaries, which look at the positive knowledge of one agent.

Corollary 1. Let $K^n_{(M,w)}(a) = \{\varphi \in \mathcal{L}^n_{pos} | (M, w) \models \Box_a\varphi\}$. Let $\varphi \in \mathcal{L}^n_{pos}$, and let $\neg\varphi$ be successful, then $K^n_{(M|\neg\varphi,w)}(a) = Cn^n_{pos}(K^n_{(M,w)}(a) \cup \{E^n\neg\varphi\})$

Propositional formulas are always successful. Therefore we do not need it as an explicit assumption for $\mathcal{L}^0_{pos}$. Also, we can use the ordinary notion of consequence. This leads to the following corollary, saying that public announcements are the same as expansions on a propositional level.

Corollary 2. For $\mathcal{L}^0_{pos}$, $K^0_{(M|\neg\varphi,w)}(a) = Cn(K^0_{(M,w)}(a) \cup \{\neg\varphi\})$

5 Conclusion and further research

In this paper we investigated to what extent public announcement and belief expansion are the same. We showed that the two notions coincide as far as positive knowledge is concerned, but it seems difficult to extend this to broader sublanguages of public announcement logic. This indicates, that even in the simple case of learning something new, one's intuitions may fail. It also poses the question which methodology is to be preferred when studying information change.

There are three obvious main directions for further research on the short term. First of all, now that we have determined a counterpart of *expansions* in DEL, a natural next enterprise is to extend DEL to a framework in which one can characterize a notion of *contraction*, and, once that is in place, a notion of *revision*. The other two directions are partly suggested by similar problems when studying 'only knowing' [18]. First, it is not entirely clear at this stage how our results generalize to, or are dependent, on specific axiomatic systems. Finally, we should expand our analysis to a real *multi-agent* framework, allowing higher order belief revision for individual agents in a larger group, unlike the *public* expansion we have modelled her. Where in our current approach the agent's knowledge can not occur in the scope of a negation, it is not clear how this generalizes to the case of more agents: maybe a formula of type $\Box_a \neg \Box_b \Box_a p$ would not disturb the results proven here.

Acknowledgements

Hans van Ditmarsch thanks the Netherlands Organisation for Scientific Research (NWO) for a grant to facilitate collaboration with Barteld Kooi.

BIBLIOGRAPHY

[1] C. E. Alchourrón, P. Gärdenfors, and D. Makinson. On the logic of theory change: partial meet functions for contraction and revision. *Journal of Symbolic Logic*, 50:510–530, 1985.

[2] R. J. Aumann and A. Brandenburger. Epistemic conditions for Nash equilibrium. *Econometrica*, 63(5):1161–1180, 1995.

[3] A. Baltag and L.S. Moss. Logics for epistemic programs. *Synthese*, 139:165–224, 2004. Knowledge, Rationality & Action 1–60.

[4] S. Benferhat, D. Dubois, H. Prade and M.A. Williams. A practical approach to revising prioritized knowledge bases. *Studia Logica*, 70(1), 2002.

[5] A. Darwiche and J. Pearl. On the logic of iterated belief revision. *Artificial Intelligence*, 89(1-2):1–29, 1997.

[6] R. Fagin, J.Y. Halpern, Y. Moses, and M.Y. Vardi. *Reasoning About Knowledge.* MIT Press, Cambridge, Massachusetts, 1995.

[7] P. Gärdenfors. *Knowledge in Flux, modeling the dynamics of epistemic states.* MIT Press, 1988.

[8] J. Gerbrandy and W. Groeneveld. Reasoning about information change. *Journal of Logic, Language, and Information*, 6:147 –196, 1997.

[9] J. Hintikka. *Knowledge and Belief, An Introduction to the Logic of the Two Notions.* Cornell University Press, Ithaca & London, 1962.

[10] J. Hintikka. Reasoning about knowledge in philosophy. In J. Y. Halpern, editor, *Proceedings of Theoretical Aspects of Reasoning About Knowledge*, pages 63–80. Morgan Kaufmann Publishers, 1986.

[11] H. Katsuno and A. Mendelzon. On the difference between updating a knowledge base and revising it. In *Proceedings of the Second International Conference on Principles of Knowledge Representation and Reasoning*, pages 387–394, 1991.

[12] J.-J.Ch. Meyer and W. van der Hoek. *Epistemic Logic for AI and Computer Science.* Cambridge University Press, Cambridge, 1995.

[13] T.A. Meyer, W.A. Labuschagne, and J. Heidema. Refined epistemic entrenchment. *Journal of Logic, Language and Information*, 9:237–259, 2000.

[14] J. A. Plaza. Logics of public communications. In M. L. Emrich, M. S. Pfeifer, M. Hadzikadic, and Z. W. Ras, editors, *Proceedings of the 4th International Symposium on Methodologies for Intelligent Systems*, pages 201–216, 1989.

[15] K. Segerberg. Two traditions in the logic of belief: bringing them together. In H. J. Ohlbach and U. Reyle, editors, *Logic, Language and Reasoning: essays in honour of Dov Gabbay*, volume 5 of *Trends in Logic*, pages 135–147. Kluwer Academic Publishers, Dordrecht, 1999.

[16] H. P. van Ditmarsch. Descriptions of game actions. *Journal of Logic, Language and Information*, 11(3):349–365, 2002.
[17] H.P. van Ditmarsch, W. van der Hoek, and B.P. Kooi. Concurrent dynamic epistemic logic. In V.F. Hendricks, K.F. Jørgensen, and S.A. Pedersen, editors, *Knowledge Contributors*, pages 45–82. Kluwer Academic Publishers, Dordrecht, 2003. Synthese Library Volume 322.
[18] J. Jaspars W. van der Hoek and E. Thijsse. Persistence and minimality in epistemic logic. *Annals of Mathematics and Artificial Intelligence*, 27(1–4):25–47, 1999 (printed 2000).

Hans van Ditmarsch

Department of Computer Science, University of Otago, PO Box 56, Dunedin 9015, New Zealand,

hans@cs.otago.ac.nz

Wiebe van der Hoek

Department of Computer Science, University of Liverpool, Liverpool L69 7ZF, United
Kingdom

wiebe@csc.liv.ac.uk

Barteld Kooi

Department of Philosophy, University of Groningen, Oude Boteringestraat 52, 9712 GL Groningen, The Netherlands

barteld@philos.rug.nl

Consistency Proofs for Systems of Multi-agent Only Knowing

ARILD WAALER

ABSTRACT. A new and natural multi-modal system of *only knowing* is proposed along with a sequent calculus. The main technical results are cut-elimination theorems and a proof of the consistency of the logic; the latter result follows from the subformula property of cut-free proofs. The system is extended to a multi-modal logic which allows the representation of belief states with confidence levels in a multi-agent context.

1 Introduction

Multi-agent belief logics can be viewed as systems designed for the reasoning about, and representation of, representations that agents use in reasoning about other agents' cognitive states. A multi-agent *only knowing* system has language constructs for representing upper and lower bounds of beliefs, which may be combined to express the exact content of an agent's belief state. The natural starting point for the design of such a system is to generalize the only knowing system of Levesque [8] to the multi-modal case. The tricky part is due to what I shall refer to as the *$\Diamond$-axiom* of Levesque's system: that $\Diamond\varphi$, "φ is logically possible", is an axiom *for each satisfiable, objective* φ ("objective" meaning that φ does not contain any modal operators). One route to generalization of this axiom goes through generalization of the semantical notion of satisfiability. This is what Halpern and Lakemeyer have attempted in a series of papers [7, 4, 5, 6], at the cost of coding the satisfiability relation into the system.

Another possibility is to express the generalized $\Diamond$-axiom syntactically. It is then natural to take the syntactical counterpart of Levesque's $\Diamond$-axiom as the starting point: $\vdash \Diamond\varphi$ if $\varphi \nvdash \bot$ (i.e. if φ is consistent), where φ is any purely Boolean formula. This seemingly circular pattern is in the single modality case completely innocent since the axiom is restricted to φs outside the modal part of the language; in particular it is easy to show that a purely Boolean φ is consistent in the system only if it is consistent in propositional logic. This constraint is clearly too restrictive in the multi-modal setting.

¿From the syntactical perspective, what seems to be appropriate is to generalize the $\Diamond$-axiom to the following: $\vdash \Diamond_k\varphi$ provided that $\varphi \nvdash \bot$, where k is an index individuating an agent, $\Diamond_k$ is a *conceivability* operator for agent k and φ is any formula in which every k-modality which occurs in it, occurs within the scope of a modality for another agent (Halpern and Lakemeyer call such a φ *k-objective*). This formulation arguably gives a cleaner and more transparent proof system than the systems studied by Halpern and Lakemeyer. Note that in the single agent case, the generalized $\Diamond$-axiom is identical to the single-agent formulation which we departed from.

However, the syntactically motivated $\Diamond$-axiom has a circular pattern, and an argument is needed in order to show that the circularity is not vicious. Note, incidentally, that consistency cannot easily be obtained by standard semantical methods. To prove soundness of the $\Diamond$-axiom one needs to recast the consistency condition into a corresponding semantical condition in terms of satisfiability and this, in turn, presupposes that completeness has already been established. The standard way of proving completeness is via maximal consistent sets, a construction which builds on consistency.

The scope of this paper is entirely syntactical. More precisely we define a multi-modal logic closed under a generalization of the syntactical form of the $\Diamond$-axiom. We introduce a sequent calculus for the proposed logic and prove cut-elimination theorems by constructive methods. A consequence of these results is that we obtain a restricted form of the subformula property: a formula φ is provable iff it has a proof using only subformulae of φ.

Modal contexts give rise to natural layers of representation in the syntax. A formula, say $\mathsf{B}_k\mathsf{B}_i\varphi$, expresses a property of agent k's representation of the beliefs of agent i (we regard agent k's representation as one distinct layer). This perspective is illuminating for the interpretation of the subformula property, since it means that the structural layers in the construction of formulae are mirrored in the proof theory. More precisely, statements about one representation layer can be proved only by means of statements about the same layer. It is immediate that every sequent calculus rule except the cut rule respects this property. Moreover, the cut-elimination results show that cut has no damaging effects. A consequence of this it that the circularity in the definition of the $\Diamond$-axiom is not vicious. Note that the introduction of a formula $\Diamond_k\varphi$ on the assumption that φ is consistent has no effect on the assumption itself as long as φ and $\Diamond_k\varphi$ belong to different representation layers. In particular, the consistency of the logic follows. Moreover, using the subformula property we show in section 3 that the system proposed in this paper is indeed equivalent to the system that Halpern and Lakemeyer claim is the correct multi-modal generalization of Levesque's system.

Since the logic for the single agent case is identical to Levesque's system, the paper also presents a sequent calculus for the propositional part of Levesque's logic. We shall also see that the cut-elimination results for the core system can be generalized to a system in which a preorder is imposed on the index set individuating the modalities. The motivation for the latter construct is to enable the representation of confidence levels for each agent; this is necessary for the representation of certain multi-modal defaults along the same lines as one can represent prioritized supernormal defaults in an extension of Levesque's system with confidence levels [9].

2 The multi-agent logic $\mathbf{L}_I$

2.1 Axiomatic system

The object language $\mathcal{L}_I$ contains a stock of propositional letters, the constants $\top$ and $\bot$, the Boolean connectives $\neg$, $\vee$, $\wedge$, $\supset$, and $\equiv$, and modal operators B_k and C_k for each k in a non-empty index set I. The intended interpretation is that the index set I represents the set of agents, B_k is a belief operator, and C_k a complementary *co-belief* operator for agent k. We define the symbol O_k by $\mathsf{O}_k\varphi = \mathsf{B}_k\varphi \wedge \mathsf{C}_k\neg\varphi$. Intuitively $\mathsf{O}_k\varphi$ means that φ is exactly what agent k believes; the construction is inspired by the "all I know" operator of Levesque [8].

Further abbreviations: b_k is $\neg\mathsf{B}_k\neg$, c_k is $\neg\mathsf{C}_k\neg$, $\Box_k\varphi$ is $\mathsf{B}_k\varphi \wedge \mathsf{C}_k\varphi$, the dual $\Diamond_k\varphi$ is $\mathsf{b}_k\varphi \vee \mathsf{c}_k\varphi$. The intended meaning of $\Box_k$ is logical necessity.

A formula is *purely Boolean* if it contains no occurrences of any modality, a *modal atom of modality k* if it is of the form $\mathsf{B}_k\psi$ or $\mathsf{C}_k\psi$ for a $k \in I$, *completely k-modalized* if it is a Boolean combination of modal atoms of modality k, and *free of modality k* if it is a Boolean combination of propositional letters and modal atoms not of modality k. φ is a *first-order formula* if, for each $k \in I$ and each subformula $\mathsf{B}_k\psi$ or $\mathsf{C}_k\psi$ in φ, ψ is free of modality k.

The *modal depth* $m(\varphi)$ of a formula φ expresses the nesting of alternating modalities in φ. Formally, the modal depth of a purely Boolean φ is 0. Otherwise, if φ is $\mathsf{B}_k\psi$ or $\mathsf{C}_k\psi$, let Ψ be the set of modal atoms which occur as subformulae in ψ. Then $m(\varphi)$ is the maximal number in $\{m(\chi) + 1 \mid \chi \in \Psi \text{ and } \chi \text{ is not } k\text{-modalized}\} \cup \{m(\chi) \mid \chi \in \Psi \text{ and } \chi \text{ is } k\text{-modalized}\}$. Otherwise, the modal depth of φ is the maximal $m(\psi)$ for a subformula ψ of φ.

The logic L_I is defined by means of the axioms and inference rules below. We write $\vdash \varphi$ if φ is theorem and $\varphi_1, \ldots, \varphi_n \vdash \psi$ if $(\varphi_1 \wedge \cdots \wedge \varphi_n) \supset \psi$ is a theorem; the provability symbol '$\vdash$' is used also for extensions of L_I and shall always be understood relative to the system at hand. A formula φ is *consistent* if $\varphi \nvdash \bot$.

Let us say that a *tautology* is any substitution instance of a formula valid in classical propositional logic such as $\Box_k\varphi \supset \Box_k\varphi$. L_I is defined as the least set that contains all tautologies, is closed under all instances of the rules

$$\frac{\varphi}{\Box_k\varphi}\ \mathrm{RN} \qquad\qquad \frac{\varphi \quad \varphi \supset \psi}{\psi}\ \mathrm{MP}$$

and contains all instances of the following schemata for each $k \in I$:

K_B: $\mathsf{B}_k(\varphi \supset \psi) \supset (\mathsf{B}_k\varphi \supset \mathsf{B}_k\psi)$
K_C: $\mathsf{C}_k(\varphi \supset \psi) \supset (\mathsf{C}_k\varphi \supset \mathsf{C}_k\psi)$
$B_\Box$: $\mathsf{B}_k\varphi \supset \Box_k\mathsf{B}_k\varphi$
$C_\Box$: $\mathsf{C}_k\varphi \supset \Box_k\mathsf{C}_k\varphi$
$\overline{B}_\Box$: $\neg\mathsf{B}_k\varphi \supset \Box_k\neg\mathsf{B}_k\varphi$
$\overline{C}_\Box$: $\neg\mathsf{C}_k\varphi \supset \Box_k\neg\mathsf{C}_k\varphi$

$\diamond$: $\quad \diamond_k \varphi$ provided $\varphi \not\vdash \bot$, φ free of modality k

The core system without the $\diamond$-axiom is called L'_I. It is easy to show that B_k and C_k are both K45 modalities. The axioms can be viewed as modality reduction rules; the main properties are captured in the following two lemmata, both of which are proved without the $\diamond$-axiom and hence hold in L'_I

LEMMA 1. *Let κ be completely k-modalized and φ be any formula. Then $\vdash \mathsf{B}_k(\kappa \vee \varphi) \equiv (\kappa \vee \mathsf{B}_k\varphi)$ and $\vdash \mathsf{C}_k(\kappa \vee \varphi) \equiv (\kappa \vee \mathsf{C}_k\varphi)$. Hence $\vdash \mathsf{B}_k\kappa \equiv (\kappa \vee \mathsf{B}_k\bot)$ and $\vdash \mathsf{C}_k\kappa \equiv (\kappa \vee \mathsf{C}_k\bot)$.*

Proof. We first show, by induction on θ, that $\vdash \theta \supset \square_k\theta$ for each completely k-modalized θ. The base case in which θ is $\mathsf{B}_k\psi$ follows from axiom $B_\square$; likewise for the other base cases (using $C_\square$, $\overline{B}_\square$, or $\overline{C}_\square$). The induction step follows from the induction hypothesis and simple modal reasoning. To establish the first equivalence we use instances of the K_B axiom, that $\vdash \kappa \supset \mathsf{B}_k\kappa$ and that $\vdash \neg\kappa \supset \mathsf{B}_k\neg\kappa$. The second equivalence is proved similarly. The latter statements follow from the former since $\vdash \kappa \equiv (\kappa \vee \bot)$. ■

LEMMA 2. *Any formula is provably equivalent to a first-order formula with the same modal depth.*

Proof. A modal atom $\mathsf{B}_k\varphi$ is equivalent to a formula of the form $\mathsf{B}_k(\psi_1 \wedge \cdots \wedge \psi_m)$ where each ψ_i is a disjunction of a formula of modality k and a formula free of modality k (either disjunct may be absent). If φ is not free of modality k, we can distribute B_k over the conjunctions and use Lemma 1 and the induction hypothesis on each $\mathsf{B}_k\psi_j$. ■

A *k-block* is a conjunction of formulae of the form $\mathsf{B}_k\psi$ or $\mathsf{C}_k\psi$, possibly negated. An I-block is a conjunction of k-blocks (for zero or more $k \in I$) and possibly also a purely Boolean formula.

LEMMA 3. *Let φ be L_I-consistent. Then there is an I-block ψ such that ψ is L_I-consistent, $\psi \supset \varphi$ is provable in L'_I and $m(\psi) = m(\varphi)$.*

Proof. By Lemma 2 φ has a first-order equivalent with the same modal depth; this formula can be put on disjunctive normal form such that each disjunct is an I-block. If there is a proof of $\bot$ from each disjunct, simple propositional reasoning yields that $\varphi \vdash \bot$. Hence, if φ is L_I-consistent, one of the disjuncts ψ in its DNF equivalent must also be consistent; ψ clearly implies φ. ∎

If φ is free of k, the $\Diamond$-axiom entails that $\Box_k\varphi$ is provable only if φ is provable; otherwise the property holds by the axioms. Hence, by RN, $\vdash \Box_k\varphi$ iff $\vdash \varphi$. The logic of $\Box_k$ is hence S5. The next lemma shows that the $\Diamond$-axiom makes B_k and C_k complementary modalities.

LEMMA 4. *Let φ and ψ be formulae free of modality k such that $\{\neg\varphi, \neg\psi\}$ is L_I-consistent. Then $\vdash \neg(\mathsf{B}_k\varphi \wedge \mathsf{C}_k\psi)$.*

Proof. By assumption, $\Diamond_k(\varphi \vee \psi)$ is an instance of the $\Diamond$-axiom. The result follows from this and the theorems $\mathsf{B}_k\varphi \supset \mathsf{B}_k(\varphi \vee \psi)$ and $\mathsf{C}_k\psi \supset \mathsf{C}_k(\varphi \vee \psi)$. ∎

Structural rules

$$\frac{\Gamma \Rightarrow \Delta}{\Gamma, \varphi \Rightarrow \Delta}\ \textsc{Lt} \qquad \frac{\Gamma \Rightarrow \Delta}{\Gamma \Rightarrow \varphi, \Delta}\ \textsc{Rt}$$

$$\frac{\Gamma, \varphi, \varphi \Rightarrow \Delta}{\Gamma, \varphi \Rightarrow \Delta}\ \textsc{Lc} \qquad \frac{\Gamma \Rightarrow \varphi, \varphi, \Delta}{\Gamma \Rightarrow \varphi, \Delta}\ \textsc{Rc}$$

Identity rules

$$\frac{}{\varphi \Rightarrow \varphi}\ \textsc{Id} \qquad \frac{\Gamma_1 \Rightarrow \varphi, \Delta_1 \quad \Gamma_2, \varphi \Rightarrow \Delta_2}{\Gamma_1, \Gamma_2 \Rightarrow \Delta_1, \Delta_2}\ \textsc{Cut}$$

Logical rules for the connectives

$$\frac{\Gamma \Rightarrow \varphi, \Delta}{\Gamma\neg\varphi \Rightarrow \Delta}\ \mathrm{L}\neg \qquad \frac{\Gamma, \varphi \Rightarrow \Delta}{\Gamma \Rightarrow \neg\varphi, \Delta}\ \mathrm{R}\neg$$

$$\frac{\Gamma, \varphi_i \Rightarrow \Delta}{\Gamma, \varphi_1 \wedge \varphi_2 \Rightarrow \Delta}\ \mathrm{L}\wedge_i\ (i = 1, 2) \qquad \frac{\Gamma_1 \Rightarrow \varphi, \Delta_1 \quad \Gamma_2 \Rightarrow \psi, \Delta_2}{\Gamma_1, \Gamma_2 \Rightarrow \varphi \wedge \psi, \Delta_1, \Delta_2}\ \mathrm{R}\wedge$$

$$\frac{\Gamma_1, \varphi \Rightarrow \Delta_1 \quad \Gamma_2, \psi \Rightarrow \Delta_2}{\Gamma_1, \Gamma_2, \varphi \vee \psi \Rightarrow \Delta_1, \Delta_2}\ \mathrm{L}\vee \qquad \frac{\Gamma \Rightarrow \varphi_i, \Delta}{\Gamma \Rightarrow \varphi_1 \vee \varphi_2, \Delta}\ \mathrm{R}\vee_i\ (i = 1, 2)$$

$$\frac{\Gamma_1 \Rightarrow \varphi, \Delta_1 \quad \Gamma_2, \psi \Rightarrow \Delta_2}{\Gamma_1, \Gamma_2, \varphi \supset \psi \Rightarrow \Delta_1, \Delta_2}\,\mathrm{L}\supset \qquad \frac{\Gamma, \varphi \Rightarrow \psi, \Delta}{\Gamma \Rightarrow \varphi \supset \psi, \Delta}\,\mathrm{R}\supset$$

$$\frac{\Gamma_1, \varphi, \psi \Rightarrow \Delta_1 \quad \Gamma_2 \Rightarrow \varphi, \psi, \Delta_2}{\Gamma_1, \Gamma_2, \varphi \equiv \psi \Rightarrow \Delta_1, \Delta_2}\,\mathrm{L}\equiv \qquad \frac{\Gamma_1, \varphi \Rightarrow \psi, \Delta_1 \quad \Gamma_2, \psi \Rightarrow \varphi, \Delta_2}{\Gamma_1, \Gamma_2 \Rightarrow \varphi \equiv \psi, \Delta_1, \Delta_2}\,\mathrm{R}\equiv$$

Logical rules for the modalities

$$\frac{\Theta^k, \Gamma \Rightarrow \Delta^k, \varphi}{\Theta^k, \Gamma^{\mathsf{B}_k} \Rightarrow \Delta^k, \mathsf{B}_k\varphi}\,\mathrm{LRB}_k \qquad \text{provided } \Delta^k, \varphi \text{ is non-empty}$$

$$\frac{\Theta^k, \Gamma \Rightarrow \Delta^k, \varphi}{\Theta^k, \Gamma^{\mathsf{C}_k} \Rightarrow \Delta^k, \mathsf{C}_k\varphi}\,\mathrm{LRC}_k \qquad \text{provided } \Delta^k, \varphi \text{ is non-empty}$$

$$\frac{}{\mathsf{B}_k\varphi, \mathsf{C}_k\psi \Rightarrow}\,\textsc{Conv} \qquad \text{provided } \{\neg\varphi, \neg\psi\} \text{ is } \mathrm{L}_I\text{-consistent}$$

The φ may be absent in both LRB_k and LRC_k. In Conv φ and ψ are both free of modality k.

Figure 1. Sequent calculus rules for L_I.

2.2 Sequent calculus

A *sequent* is an expression of the form $\Gamma \Rightarrow \Delta$, where the *antecedent* Γ and the *succedent* Δ are finite multisets of formulae, i.e. sets which may contain several occurrences of the same formula. We shall use usual conventions like Γ, φ for $\Gamma \cup \{\varphi\}$. Intuitively a sequent $\Gamma \Rightarrow \Delta$ corresponds to the formula $(\bigwedge\Gamma) \supset (\bigvee\Delta)$, which we in the following shall denote $\Gamma \rightarrow \Delta$; $(\bigwedge\Gamma)$ is $\top$ when $\Gamma = \emptyset$ and $(\bigvee\Delta)$ is $\bot$ when $\Delta = \emptyset$.

The rules of the sequent calculus for L_I are given in Figure 1. In the definition of the rules Γ^{B_k} is the set $\{\mathsf{B}_k\varphi \mid \varphi \in \Gamma\}$, Γ^{C_k} is defined correspondingly. When a set Γ is superscripted with k, like Γ^k, it contains only modal atoms of modality k.

Proofs are trees regulated by the rules whose leaves are axioms. Note that an axiom is a zero-premiss rule. Two inferences are *successive* if the conclusion of the first is a premiss of the next. In derivation figures a series of 0 or more successive structural inferences is indicated with a double line. The *height* $h(\pi)$ of a proof π is defined as the number of successive inferences (including structural inferences) on the longest branch in π. If π is an axiom, $h(\pi) = 1$.

THEOREM 5 (Correctness). $\Gamma \Rightarrow \Delta$ *is provable in the sequent calculus iff* $\vdash \Gamma \rightarrow \Delta$.

Proof. "If" is proved by routine induction on the axiomatic derivation. The axioms are provable in the sequent calculus; in particular is $\Rightarrow \Diamond_k \varphi$ provable if $\Diamond_k \varphi$ is an instance of the $\Diamond$-axiom. RN is replaced by applications of LRB_k and LRC_k, while MP is replaced by CUT. The other direction is proved by induction on a proof π of $\Gamma \Rightarrow \Delta$. All the rules for the Boolean connectives are standard. We consider the three modality rules.

Case 1: the last inference in π is

$$\frac{\Theta^k, \Gamma \Rightarrow \Delta^k, \varphi}{\Theta^k, \Gamma^{\mathsf{B}_k} \Rightarrow \Delta^k, \mathsf{B}_k \varphi}\ \text{LRB}_k$$

The rule can be used with φ absent, provided that Δ is non-empty. By induction hypothesis, $\vdash \Theta^k, \Gamma \rightarrow \Delta^k, \varphi$. By RN and modal logic, $\vdash \mathsf{B}_k(\Theta^k, \Gamma \rightarrow \Delta^k, \varphi)$. Let $\Theta^k = \{\theta_1, \ldots, \theta_m\}$ and $\Delta^k = \{\delta_1, \ldots, \delta_n\}$. There are two cases, depending on whether φ is present in the succedent of the premiss or not. Assume first that φ is present. It follows from Lemma 1 and standard modal logic that (i) $\vdash \mathsf{B}_k(\theta_1 \wedge \cdots \wedge \theta_m) \equiv (\theta_1 \wedge \cdots \wedge \theta_m) \vee \mathsf{B}_k \bot$, (ii) $\vdash \mathsf{B}_k(\delta_1 \vee \cdots \vee \delta_n \vee \varphi) \equiv (\delta_1 \vee \cdots \vee \delta_n \vee \mathsf{B}_k \varphi)$. By (i), (ii) and principles of normal modal logics it follows that $\vdash (\theta_1 \wedge \cdots \wedge \theta_m) \vee \mathsf{B}_k \bot, \Gamma^{\mathsf{B}_k} \rightarrow \Delta, \mathsf{B}_k \varphi$, from which we conclude that $\vdash \Theta^k, \Gamma^{\mathsf{B}_k} \rightarrow \Delta^k, \mathsf{B}_k \varphi$. If φ is absent, we can reason as if φ is $\bot$. We then get $\vdash (\theta_1 \wedge \cdots \wedge \theta_m) \vee \mathsf{B}_k \bot, \Gamma^{\mathsf{B}_k} \rightarrow \Delta^k, \mathsf{B}_k \bot$. Since Δ^k is non-empty, $\vdash \mathsf{B}_k \bot \rightarrow \Delta^k$. Hence $\vdash \Theta^k, \Gamma^{\mathsf{B}_k} \rightarrow \Delta^k$.

Case 2: the last inference in π is LRC_k. Same as case 1.

Case 3: the last inference in π is CONV. Apply Lemma 4. ∎

2.3 Cut-elimination theorems

The proof of the Cut-elimination theorem is inspired by Girard [3] and Shvarts [10]. As usual for such arguments, the proof breaks down into a list of cases, each of which is simple to handle. In our proof of the cut-elimination theorems we follow Gentzen's construction in the original paper [2] and replace the CUT rule by the MIX:

$$\frac{\Gamma_1 \Rightarrow (\varphi)^m, \Delta_1 \quad \Gamma_2, (\varphi)^n \Rightarrow \Delta_2}{\Gamma_1, \Gamma_2 \Rightarrow \Delta_1, \Delta_2}\ \text{MIX}$$

where $\varphi \notin \Delta_1$, $\varphi \notin \Gamma_2$, and $(\varphi)^m$ denotes m occurrences of φ. φ is called the *mix formula* of the inference. It is easy to show that a CUT can be accomplished by a MIX and structural inferences, and *vice versa*.

DEFINITION 6. The *degree* $d(\chi)$ of a formula χ is inductively defined by

(i) $d(p) = 1$, p a propositional letter,

(ii) $d(\neg\varphi) = d(\mathsf{B}_k\varphi) = d(\mathsf{C}_k\varphi) = d(\varphi) + 1$,

(iii) $d(\varphi \circ \psi) = \max\{d(\varphi), d(\psi)\} + 1$; $\circ$ a binary connective.

The degree of a MIX (CUT) inference is the degree of its mix (cut) formula. The degree $d(\pi)$ of a proof π is the maximum $d(\varphi)$ such that φ is the mix formula of a MIX inference in π *and such that the φ in the right premiss of the* MIX *is not introduced by a* CONV *inference.* Note that a mix formula introduced by a CONV inference in the right premiss of a MIX does not contribute to the degree of a proof.

LEMMA 7 (Principal lemma). *Let $\Sigma - \varphi$ be Σ with every occurrence of φ removed. Assume that π_1 is a proof of $\Sigma_1 \Rightarrow \Phi_1$ and π_2 is a proof of $\Sigma_2 \Rightarrow \Phi_2$ such that $d(\pi_1) < d(\varphi)$ and $d(\pi_2) < d(\varphi)$. Then there is a proof π of $\Sigma_1, \Sigma_2 - \varphi \Rightarrow \Phi_1 - \varphi, \Phi_2$ such that $d(\pi) < d(\varphi)$.*

Proof. Note first that if $\varphi \notin \Phi_1$, $\Phi_1 = \Phi_1 - \varphi$; π can be constructed using thinnings:

$$\genfrac{}{}{0pt}{}{\pi_1 \vdots}{\dfrac{\Sigma_1 \Rightarrow \Phi_1 - \varphi}{\Sigma_1, \Sigma_2 - \varphi \Rightarrow \Phi_1 - \varphi, \Phi_2}}$$

In particular this is the case when π_1 is an identity axiom $\psi \Rightarrow \psi$ where $\psi \neq \varphi$. In the same way we use thinnings when $\varphi \notin \Sigma_2$. We assume in the following that $\varphi \in \Phi_1$ and $\varphi \in \Sigma_2$.

Assume that the last rule r_1 of π_1 has premisses $\Sigma'_i \Rightarrow \Phi'_i$ proved by π'_i and that the last rule r_2 of π_2 has premisses $\Sigma''_j \Rightarrow \Phi''_j$ proved by π''_j. The lemma is proved by induction on $h(\pi_1) + h(\pi_2)$. There are several cases to consider, but the only interesting one is really the case in which r_1 and r_2 are logical inferences such that φ is the principal formula of r_1 in the succedent of the conclusion and φ is the principal formula of r_2 in the antecedent of the conclusion. We consider the subcase where $\varphi = \mathsf{B}_k\psi$; the case when φ is $\mathsf{C}_k\psi$ is symmetrical, and all the other cases are standard. π_1 must be as follows:

$$\dfrac{\begin{array}{c}\pi'_1\vdots\\ \Theta_1^k, \Gamma_1 \Rightarrow \Delta_1^k, (\mathsf{B}_k\psi)^m, \psi\end{array}}{\Theta_1^k, \Gamma_1^{\mathsf{B}_k} \Rightarrow \Delta_1^k, (\mathsf{B}_k\psi)^m, \mathsf{B}_k\psi}\, r_1$$

Case 1: r_2 is LRB$_k$. π_2 is then

$$\dfrac{\begin{array}{c}\pi''_1\vdots\\ \Theta_2^k, \Gamma_2, (\psi)^j, (\mathsf{B}_k\psi)^n \Rightarrow \Delta_2^k, \chi\end{array}}{\Theta_2^k, \Gamma_2^{\mathsf{B}_k}, (\mathsf{B}_k\psi)^j, (\mathsf{B}_k\psi)^n \Rightarrow \Delta_2^k, \mathsf{B}_k\chi}\, r_2$$

Apply the induction hypothesis to π'_1 and π_2; call the resulting proof π'. Then, apply the induction hypothesis to π''_1 and π_1; this gives a proof π''. The two proofs are then merged with a MIX of degree less than $d(\varphi)$ (the MIX is the uppermost inference):

$$\dfrac{\dfrac{\dfrac{\dfrac{\begin{array}{c}\pi'\vdots\\ \Theta_1^k, \Gamma_1, \Theta_2^k, \Gamma_2^{\mathsf{B}_k} \Rightarrow \Delta_1^k, \Delta_2^k, \mathsf{B}_k\chi, \psi\end{array} \qquad \begin{array}{c}\pi''\vdots\\ \Theta_1^k, \Gamma_1^{\mathsf{B}_k}, \Theta_2^k, \Gamma_2, (\psi)^j \Rightarrow \Delta_1^k, \Delta_2^k, \chi\end{array}}{\Theta_1^k, \Theta_1^k, \Theta_2^k, \Theta_2^k, \Gamma_1, \Gamma_2, \Gamma_1^{\mathsf{B}_k}, \Gamma_2^{\mathsf{B}_k} \Rightarrow \Delta_1^k, \Delta_1^k, \Delta_2^k, \Delta_2^k, \mathsf{B}_k\chi, \chi}}{\Theta_1^k, \Theta_2^k, \Gamma_1, \Gamma_2, \Gamma_1^{\mathsf{B}_k}, \Gamma_2^{\mathsf{B}_k} \Rightarrow \Delta_1^k, \Delta_2^k, \mathsf{B}_k\chi, \chi}}{\Theta_1^k, \Theta_2^k, \Gamma_1^{\mathsf{B}_k}, \Gamma_1^{\mathsf{B}_k}, \Gamma_2^{\mathsf{B}_k}, \Gamma_2^{\mathsf{B}_k} \Rightarrow \Delta_1^k, \Delta_2^k, \mathsf{B}_k\chi, \mathsf{B}_k\chi}\,\mathrm{LRB}_k}{\Theta_1^k, \Theta_2^k, \Gamma_1^{\mathsf{B}_k}, \Gamma_2^{\mathsf{B}_k} \Rightarrow \Delta_1^k, \Delta_2^k, \mathsf{B}_k\chi}$$

Case 2: π_2 is

$$\frac{}{\mathsf{B}_k\psi, \mathsf{C}_k\chi \Rightarrow}\ \textsc{Conv}$$

The induction hypothesis gives a proof π' of $\Theta_1^k, \mathsf{C}_k\chi, \Gamma_1 \Rightarrow \Delta_1^k, \psi$. π is then

$$\dfrac{\dfrac{\dfrac{\pi' \vdots}{\Theta_1^k, \mathsf{C}_k\chi, \Gamma_1 \Rightarrow \Delta_1^k, \psi}}{\Theta_1^k, \mathsf{C}_k\chi, \Gamma_1^{\mathsf{B}_k} \Rightarrow \Delta_1^k, \mathsf{B}_k\psi} \quad \dfrac{}{\mathsf{B}_k\psi, \mathsf{C}_k\chi \Rightarrow}\ \textsc{Conv}}{\dfrac{\Theta_1^k, \mathsf{C}_k\chi, \mathsf{C}_k\chi, \Gamma_1^{\mathsf{B}_k} \Rightarrow \Delta_1^k}{\Theta_1^k, \mathsf{C}_k\chi, \Gamma_1^{\mathsf{B}_k} \Rightarrow \Delta_1^k}}\ \textsc{Cut}$$

At this point a Cut is introduced in the proof. By definition of degree, this Cut inference does not increase the degree of the proof. ∎

THEOREM 8 (Basic Cut-elimination). *Let $\Gamma \Rightarrow \Delta$ be L_I-provable. Then there is a proof of $\Gamma \Rightarrow \Delta$ which contains a* Cut *inference only if the cut formula in the right premiss is introduced by* Conv.

Proof. First we can prove the following result by induction on π (the induction hypothesis uses the Principal lemma 7): Let π prove $\Gamma \Rightarrow \Delta$, and assume $d(\pi) > 0$. Then there is a proof π' of $\Gamma \Rightarrow \Delta$ such that $d(\pi') < d(\pi)$. By repeated application of this result we can transform any proof into a proof with the same end-sequent, and with degree 0. ∎

COROLLARY 9. *For each formula φ there is a first-order formula ψ such that the sequent $\Rightarrow \varphi \equiv \psi$ has a cut-free proof.*

Proof. We know from the proof of Lemma 2 that the equivalence can be proved without reference to the $\Diamond$-axiom. There is thus a sequent calculus proof of it which does not use Conv. Hence, if the proof contains instances of Mix, they can all be eliminated. ∎

LEMMA 10. *Let L be a logic containing classical logic. Assume that $\{\neg\varphi, \neg\psi\}$ is L-consistent and that $\theta_1, \ldots, \theta_n \Rightarrow \varphi$ is provable. Then there is a θ_i, $1 \leq i \leq n$, such that $\{\neg\theta_i, \neg\psi\}$ is L-consistent.*

Proof. Assume the contrary, i.e. that $\neg\psi \Rightarrow \theta_j$ is provable for each $j \leq n$. Since we have a proof of $\theta_1, \ldots, \theta_n \Rightarrow \varphi$, n applications of CUT give a proof of $\neg\psi \Rightarrow \varphi$, contradicting the consistency of $\{\neg\varphi, \neg\psi\}$. ∎

THEOREM 11 (Cut-elimination for normal form input). *Let $\Gamma \Rightarrow \Delta$ be L_I-provable and assume that all formulae in Γ, Δ are first-order. Then the sequent has a cut-free proof.*

Proof. By Theorem 8 there is a proof of $\Gamma \Rightarrow \Delta$ with degree 0. This means that any right CUT formula is introduced by CONV. Consider the last inference in the proof π.

$$\dfrac{\begin{array}{c}\pi'\vdots\\ \Theta \Rightarrow \Psi, \mathsf{B}_k\varphi\end{array} \qquad \overline{\mathsf{B}_k\varphi, \mathsf{C}_k\psi \Rightarrow}}{\Theta, \mathsf{C}_k\psi \Rightarrow \Psi}\ \text{CUT}$$

We may without loss of generality assume that Θ and Ψ contain only first-order modal atoms of modality k. The idea is to replace π' by another proof π'' and derive $\Theta, \mathsf{C}_k\psi \Rightarrow \Psi$ by thinnings from the end-sequent of π''. There are two cases. If the succedent occurrence of $\mathsf{B}_k\varphi$ in the end-sequent of π' is not traceable to an axiom (i.e. it is never used to close a branch), we can simply remove it from all sequents in π' in which it occurs. This will give us a proof π'' of $\Theta \Rightarrow \Psi$, from which $\Theta, \mathsf{C}_k\psi \Rightarrow \Psi$ follows by thinning. Otherwise, we must use LRB_k on $\mathsf{B}_k\varphi$. Assume for simplicity that this is the lowermost logical inference in π'; the argument for the general case is essentially the same. By this assumption the premiss of the inference must be of the form $\Gamma^k, \theta_1, \ldots, \theta_n \Rightarrow \Delta^k, \varphi$, where $\mathsf{B}_k\theta_1, \ldots, \mathsf{B}_k\theta_n$ are in Θ. Since $\theta_1, \ldots, \theta_n, \varphi$ are all first-order they will be lost in a further inverse application of a LR modality rule. Hence $\theta_1, \ldots, \theta_n \Rightarrow \varphi$ must be provable. Moreover, by definition of the CONV rule $\{\neg\varphi, \neg\psi\}$ is L_I-consistent and free of modality k. By Lemma 10 there is a θ_j such that $\{\neg\theta_j, \neg\psi\}$ is L_I-consistent. Thus the following is an axiom:

$$\overline{\mathsf{B}_k\theta_j, \mathsf{C}_k\psi \Rightarrow}\ \text{CONV}$$

The axiom now serves the function as π''. Since $\mathsf{B}_k\theta_j \in \Theta$, we see that $\Theta, \mathsf{C}_k\psi \Rightarrow \Psi$ follows from $\mathsf{B}_k\theta_j, \mathsf{C}_k\psi \Rightarrow$ by thinnings. By repeating this procedure, we can remove every instance of CUT. ■

2.4 Consistency results for $\mathbf{L}_I$

L_I is a *consistent* logic if there are formulae in the language which are not theorems of L_I; equivalently, if $\perp$ is not derivable. We shall now see why the cut-elimination results entail that L_I is consistent.

Let us define the *rank* of an instance $\Diamond_k\varphi$ of the $\Diamond$-axiom as the modal depth of φ. The rank of an instance of the CONV-rule with conclusion $\mathsf{B}_k\varphi, \mathsf{C}_k\psi \Rightarrow$ is the maximum of $m(\varphi)$ and $m(\psi)$.

LEMMA 12. *Let φ be L_I-provable. Then there is an L_I-proof π of φ such that the rank of every $\Diamond$-axiom in π is less than $m(\varphi)$.*

Proof. It is immediate from Corollary 9 and Theorem 11 that there is a sequent calculus proof of $\Rightarrow \varphi$ such that the rank of every of its CONV-inferences is less than $m(\varphi)$. By inspecting the proof of Theorem 5, which constructs an axiomatic proof from a sequent calculus proof, we see that each CONV-inference is mapped to a $\Diamond$-axiom with the same rank, and no other instances of the $\Diamond$-axiom are introduced. ■

THEOREM 13. *L_I is consistent.*

Proof. It is clear from Theorem 8 that the core system L'_I is consistent since there are obviously sequents without a cut-free proof in the calculus without the CONV-rule. Since the modal depth of $\perp$ is 0, it is immediate from Lemma 12 that there is no proof in L_I of it either. ■

In fact we can prove a stronger property, namely that the $\Diamond$-axiom is well-defined. To prove this, we must prove that for any instance $\Diamond_k\varphi$ of the $\Diamond$-axiom it is sound to establish the unprovability of $\neg\varphi$ in the side condition of the axiom without any reference to this instance of the $\Diamond$-axiom. However, should this not be the case, there must be a proof of $\neg\varphi$ with an instance of the $\Diamond$-axiom with a rank at least equal to $m(\neg\varphi)$, contradicting Lemma 12. Hence, there can be no circularity in proofs due to the $\Diamond$-axiom.

3 Equivalence with Halpern and Lakemeyer's logic

In an attempt to define a correct multi-modal generalization of the single-agent $\Diamond$-axiom of Levesque, Halpern and Lakemeyer [6] enrich the object language with two dual operators *Val* and *Sat*. Let us call their language $\mathcal{L}_{HL}$; note that $\mathcal{L}_{HL}$ is an extension of $\mathcal{L}_I$.

Recall that the core system L'_I is the logic over $\mathcal{L}_I$ defined in section 2.1 without the $\Diamond$-axiom. The system L_{HL} by Halpern and Lakemeyer is defined over $\mathcal{L}_{HL}$ as the weakest system which is closed under the axioms and rules of L'_I, the axioms

V1 $Val(\varphi \supset \psi) \supset (Val(\varphi) \supset Val(\psi))$

V2 $Sat(\varphi)$, provided φ is a satisfiable purely Boolean formula

V3 $(Sat(\alpha \wedge \beta_1) \wedge \cdots \wedge Sat(\alpha \wedge \beta_m) \wedge$
$Sat(\gamma \wedge \delta_1) \wedge \cdots \wedge Sat(\gamma \wedge \delta_n) \wedge Val(\alpha \vee \gamma)) \supset$
$Sat(\mathsf{B}_k\alpha \wedge \mathsf{b}_k\beta_1 \wedge \cdots \wedge \mathsf{b}_k\beta_m \wedge \mathsf{C}_k\gamma \wedge \mathsf{c}_k\delta_1 \wedge \cdots \wedge \mathsf{c}_k\delta_n)$,
provided α, γ and each β_i, δ_j are free of modality k

V4 $(Sat(\alpha) \wedge Sat(\beta)) \supset Sat(\alpha \wedge \beta)$, if α is free of modality k and β is completely k-modalized

SAT $Sat(\varphi) \supset \Diamond_k\varphi$, provided φ is free of modality k

and the following rule of inference:

$$\frac{\varphi}{Val(\varphi)}\ \mathrm{VAL}$$

The deducibility relation for L_{HL} is denoted $\vdash_{HL}$. By VAL and V1, *Val* is a normal modal operator; *Val* can in fact be shown to be S5. Moreover, *Val* and *Sat* correspond to provability and consistency in the intended way.

PROPOSITION 14. *Let φ be a formula in $\mathcal{L}_{HL}$. If $\vdash_{HL} \varphi$, then $\vdash_{HL} Val(\varphi)$. If $\nvdash_{HL} \varphi$, then $\vdash_{HL} \neg Val(\varphi)$.*

Proof. Consult Proposition 5.1 of [6]. ■

THEOREM 15. *Let φ be a formula in $\mathcal{L}_{HL}$. Then φ is a theorem of L_I if and only if it is a theorem of L_{HL}.*

Proof. We prove both directions of the theorem simultaneously by induction on the modal depth of φ. For the left-to-right direction, assume that φ is L_I-probable. By Lemma 12 there is an L_I-proof π of φ such that the modal depth of every $\Diamond$-axiom in π is less that $m(\varphi)$. By induction on the length of π we can now construct an L_{HL}-proof of φ. The only axiom in L_I which is not in L_{HL} is the $\Diamond$-axiom. If φ is purely Boolean, there are no such occurrences and we are done; this situation occurs in the base case of the outer induction. However, if $\Diamond_k\chi$ is an instance of this axiom in π we know that $m(\chi) < m(\varphi)$, that χ is free of modality k and that χ is L_I-consistent. By the induction hypothesis there is an L_{HL}-proof of $Sat(\chi)$. By axiom SAT we conclude that $\Diamond_k\chi$ is L_{HL}-provable. There is hence a proof of φ in L_{HL}.

For the right-to-left direction, it suffices, by Proposition 14, to prove that $\varphi \nvdash \bot$ entails that $\vdash_{HL} Sat(\varphi)$. So assume that φ is L_I-consistent. If φ is purely Boolean, the result follows from axiom V2 and the fact that $\varphi \nvdash \bot$ iff φ is propositionally consistent. In the inductive step we use Lemma 3 to conclude that there is an I-block ψ with $m(\psi) = m(\varphi)$ such that ψ is L_I-consistent and $\psi \supset \varphi$ is provable in the core system L'_I. ψ is the conjunction of a purely Boolean ψ_0 and k-blocks ψ_k for zero or more $k \in I$. Since ψ is L_I-consistent, each k-block ψ_k is L_I-consistent. We show that $\vdash_{HL} Sat(\psi_k)$ for each of these. If $m(\psi_k) < m(\psi)$, this follows by the induction hypothesis. Otherwise, ψ_k is of the form

$$\mathsf{B}_k\alpha \wedge \mathsf{b}_k\beta_1 \wedge \cdots \wedge \mathsf{b}_k\beta_m \wedge \mathsf{C}_k\gamma \wedge \mathsf{c}_k\delta_1 \wedge \cdots \wedge \mathsf{c}_k\delta_n,$$

where $\alpha, \beta_1, \ldots, \beta_m, \gamma, \delta_1, \ldots, \delta_n$ are formulae free of modality k. $\alpha \wedge \beta_i$ is then L_I-consistent for every $i \leq m$. For suppose not. Then $\alpha \vdash \neg\beta_i$, and so $\psi_k \vdash \mathsf{B}_k\neg\beta_i \wedge \mathsf{b}_k\beta_i$, contradicting the consistency of ψ_k. By the same argument, $\gamma \wedge \delta_j$ is consistent for every $j \leq n$. Furthermore, $\vdash \alpha \vee \gamma$. For if not, then, by Lemma 4, $\vdash \neg(\mathsf{B}_k\alpha \wedge \mathsf{C}_k\gamma)$, contradicting the consistency of ψ_k. By the induction hypothesis, $\vdash_{HL} Sat(\alpha \wedge \beta_i)$, $\vdash_{HL} Sat(\gamma \wedge \delta_j)$ and $\vdash_{HL} \alpha \vee \gamma$. By VAL, the last theorem entails that $\vdash_{HL} Val(\alpha \vee \gamma)$. By axiom V3,

$$\vdash_{HL} Sat(\mathsf{B}_k\alpha \wedge \mathsf{b}_k\beta_1 \wedge \cdots \wedge \mathsf{b}_k\beta_m \wedge \mathsf{C}_k\gamma \wedge \mathsf{c}_k\delta_1 \wedge \cdots \wedge \mathsf{c}_k\delta_n)$$

i.e. $\vdash_{HL} Sat(\psi_k)$. Repeated application of axiom V4 gives $\vdash_{HL} Sat(\psi)$.

Since $\psi \supset \varphi$ is provable in L'_I, $\vdash_{HL} \psi \supset \varphi$. Use VAL and V1 to conclude that $\vdash_{HL} Sat(\varphi)$. ■

4 Including a preorder on I

We will in this section address an interesting extension of the multi-modal logic by imposing further structure to the modalities, and show that the main results in this paper still hold.

Assume that there is a preorder $\leq$ (reflexive and transitive, not necessarily anti-symmetric) on I. Let E be the transitive, symmetric closure of $\leq$; if iEk, we say that i and k are *E-equivalent.* The equivalence class of k modulo E is denoted $\overline{k}$. Extending the terminology used previously we say that a modal atom of a modality E-equivalent to k is a *modal $\overline{k}$-atom*; the notions of formulae free of $\overline{k}$-modalities and completely $\overline{k}$-modalized formulae are defined analogously. By a first-order formula we now understand a formula in which the ψ in each subformula of the form $\mathsf{B}_k\psi$ or $\mathsf{C}_k\psi$ is free of $\overline{k}$-modalities, for any $k \in I$.

The idea is to use the equivalence classes of I modulo E to represent agents. The indices in an equivalence class then denote different confidence levels for the given agent. Hence, if k_1 and k_2 are distinct members of I, they denote two different confidence levels of beliefs; moreover, these two confidence levels pertain to the same agent if and only if k_1 and k_2 are E-equivalent.

The logic we address in this section extends L_I in two ways. First, we introduce a modality $\Box_{\overline{k}}$ for each $k \in I$, add the axiom

$$\Box_{\overline{k}}\varphi \equiv \Box_i\varphi \text{ if } iEk$$

and change the occurrence of $\Box_k$ to $\Box_{\overline{k}}$ in the definition of RN and the axioms $B_\Box$, CN, $\overline{B}_\Box$, and $\overline{C}_\Box$. Second, we change the $\Diamond$-axiom and add two persistence axioms to the system:

P_B:	$\mathsf{B}_i\varphi \supset \mathsf{B}_k\varphi,\ i \leq k$
P_C:	$\mathsf{C}_k\varphi \supset \mathsf{C}_i\varphi,\ i \leq k$
$\Diamond$:	$\Diamond_{\overline{k}}\varphi$ provided $\varphi \not\vdash \bot$, φ free of modality $\overline{k}$

The generalization of the two modality reduction lemmata 1 and 2 are easily seen to hold.

The rules for the extended sequent calculus are as in the previous sections, with the following two modifications. First, each part of the sequent in a modality rule in Figure 1 which is superscripted with a k is now superscripted with $\overline{k}$; for instance is Θ^k in LRB_k now changed to $\Theta^{\overline{k}}$ to indicate that Θ contains only modal $\overline{k}$-atoms. Second, the calculus contains in addition the following rules provided $i \leq k$:

$$\frac{\Theta^{\overline{k}}, \mathsf{B}_k\varphi \Rightarrow \Delta^{\overline{k}}}{\Theta^{\overline{k}}, \mathsf{B}_i\varphi \Rightarrow \Delta^{\overline{k}}}\ \mathrm{PB}_i \qquad \frac{\Theta^{\overline{k}}, \mathsf{C}_i\varphi \Rightarrow \Delta^{\overline{k}}}{\Theta^{\overline{k}}, \mathsf{C}_k\varphi \Rightarrow \Delta^{\overline{k}}}\ \mathrm{PC}_k$$

Correctness of the calculus is proved by extending the proof of Theorem 5 and is straightforward.

To prove cut-elimination, we must change the definition of the degree of a formula. According to Definition 6, $d(\mathsf{B}_k\varphi) = d(\mathsf{B}_i\varphi)$. This is as it should if k and i are unrelated by $\leq$; if, however, $i < k$, we need to state that the degree of $\mathsf{B}_k\varphi$ is less than the degree of $\mathsf{B}_i\varphi$. Hence we change d to d', where $d'(\chi) = d(\chi)$ if χ is not a modal atom, otherwise $d'(\mathsf{B}_k\varphi) = d(\mathsf{B}_k\varphi) + \epsilon(k)$, $\epsilon(k) < 1$. ϵ is a function satisfying $\epsilon(k) < \epsilon(i)$ iff $i < k$, and can, since $\leq$ is a preorder, easily be defined.

The two new rules cause some new cases to be checked in the proof of the Principal lemma. We consider the new subcase to the case that we have addressed for the other system. Assume r_1 and r_2 are logical inferences such that φ is the principal formula of r_1 in the succedent of the conclusion and φ is the principal formula of r_2 in the antecedent of the conclusion. Let φ be $\mathsf{B}_i\psi$ and π_1 end with an instance r_1 of LRB_i (below to the left).

$$\frac{\overset{\pi_1'}{\vdots}\quad \Theta_1^{\overline{k}}, \Gamma_1 \Rightarrow \Delta_1^{\overline{k}}, (\mathsf{B}_i\psi)^m, \psi}{\Theta_1^{\overline{k}}, \Gamma_1^{\mathsf{B}_i} \Rightarrow \Delta_1^{\overline{k}}, (\mathsf{B}_i\psi)^m, \mathsf{B}_i\psi}\ r_1 \qquad \frac{\overset{\pi_1''}{\vdots}\quad \Theta_2^{\overline{k}}, \mathsf{B}_k\psi, (\mathsf{B}_i\psi)^n \Rightarrow \Delta_2^{\overline{k}}}{\Theta_2^{\overline{k}}, \mathsf{B}_i\psi, (\mathsf{B}_i\psi)^n \Rightarrow \Delta_2^{\overline{k}}}\ r_2$$

We must check the case that π_2 ends with a PB_i, depicted above to the right. Apply the induction hypothesis to π_1' and π_2, giving π', and to π_1 and π_1'', giving π''. In the proof π below the application of LRB_i in π_1 has been changed to an instance of LRB_k. The instance of PB_i

in π_2 has been changed to many applications of PB_i (indicated by double lines in the proof):

$$\dfrac{\dfrac{\begin{array}{c}\pi'\vdots\\ \Theta_1^{\overline{k}}, \Theta_2^{\overline{k}}, \Gamma_1 \Rightarrow \Delta_1^{\overline{k}}, \Delta_2^{\overline{k}}, \psi\end{array}}{\Theta_1^{\overline{k}}, \Theta_2^{\overline{k}}, \Gamma_1^{\mathsf{B}_k} \Rightarrow \Delta_1^{\overline{k}}, \Delta_2^{\overline{k}}, \mathsf{B}_k\psi}\,\mathrm{LRB}_k \qquad \begin{array}{c}\pi''\vdots\\ \Theta_1^{\overline{k}}, \Theta_2^{\overline{k}}, \Gamma_1^{\mathsf{B}_i}, \mathsf{B}_k\psi \Rightarrow \Delta_1^{\overline{k}}, \Delta_2^{\overline{k}}\end{array}}{\Theta_1^{\overline{k}}, \Theta_1^{\overline{k}}, \Theta_2^{\overline{k}}, \Theta_2^{\overline{k}}, \Gamma_1^{\mathsf{B}_k}, \Gamma_1^{\mathsf{B}_i} \Rightarrow \Delta_1^{\overline{k}}, \Delta_1^{\overline{k}}, \Delta_2^{\overline{k}}, \Delta_2^{\overline{k}}}\,\textsc{Mix}$$

$$\begin{array}{c}
\Theta_1^{\overline{k}}, \Theta_1^{\overline{k}}, \Theta_2^{\overline{k}}, \Theta_2^{\overline{k}}, \Gamma_1^{\mathsf{B}_k}, \Gamma_1^{\mathsf{B}_i} \Rightarrow \Delta_1^{\overline{k}}, \Delta_1^{\overline{k}}, \Delta_2^{\overline{k}}, \Delta_2^{\overline{k}} \\ \hline\hline
\Theta_1^{\overline{k}}, \Theta_2^{\overline{k}}, \Gamma_1^{\mathsf{B}_k}, \Gamma_1^{\mathsf{B}_i} \Rightarrow \Delta_1^{\overline{k}}, \Delta_2^{\overline{k}} \\ \hline\hline
\Theta_1^{\overline{k}}, \Theta_2^{\overline{k}}, \Gamma_1^{\mathsf{B}_i}, \Gamma_1^{\mathsf{B}_i} \Rightarrow \Delta_1^{\overline{k}}, \Delta_2^{\overline{k}} \quad \mathrm{PB}_i \\ \hline\hline
\Theta_1^{\overline{k}}, \Theta_2^{\overline{k}}, \Gamma_1^{\mathsf{B}_i} \Rightarrow \Delta_1^{\overline{k}}, \Delta_2^{\overline{k}}
\end{array}$$

π contains a MIX on $\mathsf{B}_k\psi$. Note that $d'(\pi) < d'(\varphi)$, even though $d(\pi) = d(\varphi)$. This is why we must use d' as a measure of degree instead of d in the proof of the Principal lemma. The case in which C_k is principal in r_1 is symmetrical. Consistency of the system follows from the cut-elimination results as for L_I.

THEOREM 16. *In the sequent calculus for the extended version of L_I, a sequent is provable iff it has a proof which contains a* CUT *only if the cut formula in the right premiss is introduced by* CONV. *If every formula in the sequent is first-order the sequent is provable iff it has a cut-free proof.*

5 Application to Gricean implicatures

The system presented in section 4 is capable of representing an interesting class of Gricean implicatures covered by Gazdard in [1]. The nature of implicatures is multi-modal; they result from the audience's reflections about what an utterer meant by an utterance put forth in a particular conversational context. The representation of implicatures must have a mechanism which reflects the priority mechanism that governs Gricean implicatures.

To give an idea of the representation, let us address a simple example. Assume that U has said "p or q" and that the context of utterance is represented by the formula κ; the context typically represents what has been said in the conversation in addition to $p \vee q$. Since the representation of the context is constrained by the language of L_I, contextual information like common knowledge is excluded; this puts limitations to the formalization. If we neglect the priority mechanism we can use L_I (i.e. the system of section 2; let the index set be $\{A, U\}$ for audience and utterer.

There are two kinds of implicatures relevant for this example, both generated on the basis of the Gricean Maxim of Quantity. Scalar implicatures assume that the strength of expressions in some cases can be ordered on a linguistic scale. In particular, conjunction is stronger than disjunction. The idea is that if U says $p \vee q$ and the context does not entail that U meant $p \wedge q$, then U implicates $\neg(p \wedge q)$, i.e. that the disjunction should be taken exclusively. The mechanism behind this pragmatical inference can be formalized by

$$\mathrm{sc}(A) = \neg \mathsf{B}_U \neg(p \wedge q) \supset \mathsf{B}_A \neg \mathsf{B}_U \neg(p \wedge q).$$

The formula has the form of a supernormal default in a multi-modal language and expresses that if $p \wedge q$ it is compatible with the utterer's beliefs, then the audience will know about this. Conversely, if A does not explicitly know that U said $p \wedge q$, then A believes that U does not mean that $p \wedge q$. Using the sequent calculus it is easy to prove that if $\kappa, \mathsf{B}_U(p \vee q) \not\vdash \neg \mathsf{B}_U \neg(p \wedge q)$, then $\vdash \mathsf{O}_A(\kappa \wedge \mathsf{B}_U(p \vee q) \wedge \mathrm{sc}(A)) \equiv \mathsf{O}_A(\kappa \wedge \mathsf{B}_U(p \vee q) \wedge \mathsf{B}_U \neg(p \wedge q))$, otherwise $\vdash \mathsf{O}_A(\kappa \wedge \mathsf{B}_U(p \vee q) \wedge \mathrm{sc}(A)) \equiv \mathsf{O}_A(\kappa \wedge \mathsf{B}_U(p \vee q))$.

Another type of implicature arises from implications like $\mathsf{B}_U p \supset \mathsf{B}_U(p \vee q)$. If U says $p \vee q$ and the context does not entail that U knows which disjunct is true, then A believes that U does not know this (this is called a clausal implicature). If we let $\iota(\varphi)$ be the formula $\mathsf{B}_U \varphi \supset \mathsf{B}_A \mathsf{B}_U \varphi$, there are four clausal implicatures in this example: $\iota(p) \wedge \iota(\neg p) \wedge \iota(q) \wedge \iota(\neg q)$. Let us call this formula $\mathrm{cl}(A)$. We can now, e.g., prove that if $\kappa, \mathsf{B}_U(p \vee q), -\mathsf{B}_U p, \neg \mathsf{B}_U \neg p, \neg \mathsf{B}_U q, \neg \mathsf{B}_U \neg q \not\vdash \bot$, then $\vdash \mathsf{O}_A(\kappa \wedge \mathsf{B}_U(p \vee q) \wedge \mathrm{cl}(A)) \equiv \mathsf{O}_A(\kappa \wedge \mathsf{B}_U(p \vee q) \wedge \neg \mathsf{B}_U p \wedge \neg \mathsf{B}_U \neg p \wedge \neg \mathsf{B}_U q \wedge \neg \mathsf{B}_U \neg q)$, as expected.

In the general case there may, however, be a conflict between clausal and scalar implicatures in which case the clausal ones should be given priority [1]. In the modal language we can capture this by having two modalities for the audience, representing two confidence levels: one for the beliefs after clausal implicatures have been accounted for and one for beliefs after the scalar implicatures have been processed. We can implement this in the language introduced in section 4 by taking $I = \{A_1, A_2, U\}$ and have only one preference constraint, namely $A_1 < A_2$. The representation $\mathbf{O}_{A_1}(\kappa \wedge \mathsf{B}_U(p \vee q) \wedge \mathrm{cl}(A_1)) \wedge \mathbf{O}_{A_2}(\kappa \wedge \mathsf{B}_U(p \vee q) \wedge \mathrm{cl}(A_1) \wedge \mathrm{sc}(A_2))$ will then serve as a formalization with the desired properties. Details of this encoding will be closer addressed in a subsequent paper.

Acknowledgement

For the results in section 3 I am grateful to Bjørnar Solhaug for valuable discussions and criticism. I also thank two anonymous referees.

BIBLIOGRAPHY

[1] **Gazdard, G.** *Pragmatics: Implicatures, Presupposition and Logical Form.* Academic Press, 1979.

[2] **Gentzen, G.** Untersuchungen über das logische Schliessen. *Matematische Zeitschrift* **39** (176-210, 405-431), 1935. Translation in Szabo, M.E., *The collected papers of Gerhard Gentzen*, North-Holland, 1969.

[3] **Girard, J., Lafont, Y., Taylor, P.** *Proofs and Types.* Cambridge Tracts in Theoretical Computer Science 7, 1989.

[4] **Halpern J.Y.** Reasoning about Only Knowing with Many Agents. In *Proc. of the 11th National Conference on Artificial Intelligence (AAAI-93)*, 1993.

[5] **Halpern J.Y.** Theories of Knowledge and Ignorance for Many Agents. In *Journal of Logic and Computation* **7**:1 (79-108), 1997.

[6] **Halpern J.Y. and Lakemeyer, G.** Multi-Agent Only Knowing. In *Journal of Logic and Computation* **11**:1 (40-70), 2001.

[7] **Lakemeyer, G.** All They Know: A Study in Multi-Agent Autoepistemic Reasoning. In *Proc. of the 13th International Joint Conference on Artificial Intelligence (IJCAI-93)* (376-381), 1993.

[8] **Levesque, H. J.** All I Know: Study in Autoepistemic Logic. *Artificial Intelligence* **42** (263-309), 1990.

[9] **Lian, E., Langholm, T. and Waaler, A.** *Only Knowing with Confidence Levels: Reductions and Complexity.* JELIA 2004, Proceedings, Lecture Notes in Computer Science **3229** (2004) 500–512 '

[10] **Shvarts , G. F.** Gentzen Style Systems for K45 and K45D. In Meyer, A.R., Taitslin, M.A. (eds.), *Logic at Botik'89*, Lecture Notes in Computer Science 363, Springer-Verlag 1989.

[11] **Waaler, A.** *Logical studies in Complementary Weak S5.* Doctoral thesis, Department of Philosophy, University of Oslo, 1994.

Arild Waaler

Finnmark College and University of Oslo, Norway

arild@ifi.uio.no

Connexive Modal Logic

HEINRICH WANSING

ABSTRACT. Connexive logic is a neglected direction in non-classical logic. In the present paper, first an axiomatic system of connexive propositional logic is presented. This logic, C, is shown to be sound and complete with respect to a class of relational models. It seems that this semantics is, in fact, the first known *intuitively plausible* interpretation of a system of connexive logic. The presentation of C suggests that connexive logic is *constructive*. It is a variant of David Nelson's constructive logics with strong negation. In Nelson's logics the verification conditions of implications are dynamic, whereas all falsification conditions are static conditions of falsification on the spot. In C, both the verification and the falsification conditions of implications are dynamic. This is enough to ensure that C is connexive and can be given a comprehensible and clear interpretation in terms of information states.

In a second step, the language of the system C is extended by the modal operators $\Box$ and $\Diamond$ to obtain a connexive analogue of the smallest normal modal propositional logic K. Aiming at a connexive analogue of K that can be faithfully embedded by a modal translation into QC, quantified C, we arrive at a system that will be called CK, connexive K. The system CK is a connexive version of the constructive modal logic FSK^d characterized in [12]. CK is shown to be sound and complete with respect to relational models and to be decidable.

We shall also critically discuss the evaluation clauses for the modal operators that are induced by the *standard* translation from modal propositional logic into first-order logic. In the context of the connexive base logic, the falsification clauses of formulas $\Box A$ induced by the standard translation appear to be intuitively implausible. In any case, both syntactic duality axioms $\sim \Box A \leftrightarrow \Diamond \sim A$ and $\sim \Diamond A \leftrightarrow \Box \sim A$ fail to hold.

It seems that CK is the first system of connexive modal logic considered in the modal logic literature. This paper may therefore be seen as a contribution to establishing connexive modal logic as a respectable branch of modal logic.

1 Aristotle's Theses and Boethius' Theses

The following principle is well-known as "Aristotle's Thesis":

$$AT \quad \sim(\sim A \to A),$$

see, for example, [11]. AT is of interest, because although AT is not a theorem of classical logic, nevertheless in the history of logic, AT has sometimes been found plausible, if it is viewed as a formalization of "it is not the case that not-A implies A". In classical logic, AT is not logically equivalent with the non-theorem:

$$AT' \quad \sim(A \to \sim A).^{1}$$

But also against the background of logics in which AT and AT' fail to be logically equivalent, they would seem to constitute a pair of equally intuitive (or unintuitive) theses. There exists, thus, a tension between the supraclassicality (the fact that AT and AT' are not classically valid) and some intuitive plausibility of AT and AT'.

One may observe that, using intuitionistically acceptable means only, the pair of theses AT and AT' is equivalent in deductive power with another pair of schemata, which in accordance with previously introduced terminology are here called (Strong) "Boethius' Theses" BT and BT'. Boethius' theses are:

$$BT \quad (A \to B) \to \sim(A \to \sim B); \qquad BT' \quad (A \to \sim B) \to \sim(A \to B).$$

More precisely, BT and BT' each allow one to intuitionistically derive both AT and AT', and AT' allows one to intuitionistically derive both BT and BT', whereas AT neither classically implies BT nor classically implies BT'. Moreover, the theses BT and BT' can easily be shown to be intuitionistically equivalent. Whereas the converses

$$BT_c \quad \sim(A \to \sim B) \to (A \to B), \qquad BT'_c \quad \sim(A \to B) \to (A \to \sim B)$$

of BT and BT' are classically valid, only BT'_c is also intuitionistically valid.

2 Connexive logic

Let $\mathcal{L}$ be a language containing a unary connective $\sim$ (negation) and a binary connective $\to$ (implication). A logical system in the language $\mathcal{L}$ is called a *connexive logic*, if AT, AT', BT, and BT' are valid schemata and,

[1] Since classical propositional logic is Post complete, neither AT nor AT' can consistently be added to it.

moreover, $(A \to B) \to (B \to A)$ fails to be a valid schema.[2] The connective $\to$ in a system of connexive logic is said to be a *connexive implication.*

In [11] McCall presents an axiomatization of a system of propositional connexive logic introduced by Angell [1] in terms of certain four-valued matrices. The language of McCall's logic $CC1$ contains as primitive (notation adjusted) a unary connective $\sim$ (negation) and the binary connectives $\wedge$ (conjunction) and $\to$ (implication). Disjunction $\vee$ and equivalence $\leftrightarrow$ are defined in the usual way. The schematic axioms and the rules of $CC1$ are as follows:

$A1$ $(A \to B) \to ((B \to C) \to (A \to C))$
$A2$ $((A \to A) \to B) \to B$
$A3$ $(A \to B) \to ((A \wedge C) \to (B \wedge C))$
$A4$ $(A \wedge A) \to (B \to B)$
$A5$ $(A \wedge (B \wedge C)) \to (B \wedge (A \wedge C))$
$A6$ $(A \wedge A) \to ((A \to A) \to (A \wedge A))$
$A7$ $A \to (A \wedge (A \wedge A))$
$A8$ $((A \to \sim B) \wedge B) \to \sim A$
$A9$ $(A \wedge \sim (A \wedge \sim B)) \to B$
$A10$ $\sim (A \wedge \sim (A \wedge A))$
$A11$ $(\sim A \vee ((A \to A) \to A)) \vee (((A \to A) \vee (A \to A)) \to A)$
$A12$ $(A \to A) \to \sim (A \to \sim A)$
$R1$ modus ponens
$R2$ adjunction

Note that among these axiom schemata, only A12 is supraclassical. The system $CC1$ is characterized by the following four-valued truth tables with designated values 1 and 2:

$\sim$	
1	4
2	3
3	2
4	1

$\wedge$	1	2	3	4
1	1	2	3	4
2	1	2	4	3
3	3	3	3	4
4	4	3	4	3

$\to$	1	2	3	4
1	1	4	3	4
2	4	1	4	3
3	1	4	1	4
4	4	1	4	1

McCall emphasizes that the logic $CC1$ is only one among many possible systems satisfying the theses of Aristotle and Boethius and that he "does

[2] According to McCall [11, p. 416], a "system of connexive logic may range from one in which no proposition implies or is implied by its own negation to one in which Boethius' thesis [i.e., BT] is asserted." But, if "it is to be of any interest, it must exclude the characteristic thesis $(p \to q) \to (q \to p)$ of equivalence" [11, p. 417, notation adjusted]

not wish it to be regarded as *the* system of connexive logic" [11, p. 418]. Indeed, McCall does not suggest any intuitive interpretation of the four truth values employed in the above matrices and the fact that the values 1 and 2 are designated, and hence the semantics appears to be a purely formal method with little explanatory power.[3] In contrast to this, the system C of the present paper is meant to be a suitable candidate for the title 'the basic system of connexive logic'.

Moreover, McCall points out that $CC1$ has some properties that are difficult to justify, if the name connexive logic is meant to reflect the fact that in a valid implication $A \rightarrow B$ there exists a connection between the antecedent A and the succedent B. Axiom A4, for example, is bad in this respect. On the other hand, $CC1$ might be said to undergenerate, since $(A \wedge A) \rightarrow A$ and $A \rightarrow (A \wedge A)$ fail to be theorems of $CC1$. If the validity of Aristotle's and Boethius' theses is distinctive of connexive logics, it is not quite clear, however, how damaging the above criticism is.

More serious doubts have been cast on $CC1$ by Routley and Montgomery [22] and, as these authors have claimed, on connexive logics in general. The starting point of their criticism, however, is a certain subsystem $Z1$ of $CC1$ containing the contraposition theorem $(\sim A \rightarrow B) \rightarrow (\sim B \rightarrow A)$. This troublesome schema fails to be a theorem of the connexive logic C, defined in Section 3.

3 A basic system of connexive logic

The key observation for obtaining a connexive logic admitting a transparent and illuminating semantical characterization is simple: in the presence of the double negation introduction law, it suffices to validate both BT' and BT'_c. In other words, an interpretation of the falsification conditions of implications is called for, which deviates from the standard conditions. In Nelson's systems of constructive logic, the double negation laws hold, and the relational semantics for these logics is such that falsification and verification of formulas are dealt with separately. However, the falsification conditions of implications are the classical ones expressed by the schema $\sim (A \rightarrow B) \leftrightarrow (A \wedge \sim B)$.[4] To obtain a connexive implication, it is there-

[3] Note that the constant truth functions **1**, **2**, **3**, and **4** can be defined as follows [11, p. 421]: $\mathbf{1} := (p \rightarrow p)$, $\mathbf{2} := \sim (p \leftrightarrow \sim p)$, $\mathbf{3} := (p \leftrightarrow \sim p)$, $\mathbf{4} := \sim (p \rightarrow p)$, for some sentence letter p. Routley and Montgomery [22, p. 95] point out that $CC1$ "can be given a semantics by associating the matrix value 1 with logical necessity, value 4 with logical impossibility, value 2 with contingent truth, and value 3 with contingent falsehood. However, many anomalies result; e.g. the conjunction of two contingent truths yields a necessary truth".

[4] As a result, provable equivalence fails to be a congruence relation. Otherwise, the formula $\sim\sim (A \rightarrow B) \leftrightarrow \sim (A \wedge \sim B)$ and hence, due to the double negation and

fore enough to assume another interpretation of the falsification conditions of implications expressed by the schema $\sim (A \rightarrow B) \leftrightarrow (A \rightarrow \sim B)$.

Consider the language $\mathcal{L} := \{\wedge, \vee, \rightarrow, \sim\}$ based on the denumerable set $At_{\mathcal{L}}$ of propositional variables. Equivalence $\leftrightarrow$ is defined as usual. The schematic axioms and rules of the logic C are:

$a1$	the axioms of intuitionistic positive logic
$a2$	$\sim\sim A \leftrightarrow A$
$a3$	$\sim (A \vee B) \leftrightarrow (\sim A \wedge \sim B)$
$a4$	$\sim (A \wedge B) \leftrightarrow (\sim A \vee \sim B)$
$a5$	$\sim (A \rightarrow B) \leftrightarrow (A \rightarrow \sim B)\}$
$R1$	modus ponens

Clearly, a5 is the only supraclassical axiom of C. Deducibility in C and the consequence relation $\vdash_{\mathsf{C}}$ are defined as usual.

A C-frame is a pair $\mathcal{F} = \langle W, \leq \rangle$, where $\leq$ is a reflexive and transitive binary relation on the non-empty set W. Let $\langle W, \leq \rangle^{+}$ be the set of all $X \subseteq W$ such that if $u \in X$ and $u \leq w$, then $w \in X$. A C-model is a structure $\mathcal{M} = \langle W, \leq, v^{+}, v^{-} \rangle$, where $\langle W, \leq \rangle$ is a C-frame and v^{+} and v^{-} are valuation functions from $At_{\mathcal{L}}$ into $\langle W, \leq \rangle^{+}$. Intuitively, W is a set of information states. The function v^{+} sends an atom p to the states in W that support the truth of p, whereas v^{-} sends p to the states that support the falsity of p. $\mathcal{M} = \langle W, \leq, v^{+}, v^{-} \rangle$ is said to be the model based on the frame $\langle W, \leq \rangle$. The relations $\mathcal{M}, t \models^{+} A$ ($\mathcal{M}$ supports the truth of A at t) and $\mathcal{M}, t \models^{-} A$ ($\mathcal{M}$ supports the falsity of A at t) are inductively defined as follows:

$\mathcal{M}, t \models^{+} p$	iff	$t \in v^{+}(p)$
$\mathcal{M}, t \models^{-} p$	iff	$t \in v^{-}(p)$
$\mathcal{M}, t \models^{+} (A \wedge B)$	iff	$\mathcal{M}, t \models^{+} A$ and $\mathcal{M}, t \models^{+} B$ t
$\mathcal{M}, t \models^{-} (A \wedge B)$	iff	$\mathcal{M}, t \models^{-} A$ or $\mathcal{M}, t \models^{-} B$ t
$\mathcal{M}, t \models^{+} (A \vee B)$	iff	$\mathcal{M}, t \models^{+} A$ or $\mathcal{M}, t \models^{+} B$ t
$\mathcal{M}, t \models^{-} (A \vee B)$	iff	$\mathcal{M}, t \models^{-} A$ and $\mathcal{M}, t \models^{-} B$ t
$\mathcal{M}, t \models^{+} (A \rightarrow B)$	iff	$\forall v \geq t\ (\mathcal{M}, v \models^{+} A$ implies $\mathcal{M}, v \models^{+} B)$
$\mathcal{M}, t \models^{-} (A \rightarrow B)$	iff	$\forall v \geq t\ (\mathcal{M}, v \models^{+} A$ implies $\mathcal{M}, v \models^{-} B)$
$\mathcal{M}, t \models^{+} \sim A$	iff	$\mathcal{M}, t \models^{-} A$
$\mathcal{M}, t \models^{-} \sim A$	iff	$\mathcal{M}, t \models^{+} A$

DeMorgan laws, $(A \rightarrow B) \leftrightarrow (\sim A \vee B)$ would be provable. But this is not the case in a constructive logic, where the verification conditions of implications are those of intuitionistic implications.

Validity of a formula A in a C-model ($\mathcal{M} \models A$) and validity of A on a frame ($\mathcal{F} \models A$) are defined in the usual way. This means that if $\mathcal{M} = \langle W, \leq, v^+, v^- \rangle$ is a C-model, then $\mathcal{M} \models A$ iff for every $t \in W$, $\mathcal{M}, t \models^+ A$. $\mathcal{F} \models A$ holds iff $\mathcal{M} \models A$ for every model $\mathcal{M}$ based on $\mathcal{F}$. A formula is C-valid iff it is valid on every frame. Support of truth and support of falsity for arbitrary formulas are persistent with respect to the relation $\leq$ of possible expansion of information states. That is, for any C-model $\mathcal{M} = \langle W, \leq, v^+, v^- \rangle$, $s, t \in W$, and formula A, if $s \leq t$, then $\mathcal{M}, s \models^+ A$ implies $\mathcal{M}, t \models^+ A$ and $\mathcal{M}, s \models^- A$ implies $\mathcal{M}, t \models^- A$. It can easily be shown that a negation normal form theorem holds. Using familiar methods (see also Section 4), the following can be shown:

PROPOSITION 1. *For any $\mathcal{L}$-formula A, $\mathsf{C} \vdash A$ iff A is* C*-valid.*

PROPOSITION 2. *The logic* C *satisfies the disjunction property and the constructible falsity property. If $\mathsf{C} \vdash A \vee B$, then $\mathsf{C} \vdash A$ or $\mathsf{C} \vdash B$. If $\mathsf{C} \vdash \sim (A \wedge B)$, then $\mathsf{C} \vdash \sim A$ or $\mathsf{C} \vdash \sim B$.*

PROPOSITION 3. *The connexive logic* C *is decidable.*

It is obvious form the above presentation that C differs from Nelson's four-valued constructive logic N4 only with respect to the falsification (or support of falsity) conditions of implications. This modification is significant, as it not only turns $\rightarrow$ into a connexive implication but also leads to a constructive logic with intuitionistic implication in which the DeMorgan laws hold *and* provable equivalence is a congruence relation.

PROPOSITION 4. *The set $\{A \mid \mathsf{C} \vdash A\}$ is closed under the rule $A \leftrightarrow B$ / $C(A) \leftrightarrow C(B)$.*

4 First-order connexive logic

In this section we define a first-order extension QC of C. The system differs from the first-order extension QN4 of Nelson's four-valued constructive logic characterized in [12, Section 4.1] only with respect to the treatment of negated implications. In order to keep this paper self-contained, we shall here briefly present the axiomatization and semantic characterization of QC, following the presentation in [12].

We extend the propositional language $\mathcal{L}$ to a first-order language by adding denumerably many constant and predicate symbols. Let $T_{\mathcal{L}}$ ($CT_{\mathcal{L}}$) denote the set of $\mathcal{L}$-terms (closed $\mathcal{L}$-terms), and let $At_{\mathcal{L}}$ be the set of atomic formulas and $For_{\mathcal{L}}$ be the set of all $\mathcal{L}$-formulas. The schematic axioms and rules of QC are those of C together with:

$$\sim \exists x A \leftrightarrow \forall x \sim A; \quad \sim \forall x A \leftrightarrow \exists x \sim A$$

$$A(t) \rightarrow \exists x A(x) \text{ } (t \text{ is free for } x \text{ in } A)$$

$\forall x A(x) \to A(t)$ (t is free for x in A)

$$\frac{A \to B(x)}{A \to \forall x B(x)} \ (x \text{ not free in } A); \quad \frac{A(x) \to B}{\exists x A(x) \to B} \ (x \text{ not free in } B)$$

Deducibility in QC and the consequence relation $\vdash_{\mathsf{QC}}$ are defined in the usual way.

A QC-model is a structure $\langle W, \leq, \Delta, D, v^+, v^- \rangle$, where $\langle W, \leq \rangle$ is a C-frame; Δ is a set such that $CT_{\mathcal{L}} \subseteq \Delta \subseteq T_{\mathcal{L}}$, and D is a function from W to subsets of Δ such that (i) $CT_{\mathcal{L}} \subseteq D_u$ for all $u \in W$ and (ii) $D_u \subseteq D_t$ whenever $u \leq t$. Finally, v^+ and v^- are functions from $At_{\mathcal{L}}$ to $\langle W, \leq \rangle^+$ such that if $u \in v^{+(-)}(P(a_1, \ldots, a_n))$, then $a_1, \ldots, a_n \in D_u$. The function D assigns to every state $u \in W$ its domain D_u. If $\mathcal{M} = \langle W, \leq, \Delta, D, v^+, v^- \rangle$ is a model and $t \in W$, the notions $\mathcal{M}, t \models^+ A$ (state t supports the truth of A in $\mathcal{M}$) and $\mathcal{M}, t \models^- A$ (t supports the falsity of A in $\mathcal{M}$) are inductively defined as for C, except that in addition we have:

$\mathcal{M}, t \models^+ \forall x A(x)$	iff	$(\forall u \in W)$ if $t \leq u$, then $(\forall c \in D_u)\ \mathcal{M}, u \models^+ A(c)$
$\mathcal{M}, t \models^- \forall x A(x)$	iff	$(\exists c \in D_t)\ \mathcal{M}, t \models^- A(c)$
$\mathcal{M}, t \models^+ \exists x A(x)$	iff	$(\exists c \in D_t)\ \mathcal{M}, t \models^+ A(c)$
$\mathcal{M}, t \models^- \exists x A(x)$	iff	$(\forall u \in W)$ if $t \leq u$, then $(\forall c \in D_u)\ \mathcal{M}, u \models^- A(c)$

Again, support of truth and support of falsity for arbitrary formulas are persistent with respect to $\leq$. A formula $A \in For_{\mathcal{L}}$ is QC-valid if for any model $\mathcal{M} = \langle W, \leq, \Delta, D, v^+, v^- \rangle$ and $t \in W$, $\mathcal{M}, t \models^+ A$. Soundness can be shown by induction on the length of proofs.

PROPOSITION 5. *For any $A \in For_{\mathcal{L}}$, if* QC $\vdash A$, *then A is* QC-*valid.*

As in the case of QN4, completeness can be shown by a faithful embedding into QInt$^+$, positive first-order intuitionistic logic, see [2]. $\mathcal{L}^+$, the language of QInt$^+$, is the result of deleting $\sim$ from $\mathcal{L}$. A QInt$^+$-model is a structure $\langle W, \leq, \Delta, D, v \rangle$ defined like a QC-model, except that it has only one valuation function v from $At_{\mathcal{L}}$ to $\langle W, \leq \rangle^+$ such that if $u \in v(P(a_1, \ldots, a_n))$, then $a_1, \ldots, a_n \in D_u$.

Let $\mathcal{M} = \langle W, \leq, \Delta, D, v \rangle$ be a QInt$^+$-model, $t \in W$ and $A \in For_{\mathcal{L}^+}$. The relation $\mathcal{M}, t \models A$ (A is true at t in $\mathcal{M}$) is defined exactly as the relation $\mathcal{M}, t \models^+ A$ was defined for positive connectives and quantifies (with v^+ replaced by v in case of atomic formulas). A formula $A \in For_{\mathcal{L}^+}$ is said to be QInt$^+$-valid if for any QInt$^+$-model $\mathcal{M} = \langle W, \leq, \Delta, D, v \rangle$ and $t \in W$, A is true at t in $\mathcal{M}$.

PROPOSITION 6. *For any $A \in For_{\mathcal{L}^+}$,* QInt $\vdash A$ *iff A is* QInt$^+$-*valid.*

The proof is essentially the same as the completeness proof for full first-order intuitionistic logic [2], omitting the part concerning negation.

We say that a formula $A \in For_{\mathcal{L}}$ is in *negation normal form* (nnf), if it contains negations only in front of atomic formulas, in other words, if $\sim B$ is a subformula of A, then B is atomic.

DEFINITION 7. We define the transformation $\overline{(\cdot)}$ on $For_{\mathcal{L}}$ as follows:

1. $\overline{B} := B$, $\overline{\sim B} :=\sim B$, for $B \in At_{\mathcal{L}}$.
2. $\overline{\sim\sim A} := \overline{A}$, for a formula A.
3. $\overline{A \diamond B} := \overline{A} \diamond \overline{B}$, where A and B are formulas and $\diamond \in \{\vee, \wedge, \rightarrow\}$.
4. For any formula A and $Q \in \{\forall, \exists\}$, $\overline{QxA} := Qx\overline{A}$.
5. For any formulas A and B, $\overline{\sim (A \vee B)} := \overline{\sim A} \wedge \overline{\sim B}$, $\overline{\sim (A \wedge B)} := \overline{\sim A} \vee \overline{\sim B}$, and $\overline{\sim (A \rightarrow B)} := \overline{A} \rightarrow \overline{\sim B}$.
6. For any formula A, $\overline{\sim \forall xA} := \exists x\overline{\sim A}$ and $\overline{\sim \exists xA} := \forall x\overline{\sim A}$.

PROPOSITION 8. *For any formula A, $\overline{A}$ is in negation normal form and* $\mathsf{QC} \vdash A \leftrightarrow \overline{A}$.

We now define the positive intuitionistic first-order language $\mathcal{L}'^{+}$ by deleting $\sim$ from $\mathcal{L}$ and adding for each predicate symbol P of $\mathcal{L}$ a new predicate symbol $P^{\sim}$ of the same arity.

DEFINITION 9. The transformation $(\cdot)^*$ of formulas from $For_{\mathcal{L}}$ into $For_{\mathcal{L}'^{+}}$ is defined as follows:

1. For an atomic $P(\overline{a})$, $(P(\overline{a}))^* := P(\overline{a})$ and $(\sim P(\overline{a}))^* := P^{\sim}(\overline{a})$.
2. $(A \diamond B)^* := A^* \diamond B^*$, where A and B are formulas in nnf and $\diamond \in \{\vee, \wedge, \rightarrow\}$.
3. $(QxA)^* := QxA^*$, where A is a formula in nnf and $Q \in \{\forall, \exists\}$.
4. $(A)^* := (\overline{A})^*$ for any formula A not in nnf.

PROPOSITION 10. *For any $A \in For_{\mathcal{L}}$, $\mathsf{QC} \vdash A$ iff $\mathsf{QInt}^{+} \vdash A^*$.*

Proof. If $\mathsf{QC} \vdash A$, one can easily check by induction on the length of proofs that A^* is provable in QInt^{+}. If $\mathsf{QInt} \vdash A^*$ and $B_0, \ldots B_n = A^*$ is a proof, then $B'_0, \ldots B'_n = \overline{A}$, where B'_i is the result of replacing any subformula $P^{\sim}(\overline{a})$ by $\sim P(\overline{a})$, is a proof of the nnf of A in QC. By Proposition 8, we then have $\mathsf{QC} \vdash A$. ■

PROPOSITION 11. *For any $A \in For_{\mathcal{L}}$, $\mathsf{QC} \vdash A$ iff A is QC-valid.*

Proof. Let A_0 be a QC-valid formula. Assume that $\mathsf{QC} \nvdash A_0$. Then $\mathsf{QInt}^+ \nvdash (A_0)^*$ by Proposition 10. This means that there is a QInt^+-model $\mathcal{M} = \langle W, \leq, \Delta, D, v\rangle$ of the language $\mathcal{L}'^+$ and $t_0 \in W$ such that $\mathcal{M}, t_0 \not\models (A_0)^*$. We define a QC-model $\mathcal{M}' = \langle W, \leq, \Delta, D, v^+, v^-\rangle$ with the same W, $\leq$, and Δ. The assignment functions v^+ and v^- are defined as follows:

$$v^+(P(\overline{a})) := v(P(\overline{a})) \text{ and } v^-(P(\overline{a})) := v(P^\sim(\overline{a})).$$

Using induction on the structure of a formula, we can show that

$$\mathcal{M}, t \models A^* \text{ if and only if } \mathcal{M}', t \models^+ A$$

for any $t \in W$ and formula A in nnf. Thus, $\mathcal{M}', t_0 \not\models^+ \overline{A_0}$ and $\mathcal{M}', t_0 \not\models^+ A_0$ by Proposition 8, which conflicts with the QC-validity of A_0. ■

5 The connexive modal logic CK

Let now $\mathcal{L}_{\Box,\Diamond} := \{\wedge, \vee, \rightarrow, \sim, \Box, \Diamond\}$. To obtain a connexive analogue of the smallest normal modal logic K, we consider the translation T_x from $For_{\mathcal{L}_{\Box,\Diamond}}$ into the first-order language $\mathcal{L}$ of QC:

$$\begin{array}{rcl}
p & \overset{T_x}{\longmapsto} & P(x) \\
\sim p & \overset{T_x}{\longmapsto} & \sim P(x) \\
A * B & \overset{T_x}{\longmapsto} & T_x(A) * T_x(B),\ * \in \{\wedge, \vee, \rightarrow\} \\
\sim (A * B) & \overset{T_x}{\longmapsto} & \sim (T_x(A) * T_x(B)),\ * \in \{\wedge, \vee, \rightarrow\} \\
\Box A & \overset{T_x}{\longmapsto} & \forall y(R(x,y) \rightarrow T_y(A)) \\
\Diamond A & \overset{T_x}{\longmapsto} & \exists y(R(x,y) \wedge T_y(A)) \\
\sim\Box A & \overset{T_x}{\longmapsto} & \exists y(R(x,y) \wedge T_y(\sim A)) \\
\sim\Diamond A & \overset{T_x}{\longmapsto} & \forall y(R(x,y) \rightarrow T_y(\sim A))
\end{array}$$

Consider the following axioms and rules.

$$\begin{array}{ll}
a6 & \Box A \wedge \Box B \rightarrow \Box(A \wedge B) \\
a7 & \Box(A \rightarrow A) \\
a8 & \Diamond(A \vee B) \rightarrow \Diamond A \vee \Diamond B \\
a9 & \Diamond(A - B) \rightarrow (\Box A \rightarrow \Diamond B) \\
a10 & (\Diamond A - \Box B) \rightarrow \Box(A \rightarrow B) \\
a11 & \sim\Box A \rightarrow \Diamond \sim A \\
a12 & \sim\Diamond A \leftrightarrow \Box \sim A \\
R_\Box & A \rightarrow B \,/\, \Box A \rightarrow \Box B \\
R_\Diamond & A \rightarrow B \,/\, \Diamond A \rightarrow \Diamond B
\end{array}$$

The connexive modal logic CK is defined as the deductive closure of axioms a1–a12 under the rules $R1$, $R_\Box$, and $R_\Diamond$. If Δ is a set of $\mathcal{L}_{\Box,\Diamond}$-formulas and A an $\mathcal{L}_{\Box,\Diamond}$-formula, the relation $\Delta \vdash A$ holds iff A belongs to the closure of $\mathsf{CK} \cup \Delta$ under $R1$. Axioms a9 and a10 ensure that $\Box$ and $\Diamond$ can be interpreted semantically with respect to a single accessibility relation R (instead of a pair of independent accessibility relations $R_\Box$ and $R_\Diamond$) (see [4], [13], [25]) and that CK can be faithfully embedded into QC.

A CK-frame is a triple $\mathcal{F} = \langle W, \leq, R\rangle$, where $\langle W, \leq\rangle$ is a C-frame and R is a binary relation on W such that:

1. $\leq$ is reflexive and transitive;
2. $\leq^{-1} \circ R \subseteq R \circ \leq^{-1}$;
3. $R \circ \leq \; \subseteq \; \leq \circ R$.

A CK-model is a structure $\mathcal{M} = \langle W, \leq, R, v^+, v^-\rangle$, where $\langle W, \leq, R\rangle$ is a CK-frame and v^+ and v^- are valuation functions from $At_{\mathcal{L}}$ into $\langle W, \leq\rangle^+$. The relations $\mathcal{M}, t \models^+ A$ and $\mathcal{M}, t \models^- A$ are inductively defined as for C, except that in addition we have:

$$\begin{array}{lll} \mathcal{M}, t \models^+ \Box A & \text{iff} & \forall u \geq t \, \forall v (uRv \text{ implies } \mathcal{M}, v \models^+ A) \\ \mathcal{M}, t \models^- \Box A & \text{iff} & \exists u (tRu \text{ and } \mathcal{M}, u \models^- A) \end{array}$$

$$\begin{array}{lll} \mathcal{M}, t \models^+ \Diamond A & \text{iff} & \exists u (tRu \text{ and } \mathcal{M}, u \models^+ A) \\ \mathcal{M}, t \models^- \Diamond A & \text{iff} & \forall u \geq t \, \forall v (uRv \text{ implies } \mathcal{M}, v \models^- A) \end{array}$$

A discussion of the support of truth conditions for the modal operators can be found in [24]. Support of truth and support of falsity are persistent with respect to the relation $\leq$ of possible expansion of information states. A formula A is valid in a model $\mathcal{M} = \langle W, \leq, R, v^+, v^-\rangle$ ($\mathcal{M} \models A$) if $\mathcal{M}, t \models^+ A$ for all $t \in W$; A is said to be valid on a frame $\mathcal{F}$ ($\mathcal{F} \models A$) if A is valid in any model based on $\mathcal{F}$.

PROPOSITION 12. *For any $\mathcal{L}$-formula A, $\mathsf{CK} \vdash A$ iff A is valid on every CK-frame.*

The proof is basically the completeness proof for the constructive modal logic FSK^d, see [13]. A canonical model is defined and shown to satisfy the above conditions 1.–3. on its underlying frame.

DEFINITION 13. We define the transformation $\overline{(\cdot)}$ on $For_{\mathcal{L}_{\Box,\Diamond}}$ as in Definition 7 except that the clause for quantified formulas is replaced by the following:

6. For any $\mathcal{L}_{\Box,\Diamond}$-formula A, $\overline{\sim \Box A} := \Diamond \overline{\sim A}$ and $\overline{\sim \Diamond A} := \Box \overline{\sim A}$.

PROPOSITION 14. *For any $\mathcal{L}_{\Box,\Diamond}$-formula A, $\overline{A}$ is in negation normal form and $\mathsf{CK} \vdash A \leftrightarrow \overline{A}$.*

This negation normal form theorem allows CK to be rewritten as FS^+, the positive fragment of Fischer Servi's intuitionistic modal logic FS, see [3], [4], [5], [24].

PROPOSITION 15. *The translation T_x is a faithful embedding of* CK *into* QC.

Proof. Observe that the translation of $\mathcal{L}_{\Box,\Diamond}$-formulas in negation normal form into the language $\mathcal{L}_{\Box,\Diamond}{}'^{+}$ that maps every propositional variable p to itself, sends every negated propositional variable $\sim p$ to a fresh propositional variable p', and commutes with the positive connectives is a faithful embedding of CK into FS^+. This is enough to establish the claim, since FS^+ is faithfully embedded by the standard translation, coinciding on negation-free formulas with T_x, into $\mathsf{QInt}^+ = \mathsf{QC}^+$, positive QK. ■

Since FS is decidable, we obtain the following:

COROLLARY 16. *The connexive modal logic* CK *is decidable.*

PROPOSITION 17. *The logic* CK *satisfies the disjunction property and the constructible falsity property.*

PROPOSITION 18. *The set $\{A \mid \mathsf{CK} \vdash A\}$ is closed under the rule $A \leftrightarrow B$ / $C(A) \leftrightarrow C(B)$.*

Note hat T_x differs from the standard translation ST_x from $For_{\mathcal{L}_{\Box,\Diamond}}$ into $For_{\mathcal{L}}$:

$$\begin{array}{rcl} p & \overset{ST_x}{\longmapsto} & P(x) \\ \sim A & \overset{ST_x}{\longmapsto} & \sim ST_x(A) \\ A * B & \overset{ST_x}{\longmapsto} & ST_x(A) * ST_x(B), \ * \in \{\wedge, \vee, \rightarrow\} \\ \Box A & \overset{ST_x}{\longmapsto} & \forall y(R(x,y) \rightarrow ST_y(A)) \\ \Diamond A & \overset{ST_x}{\longmapsto} & \exists y(R(x,y) \wedge ST_y(A)) \end{array}$$

Whereas T_x guarantees the syntactic duality between $\Box$ and $\Diamond$, the standard translation of $\sim \Box A$ is equivalent with $\exists y(R(x,y) \rightarrow \sim ST_y(A))$, which is not equivalent with the standard translation of $\Diamond \sim A$. Also $\sim \Diamond A$ and $\Box \sim A$ are not equivalent under ST_x. The standard translation of $\sim \Diamond A$ yields a formula equivalent with $\forall y(\sim R(x,y) \vee \sim ST_y(A))$. The support of falsity clauses for modal formulas induced by ST_x are:

$$\begin{array}{lll} \mathcal{M}, t \models^- \Diamond A & \text{iff} & \forall s \geq t\, \forall u(sR^{\sim}u \text{ or } \mathcal{M}, u \models^- A) \\ \mathcal{M}, t \models^- \Box A & \text{iff} & \exists s \forall u \geq s(tRs \text{ implies } \mathcal{M}, u \models^- A) \end{array}$$

where $R^{\sim}$ is an impossibility relation. The support of falsity clause for formulas $\Diamond A$ is plausible: if for every expansion s of information state t, every state u is such that either u is impossible relative to s or u supports the falsity of A, then t supports the falsity of $\Diamond A$. The support of falsity clause for $\Box A$, however, appears to be much more problematic. To support the falsity of $\Box A$, it is enough for a state t that there exists one inaccessible state, which is not very convincing.

6 Another system of connexive logic

Axiom a5 calls for some explanation. Whereas the direction from right to left can be justified by rejecting the view that if A implies B and A is inconsistent, A implies any formula, in particular B, the direction from left to right seems rather strong. If the verification conditions of implications are dynamic, then a5 indicates that the falsification conditions of implications are dynamic as well. The falsity of $(A \to B)$ thus implies that if A is true, B is *false*. Yet, one might wonder why it is not required that the falsity of $(A \to B)$ implies that if if A is true, B is *not true*. This cannot be expressed in a language with just one negation $\sim$ expressing falsity instead of absence of truth (classically at the state of evaluation or intuitionistically at all related states). If one adds to C the further axiom $\sim A \to (A \to B)$ to obtain a connexive variant of Nelson's three-valued logic $\mathsf{N3}$, intuitionistic negation $\neg$ is definable by setting: $\neg A := A \to\sim A$. Then a5 could be replaced by

$$\text{a5}' \quad \sim (A \to B) \leftrightarrow (A \to \neg B).$$

The resulting system is another system of connexive logic with a transparent semantics and an intuitive interpretation in terms of information states. It satisfies AT, AT', BT, and BT', because $A \to \neg \sim A$ and $\sim A \to \neg A$ are theorems. For BT, for example, we have:

1.	$A \to B$	assumption
2.	$B \to \neg \sim B$	theorem
3.	$A \to \neg \sim B$	1., 2., transitivity of $\to$
4.	$(A \to \neg \sim B) \to \sim (A \to \sim B)$	a5′
5.	$\sim (A \to \sim B)$	3., 4., R1
6.	$(A \to B) \to \sim (A \to \sim B)$	1., 5., deduction theorem

7 Systems of consequential implication

Aristotle's and Boethius' Theses express, as it seems, some pre-theoretical intuitions about meaning relations between negation and implication. But is not clear that a language must contain only one negation operation and only one implication. In the previous section we have encountered a system with two negations, and the language of systems of *consequential implication* comprises two implication connectives together with one negation, see [14], [15], [16], [17]. In [18], the notion of a normal system of analytic consequential implication is defined. The smallest such normal consequential logic that satisfies AT is called CI. Alternatively, CI can be characterized as the smallest normal system that satisfies Weak Boethius' Thesis:

$$(A \to B) \supset \neg(A \to \neg B),$$

where $\to$ is consequential implication, $\supset$ is material implication, and $\neg$ is classical negation.

Pizzi and Williamson show that CI can be faithfully embedded into the normal modal logic KD, and vice versa. Analytic consequential implication is interpreted according to the following translation function φ:

$$\varphi(A \to B) = \Box(\varphi A \supset \varphi B) \wedge (\Box\varphi B \supset \Box\varphi A) \wedge (\Diamond\varphi B \supset \Diamond\varphi A)$$

As Pizzi and Williamson [18, p. 571] point out, their investigation is a "contribution to the modal treatment of logics intermediate between logics of consequential implication and connexive logics." They emphasize a difficulty of regarding consequential implication as a genuine implication connective by showing that in any normal system of consequential logic that admits modus ponens for consequential implication and contains BT, the following formulas are provable:

$$(a)\ \ (A \to B) \equiv (B \to A), \ \ (b)\ \ (A \to B) \equiv \neg(A \to \neg B),$$

where $\equiv$ is classical equivalence. Since $(A \to B) \leftrightarrow \sim(A \to \sim B)$ is a theorem of C, the more problematic fact, from the point of view of the present paper, is the provability of (a). Pizzi and Williamson also show that in any normal system of consequential logic that contains BT, the formula $(A \to B) \equiv (A \equiv B)$ is provable, if $(A \to B) \supset (A \supset B)$ is provable, in other words, consequential implication collapses into classical equivalence if $(A \to B) \supset (A \supset B)$ is provable.

8 Subconnexive logic and the logic of scientific research

In the literature, there is a discussion about whether some system of constructive logic may be regarded as the logic of scientific research. According to Grzegorczyk [6], intuitionistic logic may be viewed as the logic of scientific research, understood positivistically as the logic of verification. However, intuitionistic falsity, as expressed by intuitionistic negation, is still a rather weak notion of falsity. The intuitionistic negation of a formula A is verified at an information state u iff A is *not true* at all possible extensions of u. Gurevich [7] criticizes this weak notion of falsity and suggests taking direct falsification seriously by admitting falsification of atomic formulas on the spot. These considerations lead him to Nelson's three-valued logic N3 with strong, constructive negation. The system N3 is a logic of scientific research in which both truth and falsity are understood positivistically in the sense that both truth and falsity are persistent with respect to the relation of possible growth of information states. Moreover, truth and falsity are mutually exclusive: there is no information state u and no formula A such that both A and its strong negation are true at u. In [26] is has been pointed out that Gurevich's emphasis on falsification may be seen to have a precursor in Popper's idea of falsification in the natural sciences, see [19], [20].

In N3 and N4, positive and negative information are treated on a par, by distinguishing a support of truth relation $\models^+$ and a support of falsity relation $\models^-$. Whereas the persistence with respect to $\models^-$ is compatible with Popper's notion of falsification, persistence of positive information is incompatible with Popper's notion of confirmation. According to Popper, empirical hypotheses may be falsified by observation, measurement, and suitable experiments. This means falsification on the spot, justifying, in Popper's view, a dismissal of the statement under consideration henceforth. In contrast to this, confirmation is only preliminary confirmation. It may happen that a future state falsifies the hypothesis in question, so that confirmation at the given state does not establish the hypothesis once and for all. It does not help to consider the logic resulting from N3 by giving up the persistence condition for support of truth of atomic formulas. In this subconstructive logic, *falsification*, too, is no longer persistent. Consider simple law-like hypotheses like $\forall x(S(x) \rightarrow B(x))$. If strong negation embodies the notion of falsification, then $\forall x(S(x) \rightarrow B(x))$ is falsified at a state u iff for some element a from the individual domain of u, $(S(a) \rightarrow B(a))$ is falsified at u iff $S(a)$ is verified at u and $B(a)$ is falsified at u. Although falsification is falsification on the spot, falsification in *Nelson's systems* involves *persistent* verification. But if persistence of verification is given up, so that verification may be regarded as confirmation in Popper's sense, then, if for some a, $S(a)$ is confirmed and $B(a)$ is falsified, falsification at all future states is not guaranteed. This problem of failure of persistence of negative information, of course, arises for the falsification of other kinds of compound formulas, too.[5]

The problem of identifying a formal system that captures Popper's con-

[5] In [23], it has been suggested to consider a certain system of implication-free logic as Popper's logic of scientific research. In this system, called falsificationist logic, the

ception of the logic of scientific research looses much of its attraction, when it is recognized that Popper's model of scientific inquiry is an outdated and abandoned model. It is well-know that Popper's much celebrated distinction between confirmation and falsification of scientific theories has been vigorously criticized by many philosopher's of science like, for example, Kuhn [9] and Lakatos [10]. Since virtually every empirical theory is confronted with some kind of anomalies, Popper's thesis that anomalies falsify a theory once and for all would have the highly undesirable consequence that there is no non-falsified empirical theory. There simply is nothing like persistent direct falsification in the empirical sciences. Moreover, and maybe even more importantly, there is little reason to assume that the development of empirical or non-empirical scientific theories is adequately described by a system of propositional or first-order *logic* instead of a suitable first- or higher-order *theory*.

Whether or not the system QC or the sub-connexive logic obtained from QC by giving up both persistence of support of truth and persistence of support of falsity of atomic information are logics of scientific research, these systems provide an interesting model of *information processing*. If the informational interpretation of constructive logics (see [26] and references therein) is taken seriously as a model of information processing, then it is conspicuous that although support of truth and support of falsity in Nelson's logics are treated on a par, there still is an asymmetry between verification and falsification: the former may be dynamic, the latter is static. The support of falsity conditions of implications in C are one way of overcoming this asymmetry.[6] Interestingly, this modification of the classical falsity conditions of implications gives rise to a system of connexive logic. As a system of constructive logic with *strong* negation, QC may also play a role in investigations of contrariety, see [27], [28]. More specifically, CK may play a role in investigations of necessity versus unnecessity and possibility

evaluation clauses for negation are as follows (notation adjusted):

$$u \models^+ \sim A \quad \text{iff} \quad (\exists w \geq u)\, w \models^- A; \qquad u \models^- \sim A \quad \text{iff} \quad (\forall w \geq u)\, w \models^+ A.$$

Moreover, the verification of atomic information is taken to be backward-persistent: if $u \models^+ p$ and $u \geq w$, then $w \models^+ p$.

[6]Another suggestion for rendering the evaluation of negated dynamic implications dynamic has been made independently of the present paper in [8], where N. Kamide defines a sequent calculus CLS, in which the following equivalence is provable (notation adjusted):

$$\sim (A \to B) \leftrightarrow (\sim A \to \sim B).$$

Kamide points out that "the substitution theorem with respect to $\leftrightarrow$ does not hold". The treatment of negated implications is motivated by the idea of treating $\sim$ as a modal operator satisfying the K axiom schema, leading to the left-to-right direction of the displayed formula.

versus impossibility.

Acknowledgements

The author would like to thank the anonymous referees for their instructive comments and Sergei Odintsov for agreeing to use here material from the joint papers [12] and [13].

BIBLIOGRAPHY

[1] R. Angell, A Propositional Logic with Subjunctive Conditionals, *Journal of Symbolic Logic* 27 (1962), 327–343.
[2] D. van Dalen, Intuitionistic Logic, in: D. Gabbay and F. Guenthner (eds.), *Handbook of Philosophical Logic Vol. III*, Reidel, Dordrecht, 1986, 225–339.
[3] G. Fischer Servi, Axiomatizations for some Intuitionistic Modal Logics, *Red. Sem. Mat. Universi. Politec. Torino* 42 (1984), 179–194.
[4] D. Gabbay, A. Kurucz, F. Wolter, and M. Zakharyaschev, *Many-Dimensional Modal Logics: Theory and Applications* , Elsevier, Amsterdam, 2003.
[5] C. Grefe, Fischer Servi's Intuitionistic Modal Logic has the Finite Model Property, in: M. Kracht et al. (eds.), *Advances in Modal logic. Vol. 1.*, CSLI Lecture Notes 87, 1998, 85–98.
[6] A. Grzegorczyk, A philosophically plausible formal interpretation of intuitionistic logic, *Indagationes Mathematicae* 26 (1964), 596–601.
[7] Y. Gurevich, Intuitionistic logic with strong negation, *Studia Logica* 36 (1977), 49–59.
[8] N. Kamide, Gentzen-Type Methods for Bilattice Negation, manuscript, 2004, to appear in: *Studia Logica.*
[9] T. Kuhn, *The Structure of Scientific Revolutions*, Chicago, Chicago UP, 1962.
[10] I. Lakatos, Falsification and the Methodology of Scientific Research Programmes, in: I. Lakatos and A. Musgrave (eds.), *Criticism and the Growth of Knowledge*, Cambridge, Cambridge UP, 1970, 91–196.
[11] S. McCall, Connexive Implication, *Journal of Symbolic Logic* 31 (1966), 415–433.
[12] S.P. Odintsov and H. Wansing, Inconsistency-tolerant Description Logic: Motivation and Basic Systems, in: V. Hendricks and J. Malinowski (eds.), *Trends in Logic: 50 Years of Studia Logica*, Kluwer Academic Publishers, Dordrecht, 2003, 287–321.
[13] S.P. Odintsov and H. Wansing, Constructive Predicate Logic and Constructive Modal Logic. Formal Duality versus Semantical Duality, in: V. Hendricks et al. (eds.), *First-Order Logic Revisited*, Logos Verlag, Berlin, 2004, 269–286.
[14] C. Pizzi, Boethius' Thesis and Conditional Logic, *Journal of Philosophical Logic* 6 (1977), 283–302.
[15] C. Pizzi, Decision Procedures for Logics of Consequential Implication, *Notre Dame Journal of Formal Logic* 32 (1991), 618–636.
[16] C. Pizzi, Consequential Implication: A Correction *Notre Dame Journal of Formal Logic* 34 (1993), 621–624.
[17] C. Pizzi, Weak vs. Strong Boethius' Thesis: A Problem in the Analysis of Consequential Implication, in: A. Ursini and P. Aglinanò (eds.), *Logic and Algebra*, Marcel Dekker, New York, 1996, 647–654.
[18] C. Pizzi and T. Williamson, Strong Boethius' Thesis and Consequential Implication, *Journal of Philosophical Logic* 26 (1997), 569–588.
[19] K. Popper, *Logik der Forschung*, Springer Verlag, Wien, 1934.
[20] K. Popper, *Conjectures and Refutations*, Routledge and Kegan Paul, London, 1963.
[21] R. Routley, Semantical Analyses of Propositional Systems of Fitch and Nelson, *Studia Logica*, 33 (1974), 283–298.
[22] R. Routley and H. Montgomery, On Systems Containing Aritotle's Thesis, *Journal of Symbolic Logic* 33 (1968), 82–96.

[23] Y. Shramko, The Logic of Scientific Research, TU Dresden, Institute of Philosophy, 2004, Dresden Preprints in Theoretical Philosophy and Philosophical Logic 7-2004.
[24] A. Simpson, *The Proof Theory and Semantics of Intuitionistic Modal Logics*, PhD Thesis, University of Edinburg, 1994.
[25] V. Sotirov, Modal Theories With Intuitionistic Logic, in: *Mathematical Logic. Proceedings of the Conference on Mathematical Logic, Dedicated to the Memory of A.A. Markov (1903-1979), Sofia, September 22-23, 1980*, Sofia, 1984, 139–171.
[26] H. Wansing, *The Logic of Information Structures*, Springer Lecture Notes in AI 681, Springer Verlag, Berlin, 1993.
[27] H. Wansing, Negation, in L. Goble (ed.), *Blackwell's Guide to Philosophical Logic*, Basil Blackwell Publishers, Cambridge/MA., 2001, 415–436.
[28] H. Wansing, Widersprüchlichkeit und Kontrarität. Priest über Negation, in: B. Christiansen and U. Scheffler (eds.), *Was folgt. Themen zu Wessel*, Logos Verlag, Berlin, 2004, 251–267.

Heinrich Wansing
Dresden University of Technology
Institute of Philosophy
01062 Dresden
Germany
E-mail: Heinrich.Wansing@mailbox.tu-dresden.de

INDEX

www.ingramcontent.com/pod-product-compliance
Ingram Content Group UK Ltd.
Pitfield, Milton Keynes, MK11 3LW, UK
UKHW021316180426
11947UKWH00015B/1261

9 781904 987222